D0946963

Feathered Dragons

LIFE OF THE PAST
James O. Farlow, Editor

Philip J. Currie, Eva B. Koppelhus,
Martin A. Shugar, and Joanna L. Wright, Editors

Feathered Dragons

Studies on the Transition from Dinosaurs to Birds

Indiana University Press
Bloomington & Indianapolis

This book is a publication of
Indiana University Press
601 North Morton Street
Bloomington, IN 47404-3797 U.S.
http://iupress.indiana.edu

Telephone orders 800-842-6796
Fax orders 812-855-7931
Orders by e-mail iuporder@indiana.edu

The paper used in this publication meets the minimum requirements of American National Standard for Information Sciences—Permanence of Paper for Printed Library Materials, ANSI Z39.48-1984.

Manufactured in the United States of America

Library of Congress Cataloging-in-Publication Data

Feathered dragons: studies on the transition from dinosaurs to birds / Philip J. Currie . . . [et al.], editors.
p. cm. —(Life of the past)
Includes bibliographical references and index.
ISBN 0-253-34373-9 (cloth: alk. paper)
1. Dinosaurs. 2. Birds, Fossil. 3. Birds—Evolution. I. Currie, Philip J. II. Series.
QE861.4.F32 2004
567.9—dc22 2003019035

1 2 3 4 5 09 08 07 06 05 04

This book is dedicated to
Michael Feinberg,
whose unwavering commitment and financial support made this project and volume possible.

CONTENTS

Contributors

Robert T. Bakker, 1447 Sumac Avenue, Boulder, CO 80304, U.S.

Gary Bir, 1447 Sumac Avenue, Boulder, CO 80304, U.S.

David A. Burnham, Natural History Museum and Biodiversity Research Center, University of Kansas, Lawrence, KS 66045-2454, U.S.

Sankar Chatterjee, Museum of Texas Tech University, Box 43191, Lubbock, TX 79409-3191, U.S.

Luis M. Chiappe, Department of Vertebrate Paleontology, Natural History Museum of Los Angeles County, 900 Exposition Boulevard, Los Angeles, CA 90007, U.S.

James M. Clark, Department of Biological Sciences, George Washington University, Washington, DC 20052, U.S.

Philip J. Currie, Royal Tyrrell Museum of Palaeontology, Box 7500, Drumheller, AB T0J 0Y0, Canada.

Stephen J. Godfrey, Calvert Marine Museum, P.O. Box 97, Solomons, MD 20688, U.S.

Gerald Grellet-Tinner, Department of Vertebrate Paleontology, Natural History Museum of Los Angeles County, 900 Exposition Boulevard, Los Angeles, CA 90007, U.S.; and Department of Earth Sciences, University of Southern California, Los Angeles, CA 90007, U.S.

Thomas P. Hopp, Protein Research Laboratories, Inc., 4842 51st Avenue SW, Seattle, WA 98116, U.S.

Frankie D. Jackson, Department of Earth Sciences, Montana State University, Bozeman, MT 59717, U.S.

Eva B. Koppelhus, Royal Tyrrell Museum of Palaeontology, Box 7500, Drumheller, AB T0J 0Y0, Canada.

Peter J. Makovicky, Department of Geology, Field Museum of Natural History, 1400 S. Lake Shore Drive, Chicago, IL 60605-2496, U.S.

Mark A. Norell, Division of Paleontology, American Museum of Natural History, 79th Street at Central Park West, New York, NY 10024-5192, U.S.

Fernando E. Novas, Museo Argentino de Ciencias Naturales, Av. Angel Gallardo 470, Buenos Aires (1405), Argentina.

Mark J. Orsen, Protein Research Laboratories, Inc., 4842 51st Avenue SW, Seattle, WA 98116, U.S.

Gregory J. Retallack, Department of Geological Sciences, University of Oregon, Eugene, OR 97403-1272, U.S.

Dale A. Russell, North Carolina Museum of Natural Sciences, and Department of Marine, Earth and Atmospheric Sciences, North Carolina State University, Raleigh, NC 27695-8202, U.S.

Martin A. Shugar, M.D., 3850 Hollywood Blvd., Suite 401, Hollywood, FL 33021, U.S.

R. J. Templin, 2212 Aster Street, Ottawa, ON K1H 6R6, Canada.

David J. Varricchio, Museum of the Rockies, Montana State University, Bozeman, MT 59715, U.S.

Peter Wellnhofer, Bayerische Staatssammlung für Paläontologie und historische Geologie, Richard-Wagner-Str. 10, 80333 München, Germany.

Joanna L. Wright, University of Colorado at Denver, Department of Geology, Campus Box 172, PO Box 173364, Denver, CO 80217-3364, U.S.

Foreword

This book was conceived during the Florida Symposium on Dinosaur-Bird Evolution, which was held on April 7th and 8th, 2000, at the Greater Fort Lauderdale–Broward County Convention Center. This meeting fulfilled a promise made to Dr. John Ostrom in October 1997, following his first examination of what would become *Bambiraptor feinbergi.* At that time, he stressed the importance of an international conference to showcase this treasure, and to record its existence as a new dinosaurian taxon. This was particularly important in the context of the discovery of the first "feathered" dinosaurs, reports of which were just starting to flow out of China. Appropriately, *Bambiraptor feinbergi* and several of the Chinese specimens were on display at the meetings, which became a forum for discussion of the evolution of birds. Dr. Ostrom was the guest of honor, and more than 35 talks were delivered to large, enthusiastic crowds.

The year 2000 saw more than the turnover of the centuries and the establishment of *Bambiraptor feinbergi.* Meat-eating theropod dinosaurs had been recognized as potential ancestors of birds (Huxley 1868) ever since the discovery of *Archaeopteryx* and *Compsognathus* in the middle of the nineteenth century. But the remains of these animals were rare, usually incomplete, and often poorly preserved. A revolution began in 1964 with the discovery of *Deinonychus,* an animal revealed to the world five years later by John Ostrom (Ostrom 1969). The work of Ostrom and others showed startling similarities between dinosaurs and birds. With each new small theropod discovery, the ties became stronger. The spread of phylogenetic systematics in vertebrate paleontology also had its role, and the seminal paper by Gauthier (1986) was a watershed in scientific acceptance of dinosaurs as the most logical ancestors of birds.

Public interest in the origin of birds grew slowly throughout the 1970s and 1980s, and exploded with the release of the movie *Jurassic Park* in 1993. In spite of the fact that the message of dinosaur-bird relationships was subtle (almost subliminal) in the movie, the idea had never before been widely known outside of the scientific community. The public was finally ready for the next surprising discovery—a dinosaur with feathers!

When the existence of *Sinosauropteryx* (Ji and Ji 1996) was announced in 1996, the concept of feathered dinosaurs had already been around for more than twenty years. Following the evidence of warm-bloodedness in small theropods and their relationship to birds, it was logical to hypothesize that feathers may have evolved initially as a form of body insulation. However, even those of us who believed the logic of the idea never really expected to be lucky enough to see feathers preserved on a dinosaur. Furthermore, all scientists had not (and still have not) accepted the concept of birds being phylogenetically nested within the Dinosauria. The last years of the twentieth century saw one of the greatest controversies in vertebrate paleontology peak in scientific and public debate. This was the atmosphere that enveloped the participants of the Florida Symposium as proponents of all sides of the debates concerning the origin of birds, the origin of feathers, and the origin of flight gathered to examine *Bambiraptor,* a remarkable new specimen from Montana.

The onslaught of new Chinese feathered dinosaurs has continued over the intervening years, and the level of controversy over bird origins has waned. As more people accept birds as the living representatives of the Dinosauria, the focus of research has been shifting to the evolution of feathers and avian flight. Great ideas are like architectural plans; once accepted, the real work begins. Just as bricks must be laid to form the foundation of a building, basic concepts must be tested before further scientific theories are advanced. As paleontologists follow through with the nuts-and-bolts work of description and analysis, new discoveries will constantly force the modification of the framework. The ensuing debates ensure that every decision made is well considered. And from this process emerges a greater understanding of the complexities of life—both past and present.

References

Gauthier, J. 1986. Saurischian monophyly and the origin of birds. *California Academy of Sciences, Memoirs* 8: 1–55.

Huxley, T. H. 1868. On the animals which are most nearly intermediate between birds and reptiles. *Annals and Magazine of Natural History* 2 (4th series): 66–75.

Ji Q. and Ji S.-A. 1996. On the discovery of the earliest bird fossil in China and the origin of birds. *Chinese Geology* 233: 30–33 (in Chinese).

Ostrom, J. H. 1969. Osteology of *Deinonychus antirrhopus,* an unusual theropod from the Lower Cretaceous of Montana. *Peabody Museum of Natural History, Bulletin* 30: 1–165.

Acknowledgments

A book like this comes together only because of the hard work of many people, including Darren Tanke (Royal Tyrrell Museum of Palaeontology) and graduate students at the University of Calgary (Sandra Jasinoski, Michael Ryan, Tanya Samman, Tamaki Sato, Eric Snively, and Darla Zelenitsky).

Introduction: Dinosaurs Acting Like Birds, and Vice Versa

An Homage to the Reverend Edward Hitchcock, First Director of the Massachusetts Geological Survey

Robert T. Bakker

So, who started it all?

The dinosaur-bird link is a hot topic in the popular science media. Even "Jurassic Park III" tried to jump on the avian-dino bandwagon by making a brave attempt to adorn *Velociraptor* with a feathery hairpiece. (The result looked like a roadrunner's toupee—don't blame the effects-artists; it's notoriously difficult to render a full body covering of feathers in computer-graphic animation, so we'll have to wait for "JP IV" for a more thoroughly rendered avian pelage.) My local toy and stuffed-animal store offers half a dozen plastic dromaeosaurid models, each with a texture that expresses some sort of bird-style skin covering.

Who gets the credit—or blame—for the fowl-image of dromaeosaurids in particular and dinosaurs in general? Actually, there are three separate issues discussed in this and similar books and symposia: 1) Are dinosaurs birds, in a phyletic-taxonomic sense? 2) Did dinosaurs act like birds, in physiology and ecology and behavior? 3) How did birdlike dinosaurs fit into the long-term evolutionary interactions that shaped the history of life? The Dinofest symposium and other recent discussions have developed arguments in the first two conceptual arenas; the third question has attracted surprisingly little attention—far less than in the 19th century, as I'll show a little later. Many reporters in the popular media have conflated the three issues, but they are really, really different. It's possible to believe that birds are the descendants of dinosaurs and yet insist that most dinosaurs were cold-blooded "good reptiles." Some of the most valiant champions of the dinosaur-bird phyletic link ridiculed the notion of dinosaurs having feathers and being hot-blooded. Sir Richard Owen took the opposite tack—he nurtured the idea that dinosaur physiology was advanced, and that his Subclass Dinosauria included the reptiles who made the "closest approach to the viviparous quadrupeds," e.g., mammals. However, inexplicably, he opposed any phylogenetic connections between dinosaurs and birds.

So, who was the first to explore the birdness of dinosaurs in all three senses? In recent symposia, many references were made to Ostrom's 1973 paper on the *Archaeopteryx*-theropod connection or to my 1975 *Scientific American* article "Dinosaur Renaissance." However, the triple-treatment of dino-bird topics goes much further back. Thomas Henry Huxley often is fingered as the culprit. And it is true that in 1867 Oxford's John Phillips prodded Huxley to rework megalosaur bones into a rough approximation of the modern view (Phillips 1871). But the Phillips-Huxley duo was a latecomer to the dinosaur-bird problem. The first publication placing birds and dinosaurs in their correct mutual relations came in 1836, five years *before* Owen coined his memorable term "Dinosauria." It was written by the Father of Dinosaurology, first Director of the Massachusetts Geological Survey, President of Amherst College, and the most influential Congregationalist pastor and theologian in the first half of the 19th century: the Reverend Edward Hitchcock. Hitchcock is a paleontological paradox. The earliest and in many ways the best mind in dinosaur science, he was a fervent anti-evolutionist and never used the term "dinosaur," not even in the late 1850s. My inaugural publication was "The superiority of dinosaurs" in 1968, where I argued that dinosaur physiology had to be "higher" than a lizard's because dinosaurs had suppressed mammal evolution for 165 million years. I didn't know it then, but this style of thinking was first perfected by Hitchcock and his colleagues.

Hitchcock—and only Hitchcock—deserves the rubric "Father of Jurassic Park." And in the 21st century the good Reverend's reputation most definitely deserves a makeover. The dino books I grew up with dismissed Hitchcock as a bungling footprint specialist who deluded himself into believing that dinosaur tracks in the Connecticut Valley red beds were made by nonexistent Jurassic birds (Colbert 1945). "Tsk, tsk . . . poor Hitchcock, he never realized his 'Ornithichnites' were imprints left by early theropod and ornithopod dinosaurs . . ." is the sort of comment I heard often. Yes, indeed, Hitchcock never used the word "dinosaur" when discussing his beloved bird-tracks. But that was not because he was a naïve paleo-bumpkin. Hitchcock didn't apply the "dinosaur" label to his tracks because the finest minds of European paleontology had reconstructed dinosaur feet and body form in a totally erroneous fashion. Standard texts before 1870 made dinosaurs look like hybrid crocodiles/monitor lizards/bears/rhinos who padded around on flat-footed paws. Hitchcock knew that his Early Jurassic trackmakers were digitigrade and fundamentally avian, not plantigrade and fundamentally reptilian or mammalian.

To understand how Hitchcock was right and the more famous

dinosaur-paleontologists of his day were wrong, we need only to survey the restorations of the Dinosauria presented to the public before 1870. The standard view of dinosaurs had been built up from Jurassic megalosaurs and Cretaceous iguanodonts. All European opinion seemed to agree—dinosaur feet and legs were mostly of ursine and crocodilian design, with some rhino thrown in (Dean 1999; Mantell 1844). Hardly any osteologists detected strong avian influence. The bear-lizard-croc-rhino view of dinosaurs resulted from selected reading of skeletal parts combined with sophomoric mistakes. Back in the 1810s and '20s, in the original excavation of megalosaurian elements from Stonesfield, England, a huge birdlike ilium had been found. The young Oxford don William Buckland noted the avian geometry. Alas, he was overruled by Cuvier and Owen, who insisted that the ilium really was from the other end of the creature. Consequently, the ilium was inserted into the unfortunate megalosaur's shoulder region as a coracoid, foisting on the beast a grotesquely misproportioned shoulder region. A slender ischium was also misappropriated into the outsized pectoral girdle, being forced to serve as a giant "clavicle" (Mantell 1844; Rupke 1983; Phillips 1871). The first Jurassic Park special-effects extravaganza was set up in 1853 outside the Crystal Palace. Life-size megalosaur sculptures bared their ceramic teeth at magnificently lumpy iguanodonts. Every dinosaur was outfitted with lizardoid-pentadactyl paws and muscle-bound forequarters of Schwarzeneggerian proportions. Shoulders and arms were as thick as or thicker than thigh and calf (Gordon 1894; Rupke 1994).

As the finest European fossil-savants were committing anatomical malpractice, Hitchcock was reconstructing true dinosaur posture and gait and decoded the dinosaur digital formulae from tracks preserved in the red beds of the Connecticut Valley. These rift-valley sediments the Reverend took to be Early Jurassic, a bit earlier than the most famous Jurassic bone locality, the megalosaur- and mammal-bearing beds of the English Stonesfield Slate. (The Reverend was correct about the age; the richest track-yielding strata in the Connecticut Valley are Liassic, with a few good footprint sites from the latest Triassic.) For most of his career, Hitchcock had little bone material from the red beds. The best specimens were partial skeletons unearthed during excavation for a well by the rather imprecise technique of exploding a keg of black powder (Lull 1915). Shattered limb bones showed a peculiar hollowness in the core—an almost avian hollowness—combined with a long tail and crocodilian features in the limb bones. These exploded fossils would become the type sample of *Anchisaurus colurus,* a gazelle-sized prosauropod dinosaur. But they offered Hitchcock little help in figuring out the identity of the trackmakers.

In fact, Hitchcock didn't need help from bones. He reconstructed the fore and hind paws of Jurassic dinosaurs directly from the footprints with extraordinary sagacity. I find Hitchcock's inaugural 1836 monograph awe-inspiring. His observations are detailed, his reasoning tight. This one publication made paleoichnology a robust discipline, a window into the grand succession of dominant life forms in the terrestrial sphere.

Hitchcock's first inference was that his trackmakers kept their heels high off the ground, bird-style, when walking and running. Nearly all tracks were imprints of the pads beneath the toe bones, with no marks left by the sole below metatarsals or upper tarsals. Only a few tracks showed the backside of the Achilles tendon and the ankle region, and these prints were from animals hunkered down. The second Hitchcockian conclusion was that his animals walked and ran almost entirely on their hind feet—handprints were found in only a few cases. The third inference was that the trackmakers moved with long-legged strides on limbs set close to the body midline: the trackways were exceedingly narrow, with little space between right and left track-paths, and the distance between successive footfalls was huge. Hitchcock deduced that the metatarsal region was long and narrow, because the three main toes were close to one another where they attached to the ankle. All these deductions meant that the Connecticut Valley trackways could not be made by a lizard or alligator or shuffle-footed mammal such as a bear or raccoon. And even more certainly, the footprints could not be the product of a front-heavy megalosaur or iguanodont of the build universally favored by European dinosaur specialists.

Hitchcock went on to crack the phalangeal code of his dinosaurs by observing that the biggest, most distinct pads would lie beneath joints near where two toe bones came together. And that assumption allowed him to write down the toe-bone formula in his Ornithichnites animals. Toe I, the hallux, the innermost digit, had two phalanges. This toe was the shortest and often failed to impress the substrate, but Hitchcock was pretty sure that most of his Ornithichnite animals carried a hallux. Toe II had three phalanges. Toe III, four; toe IV, five. Toe V, the pinkie toe, left no marks ever and must have had only vestigial phalanges. An extra pad behind toes IV and II flummoxed the Reverend for a while, until he realized that this bump was a cushion beneath the metatarsal bundle. The toe formula Hitchcock calculated was 2-3-4-5-0, most emphatically not the basic mammalian 2-3-3-3-3.

Crocodile hind feet were a numerical match for Ornithichnites. But crocs and gators had been eliminated on account of being splay-legged and flat-footed. Plus, as Hitchcock observed, the pedal symmetry of his trackmakers was non-crocodilian. Crocodilian hind feet have Toes II–III–IV subequal. Hitchcock's Jurassic animals walked on a foot centered on Toe III, with the adjacent II and IV distinctly shorter. The inner toe, too, was unlike that of any reptile—it was turned inward at a strong angle to Toe III. The in-turned I was suggestive of the bird digit I, though Hitchcock recognized that most birds have a longer hallux that is turned even further backward. The symmetry of the hind paw, the digital formula, and the hallux all fit the Class Aves nearly perfectly. The quantified paleo-podiatry added overwhelming proof that no Owen-Cuvier dinosaur could have squeezed its feet into the Reverend's tracks.

Hitchcock's first paper of 1836 contained behavioral observations as well as anatomy. The great strides of his Jurassic avians demonstrated high speeds. And several of his trackmakers were social—their footprints occurred in groups where all the individuals were walking in parallel, close together. The Reverend drew vivid word portraits of the Jurassic ecosystem: great flocks of giant ground birds contended for dominance with each other on the river-edge floodplains and along the wide banks of warm lakes.

Hitchcock was not content to place his Jurassic trackmakers within the Class-structure of 19th-century zoology. He struggled to understand where Ornithichnite animals fit in the entire flow of Creation. Rabble-rousing theories of transmutation were in the air in the 1830s. Oken, Lamarck, and Geoffroy St. Hilaire and their students were preaching evolutionary progressivism, arguing that fossils demonstrated a general rise in complexity from fish to reptile-oids to mammals to man. To Hitchcock, these scenarios smacked of impious materialism. He hated them. There was a God-fearing alternative: the elegantly reasoned theory of serial Creation (Desmond 1989; Rupke 1983, 1994). It went like this:

1) Vertebrates are arrayed up and down a single *Scala Naturae*. Birds were the second hot-blooded Class, just a rung lower in the *Scala* than Class Mammalia; both were higher than typical, lizardoid reptiles, which in turn were higher than froggish batrachians. The "higher" the Class, the more temperate their ideal environment. Frogs and sluggish reptiles preferred hothouse habitats of great heat and humidity. The highest mammals—people—needed a brisk, temperate climate, like that of Western Europe or New England.

2) The globe began as a molten mass and then cooled off gradually over many millennia. Experiments carried out by Comte de Buffon, the "Inspector of the King's Lumber," in the mid-18th century seemed definitive (Buffon's title reflected his duties in forestry, not urology). He constructed balls of a constant diameter, composed of various combinations of metals and stone, and then heated them in an oven. By measuring the rate of cooling, he produced the first quantitative calculation of the earth's age (Roger 1997). "Brass Balls Buffon" rattled the cage of contemporary clerics with his published estimate of many tens of thousands of years. Though Buffon was suspected of heresy, serial creationists were convinced that earth history was a long story of climatic amelioration.

3) As earth climate went from soggy and torrid to dry and temperate, the global environment became suitable for higher and higher Classes. And every time the changed climate required an upgrade in physiology, the Creator inserted the correct Class of species.

4) The Coal Age was too hot and humid for any life-form higher than a frogoid amphibian or lowly lizard.

5) The Mesozoic was drier and more invigorating, suitable for high-Class reptiles like dinosaurs which were outfitted with advanced anatomical design.

6) The Cenozoic ushered in a modernized climate that demanded the brightest and most hot-blooded Classes, advanced birds and mammals.

Implicit in this program for progress was the emphasis on the large land vertebrates. Lower Classes could continue as small-bodied species after the climate had changed enough to favor replacement of giant species by the next higher Class. And so it was OK to let lizards and frogs scuttle and hop around the feet of lions and tigers and bears and mammoths in the final Creation days. It was not OK to have gargantuan lizards or frog-oids competing with large mammals in the Cenozoic. And it was not OK to let archaic cold-bloods rule with semi-warm-blooded dinosaurs in the Jurassic.

If we selectively eliminate the theology, this theory of Progress is not a bad description of Laurasian habitat evolution since the Carboniferous. Coal swamps across Western Europe and the U.S. were as wet and warm as modern-day Borneo. The Jurassic, on average, had much drier summers, and from the Cretaceous to the Recent there was an overall trend toward colder winters. In serial Creation, each time-slice was a balanced ecological system. Plants were fitted to their soils—giant fern foliage and horsetails for the Coal Age swamps, conifers and cycads for the better drained Jurassic landscape, grasses for the late Cenozoic. Each period had herbivores fitted to browse on the dominant plants, and top predators to keep the hay-burners from overpopulating.

At first, Hitchcock was skeptical of creationist ladders through time. He pointed to his giant Jurassic birds as Progress debunkers. Huge avian species, belonging to the second-highest Class, shouldn't have appeared until the later Cenozoic, much later than the Connecticut Valley red beds. And Hitchcock used the Ornithichnites to cast doubt on the value of the fossil bone record. Since no one had dug up giant bird bones from the Jurassic, and his trackmakers clearly were birds, osseous paleontology failed to document this important epoch of vertebrate evolution.

But a decade and a half later, in his delightful book *The Religion of Geology* (1851) and his great, final monograph (1858), Hitchcock displayed a conversion to a progressive serial-Creation view of earth history. The stratigraphic sequence of land vertebrates, especially *large* land vertebrates, did climb a ladder of progress. Carboniferous–Permian–Early Triassic times were ruled by gigantic frog-oids (labyrinthodonts) and lowly reptiles. Jurassic-Cretaceous epochs were the realm of distinctly higher Classes and subclasses. Owen's dinosaurs were technically reptilian but they were the highest of their Class, with hearts and legs far superior to those of lizards. Alongside these very high Reptilia lived the earliest mammals, their tiny jaws surprisingly common in the Jurassic sites of Stonesfield and Purbeck. Though they were definitely mammalian, the Jurassic jaws appeared to be closest to marsupials of possum grade, and possums and marsupials in general were known to be the lowest of their Class. Low-Class, marsupialoid mammals seemed compatible with high-Class Reptilia within the context of the Jurassic climate (Desmond 1989; Rupke 1983; Mantell 1844).

How about Hitchcock's Jurassic birds? They were not as stratigraphically precocious as he had believed. Certainly his Jurassic trackmakers were avian in many features; nevertheless, the Reverend could detect a mixture of Class characteristics. On one hand, the Ornithichnites had avian foot pads and long-striding, ostrich-like hind legs. On the other hand, there were strong whiffs of batrachian features mingled with the geometry of Australian marsupials, plus design elements of lizards and crocodiles.

The two most stunning ways in which his Jurassic birds differed from all modern avians were their hands and tails. All modern-day birds had stiff, simplified hands, the basic plan being fingers II and III fused up to provide a firm support for flight feathers. Hitchcock could prove that some—maybe many—of the Ornithichnite animals carried a primitive hand with the gestalt of a marsupial or a frog or a reptile. *Anomoepus* was the trackmaker genus that vouchsafed the clearest hand prints. (The manus we now know would fit the Como Jurassic *Othnielia* or *Drinker* quite exactly.) Hitchcock was struck by two features. The hand was tiny compared to the hind paw, about one-tenth the area, a discrepancy that would fit a kangaroo-like marsupialoid or a hybrid possum-anuran. And, even more important, *Anomoepus* hand was pentadactyl with five separate, short, padded fingers. Three of them, the innermost digits I–III, carried claws. The mud film on the trackway slab would have preserved the imprint of flight feathers; there were no such marks. "These hands did not carry feathers," was Hitchcock's conclusion. Instead, the *Anomoepus* hand was a good match for a frog forepaw, and an even better match for a crocodilian manus or a kangaroo hand. He suspected that other Ornithichnite-animals —maybe the majestic *Grandipus*—were equipped with similar free-fingered forepaws (Hitchcock 1858).

Tail traces provided the Reverend with the other batch of mixed-Class characters in the Ornithichnite animals. The *Anomoepus* had a short, muscular tail, Hitchcock concluded from a depression on footprint slabs between the thighs (the identity of the organ that made the bump is still unclear—Lull 1915). His *Grandipus* seemed to have a much longer, more crocodilian caudal appendage, because there was a sinuous, continuous tail trace left along the trackway midline. Long tails barred a species from Class Aves, at least from Aves in the modern zoological sense. Yet Hitchcock couldn't get over the extraordinary resemblance between bird feet and the hind paws of *Grandipus* and *Anomoepus*. The combination of evidence from hands and tail led Hitchcock to infer that Jurassic Ornithichnite animals were not simply giant versions of turkeys and egrets; the Connecticut trackmakers were dominant land vertebrates whose design blurred Class boundaries.

The Class confusion in his great monograph of 1858 exhilarated the Reverend. The Past expanded the definition of modern-day zoological categories. Hitchcock's Yale colleague W. D. Dana was playing with parallel ideas of transitions from one Class to another (Dana 1863; Gould 1998). The duckbill platypus combined mammalian hair and milk glands with avian/reptilian eggs and bill. The African lungfish co-mingled batrachian lungs with fish gills and fins. Dinosaurs, as restored according to the Crystal Palace blueprint, had mammalian legs and viscera added to a basic crocodile jaw and tail. Hitchcock's *Anomoepus* and *Grandipus* reinforced Dana's hope to find links between typical birds and frog/lizard/crocodile. Dana did not see such transitions as proof of evolution, but rather as evidence for the completeness of the Creation plan. Dana's Class linkages, when added to serial Creation, made sense of the Jurassic Ornithichnite animals. The Jurassic was still too early, too warm, for the mature expression of the avian body plan, just as it was too stuffy for any mammal higher than an opossum. Ah, but Jurassic habitats were perfect for a possum-frog-croc-bird subclass.

One of the greatest ironies of 19th-century geology is that the loudest *opponents* of paleontological Progress in the 1850s were the very men who would inspire Darwin and Darwinism. Charles Lyell and a young Thomas Henry Huxley scoffed at any earth history that taught a simple, directional sequence of frog–lizard–possum–advanced mammal (Desmond 1989, 1994). Lyell championed a steady-state globe, where cycles of climate change would bring Classes and then yank them out again. The most famous example was his prediction that when tropical water returned to England, ichthyosaurs would cavort in the Thames. According to Lyell and Huxley, any appearance of Class progress was an artifact of poor sampling by the fossil record. Jurassic mammals were probably not restricted to low-Class opossums, because away from the lowlands, big-brained mammoths and rhinos may have gamboled on breezy mid Mesozoic hillsides. Lyell saw the Ornithichnite animals as proof that 100-percent modern birds existed far back in the Mesozoic, and he never grappled with the anatomical chimeras so exciting to the Reverend Hitchcock. It was the cadre of serial creationists, Hitchcock, Aggasiz, Miller, Buckland, and Mantell, who championed a coherent, directional script for the history of vertebrate life on the terrestrial sphere.

Historians of science often chuckle at serial creationists and at Dana's search for Class transitions (Gould 1998). That's a bit unfair, because the Hitchcock-Dana view of Deep Time was ecologically more nuanced than that of many early evolutionary thinkers, the young Huxley, for example. Serial creationists who discovered intermediate time-slices in fact provided much of the data that Nature did operate through transmutation. Even as early as 1836, the fossil evidence for the progressive history of *big* vertebrates was pretty good. By the late 1850s, it was far-fetched to believe that elephants went back as far as the Jurassic and increasingly improbable that perfectly modern birds had stalked over the Connecticut Valley mudflats that became the red beds. More fossil discoveries were filling the breaches between Classes. In the 1840s and '50s, Richard Owen was beginning to describe a half dozen fossil intermediates between normal reptiles and possum-grade mammals, all dug from the Late Permian and Early Triassic of the South African Karoo. Owen's protomammals ruled as Top Herbivore and Top Predator where they should—after the primitive reptiles of the Early Permian, before the advent of primitive mammals and marsupialoid birds in the latest Triassic and Jurassic (Desmond 1989; Bowler 1976, 1989; Rupke 1994).

When the first edition of *The Origin of Species* hit the bookstores in 1859, the public had been prepared for new transmutationist theories by a half century of wondrous paleontological discoveries. Darwin did provide a new and plausible mechanism to explain the process of evolution. The serial creationists and Class-defying anatomists had piled up the hard data that proved evolution had actually occurred. Serial Creation became harder and harder to swallow as theology because of two progressive accumulations of data points: 1) The number of time-slices with their distinct faunas multiplied, so hundreds of discrete Creations were needed if transmutation was not allowed; Hitchcock saw six separate creation-events in 1851, but by 1871 the Jurassic alone was divided into a dozen faunal ages, and the entire Mesozoic required thirty or forty individual creation-injections. 2) Transitional fossils made classes and orders closer and closer to each other, making transmutation much more likely; the many intermediates between primitive reptiles and mammals, and between reptiles and birds, rendered obsolete the old-fashioned notions of unbridgeable gaps between classes. Darwinian evolution became the only mechanism that was intellectually acceptable.

Archaeopteryx brought serial creationists and the new evolutionists together. Darwin was ecstatic when the first skeleton was announced. Here was a near perfect Class transition—a Late Jurassic animal that bridged what many had insisted was an unbridgeable gap between Reptilia and Aves. Dana was delighted too; he had postulated just such Class-to-Class links in his Creation scheme. In a curious turn of events, it was the serial creationists who made the connection between *Archaeopteryx* and the Connecticut Valley trackmakers. In 1864, the Reverend Hitchcock's son, Edward Jr., included comments on the German Jurassic bird in an update on his father's great monograph. Also included was Richard Owen's description of the best Connecticut Valley skeleton, the second specimen excavated by gunpowder. Edward Hitchcock Jr. embraced *Archaeopteryx* as a splendid confirmation for his father's life work. *Archaeopteryx* was a bird—it had flight feathers—but its hands carried three separate fingers, each tipped with claws, a real-life fulfillment of Edward Sr.'s vision of Jurassic birds with crocodile-like hands. The elder Hitchcock had reconstructed *Grandipus* as a great avian-shaped body and legs attached to a long, reptilian tail. And *Archaeopteryx* proudly carried a long, flexible, bony tail quite capable of making a *Grandipus* caudal impression. We must date the modern view of theropod dinosaurs from this 1864 publication. All the bipedal dinosaurs of the Connecticut Valley red beds, from bantamweight *Grallator* to two-ton *Grandipus,* could be imagined as variations on the body plan exhibited by *Archaeopteryx*.

A few years later, in the Oxford Museum, Phillips would lead Huxley into recognizing megalosaurs as bipeds full of avian features, while at the same time Cope was recognizing bird patterns in his Cretaceous carnivore *Laelaps* and the hadrosaur *Ornithotarsus* (Cope 1866, 1869). Soon the evolutionary descent of birds from dinosaurs was being promoted at Yale and in encyclopedias and textbooks of comparative anatomy (Huxley 1881; Bowler 1976, 1989). Cope and Marsh recognized that Hitchcock's Ornithichnites were truly relicts of protobirds, dinosaurs with avian foot pads and posture.

All of us 21st-century students of the dinosaur-bird relations owe a tip of the hat toward the Reverend Edward Hitchcock. His research program is well worth emulating: study the form and function of every joint, every bone element; put them in an ecological and behavioral context; then, ponder how the forces operating in Deep Time shaped the extinction of great categories and the arrival of new ones. Darwin's natural selection did operate when the Ornithichnite animals met possum-like creatures. Small theropods—*Bambiraptor* is a superb example—are the dinosaurs most likely to have interacted with Mesozoic Mammalia. Buckland's mid-Jurassic possumoids probably were chased by coyote-sized theropods. Hitchcock's Early Jurassic *Grallator* must have pursued late-surviving mammal-like reptiles and other early furry prey. On average, theropod predation, both real and potential, did suppress the evolution of large mammals until the end of the Cretaceous. What gave the theropod dinosaurs the advantage? Surely Hitchcock was right here—it was avian feet carried on avian limbs attached to a basic avian body plan.

References

Bakker, R. T. 1968. The superiority of dinosaurs. *Discovery* (Yale) 23: 11–23.

———. 1975. Dinosaur renaissance. *Scientific American* 232: 57–78.

Bowler, P. 1976. *Fossils and progress.* New York: Science History Publications.

———. 1989. *The invention of progress.* Oxford: Blackwell.

Colbert, E. H. 1945. *The dinosaur book.* New York: McGraw Hill.

Cope, E. D. 1866. Discovery of a gigantic carnivorous dinosaur in the Cretaceous of New Jersey. *Academy of Sciences Philadelphia Proceedings 1866:* 275–279.

———. 1869. Remarks on *Holops brevispinus, Ornithotarsus immanis,* and *Macrurosaurus proriger. Academy of Sciences Philadelphia Proceedings* 21: 123–125.

Dana, J. D. 1863. On the parallel relations of the classes of vertebrates, and on some characteristics of the reptilian birds. *American Journal of Science* 36: 315–321.

Dean, D. R. 1999. *Gideon Mantell and the discovery of dinosaurs.* New York: Cambridge.

Desmond, A. 1989. *The politics of evolution.* Chicago: University of Chicago Press.

———. 1994. *Huxley.* Reading: Addison-Wesley.

Gordon, E. 1894. *The life and correspondence of William Buckland.* New York: Appleton.

Gould, S. J. 1998. *Leonardo's mountain of clams and the diet of worms.* New York: Harmony.

Hitchcock, E. 1836. Ornithichnology—description of the foot marks of birds (Ornithichnites) on New Red sandstone in Massachusetts. *American Journal of Science* 29: 305–339.

———. 1851. *The religion of geology.* Boston: Philips, Sampson and Co.
———. 1858. *Ichnology of New England.* Boston: Wright and Porter.
Hitchcock, E., Jr. 1865. *Supplement to the ichnology of New England.* Boston: Wright and Porter.
Huxley, T. H. 1881. *The anatomy of vertebrated animals.* New York: Appleton.
Lull, R. S. 1915. Triassic life of the Connecticut Valley. *Connecticut Geological and Natural History Survey Bulletin* 81: 1–331.
Mantell, G. 1844. *The medals of creation.* London: R. Clay.
Ostrom, J. H. 1973. The ancestry of birds. *Nature* 242: 136–138.
Phillips, J. 1871. *Geology of Oxford.* Oxford: Clarendon.
Roger, J. 1997. *Buffon.* Ithaca, N.Y.: Cornell University Press.
Rupke, N. 1983. *The great chain of history.* Oxford: Clarendon.
———. 1994. *Richard Owen, Victorian naturalist.* New Haven, Conn.: Yale University Press.

Section I. The Setting

1. The Dinosaurian Setting of Primitive Asian Birds

Dale A. Russell

Abstract

Endemism in Central Asia appears to have been strongest during a 30–40-million-year interval corresponding to middle Mesozoic time, when the abundant presence of mamenchisaurs implies productive and broadly available low-level fodder. Transitional endemic/exotic assemblages during basal Cretaceous time may reflect the operation of poorly constrained paleogeographic and paleoenvironmental factors. Central Asian dune fields supported a strange ecosystem dominated by small dinosaurs. Nearly 17 percent of the total world collection of dinosaur skeletons has been recovered from fossil (Djadokhta and equivalent) dune fields. The latter assemblage generally resembles those (without dinosaurs) that survived Cretaceous-Tertiary extinctions in North America. Floodplain (Nemegt) assemblages were more diverse on a familial level than contemporary dinosaurian assemblages in North America. Both carnivorous and herbivorous dinosaurs were apparently more restricted in their dietary preferences than are modern mammals. Relative to the record of Tertiary mammals, that of dinosaurs is strikingly incomplete. Although the phylogenetic origin of birds is rapidly becoming clarified, dinosaurian proxy evidence is inadequate to constrain the biogeographic origin of birds.

Introduction

Today, birds intimately mingle with large, flightless vertebrates. It is reasonable to suppose that they did so during Mesozoic time, yet their remains are usually preserved apart from those of their larger contemporaries in the fossil record. Presumably their delicate skeletons were destroyed by fluvial or digestive processes, but were preserved intact in fine-grained lacustrine sediments where wind and water seldom carried robust skeletal parts. In the past decade, superb skeletons of Mesozoic birds have been recovered from lake deposits in Central Asia. Many excellent phylogenetic studies link their ancestry to their non-volant theropod cousins. This short essay describes the dinosaurian context of primitive Central Asian birds. This context is compared to the dinosaurian record in North America to highlight similarities and differences.

Due to the tectonic history of the western regions of the continent and the industry of collectors, the dinosaur assemblages of the Jurassic and Cretaceous of North America presently constitute a reference sequence against which dinosaur assemblages elsewhere in the world may be compared. The dinosaurs of the Kayenta, Morrison, Dinosaur Park, and Hell Creek formations were found to document major phases in the evolution of Mesozoic terrestrial life. During the latter part of the 20th century, the dinosaurian record has also been vigorously explored in Central Asia, such that the assemblages of the Lufeng, Shaximiao, Yixian, Djadokhta, and Nemegt formations are becoming as well studied as their North American counterparts. A few impressionistic comments on Asian dinosaur assemblages are presented in the hope that they will serve to stimulate discussions of terrestrial ecosystems populated by non-avian dinosaurs and ancestral birds.

Biogeography

The question of whether or not Central Asia was biogeographically isolated during middle Mesozoic time has long been a matter of speculation (e.g., Russell 1993; Hunt et al. 1994; Upchurch 1994, 1995; Lucas 1996a; Evans et al. 1998; Wilson and Sereno 1998). The immediate theropod progenitors of *Archaeopteryx* have not been identified in the Northern Hemisphere, and the diverse but fragmentary remains of the relatively larger "spinosauroids" (basal tetanurans) and "allosauroids" have not yet yielded sufficient information to disclose the biogeographic roots of Jurassic theropods (Allain 2001; T. Holtz, pers. comm. 2001). Indications of isolation may be reflected in the composition of the Lufeng, Shaximiao, and Yixian assemblages.

Lufeng Assemblage

The Lufeng Formation in Yunnan, southern China, is generally considered to be of Liassic, and probably of Sinemurian age (Luo and Wu 1994; Lucas 1996a,c), as is its approximate correlate, the Kayenta Formation in North America (Sues et al. 1994). That Central Asia was then part of Pangea is indicated by intercontinental distributions of Lufeng non-dinosaurian vertebrates (including sphenodontians, crocodylomorphs, tritylodonts, and mammals, Luo and Wu 1994; Sues et al. 1994) and by the occurrence of the ceratosaur *Dilophosaurus* (but see Lamanna et al. 1998) and basal thyreophoran *Scelidosaurus* in both the Lufeng Formation and the Kayenta Formation of Arizona (Lucas 1996a,b).

Shaximiao Assemblages

The strangeness of middle Mesozoic dinosaurs from the Red Basin of Sichuan, in central China, became clearly apparent with the description of *Mamenchisaurus* (Young and Chao 1972). Mamenchisaurids paralleled the North American diplodocids in their long, horizontal necks and split chevrons, but differed from them in not having their spatulate teeth and elongated cervical ribs (Russell and Zheng 1993). The characteristic teeth and vertebrae of diplodocoids have not been recognized by this author (personal observation) in collections of Jurassic sauropods from China. Dinosaurs from the upper Shaximiao assemblage (approximately of Late Jurassic age) are distinct from those of the Morrison and Tendaguru Formations (Kimmeridgian-Tithonian) on at least a generic level (Russell 1993).

The age of the lower Shaximiao Formation is considered to be approximately "middle Jurassic" by Russell (1993) or "tentatively Bajocian" by Lucas (1996a,c), citing correlations based on fossils of fresh water microorganisms. Its assemblage also shows evidence of generic-level endemism (Russell 1993), and contains abundant skeletal remains of the mamenchisaurid *Omeisaurus*. A Bathonian-Callovian sauropod from Morocco (*Atlasaurus*, Monbaron et al. 1999) appears to be much more closely related to *Brachiosaurus* than to *Omeisaurus*, supporting the hypothesis that Central Asia was separated from Pangea by an ecologic or topographic (oceanic) barrier.

Stegosaurs and sauropods evidently originated before the isolation of Central Asia, for both groups have Pangean distributions and are well represented in Shaximiao assemblages. In view of these circumstances and the endemism of the lower Shaximiao assemblage, it is possible that barriers to intercontinental interchange were present before the end of the Liassic.

Yixian Assemblage

Russell (1993) postulated isolation to have ended during Aptian time, based on the appearance of ornithomimids, dromaeosaurs, and iguanodonts in Central Asia, and the disappearance of mamenchisaurs and stegosaurs soon thereafter. The "replacing" fauna was also present in North America (Kirkland et al. 1997), but its area of origin is unclear. Luo (1999) concurred, noting the presence of relict or endemic dinosaurs (coelurosaurs, therizinosaurs, and psittacosaurs) in the somewhat older Yixian Formation of Liaoning, in northeastern China, in a horizon that has been radiometrically dated at 124 million years, or middle Barremian time (Swisher et al. 1999). The Yixian has recently become world-famous for its skeletons of feathered dinosaurs. However, a dromaeosaur (*Sinornithosaurus*, Xu et al. 1999) which would appear to belong to the younger "replacing" fauna has also been identified in the Yixian.

In the Mazongshan region of Gansu, fluviolacustrine strata dated as Barremian (Xinminbao Group, You et al. 1999) contain remains of mamenchisaurs, therizinosaurs, and psittacosaurs of an older, endemic aspect, as well as sauropods and a ceratopsian (*Archaeoceratops,* Dong 1997) apparently belonging to the "replacing" fauna. It seems likely that a transitional assemblage of dinosaurs was already present in northern China by Barremian time (Tang et al. 2001). Recently, a rich microvertebrate bone bed dominated by remains of fresh-water aquatic vertebrates from the Tetori Group in Japan (probably of basal Cretaceous age) has provided evidence suggesting connections with Europe across the Turgai Straits as early as Late Jurassic time (Evans et al. 1998). Dinosaurian remains from the basal Cretaceous of Japan also indicate the presence of tyrannosaurs and representatives of the oviraptorosaur-therizinosauroid clade, and that the alternatives of local origin versus immigration may be difficult to distinguish (Manabe et al. 2000). By Cenomanian-Turonian time, vertebrate assemblages in the southwestern United States show strong evidence of Asian affinities (Cifelli et al. 1997; Kirkland et al. 1997; Kirkland and Wolfe 2001).

In Inner Mongolia, fluviolacustrine strata of the Iren Dabasu Formation contain a dinosaurian assemblage of post-Cenomanian, pre-Maastrichtian age that is older than, but generally similar to that of the Nemegt Formation (Currie and Eberth 1993). By Nemegt time, the presence of titanosaurs in Mongolia (Rogers and Forster 2001), as well as faunistic (Prasad and Sahni 1999) and tectonic (Yin and Harrison 2000) evidence, suggests the possibility of faunal links to Gondwana continents. *Nemegtosaurus*-like sauropods also occur in the Hauterivian-Barremian of southeast Asia, where no remains of mamenchisaurs have been identified (Buffetaut and Suteethorn 1999). There is much to learn about the biogeographic history of terrains in the eastern Tethyan region.

Thus, the case for a mid-Mesozoic isolation of Central Asia appears to be strongest for the poorly correlated assemblages of the Shaximiao Formation, conventionally considered to be of middle and upper Jurassic age. A span of time consistent with the origin and dominance of these assemblages might be on the order of 30–40 million years, and centered on late middle Jurassic time (Gradstein et al. 1994). The apparently transitional nature of Early Cretaceous terrestrial faunas may reflect paleoenvironmental as well as paleogeographic factors. It remains to be seen whether or not the progenitors of birds were present in Asia when it was isolated.

Ecology

The dinosaurs of Central Asia inhabited the interior of a continent apparently far removed from lowland environments near sea level such as those in which the Jurassic-Cretaceous record of North America is preserved (Hicks et al. 1999). They could thus be expected to differ from their North American counterparts in reflecting the effects of continental climates and abrupt changes in topography.

Shaximiao Assemblages (Long-Necked Sauropods)

The skeletal architecture of terrestrial herbivores reflects the architecture of the plants upon which they feed. The giraffoid structure of the skeleton of *Brachiosaurus* (Janensch 1950) suggests the animal browsed on arborescent vegetation. It has been demonstrated that diplodocids (*Apatosaurus* and *Diplodocus*, Stevens and Parrish 1999) fed on low-level vegetation. *Brachiosaurus* is the dominant herbivore in the Tendaguru assemblage of Tanzania (Russell et al. 1980) and it may be presumed that trees were common in its local environment. In their general skeletal structure *Mamenchisaurus* (Young and Chao 1972) and *Omeisaurus* (He et al. 1988) resemble long-necked diplodocids, although their jaws and dentition resemble those of *Brachiosaurus* (Russell and Zheng 1993). The relative abundance of mamenchisaurid remains in the Jurassic of China is accordingly suggestive of widespread bushy vegetation.

Djadokhta Assemblage (Dune Dinosaurs)

Within the heartland of Asia lived an ecosystem dominated by small dinosaurs that was so strange as to be almost otherworldly in comparison with lowland dinosaur ecosystems in North America. Already recognizable in arid environments by late Early Cretaceous time (Khukhtykian, Jerzykiewicz and Russell 1991; Hicks et al. 1999) and along rift lakes (see Yixian Formation, above), it was typically preserved in dune field environments (Djadokhta Formation and equivalents) widely spread across the Mongolias during the latter part of the Late Cretaceous (Dashzeveg et al. 1995; Watabe and Fastovsky 1999). Nearly three-quarters of dinosaurian specimens recovered belonged to protoceratopsids, and the remainder are approximately equally divided between small theropods and ankylosaurids (Dashzeveg et al. 1995). They are of a size range comparable to that represented by barnyard animals, such as cats, chickens, geese, turkeys, dogs, and pigs. Teeth and eroded skeletal elements of larger dinosaurs (Osmólska 1980; Jerzykiewicz et al. 1993; Makovicky and Norell 1998) were only rarely transported into the dune region in streams. No nests of larger dinosaurs have yet been described.

The Djadokhta assemblage (and its ecological antecedents) is also characterized by the abundance and diversity of lizard and mammalian remains. No other assemblage of small dinosaurs is comparably well known elsewhere in the world, although isolated teeth of microdinosaurs occur in some diversity in the middle Cretaceous of Morocco and the Iberian Peninsula (Sigogneau-Russell et al. 1998).

Most of the skeletons of small dinosaurs in this assemblage are exceptionally well preserved, frequently in the positions in which they died. The high degree of articulation associated with burrows suggests a scarcity of vertebrate and abundance of arthropod scavengers (cf. Jerzykiewicz et al. 1993; Norell et al. 1995; Fastovsky et al. 1997; Loope et al. 1998). Given the abundance of skeletons of small animals (lizards, mammals), it would appear that the lower size limit for dinosaurs is well constrained in the Djadokhta assemblage; if smaller dinosaurs were present they too would have been preserved (the smaller Djadokhta coelurosaurs varied between 1.5 and 2 m in length, Paul 1988; the smallest known adult dinosaur, from the Early Cretaceous of Liaoning, China, was about 40 cm long, Xu et al. 2000; an articulated bird skeleton from the Djadokhta is smaller than that of the smallest known dinosaur adult, Norell and Clarke 2001). More than 350 specimens of dinosaurs have been counted in Djadokhta-facies sediments (Dashzeveg et al. 1995), representing nearly 17 percent of the total number of associated dinosaur specimens collected from around the world (N = 2,100, Dodson 1990). In view of the small size of the dinosaurs and the ease with which they can be excavated, Djadokhta strata may soon rival those of the Dinosaur Park Formation, Alberta, in the total number of dinosaur specimens yielded (cf. Béland and Russell 1978).

Two comparably well-collected dinosaurian assemblages in North America have produced, respectively, 37 (Dinosaur Park Formation) and 53 (Morrison Formation) species of dinosaurs. It should be noted that the Morrison Formation is distributed over a much broader area and encompasses a greater span of time (about 7 million years, Kowallis et al. 1998) than the Dinosaur Park Formation (about 1.5 million years, Eberth and Hamblin 1993). However, Djadokhta strata have produced only 13 species (see table 1.2). A paradox is presented by low diversity, implying biologically stressed environments (Jerzykiewicz et al. 1993), and high fossil recovery, implying biologically productive environments (Fastovsky et al. 1997). However, high water tables and streams developed during mesic intervals (Loope 1998), suggesting the possibility that large populations could have coincided with relatively humid conditions.

Various predatory dinosaurs may have been adapted to feed upon smaller vertebrates of either diurnal (lizard, bird) or nocturnal (mammal) habits. Abundant and morphologically varied burrows and pupal chambers, as well as invertebrate-bored dinosaur skeletons, imply moist sand rich in organic material and large populations of scavenging arthropods (probably beetles, e.g., Jerzykiewicz et al. 1993; Johnson et al. 1996; Fastovsky et al. 1997; Watabe and Fastovsky 1999). Oviraptorids may have extracted insects and lizards from their sandy burrows with long, clawed fingers. These animals nested in the dune fields, and four out of 17 skeletons of adult oviraptorids were found sitting on nests (Clark et al. 1998; Clark 1999). A variety of egg types found in Djadokhta sediments (Jerzykiewicz et al. 1993) may have provided a seasonally abundant source of food. The long legs and powerful, single-digit brachial "hooks" of theropod-mimic birds (alvarezsaurids, *Mononykus, Shuvuuia,* Chiappe et al. 1998) constitute an extraordinary combination of structures. It is, however, possible to visualize an alvarezsaur thrusting its "hooks" into a small egg and scooting away to consume its immobile prey safely elsewhere.

The little dinosaurs of the dune fields of Central Asia possessed body forms that might be seen as transitional between those of Triassic archosaurs and Tertiary mammals. There is little evidence of the gigantism that elsewhere characterized terrestrial Mesozoic faunas. On the whole, the Djadokhta vertebrate assemblage (apart from the dinosaurs) is similar to the assemblage that survived the terminal Cretaceous extinctions in the western interior of North America (cf. Sheehan and Fastovsky 1992). A conundrum is thereby posed: if a worldwide regression produced the changes observed at the boundary (cf. Archibald 1996), why did small dinosaurs adapted to regions at some distance from strand lines not survive?

Nemegt Assemblage (Riparian Tyrannosaurs)

In southern Mongolia escarpments of the Nemegt Formation, in striking contrast to those of the dune (Djadokhta) facies, have yielded few remains of microvertebrates (Jerzykiewicz and Russell 1991). Only two of the small dune-dwelling theropods (*Oviraptor, Saurornithoides*) are known to have been sufficiently adaptable to tolerate fluviolacustrine environments of the Nemegt. The wetlands that accommodated the Nemegt assemblage may have been confined to river valleys and small lakes within a semi-arid to arid environment (Jerzykiewicz 1995). Temporal relationships between fluviolacustrine and dune facies are at least to some degree ambiguous (Dashzeveg et al. 1995; Fastovsky et al. 1998). Differences between their respective dinosaurian assemblages may be due more to local environment than to time (contra Russell in Jerzykiewicz and Russell 1991).

Fossil wood and turtle remains are, however, abundantly preserved in the Nemegt, and the facies has yielded the largest Late Cretaceous megafauna known from Asia (Jerzykiewicz and Russell 1991). Overall, the fauna was nearly as diverse as contemporary, well-collected assemblages from North America (tables 1.1–1.3). At the family level it was marginally more diverse (17 versus 16 families in two North American assemblages, table 1.3). Ecological differences from the North American assemblages may also be reflected in the absence of nodosaurs, large ceratopsians, and *Edmontosaurus*-like hadrosaurs (possibly preferring more coastal environments), and the presence of therizinosaurs and a gigantic ornithomimosaur (possibly preferring more inland environments).

Using regressions in Anderson et al. (1985), skeletal measurements of the two Nemegt hadrosaurs (*Barsboldia,* Maryanska, and Osmólska 1981b; *Saurolophus,* Maryanska and Osmólska 1981a, 1984) indicate weights well in excess of 10 metric tons. *Shantungosaurus* from the Late Cretaceous of Shandong (Hu 1973) was probably even more ponderous. North American hadrosaurs typically weighed between 2 and 4 metric tons (Anderson et al. 1985). At least two varieties of large sauropods were also present in Mongolia, neither of which was closely related to the contemporary North American titanosaur (*Alamosaurus*). The Nemegt was located at approximately the same paleolatitude as northern Colorado (C. Scotese, pers. comm. 1999), so that differing sauropods may have been the result of immigration from different tropical source lands (respectively southern Asia and neotropical America). The large size of Nemegt herbivores suggests relatively high levels of primary productivity.

In the Late Cretaceous of North America, the distributions of dinosaurian herbivores are much more restricted than those of modern large mammals (Lehman 1997, 2001). In coastal plain environments, changes at the species and generic levels in dominant herbivores can occur within 10 degrees of latitude (1,000 km), and within 100 km

TABLE 1.1.
Distribution of Late Cretaceous families of dinosaurs

Family	**(1)***	**(2)**	**(3)**
Dromaeosauridae	x	x	x
Troodontidae	x	x	x
Oviraptoridae	x		
Caenagnathidae	x	x	x
Avimimidae	x		
Ornithomimidae	x	x	x
Family undet. (*Bagaraatan*)			x
Tyrannosauridae	x	x	x
Therizinosauridae	x		
Deinocheiridae	x		
Titanosauromorphs	x	x	x
Thescelosauridae		x	x
Hypsilophodontidae		x	
Homalocephalidae	x	x	x
Pachycephalosauridae	x	x	x
Protoceratopsidae	x	x	x
Ceratopsidae		x	x
Ankylosauridae	x	x	x
Nodosauridae		x	x
Lambeosaurinae	x	x	
Hadrosaurinae	x	x	x
Total	**17**	**16**	**14**

*Abbreviations: (1) Mongolia, (2) North America, (3) terminal Cretaceous "*Triceratops* fauna."

from higher to lower elevations across an aggrading coastal plain (Brinkman et al. 1998). It is likely that even more abrupt transitions occurred between regions inhabited by small protoceratopsids and large hadrosaurs in the Late Cretaceous of Central Asia. Dinosaurian herbivores were much more restricted in their vegetational (dietary) preferences than large mammalian herbivores were (Banfield 1974).

Carnivory in modern terrestrial vertebrates is usually directed toward prey animals that are smaller than their predators (see regressions in Peters 1986). Bone fragments preserved in two coprolites ascribed to tyrannosaurs (both from Canada) indicate smaller, subadult herbivore prey (Chin et al. 1998, 1999). Circumstantial evidence suggestive of active carnivory rather than scavenging in tyrannosaurs is the rapid liquefaction (2–4 days) of carcasses of large prey animals (elephants) through bacterial decay (Coe 1978; personal observation). Hot, liquefied prey tissues seem incompatible with the feeding apparatus of tyrannosaurids, and it is here assumed that the Nemegt tyrannosaur (*Tarbosaurus*) preyed upon living, subadult hadrosaurs. However, skeletons of *Tarbosaurus* are preserved about as abundantly as those of the principal herbivore *Saurolophus* (Osmólska 1980), indicating that

TABLE 1.2
Asian dinosaur assemblages

Djadokhta assemblage—After Jerzykiewicz and Russell (1991), with citations of additional records contained within parentheses. Taxa considered allochthonous to the assemblage are indicated by *.

Theropoda, family undetermined (Elzanowski and Wellnhofer 1993)
    Archaeornithoides deinosauriscus

Theropoda
    *Genus undetermined (Jerzykiewicz et al. 1993)

Cf. Tyrannosauridae
    *Genus undetermined

Ornithomimidae
    *Undescribed genus (Makovicky and Norell 1998)

Avimimidae
    Avimimus portentosis

Oviraptoridae
    Oviraptor philoceratops
    Citipati osmolskae (Clark et al. 2001)
    Khaan mckennai (Clark et al. 2001)

Troodontidae
    Saurornithoides mongoliensis
    Byronosaurus jaffei (Norell et al. 2000)

Dromaeosauridae
    Velociraptor mongoliensis

Titanosauria
    *cf. *Nemegtosaurus* sp.

Hadrosauridae
    *Genus undetermined

Homalocephalidae
    Goyocephale lattimorei

Protoceratopsidae
    Bagaceratops sp.
    (Dashzeveg et al. 1995; Jerzykiewicz et al. 1993)
    Protoceratops andrewsi
    Udanoceratops tschizhovi
    (Jerzykiewicz et al. 1993)

Ankylosauridae
    Pinacosaurus grangeri

Nemegt assemblage—After Jerzykiewicz and Russell (1991), with citations of additional records contained within parentheses.

Tetanurae, family undetermined
    Bagaraatan ostromi (Osmólska 1996)

Tyrannosauridae
    Alioramus remotus
    Tarbosaurus bataar (including *Maleevosaurus novojilovi*, Carr 1999)

Ornithomimidae
    Anserimimus planinychus
    Gallimimus bullatus

Deinocheiridae
    Deinocheirus mirificus

Caenagnathidae (Sues 1997)
    Elmisaurus rarus

Avimimidae
    Avimimus portentosus

Oviraptorosauria, family undetermined
    Nomingia gobiensis (Barsbold et al. 2000)

Oviraptoridae
    Oviraptor mongoliensis

Therizinosauridae
    Therizinosaurus cheloniformis

Troodontidae
    Borogovia gracilicrus
    Saurornithoides junior
    Tochisaurus nemegtensis
    (Kurzanov and Osmólska 1991)

Dromaeosauridae
    Adasaurus mongoliensis
    ? *Velociraptor* sp.

Titanosauria (Rogers and Forster 2001)
    Nemegtosaurus mongoliensis
    Opisthocoelicaudia skarzynskii

Hadrosauridae (Hadrosaurinae)
    Saurolophus angustirostris

Hadrosauridae (Lambeosaurinae)
    Barsboldia sicinskii

Homalocephalidae
    Homalocephale calathocercos

Pachycephalosauridae
    Prenocephale prenes

Ankylosauridae
    Tarchia gigantea

TABLE 1.3
North American dinosaur assemblages

Morrison assemblage—Kimmeridgian–Tithonian, after Chure et al. 1998, with citation of additional record contained within parentheses.

Coeluridae
- *Coelurus fragilis*
- *Ornitholestes hermanni*
- Undescribed species

? Troodontidae
- *Koparion douglassi*

Megalosauridae
- *Torvosaurus tanneri*

Ceratosauridae
- *Ceratosaurus nasicornis*
- *Ceratosaurus,* sp. nov.

Allosauridae
- *Allosaurus atrox*
- *Allosaurus fragilis*
- *Saurophaganax maximus*
- *Allosaurid,* sp. nov.

? Tyrannosauridae
- *Stokesosaurus clevelandi*

Family incertae sedis
- *Marshosaurus bicentesimus*
- *Elaphrosaurus* sp.
- Dromaeosaurid-like teeth
- Small maniraptoran ulna

Cetiosauridae
- *Haplocanthosaurus delfsi*
- *Haplocanthosaurus priscus*
- Cetiosaurid, unnamed (= "*Morosaurus*" *agilis*)
- Cetiosaurid, unnamed (= "*Apatosaurus*" *minimus*)

Brachiosauridae
- *Brachiosaurus altithorax*

Camarasauridae
- *Camarasaurus grandis*
- *Camarasaurus lentus*
- *Camarasaurus lewisi*
- *Camarasaurus supremus*

Diplodocidae
- *Amphicoelis altus*
- *Apatosaurus ajax*
- *Apatosaurus excelsus*
- *Apatosaurus louisae*
- *Barosaurus lentus*
- *Diplodocus carnegii*
- *Diplodocus lacustris*
- *Diplodocus longus*
- *Diplodocus hayi*
- *Dyslocosaurus polyonychius*
- *Dystrophaeus viaemalae*
- *Seismosaurus hallorum*
- *Supersaurus vivianae* (*Dystylosaurus edwini*)

Fabrosauridae
- *Echinodon* sp.

? Heterodontosauridae
- Undescribed species

Hypsilophodontidae
- *Drinker nisti*
- *Othnielia rex*

Dryosauridae
- *Dryosaurus altus*

Camptosauridae
- *Camptosaurus amplus*
- *Camptosaurus depressus*
- *Camptosaurus dispar*

? Iguanodontidae
- cf. *Iguanodon*, sp.

Stegosauridae
- *Stegosaurus armatus*
- *Stegosaurus stenops*
- *Hesperosaurus mjosi*
- *"Stegosaurus" longispinus*

Nodosauridae
- *Mymoorapelta maysi*

Ankylosauridae
- *Gargoyleosaurus parkpinorum* (Carpenter et al. 1998)

Family incertae sedis
- Undescribed ankylosaur

Dinosaur Park assemblage—Campanian, after Weishampel (1990), with citations of additional records contained within parentheses.

Tyrannosauridae
- *Aublysodon mirandus* (Currie et al. 1990)
- *Albertosaurus libratus*
- *Daspletosaurus*, new species (P. J. Currie, pers. comm. 1999)

Ornithomimidae
- *Dromiceomimus samueli*
- *Ornithomimus edmontonensis*
- *Struthiomimus altus*

TABLE 1.3 *(cont.)*
North American dinosaur assemblages

Caenagnathidae (Sues 1997)
- *Chirostenotes pergracilis*
- *Chirostenotes elegans*

Troodontidae
- *Troodon formosus*

Dromaeosauridae
- *Dromaeosaurus albertensis*
- *Saurornitholestes langstoni*

Maniraptora, family undetermined (Baszio 1997)
- *Ricardoestesia gilmorei*

Hadrosauridae (Hadrosaurinae)
- *Brachylophosaurus canadensis*
- *Gryposaurus notabilis*
- *Kritosaurus incurvimanus*
- *Prosaurolophus maximus*
- cf. *Maiasaura* sp. (P. J. Currie, pers. comm. 1999)

Hadrosauridae (Lambeosaurinae)
- *Corythosaurus casuarius*
- *Lambeosaurus lambei*
- *Lambeosaurus magnicristatus*
- *Parasaurolophus walkeri*

Pachycephalosauridae
- *Gravitholus albertae*
- *Pachycephalosaurus* sp.
- *Stegoceras validum*
- *Ornatotholus browni*

Protoceratopsidae (Ryan and Currie 1998)
- Cf. *Leptoceratops* sp.

Ceratopsidae (Chasmosaurinae)
- *Chasmosaurus belli*
- *Chasmosaurus canadensis*
- *Chasmosaurus russelli*
- *Chasmosaurus*, new species (Holmes et al. 1999; P. J. Currie, pers. comm. 1990)

Ceratopsidae (Centrosaurinae)
- *Centrosaurus apertus*
- *Monoclonius crassus*
- *Styracosaurus albertensis*

Nodosauridae
- *Edmontonia longiceps*
- *Edmontonia rugosidens*
- *Panoplosaurus mirus*

Ankylosauridae
- *Euoplocephalus tutus*

Hell Creek assemblage—Maastrichtian, after Russell and Manabe (2002).

Tyrannosauridae
- *Aublysodon* cf. *A. mirandus*
- *Tyrannosaurus rex* (including *Nanotyrannus lancensis*, Carr 1999)

Ornithomimidae
- *Ornithomimus* sp. (Triebold 1997)
- *Struthiomimus* sp. (Triebold 1997)

Dromaeosauridae
- *Dromaeosaurus albertensis* (Baszio 1997)
- cf. *Saurornitholestes langstoni* (Baszio 1997)

Caenagnathidae
- *Caenagnathus* sp. (Currie et al. 1993)

Troodontidae
- *Troodon formosus*

Maniraptora, family undetermined
- *Ricardoestesia gilmorei*
- *Ricardoestesia*, undescribed species (Baszio 1997)

Theropoda, family undetermined
- *Paronychodon lacustris* (Baszio 1997)

Nodosauridae
- *Edmontonia* sp.

Ankylosauridae
- *Ankylosaurus magniventris*

Thescelosauridae
- *Bugenasaura infernalis* (Galton 1995)
- *Thescelosaurus neglectus*
- *Thescelosaurus garbanii* (Morris 1976)

Hadrosauridae
- *Anatotitan copei*
- *Edmontosaurus annectens*

Pachycephalosauridae
- cf. *Homalocephalidae,* new genus and species (H. Galiano, pers. comm. 2001)
- *Pachycephalosaurus wyomingensis*
- *Stegoceras validum*
- *Stygimoloch spinifer*

Ceratopsidae
- *Torosaurus latus*
- *Triceratops horridus*
- *Triceratops prorsus* (Forster 1996)

TABLE 1.4
Herbivore/carnivore species ratios

Dinosaur assemblages—see Tables 1.2 and 1.3. Carnivores include all theropods except therizinosaurs. Gastroliths are associated with the skeletons of some Asian coelurosaurs (Kobayashi et al. 1999), but have not been found with those of their North American counterparts. Remaining dinosaurs are considered to have been herbivores.

Nemegt: 8/15		0.53
Djadokhta: 5/8		0.63
Hell Creek: 14/11		1.27
Dinosaur Park: 25/12	2.08	
Morrison: 37/16	2.31	

Mammal North American Provincial Ages—After Savage and Russell (1983); carnivores are mammals presumed to have had carnivorous habits (including marsupials and "creodonts"); pantolestans and apatotheres are considered herbivores; insectivorous mammals, bats and marine mammals have been excluded. Ages span Eocene to Pleistocene time, when dentitions adapted to carnivory are relatively easy to recognize in mammals.

Rancholabrean: 213/50	4.26	
Irvingtonian: 140/39	3.59	
Blancan: 91/23		3.96
Hemphillian: 167/50		3.34
Clarendonian: 164/54	3.04	
Barstovian: 191/56		3.41
Hemingfordian: 144/36	4.00	
Late Arikareean: 176/53	3.32	
Early Arikareean: 98/16	6.13	
Whitneyan: 75/24	3.13	
Orellan: 109/23	4.04	
Chadronian: 176/38		4.63
Duchesnean: (samples too small)		
Uintan: 175/21		8.33
Wasatchian-Lostcabinian: 81/35	2.31	
Lysitean: 56/19	2.95	
Graybullian: 69/38	1.79	

Modern mammalian assemblages—After Banfield (1974), and Kingdom (1997).

Canada: 82/26		3.15
Africa: 489/73		6.70

the Nemegt environment was a predator trap. Additional support for this conclusion lies in the generally greater number of carnivore than herbivore skeletons in the Nemegt (Osmólska 1980), and the fact that the ratio of herbivore/carnivore dinosaur species is 0.53, the smallest ratio in five major dinosaurian assemblages (table 1.4).

Were carnivorous dinosaurs' prey preferences more restricted to special species than carnivorous mammals were? That this may have been so is suggested by herbivore/carnivore species ratios (table 1.4), which in the five dinosaurian assemblages equals approximately 1.4 (minus and plus one standard deviation: 0.6 to 2.2), and in Cenozoic mammals the ratio equals approximately 3.9 (minus and plus one standard deviation: 2.4 to 5.3). Thus for every carnivorous dinosaur species there were approximately 1.4 herbivore species, but in Cenozoic mammalian assemblages there were approximately 4 herbivore species. Ratios may be affected by other factors: They decline with increasing area sampled (Van Valkenburg and Janis 1993), and may vary latitudinally (cf. table 1.4: ratios in Canadian [3.2] and African [6.7] mammalian faunas). A ratio of approximately 2.3 in the Jurassic Morrison Formation enters into the range of those for Eocene mammals. Nevertheless, the possibility remains that both herbivorous and carnivorous dinosaurs were generally more restricted in their trophic preferences than are modern mammals.

Conclusions

The most striking result of this brief review of Jurassic-Cretaceous dinosaur assemblages of Central Asia is the incompleteness of information on the global record of dinosaurs. The tabulations in Table 1.4 are based on five of the most completely known assemblages of dinosaurs in the world, averaging 30 species each, and 16 assemblages of mammals from North America, averaging nearly 170 species each. The dinosaurian assemblages were selected from an interval of 145 million years and the mammalian assemblages from one of 55 million years. It is probably fair to conclude that the dinosaurian record is less understood than the mammalian record by about an order of magnitude. That two of the five most completely known dinosaurian assemblages occur in Asia suggests the extent to which our knowledge of Asian dinosaurs augments our understanding of the dinosaurian era.

In spite of the obscurity which shrouds Mesozoic terrestrial life, it does seem that Central Asia experienced an interval of isolation during mid-Mesozoic time that may have been as profound as that in which Australia was immersed for most of Cenozoic time. Differences between dune-dominated and river-dominated environments in the Mongolias produced herbivore assemblages respectively dominated by small protoceratopsids and gigantic hadrosaurs. Such a marked difference in herbivores, together with analogous differences correlated with environments in North America during Late Cretaceous time, suggests that dinosaurian herbivores may have been less tolerant of differences in regional vegetation than are contemporary mammalian herbivores. In parallel fashion, the greater diversity of dinosaurian predators relative to that of dinosaurian herbivores suggests that they too may have been more narrowly adapted to feeding on specific prey animals than are mammals. In such manner, the Asian record deepens our appreciation of the dinosaurian world.

Many other questions remain, and more will surely follow as older questions are resolved. The mid-Mesozoic isolation of Central Asia must be constrained more precisely, both in extent and duration. Where was the place of origin of Asian-American Cretaceous dinosaurs? What was the biogeographic source of birds? Were "dune dinosaurs" only an Asian phenomenon, or might similar ecosystems be preserved, undiscovered, in the middle Cretaceous of Africa? How was it that small dinosaurs adapted to stressed environments in Central Asia were unable to survive the stresses associated with the Cretaceous-Tertiary extinctions? With such a high family-level diversity preserved in one riparian environment in Mongolia, how many unknown varieties of dinosaurs inhabited the vast area and diverse habitats of Asia during Late Cretaceous time? Finally, what were the trophic relationships between advanced carnivorous coelurosaurs and archaic carnivorous birds? It is easy to visualize early Cretaceous dromaeosaurs as effective avian predators, but some dromaeosaurs were outweighed by some contemporaneous birds by a factor of over three (Xu et al. 2000; Zhou and Zhang 2002).

The mysteries of ancient Asia will attract explorers of the dinosaurian world and students of the evolution of avian form (MacCready 1985) for many generations to come.

Acknowledgments. The author thanks Dr. Dan Chure, Dr. Philip Currie, and Dr. Hans-Dieter Sues for their assistance in reviewing and completing the lists of taxa in various dinosaur assemblages. He is also grateful to Dr. Paul Barrett, Dr. David Fastovsky, Dr. Thomas Holtz, Dr. Makoto Manabe, Dr. Paul Upchurch, and Dr. Zhe-xi Luo for sharing with him their opinions on the temporal and environmental context of Asian assemblages. Particular thanks are due to Dr. James Kirkland for his careful review of the manuscript. The immediate and detailed knowledge of generous colleagues is the most valuable data retrieval resource of the information age.

References

Allain, R. 2001. Redescription de *Streptospondylus altdorfensis,* le dinosaure théropode de Cuvier, du Jurassique de Normandie. *Geodiversitas* 23: 349–367.

Anderson, J. F., A. Hall-Martin, and D. A. Russell. 1985. Long-bone circumference and weight in mammals, birds and dinosaurs. *Journal of Zoology (London)* 207: 53–61.

Archibald, J. D. 1996. *Dinosaur extinction and the end of an era: what the fossils say.* New York: Columbia University Press.

Banfield, A. W. F. 1974. *The Mammals of Canada.* National Museum of Natural Sciences (Canada). Toronto: University of Toronto Press.

Barsbold, R., H. Osmólska, M. Watabe, P. J. Currie, and K. Tsogtbaatar. 2000. A new oviraptorosaur (Dinosauria, Theropoda) from Mongo-

lia: The first dinosaur with a pygostyle. *Acta Palaeontologica Polonica* 25: 97–106.
Baszio, S. 1997. Investigations on Canadian dinosaurs. *Courier Forschungs-institut Senckenberg* 196: 1–77.
Béland, P., and D. A. Russell. 1978. Paleoecology of Dinosaur Provincial Park (Cretaceous), Alberta, interpreted from the distribution of articulated vertebrate remains. *Canadian Journal of Earth Sciences* 15: 1012–1024.
Brinkman, D. B., M. J. Ryan, and D. A. Eberth. 1998. The paleogeographic and stratigraphic distribution of ceratopsids (Ornithischia) in the Upper Judith River Group of western Canada. *Palaios* 13: 160–169.
Buffetaut, E., and V. Suteethorn. 1999. The dinosaur fauna of the Sao Khua Formation of Thailand and the beginning of the Cretaceous radiation of dinosaurs in Asia. *Palaeogeography, Palaeoclimatology, Palaeoecology* 150: 13–23.
Carpenter, K., C. Miles, and K. Cloward. 1998. First known skull of a Jurassic ankylosaur (Dinosauria). *Nature* 393: 782–783.
Carr, T. D. 1999. Craniofacial ontogeny in Tyrannosauridae (Dinosauria, Coelurosauria). *Journal of Vertebrate Paleontology* 19: 497–520.
Chiappe, L. M., M. A. Norell, and J. M. Clark. 1998. The skull of a relative of the stem-group bird *Mononykus*. *Nature* 392: 275–278.
Chin, K., T. T. Tokaryk, G. M. Erickson, and L. C. Calk. 1998. A king-sized theropod coprolite. *Nature* 393: 680–682.
Chin, K., D. A. Eberth, and W. J. Sloboda. 1999. Exceptional soft-tissue preservation in a theropod coprolite from the Upper Cretaceous Dinosaur Park Formation of Alberta. *Journal of Vertebrate Paleontology* 19: 37–38A.
Chure, D. J., K. Carpenter, R. Litwin, S. Hasiotis, and E. Evanoff. 1998. The fauna and flora of the Morrison Formation. *Modern Geology* 23: 507–537.
Cifelli, R. L., J. L. Kirkland, A. Weil, A. R. Deinos, and B. J. Kowallis. 1997. High precision Ar^{40}/Ar^{39} geochronology and the advent of North America's Late Cretaceous terrestrial fauna. *Proceedings National Academy of Sciences U.S.* 94: 11163–11167.
Clark, J. M. 1999. An oviraptorid skeleton from the Late Cretaceous of Ukhaa Tolgod, Mongolia, preserved in an avianlike brooding position over an oviraptorid nest. *American Museum Novitates* 3269: 1–36.
Clark, J. M., M. A. Norell, and L. M. Chiappe. 1998. A "brooding" oviraptorid from the Late Cretaceous of Mongolia and its avian characters. *Journal of Vertebrate Paleontology* 18: 34A.
Clark, J. M., M. A. Norell, and R. Barsbold. 2001. Two new oviraptorids (Theropoda: Oviraptorosauria), Upper Cretaceous Djadochta Formation, Ukhaa Tolgod, Mongolia. *Journal of Vertebrate Paleontology* 21: 209–213.
Coe, M. J. 1978. The decomposition of elephant carcasses in the Tsavo (East) National Park, Kenya. *Journal of Arid Environments* 1: 71–86.
Currie, P. J., J. K. Rigby, Jr., and R. E. Sloan. 1990. Theropod teeth from the Judith River Formation of southern Alberta, Canada. In K. Carpenter and P. J. Currie (eds.), *Dinosaur systematics: approaches and perspectives*, pp. 107–125. Cambridge: Cambridge University Press.
Currie, P. J. and D. A. Eberth. 1993. Palaeontology, sedimentology and palaeoecology of the Iren Dabasu Formation (Upper Cretaceous), Inner Mongolia, People's Republic of China. *Cretaceous Research* 14: 127–144.

Currie, P. J., S. J. Godfrey, and L. Nessov. 1993. New caenagnathid (Dinosauria: Theropoda) specimens from the Upper Cretaceous of North America and Asia. *Canadian Journal of Earth Sciences* 30: 2255–2272.

Dashzeveg, D., M. J. Novacek, M. A. Norell, J. M. Clark, L. M. Chiappe, A. Davidson, M. C. McKenna, L. Dingus, C. Swisher, and P. Altangerel. 1995. Extraordinary preservation in a new vertebrate assemblage from the Late Cretaceous of Mongolia. *Nature* 374: 446–449.

Dodson, P. E. 1990. Counting dinosaurs: how many kinds were there? *Proceedings National Academy of Sciences U.S.* 87: 7608–7612.

Dong, Z.-M. 1997. *Sino-Japanese Silk Road dinosaur expedition.* Beijing: China Ocean Press.

Eberth, D. A., and A. P. Hamblin. 1993. Tectonic, stratigraphic and sedimentologic significance of a regional discontinuity in the upper Judith River Group (Belly River wedge) of southern Alberta, Saskatchewan, and northern Montana. *Canadian Journal of Earth Sciences* 30: 174–200.

Evans, S. E., M. Manabe, E. Cook, R. Hirayama, S. Isaji, C. J. Nicholas, D. Unwin, and Y. Yabumoto. 1998. An Early Cretaceous assemblage from Gifu Prefecture, Japan. *Bulletin of the New Mexico Museum of Natural History and Science* 14: 183–186.

Fastovsky, D. E., D. Badamgarav, H. Ishimoto, M. Watabe, and D. B. Weishampel. 1997. The paleoenvironments of Tugrikin-Shireh (Gobi Desert, Mongolia) and aspects of the taphonomy and paleoecology of *Protoceratops* (Dinosauria: Ornithischia). *Palaios* 12: 59–70.

Fastovsky, D. E., M. Watabe, and D. Badamgarav. 1998. Late Cretaceous dinosaur-bearing paleoenvironments, Gobi Desert, Mongolia. In D. L. Wolberg, D. Gittis, S. Miller, L. Carey, and A. Raynor (eds.), *Dinofest International,* pp. 14–15. Philadelphia: The Academy of Natural Sciences.

Forster, C. A. 1996. Species resolution in *Triceratops:* cladistic and morphometric approaches. *Journal of Vertebrate Paleontology* 16: 259–270.

Galton, P. M. 1995. The species of the basal hypsilophodont dinosaur *Thescelosaurus* Gilmore (Ornithischia: Ornithopoda) from the Late Cretaceous of North America. *Neues Jahrbuch für Paläontologie, Abhandlungen* 198: 297–311.

Gradstein, F., F. P. Agterberg, J. G. Ogg, J. Hardenbol, P. van Veen, J. Thierry, and Z.-H. Huang. 1994. A Mesozoic time scale. *Journal of Geophysical Research* 99, B12: 24,051–24,074.

He, X.-L., K. Li, and K.-J. Cai. 1988. *Sauropod dinosaurs (2) Omeisaurus tianfuensis. The Middle Jurassic dinosaur fauna from Dashanpu, Zigong, Sichuan.* Chengdu: Sichuan Publishing House of Science and Technology (in Chinese with English summary).

Hicks, J. F., D. L. Brinkman, D. J. Nichols, and M. Watabe. 1999. Paleomagnetic and palynologic analyses of Albian to Santonian strata at Bayn Shireh, Burkhant, and Khuren Dukh, eastern Gobi Desert, Mongolia. *Cretaceous Research* 20: 829–850.

Holmes, R., K. Shepherd, and C. Forster. 1999. An unusual chasmosaurine ceratopsid from Alberta. *Journal of Vertebrate Paleontology* 19: 52A.

Hu, C. 1973. A new hadrosaur from the Cretaceous of Chucheng, Shantung. *Acta Geologica Sinica* 2: 179–202 (in Chinese with English summary).

Hunt, A. P., M. G. Lockley, S. G. Lucas, and C. A. Meyer. 1994. The global sauropod record. *Gaia* 10: 261–279.

Janensch, W. 1950. Die Skelettrekonstruktion von *Brachiosaurus brancai. Palaeontographica* (Supplement 7) 3: 97–103.
Jerzykiewicz, T. 1995. Cretaceous vertebrate-bearing strata of the Gobi and Ordos basins—a demise of the Central Asian lacustrine dinosaur habitat. In K. H. Chang and S. O. Park (eds.), *Environment and tectonic history of East and South Asia,* pp. 233–256. Taegu: Proceedings of the 15th International Symposium of Kyungpook National University.
Jerzykiewicz, T., and D. A. Russell. 1991. Late Mesozoic stratigraphy and vertebrates of the Gobi Basin. *Cretaceous Research* 12: 345–377.
Jerzykiewicz, T., P. J. Currie, D. A. Eberth, P. A. Johnson, E. H. Koster, and J.-J. Zheng. 1993. Djadokhta Formation correlative strata in Chinese Inner Mongolia: an overview of the stratigraphy, sedimentary geology, and paleontology and comparisons with the type locality in the pre-Altai Gobi. *Canadian Journal of Earth Sciences* 30: 2180–2195.
Johnson, P. A., D. A. Eberth, and P. K. Anderson. 1996. Alleged vertebrate eggs from Upper Cretaceous redbeds, Gobi Desert, are fossil insect (Coleoptera) pupal chambers: *Fictovichnus* new ichnogenus. *Canadian Journal of Earth Sciences* 33: 511–525.
Kingdom, J. 1997. *The Kingdom field guide to African mammals.* London: Academic Press.
Kirkland, J. L., B. Britt, D. L. Burge, K. Carpenter, R. Cifelli, F. DeCourten, J. Eaton, S. Hasiotis, and T. Lawton. 1997. Lower to middle Cretaceous dinosaur faunas of the central Colorado Plateau: a key to understanding 35 million years of tectonics, sedimentology, evolution, and biogeography. *Brigham Young University Geology Studies* 42: 69–103.
Kirkland, J. L., and D. G. Wolfe. 2001. First definitive therizinosaurid (Dinosauria: Theropoda) from North America. *Journal of Vertebrate Paleontology* 21: 410–414.
Kobayashi, Y., J.-C. Lu, Z.-M. Dong, R. Barsbold, Y. Azuma, and Y. Tomida. 1999. Herbivorous diet in an ornithomimid dinosaur. *Nature* 402: 480–481.
Kowalis, B. J., E. H. Christiansen, A. L. Deino, F. Peterson, C. E. Turner, M. J. Kunk, and J. D. Obradovich. 1998. The age of the Morrison Formation. *Modern Geology* 22: 235–260.
Kurzanov, S. M., and H. Osmólska. 1991. *Tochisaurus nemegtensis* gen. et sp. n., a new troodontid (Dinosauria, Theropoda) from Mongolia. *Acta Palaeontologia Polonica* 36: 69–76.
Lamanna, M. C., J. B. Smith, H.-L You, T. R. Holtz, and P. Dodson. 1998. A reassessment of the Chinese theropod dinosaur *Dilophosaurus sinensis. Journal of Vertebrate Paleontology* 18: 57A.
Lehman, T. H. 1997. Late Campanian dinosaur biogeography in the western interior of North America. In D. L. Wolberg, E. Stump, and G. D. Rosenberg (eds.), *Dinofest International,* pp. 223–240. Philadelphia: The Academy of Natural Sciences.
———. 2001. Late Cretaceous dinosaur provinciality. In D. H. Tanke and K. Carpenter (eds.), *Mesozoic vertebrate life: new research inspired by the paleontology of Philip J. Currie,* pp. 310–328. Bloomington: Indiana University Press.
Loope, D. B., L. Dingus, C. C. Swisher, and C. Minjin. 1998. Life and death in a Late Cretaceous dune field, Nemegt Basin, Mongolia. *Geology* 26: 27–30.
Lucas, S. G. 1996a. Vertebrate biochronology of the Jurassic of China. *Museum of Northern Arizona Bulletin* 60: 23–33.

———. 1996b. The thyreophoran dinosaur *Scelidosaurus* from the Lower Jurassic Lufeng Formation, Yunan, China. *Museum of Northern Arizona Bulletin* 60: 81–85.

———. 1996c. Vertebrate biochronology of the Mesozoic of China. *Memoirs of Beijing Natural History Museum* 55: 109–148.

Luo, Z. 1999. A refugium for relicts. *Nature* 400: 23–24.

Luo, Z., and X.-C. Wu. 1994. The small tetrapods of the Lower Lufeng Formation, Yunnan, China. In N. C. Fraser and H. D. Sues (eds.), *In the shadow of the dinosaurs,* pp. 251–270. Cambridge: Cambridge University Press.

MacCready, P. E. 1985. Natural and artificial flying machines. *Technical Soaring* 9(3): 56–63.

Makovicky, P. J., and M. A. Norell. 1998. A partial ornithomimid braincase from Ukhaa Tolgod (Upper Cretaceous, Mongolia). *American Museum Novitates* 3247: 1–16.

Manabe, M., P. M. Barrett, and S. Isaji. 2000. A refugium for relicts? *Nature* 404: 953.

Maryanska, T., and H. Osmólska. 1981a. Cranial anatomy of *Saurolophus angustirostris* with comments on the Asian Hadrosauridae (Dinosauria). *Palaeontologia Polonica* 42: 5–24.

———. 1981b. First lambeosaurine dinosaur from Nemegt Formation, Upper Cretaceous, Mongolia. *Acta Palaeontologica Polonica* 26: 243–255.

———. 1984. Postcranial anatomy of *Saurolophus angustirostris* with comments on other hadrosaurs. *Palaeontologia Polonica* 46: 119–141.

Monbaron, M., D. A. Russell, and P. Taquet. 1999. *Atlasaurus imelakei* n.g., n.sp., a brachiosaurid-like sauropod from the Middle Jurassic of Morocco. *Comptes Rendus de l'Academie des Sciences (Paris),* Série II. 329, n°7, fasicule a, *Sciences de la terre et des planètes:* 519–526.

Morris, W. J. 1976. Hysilophodont dinosaurs: a new species and comments on their systematics. In C. S. Churcher (ed.), *Athlon: essays on paleobiology in honour of Loris Shano Russell,* pp. 93–113. Royal Ontario Museum Life Sciences, Miscellaneous Publications.

Norell, M. A., J. M. Clark, L. M. Chiappe, and D. Dashzeveg. 1995. A nesting dinosaur. *Nature* 378: 774–776.

Norell, M. A., P. J. Makovicky, and J. M. Clark. 2000. A new troodontid theropod from Ukhaa Tolgod, Mongolia. *Journal of Vertebrate Paleontology* 20: 7–11.

Norell, M. A., and J. A. Clarke. 2001. Fossil that fills a critical gap in avian evolution. *Nature* 409: 181–184.

Osmólska, H. 1980. The Late Cretaceous vertebrate assemblages of the Gobi Desert, Mongolia. *Mémoires Société Géologique de France,* nouvelle série, 139: 145–150.

———. 1996. An unusual theropod dinosaur from the Late Cretaceous Nemegt Formation of Mongolia. *Acta Palaeontologica Polonica* 41: 1–38.

Paul, G. S. 1988. *Predatory dinosaurs of the world.* New York Academy of Sciences. New York: Simon and Schuster.

Peters, R. H. 1986. *The ecological implications of body size.* Cambridge: Cambridge University Press.

Prasad, G. V. R., and A. Sahni. 1999. Were there size constraints on biotic exchanges during the northward drift of the Indian plate? *Proceedings of the Indian National Science Academy* 65A: 377–396.

Rogers, K. C., and C. A. Forster. 2001. The last of the dinosaur titans: a new sauropod from Madagascar. *Nature* 412: 530–534.

Russell, D. A. 1993. The role of Central Asia in dinosaurian biogeography. *Canadian Journal of Earth Sciences* 30: 2002–2012.

Russell, D. A., P. Béland, and J. S. McIntosh. 1980. Paleoecology of the dinosaurs of Tendaguru (Tanzania). *Mémoires Société géologique de France,* nouvelle série, 139: 169–175.

Russell, D. A., and Z. Zheng. 1993. A large mamenchisaurid from the Junggar Basin, Xinjiang, People's Republic of China. *Canadian Journal of Earth Sciences* 30: 2082–2095.

Russell, D. A., and M. Manabe. 2002. A synopsis of the Hell Creek (uppermost Cretaceous) dinosaur assemblage. In J. H. Hartman, K. R. Johnson, and D. J. Nichols (eds.), *The Hell Creek Formation and the Cretaceous-Tertiary boundary in the northern Great Plains—an integrated continental record of the end of the Cretaceous. Geological Society of America* Special Paper 361: 169–176.

Ryan, M. J., and P. J. Currie. 1998. First report of protoceratopsians (Neoceratopsia) from the Late Cretaceous Judith River Group, Alberta, Canada. *Canadian Journal of Earth Sciences* 35: 820–826.

Savage, D. E., and D. E. Russell. 1983. *Mammalian paleofaunas of the world.* London: Addison-Wesley.

Sheehan, P. M., and D. E. Fastovsky. 1992. Major extinctions of land-dwelling vertebrates at the Cretaceous-Tertiary boundary, eastern Montana. *Geology* 20: 556–560.

Sigogneau-Russell, D., S. E. Evans, J. F. Levine, and D. A. Russell. 1998. The Early Cretaceous microvertebrate locality of Anoual, Morocco: a glimpse at the small vertebrate assemblages of Africa. *Bulletin of the New Mexico Museum of Natural History and Science* 14: 177–181.

Stevens, K. A., and J. M. Parrish. 1999. Neck posture and feeding habits of two Jurassic sauropod dinosaurs. *Science* 284: 798–800.

Sues, H.-D. 1997. On *Chirostenotes,* a Late Cretaceous oviraptorosaur (Dinosauria: Theropoda) from western North America. *Journal of Vertebrate Paleontology* 17: 698–716.

Sues, H.-D., J. M. Clark, and F. A. Jenkins. 1994. A review of the Early Jurassic tetrapods from the Glen Canyon Group of the American southwest. In N. C. Fraser and H. D. Sues (eds.), *In the shadow of the dinosaurs,* pp. 284–294. Cambridge: Cambridge University Press.

Swisher, C. C., Y.-Q. Wang, X.-L. Wang, X. Xu, and Y. Wang. 1999. Cretaceous age for the feathered dinosaurs of Liaoning, China. *Nature* 400: 58–61.

Tang, F., Z.-X. Luo, Z.-H. Zhou, H.-L. You, J. A. Georgi, Z.-L. Tang, and X.-Z. Wang. 2001. Biostratigraphy and palaeoenvironment of the dinosaur-bearing sediments in Lower Cretaceous of Mazongshan area, Gansu Province, China. *Cretaceous Research* 22: 115–129.

Triebold, M. 1997. The Sandy Site: small dinosaurs from the Hell Creek Formation of South Dakota. In D. L. Wolberg, E. Stump, and G. D. Rosenberg (eds.), *Dinofest International,* pp. 245–248. Philadelphia: The Academy of Natural Sciences.

Upchurch, P. 1994. Sauropod phylogeny and palaeoecology. *Gaia* 10: 249–260.

———. 1995. The evolutionary history of sauropod dinosaurs. *Philosophical Transactions of the Royal Society of London,* Series B 349: 365–390.

Van Valkenburgh, B., and C. M. Janis. 1993. Historical diversity patterns in North America large herbivores and carnivores. In R. E. Ricklets and D. Schulter (eds.), *Species diversity in ecological communities*, pp. 330–340. Chicago: University of Chicago Press.

Watabe, M., and D. Fastovsky. 1999. Diverse dinosaur habitats of Djadokhta Age (Late Cretaceous) in the Gobi Desert, Mongolia. *Journal of Vertebrate Paleontology* 19: 83A.

Weishampel, D. B. 1990. Dinosaur distribution. In D. B. Weishampel, P. Dodson, and H. Osmólska (eds.), *The Dinosauria*, pp. 63–139. Berkeley: University of California Press.

Wilson, J. A., and P. C. Sereno. 1998. Early evolution and higher-level phylogeny of sauropod dinosaurs. *Society of Vertebrate Paleontology*, Memoir 5.

Xu X., Wang X.-L., and Wu X.-C. 1999. A dromaeosaurid dinosaur with a filamentous integument from the Yixian Formation of China. *Nature* 401: 262–266.

Xu X., Zhou Z.-H., and Wang X.-L. 2000. The smallest known non-avian theropod dinosaur. *Nature* 408: 705–708.

Yin A., and T. M. Harrison 2000. Geologic origin of the Himalayan-Tibetan orogen. *Annual Review of Earth and Planetary Sciences* 28: 211–280.

You H.-L., P. Dodson, Luo Z.-X., Dong Z.-M., and Y. Azuma. 1999. *Archaeoceratops* and other dinosaurs from the late Early Cretaceous of the Mazongshan Area, northwest China. *Journal of Vertebrate Paleontology* 19: 86A.

Young, C. C., and H. Chao. 1972. *Mamenchisaurus hochuanensis* sp. nov. *Institute of Vertebrate Paleontology and Palaeoanthropology Monograph, Series A*, 8: 1–30 (in Chinese).

Zhou Z.-H., and Zhang F.-C. 2002. Largest bird from the Early Cretaceous and its implications for the earliest avian diversification. *Naturwissenschaften* 89: 34–38.

2. End-Cretaceous Acid Rain as a Selective Extinction Mechanism between Birds and Dinosaurs

GREGORY J. RETALLACK

Abstract

Acid would have been a consequence of catastrophic events postulated for the Cretaceous-Tertiary boundary: nitric acid from atmospheric shock by bolides and from burning of trees; sulfuric acid from volcanic aerosols and from impact vaporization of evaporites; hydrochloric acid from volcanic aerosols; and carbonic acid from carbon dioxide of volcanoes, fires, and methane-hydrate release. Pedoassays for buffering of soil acid above pH 4 are indicated by the clayey, little-leached nature of latest Cretaceous and earliest Tertiary paleosols in Montana. Chemoassay of 2.7×10^{11} to 4.6×10^{17} moles of acid has been estimated from base-cation leaching of Cretaceous-Tertiary boundary beds and paleosols in eastern Montana, and these estimates are compatible with other independent chemoassays. Marine bioassay of pH suppression to no less than 7.6, allowing survival of coccolithophores, foraminifera, and dinoflagellates, requires a total acid load of less than 5×10^{16} moles. Similar limits come from a non-marine bioassay of pH suppression to less than 5.5 but no less than pH 4, allowing survival of amphibians and fish, but strong extinctions of non-marine mollusks in Montana. Acidification also may have been responsible for heavy extinctions among evergreen angiosperms. Vegetation browning would have been difficult for herbivorous dinosaurs and their predators, but less problematic for small insectivorous and detritivorous mammals and birds. Acid rain may have been an important agent of selective mortality and extinction across the Cretaceous-Tertiary boundary.

Introduction

Although differences between birds and dinosaurs have been blurred by recent fossil finds (Qiang et al. 1998; Benton 1999; Burnham et al. 2000) and by cladistic theory (Chiappe 1995; Padian 1998), birds are still with us, but dinosaurs have been gone for 65 million years. Why did birds survive mass extinction at the Cretaceous-Tertiary boundary, and not dinosaurs? The general question of differential extinction is critical to understanding the role of catastrophic events in evolution (Jablonski 1996; Sheehan et al. 1996). What, if anything, do mass extinctions select? Although survival could have been just a matter of luck, or due to tolerance of impact winter, disease, allergies, or a long list of other potential causes of end-Cretaceous extinctions (Bakker 1986; Archibald 1996), this chapter examines just one proposition: that differential extinctions at the end of the Cretaceous were the result of acid generated by massive asteroid impact.

Catastrophic impact of a large bolide at the Cretaceous-Tertiary boundary is now established beyond scientific doubt in many localities and by many lines of evidence (fig. 2.1): iridium anomalies (Alvarez et al. 1980; Orth et al. 1990; Claeys et al. 2002); carbon and carbon isotopic anomalies (Wolbach et al. 1988; Arthur et al. 1987; Arens and Jahren 2000); nitrogen and nitrogen isotopic anomalies (Gilmour et al. 1990); shocked quartz (Bohor 1990; Izett 1990; Claeys et al. 2002); stishovite (McHone et al. 1989); nanometer-sized diamonds (Carlisle and Braman 1991); dramatic changes in fossil plants, including transient abundance of fern spores (Wolfe and Upchurch 1987; Nichols et al. 1990; Johnson and Hickey 1990); and a large impact crater in the Yucatan (Sharpton et al. 1993, 1996; Kring 1995; Morgan et al. 1997). Also at this time, there were submarine landslides around the North Atlantic Ocean (Norris and Firth 2002), and flood basalt eruptions of the Deccan Traps in India (Duncan and Pyle 1988; Courtillot et al. 1990), perhaps stimulated by impact seismicity (Boslough et al. 1996; Chatterjee 1997), as well as widespread wildfires set by the impact fireball, and later ignition of browned vegetation (Wolbach et al. 1988; Heymann et al. 1996).

Acid is a likely consequence of all these events: nitric acid from atmospheric shock by the bolide and from burning of trees (Zahnle 1990); sulfuric acid from volcanic aerosols and impact vaporization of evaporites (Sigurdsson et al. 1992; Brett 1992; Yang et al. 1996); hydrochloric acid from volcanic aerosols (Caldeira and Rampino 1990); and carbonic acid from carbon dioxide of volcanoes, fires, and methane oxidation (Wolbach et al. 1988; Tinus and Roddy 1990; Ivany and Salawich 1993). All this acid should have left records in paleosols (pedoassay), in boundary beds (chemoassay), and in the differential extinction of acid-sensitive organisms (bioassay). These three methods of assessing environmental acidification will each be considered in turn.

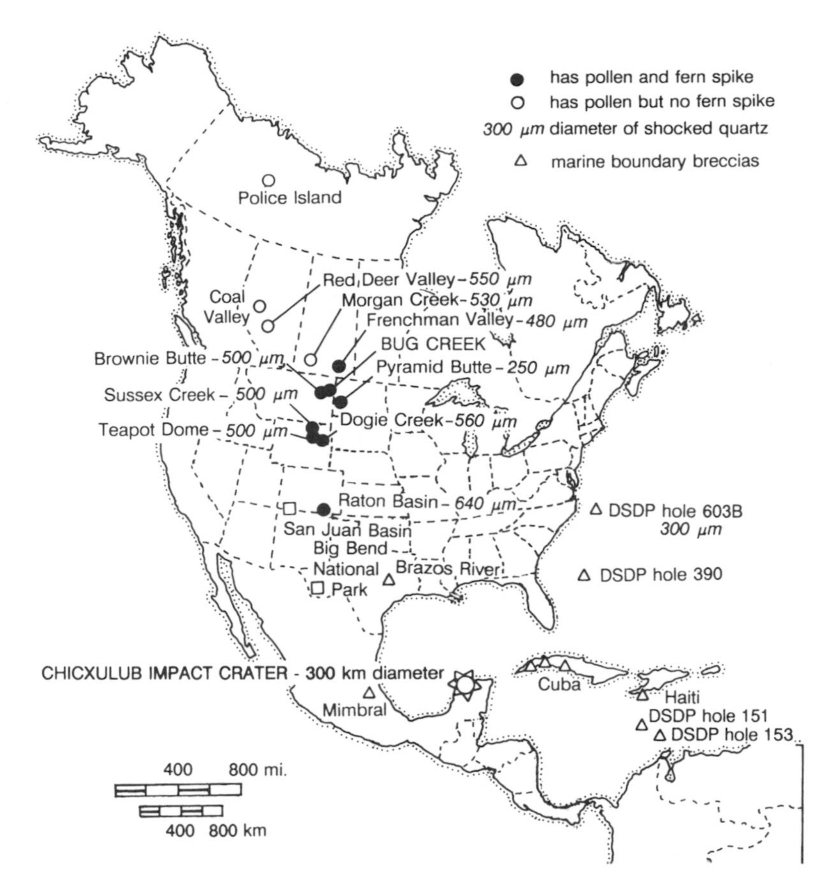

Figure 2.1. Locations of Bug Creek, Brownie Butte, and other important Cretaceous-Tertiary boundary sites in North America.

Figure 2.2. (below) The Cretaceous-Tertiary boundary paleosol sequence in Bug Creek, Montana.

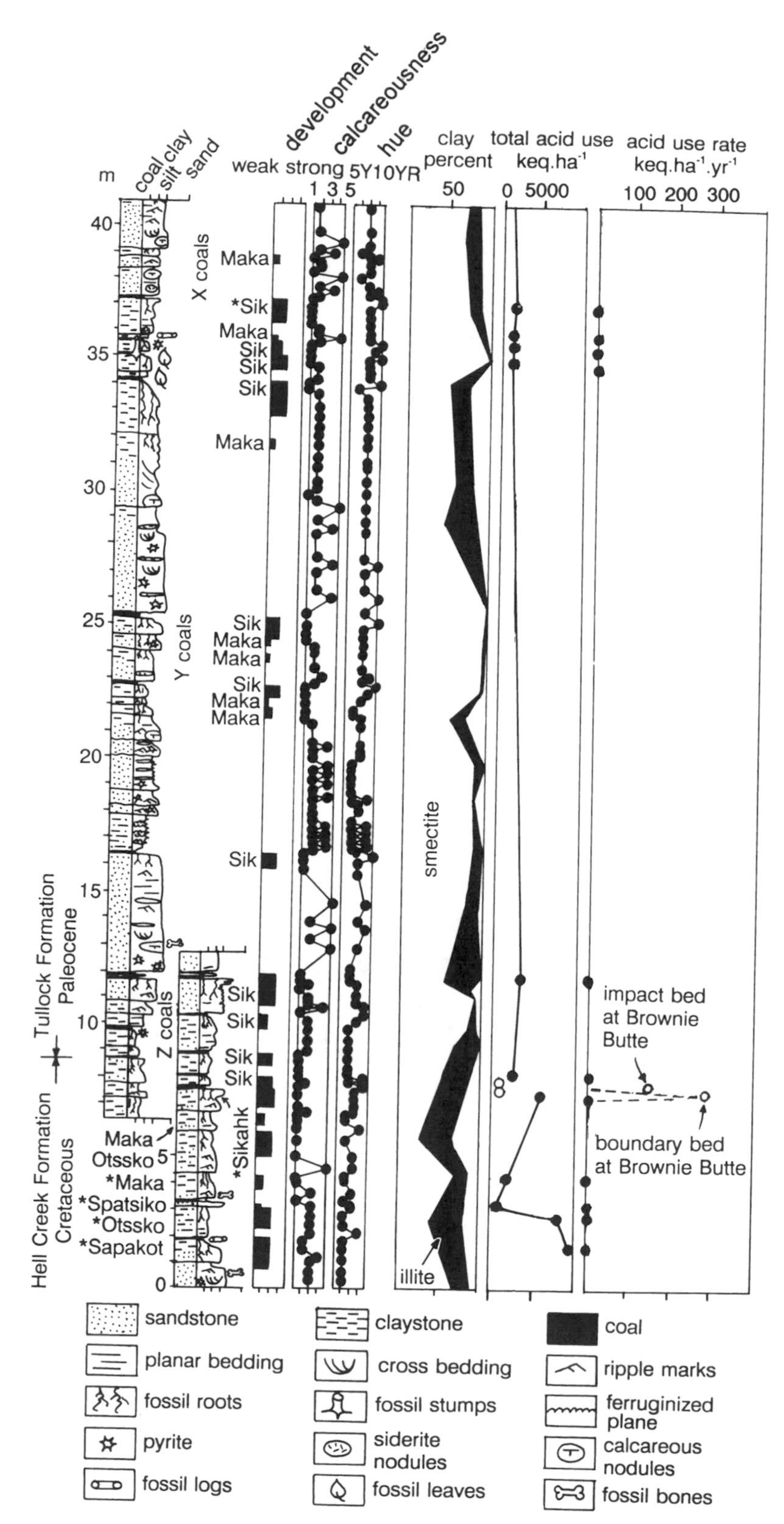

Figure 2.3. Measured section and acid use of paleosols across the Cretaceous-Tertiary boundary in Montana. Positions of individual paleosols are shown by black boxes in the development column, with width proportional to physical (not chemical) development. (Calcareousness was estimated by field application of dilute acid and hue from Munsell charts. Clay mineral proportions from Bell 1965, acid use from Retallack 1996.)

Pedoassay

Paleosols in the Bug Creek area of eastern Montana are a remarkably fossiliferous record of the Cretaceous-Tertiary boundary (figs. 2.1–2.3; Retallack et al. 1987; Retallack 1994). Only a weak iridium anomaly and no distinctive boundary beds have been found in Bug Creek (Smit et al. 1987; Rigby and Rigby 1990; Swisher et al. 1993), but the Cretaceous-Tertiary boundary can be located there by means of unusually abundant fern spores, fossil plant extinctions (Hotton 1988), and a carbon isotopic excursion (Arens and Jahren 2000). The last known dinosaur bone in place within a paleosol in this area is a poorly preserved *Triceratops* vertebral centrum 1 m below the palynologically defined extinction level, but bone is not unusually rare in paleosols 3 m below the extinction zone within the lower Z coal (Sheehan et al. 2000; contra Williams 1994). Dinosaur bone also is found in sandstone paleochannels at higher stratigraphic levels, where it was probably redeposited (Lofgren 1995; Archibald 1996). Although a case has been made for gradual extinction of dinosaurs (Sloan et al. 1986) and even for modest rather than mass extinction in this area (Archibald and Bryant 1990; Archibald 1996), comprehensive collections from nearby regions of Montana and North Dakota have shown that dinosaur extinction was abrupt at the level of the plant extinctions (Bryant 1989; Sheehan et al. 1991; Pearson et al. 1999; Sheehan et al. 2000). The dramatic plant extinctions revealed by pollen and spores (Hotton 1988; Nichols and Fleming 1990) are now supported by massive collections of fossil plants through this same region (Johnson and Hickey 1990). The "zone of death" revealed by fossil fern spores and the carbon isotopic excursion in Bug Creek (Hotton 1988; Arens and Jahren 2000) is a carbonaceous surface of a moderately developed paleosol (figs. 2.2, 2.3: Sikahk pedotype of Retallack 1994). Thin layers of impact ejecta found at many Cretaceous-Tertiary boundary sections (Bohor 1990; Izett 1990) were not found in Bug Creek, where they were presumably eroded away or mixed by the action of later roots and burrows of the Sikahk paleosol.

The gray and brown clayey paleosols and black lignitic paleosols of the upper Hell Creek and lower Tullock Formations in Bug Creek are representative of paleosols in these somber-colored formations over a wide area of the northern High Plains from Buffalo in South Dakota, to Marmarth in North Dakota, and Jordan in Montana, a distance of some 400 km (Fastovsky and McSweeney 1987; McSweeney and Fastovsky 1987). The lignitic facies of the Tullock Formation began accumulating well before the Cretaceous-Tertiary boundary in South Dakota, which was closer to the sea (Pearson et al. 1999), but only a little before the boundary in Bug Creek (Swisher et al. 1993). Thus the Cretaceous-Tertiary boundary fell at a time of local marine transgression, not the regression that has been advocated as a global explanation for end-Cretaceous events by Hallam (1987). The peaty paleosols of Montana represent swamps of water pine (*Glyptostrobus*) and dawn redwood (*Metasequoia*), which became much more widespread during the Early Paleocene than Late Cretaceous. Even well-drained paleosols of the Hell Creek and Tullock Formations formed on terraces only a few meters above the water table in seasonally wet lowland floodplains. Mean annual rainfall in the latest Cretaceous as estimated from paleosols was 900–1,200 mm per annum, with an increase in the earliest Paleocene to more than 1,200 mm (Retallack 1994). These estimates from paleosols are compatible with paleobotanical estimates for a postapocalyptic humid greenhouse: latest Cretaceous Lance flora indicates mean annual temperature (MAT) of 16.2°C and mean annual precipitation (MAP) of 700–800 mm, earliest Paleocene Brownie Butte flora indicates MAT 27.4°C and MAP >3,500 mm, and later Paleocene Clareton flora indicates MAT 19.8°C and MAP 2,500 mm (Wolfe 1990). In these floodplain terrace soils, diverse Late Cretaceous broadleaf forests, with dicots such as *"Cissites" marginatus* and *Dombeyopsis trivialis,* were replaced in the earliest Paleocene with less diverse forests of dicots such as *"Populus" nebrascensis* and *Cercidiphyllum genetrix* (Johnson and Hickey 1990). The great diversity of dinosaurs in Late Cretaceous paleosols is completely missing from earliest Paleocene paleosols, which contain champsosaurs and other survivors of the extinction (Archibald and Bryant 1990; Archibald 1996). A striking feature of the paleosol record is the general similarity of Late Cretaceous and Early Paleocene paleosols, despite the very different species of plants and animals that they supported (Retallack 1994).

Paleosols at the Cretaceous-Tertiary boundary in Bug Creek were examined in order to quantitatively assess acidification, compared with paleosols in the same sequence above and below the boundary. These analyses aimed to determine depletion of acid-soluble bases, such as the alkaline earth (Ca, Mg) and alkali (Na, K) elements (Retallack 1996). Ordinary weathering is a process of acidification, and an acid rain signature can be detected only as an unusually severe depletion of weatherable bases. Thus acidification in the paleosol at the Cretaceous-Tertiary boundary needs to be compared with background acidification due to normal weathering of paleosols above and below the boundary. These analyses of major elements (fig. 2.4) and of rare earth trace elements (fig. 2.5) show significant acidification of the boundary paleosol compared within those stratigraphically above and below, but not so profound as would indicate strong acids. Barium/strontium and base/alumina molar ratios should be particularly sensitive to acidification, and these also fail to show profound perturbation (fig. 2.5). Nor do the metals Cu, Ni, and Zn show marked anomalies (fig. 2.4), as might be predicted from strong acidification or influx of meteoritic material (Davenport et al. 1990). The total amount of acid consumed by mineral horizons of four Cretaceous paleosols averaged 5,297 +/< 3,758 keq.ha^{-1} acid (error of 1Σ), but nine Paleocene paleosols used 2,069 +/< 1,481 keq.ha^{-1} (Retallack 1996). Estimates of the rate of acid consumption also were made, using maximum values for duration of ancient soil formation estimated by comparison with studies of morphological (*not chemical*) differentiation of Quaternary soils. These calculated minimal rates of acid consumption of Late Cretaceous and Early Paleocene paleosols are not appreciably different from each other

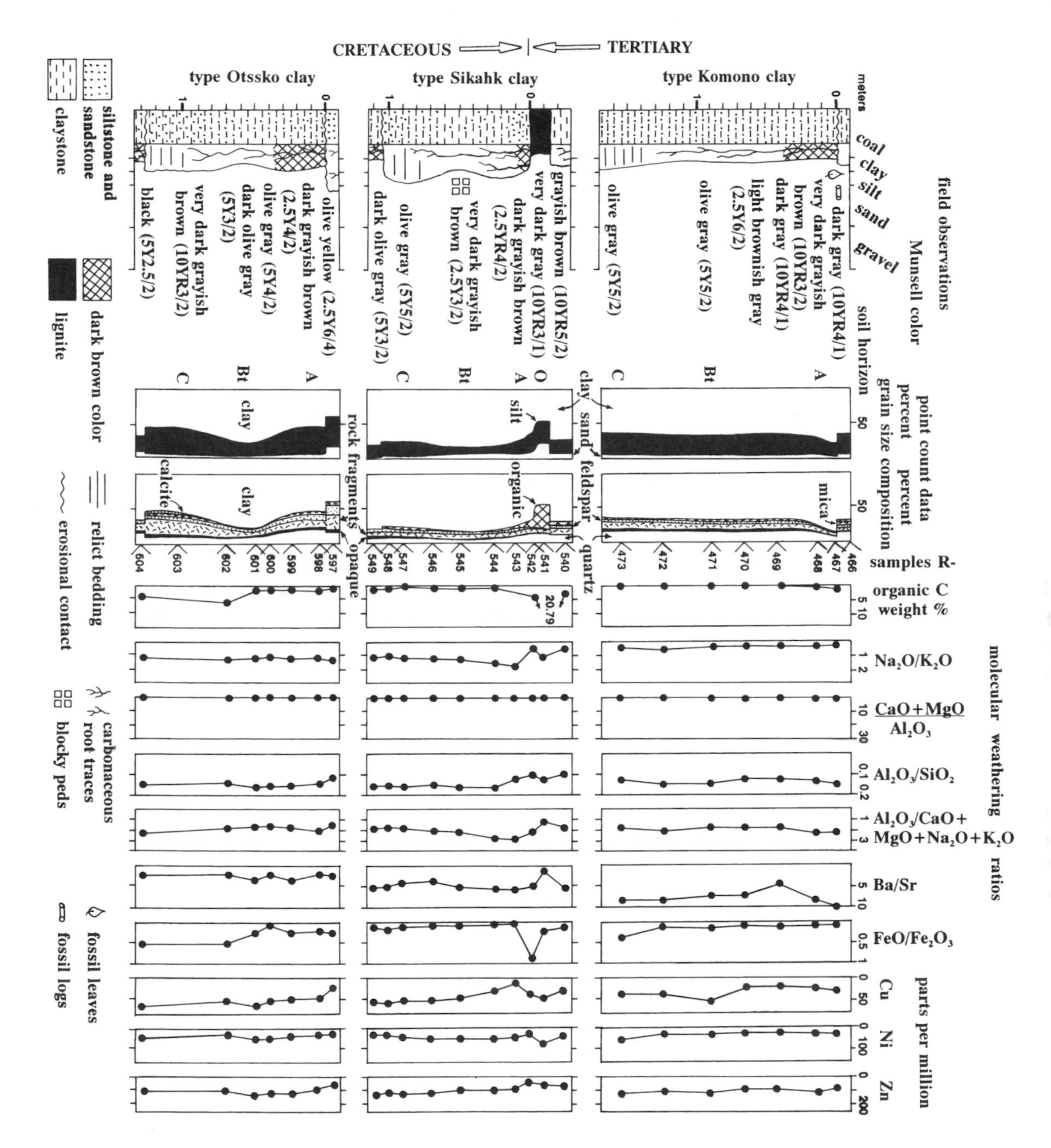

Figure 2.4. Chemical depth functions of paleosols at, above, and below the Cretaceous-Tertiary boundary in Bug Creek, Montana, showing little significant difference and no indications of podzolization at the boundary.

or from Holocene soils (Fölster 1985), which generally fall between limits of 0.2–2.3 $keq.ha^{-1}.yr^{-1}$. The four latest Cretaceous paleosols had an average rate of acid consumption of 2.0 +/< 1.7 $keq.ha^{-1}.yr^{-1}$ and 9 earliest Paleocene paleosols of 0.9 +/< 0.4 $keq.ha^{-1}.yr^{-1}$. Such calculations for 2 paleosols at the boundary in Bug Creek give an average consumption of 6,585 +/< 199 $keq.ha^{-1}$. These paleosols show profile differentiation and little remaining relict bedding compatible with some

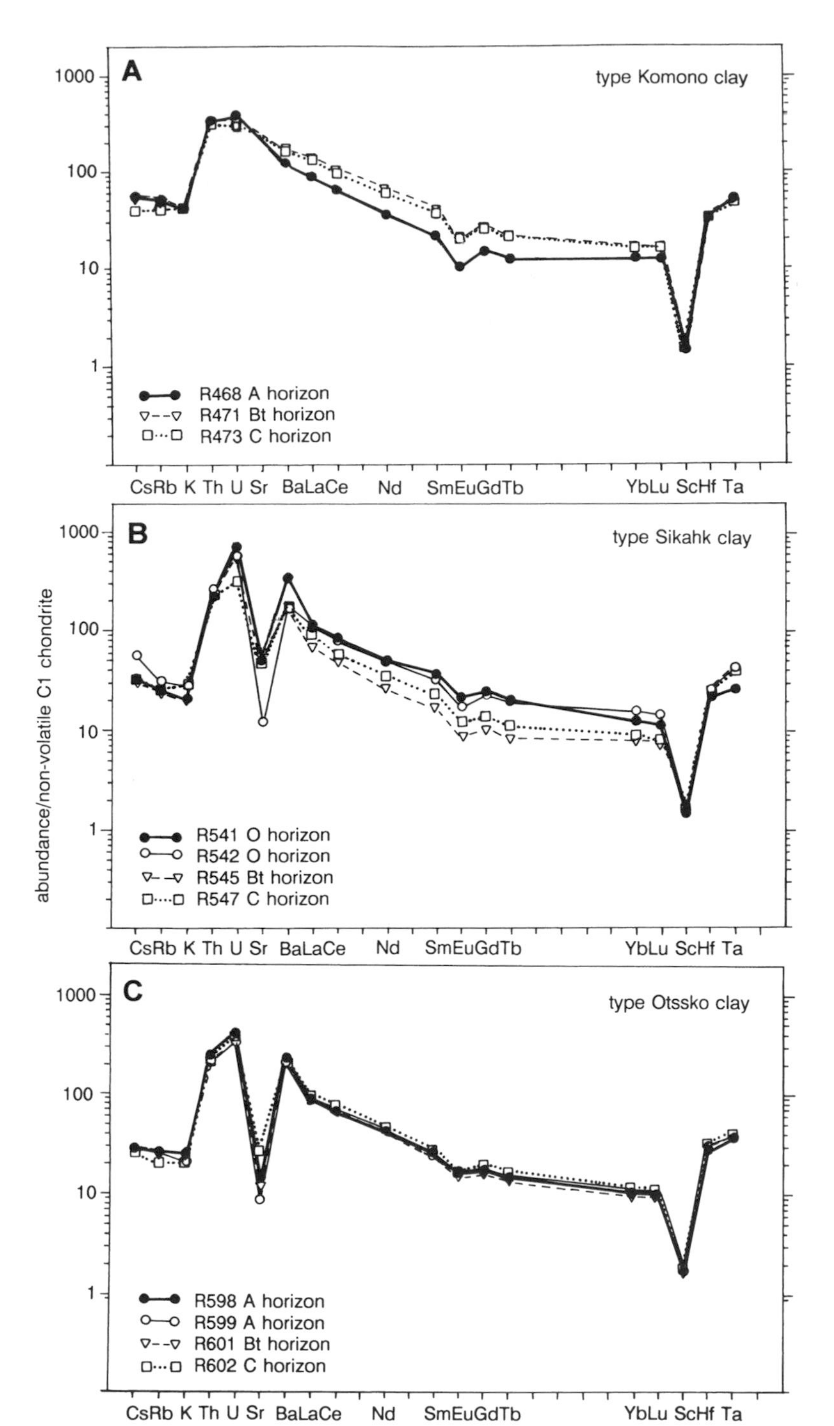

Figure 2.5. REE analyses normalized to North American shales of paleosols at, above and below the Cretaceous-Tertiary boundary in Bug Creek, Montana, showing slight, but not significant, spread of values in the boundary paleosol attributable to acid rain.

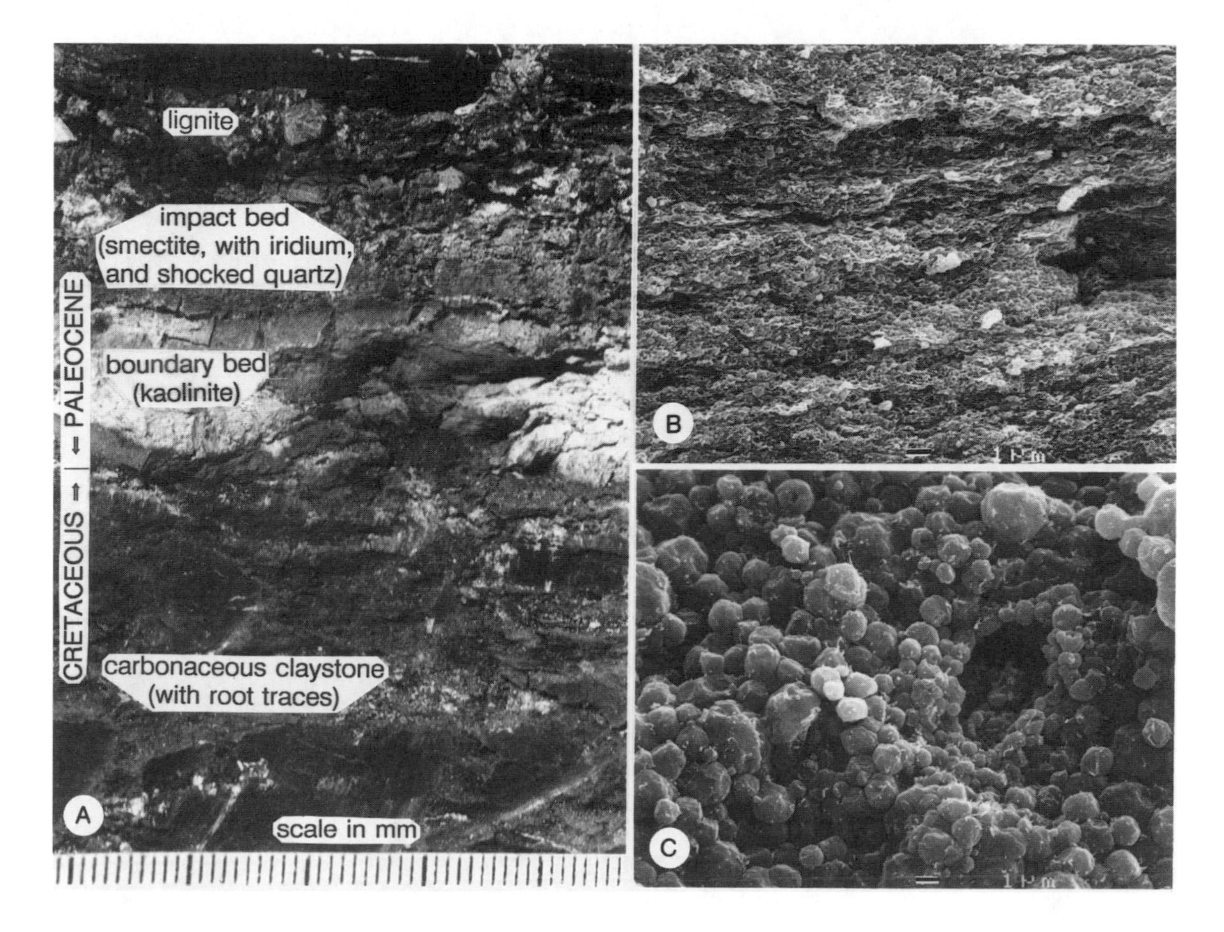

Figure 2.6. Cretaceous-Tertiary boundary beds at Brownie Butte, Montana, with their very different microstructures visible under SEM (B and C): A, field view of boundary bed and impact bed between terminal Cretaceous paleosol below and Paleocene lignite above; B, layered microstructure of impact bed under SEM; C, pelletoidal microstructure of boundary bed under SEM.

15,000 years of soil formation, which would give a rate of acid consumption of 0.2 +/< 0.006 $keq.ha^{-1}.yr^{-1}$. These estimates are permissive of either high acid load over a short time or of low acid load over a long time. They are highly dependent on time scales of observation, which are insufficiently resolved in Bug Creek. Nevertheless, these estimates do indicate acid consumption and place upper limits on amounts of acid involved.

Limits on acidification also come from the lack of petrographic or chemical evidence of podzolization in the boundary paleosols of Montana (Retallack 1994). Podzolization is a soil-forming process that creates quartz-rich soils locally cemented by iron oxides and organic matter (Spodosols), and it occurs at pH less than 4 (Lündstrom et al. 2000). Podzolization is also negated by a clay mineral study (Bell 1965) which found smectite abundant in paleosols above, below, and at the Cretaceous-Tertiary boundary (fig. 2.3). Base-rich clays such as smectite, as well as other chemical and petrographic data, are evidence that Late Cretaceous and Early Paleocene paleosols were Entisols, Inceptisols, Histosols, Ultisols, and Alfisols (Retallack 1994), not Spodosols. Podzolization involves not only destruction of clay, but also release of toxic metal cations (particularly of aluminum), which can affect organ-

TABLE 2.1
Estimates of pH in soils and waters at the Cretaceous-Tertiary boundary

pH Constraint	Evidence	Author
2.0–4.0	theoretical impact-generated acid	Prinn and Fegley 1987
3.5–10.0	freshwater fish in Montana	Archibald 1996
4.0–10.0	amphibians in Montana	Archibald 1996
4.0–10.0	non-calcareous Alfisols in Montana (unpodzolized or salinized)	Retallack 1994
<5.5	freshwater clams and snails in Montana	Hartman 1998
5.4–8.2	marine foraminifera	d'Hondt et al. 1994
6.0–8.2	marine coccolithophorids	d'Hondt et al. 1994
7.6–8.2	marine dinoflagellates	d'Hondt et al. 1994

Note: Estimates accepted here are pH 4.0–5.5 for soil and groundwater in Montana and pH 7.6–8.2 for average ocean water.

isms more severely than acid alone (Howells 1995). The lack of evidence for podzolization is thus an important limit to acid load and other toxicity experienced by Cretaceous-Tertiary boundary paleosols in Montana. Thus some past theoretical estimates of acidification at the Cretaceous-Tertiary boundary of pH 2–4 are unreasonable (table 2.1).

Chemoassay

The key to quantitative assessment of acidification at the Cretaceous-Tertiary boundary is finding a geological setting in which the timing of acid input is well constrained, as for example between the various ejecta layers from Chicxulub Crater (Alvarez et al. 1995). At Brownie Butte, Montana (fig. 2.1), the Cretaceous-Tertiary meteoritic ejecta include an impact bed, which is 1 cm thick, gray, smectitic, and layered, with shocked quartz and an iridium anomaly (fig. 2.6). It lies directly above the boundary bed, which is 2 cm thick, pink to white, kaolinitic, microspherulitic, and vuggy. These two distinctive thin beds have been discovered at 30 sites from Alberta and Saskatchewan south to New Mexico at the radiometrically and palynologically determined Cretaceous-Tertiary boundary (Bohor 1990; Izett 1990). The boundary bed at Brownie Butte has been interpreted as a paleosol with roots (Fastovsky et al. 1989), but the carbonaceous structures in the boundary bed are plant stalks, 5 mm or more in diameter, and frayed upwards (fig. 2.4). They lack the fine rootlets and downward taper of roots. The boundary bed has a trace element chemical composition most like glassy ejecta from an early ejecta blanket of melt, shocked rocks, and admixed seawater. Shocked quartz is rare in the boundary bed, but more common in the impact bed, which was probably deposited within hours by fallout from a warm fireball (Alvarez et al. 1995). Significantly, the thin boundary bed consumed much more acid than the overlying impact bed. Even if both beds were subsequently acidified further within the peaty paleosol, as argued for other kaolinitic coal partings (Staub and Cohen 1978; Demchuck and Nelson-Glatiotis 1993), their differential leaching gives a minimal acid titer. Vigorous early neutralization of hot acid by silicate ejecta may explain both the distinctive kaolinitic composition and the microspherulitic and vuggy texture of the boundary bed, in which most shocked quartz has been destroyed by profound chemical leaching (Retallack 1996).

Quantification of acid consumption of the boundary bed compared with the overlying impact bed relies on procedures developed for studies of modern soil acidification (Fölster 1985). The loss of basic cations can be used to calculate the moles of hydronium consumed from weight percent analytical values and bulk density compared with parent materials of Cretaceous-Tertiary boundary beds and paleosols (Retallack 1996). The boundary claystone is more acidified by at least 5.4 $keq.ha^{-1}$ than the sharply overlying, well-bedded, smectitic impact layer. This is an enormous amount of acid for the mere hours to days between accumulation of boundary and impact beds envisaged (Alvarez et al. 1995). For comparison, after the experimental application of rain of pH 3.5, a modern soil from near Unadilla in upstate New York maintained a pH of 4.1 in mineral horizons and lost 7.8 $keq.ha^{-1}. yr^{-1}$ from these horizons (Cronan 1985), which is comparable to the loss estimated here for the boundary bed in Montana and about three times the loss from weak acids (Fölster 1985). Thus strong acids and acid rain are indicated at the Cretaceous-Tertiary boundary, rather than only ordinary weathering or weak acids such as carbonic acid.

Such calculations can also be done with a variety of other conceivable parent materials for the kaolinitic boundary bed: local Cretaceous and Paleocene sediments (Retallack 1994), melt rock from Chicxulub Crater, Mexico (Hildebrand et al. 1991), and impact glasses from Beloc, Haiti (Sigurdsson et al. 1991, 1992) and Mimbral, Mexico (Smit et al. 1992). These various calculations give estimates of from 2.7×10^{11} to 4.6×10^{17} moles of acid produced by the terminal Cretaceous impact within hours between settling of the boundary and impact beds (Retallack 1996).

These estimates are compatible with completely independent chemoassays derived from the anomalous enrichment of $^{87}Sr/^{86}Sr$ in marine foraminifera at the Cretaceous-Tertiary boundary (MacDougall 1988; Martin and MacDougall 1991). This spike in strontium of continental origin is best explained as a product of increased acid-induced weathering of ^{87}Sr on land. Scaling up to this anomaly from modern deliveries of ^{87}Sr to the ocean gives a spike of 3.8×10^{15} moles of acid at the Cretaceous-Tertiary boundary (Vonhof and Smit 1997).

TABLE 2.2
Estimates of total acid produced by Cretaceous-Tertiary boundary impact

Acid Production	Basis of Estimate	Author
6×10^{14} to 1×10^{15} moles NO	impact shocking atmospheric N_2	Prinn and Fegley 1987
1×10^{14} to 1×10^{15} moles HNO_3	impact shocking atmospheric N_2	Zahnle 1990
1×10^{16} moles SO_2	impact vaporization of sulfate evaporites	Morgan et al. 1997
2×10^{14} to 1.3×10^{17} moles H_2SO_4	impact vaporization of sulfate evaporites	Brett 1992; Sigurdsson et al. 1992
2–20 $\times 10^{16}$ equivalents acid	impact vaporization of evaporites	Yang et al. 1996
$<1 \times 10^{16}$ moles H_2SO_4 or $<4 \times 10^{15}$ moles HNO_3	preservation of marine microplankton	d'Hondt et al. 1994
3.8×10^{15} equivalents acid	$^{87}Sr/^{86}Sr$ perturbation at boundary in marine rocks	Vonhof and Smit 1997
2.7×10^{11} to 4.6×10^{17} equivalents acid	acidification of non-marine boundary beds, Montana	Retallack 1996
4,511±330 ppmV CO_2	stomatal index of fossil ginkgo leaves	Retallack 2001

Note: Estimates accepted here are less than 10^{16} equivalents of all acids combined.

These chemoassays from the boundary beds and $^{87}Sr/^{86}Sr$ anomaly support some, but not all, theoretical estimates of acid production by impact and its attendant effects (tables 2.1, 2.2). Theoretical estimates on the production of NO_2 by a bolide capable of creating Cretaceous-Tertiary iridium anomalies have varied from 1×10^{14} to 1.2×10^{17} moles (Lewis et al. 1982; Prinn and Fegley 1987; Zahnle 1990) or some 2–2,350 keq.ha^{-1} of the earth's surface area. An additional source of acid on short time scales is vaporization of anhydrite evaporites under the impact crater of Chicxulub, Mexico. This may have produced 4×10^{17} to 1.3×10^{19} g SO_2 (Brett 1992; Sigurdsson et al. 1992), which is 6.2×10^{15} to 2.0×10^{17} moles, or 254–7,840 keq.ha^{-1} globally. Wildfires would produce comparable amounts of NO and CO_2 (Zahnle 1990), perhaps focused at the boundary (Wolbach et al. 1988; Tinus and Roddy 1990). An additional estimate from hypothetical oceanic titration (d'Hondt et al. 1994) is a total acid load of no more than 5×10^{16} moles, or 980 keq.ha^{-1} globally. The chemoassays already presented here indicate that lower estimates of acid load are reasonable, but the higher estimates are excessive. Nitrogen and nitrogen isotopic studies (Gilmour et al. 1990) and sulfur and sulfur isotopic studies (Yoder et al. 1995) of the Cretaceous-Tertiary boundary are evidence that both nitric and sulfuric acids were involved. A combination of both acids not exceeding 10^{16} moles seen from the boundary beds in Montana is further indication

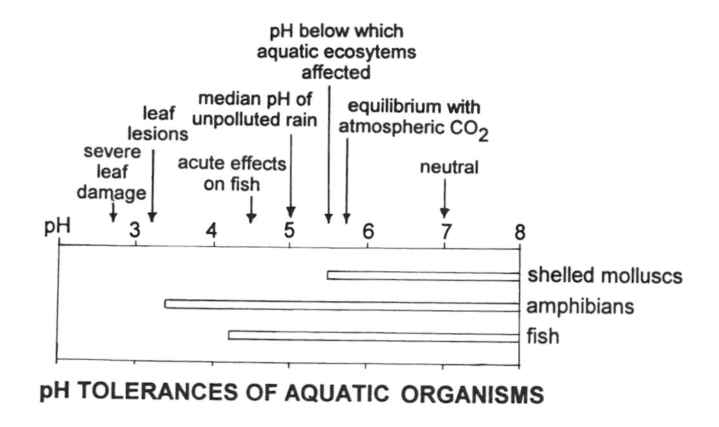

Figure 2.7. Tolerances of aquatic organisms for different pH levels. (After Howells 1995)

that estimates of the load of each acid approaching this value are also excessive, unless there was some other buffer of acidity such as scrubbing by silicate dust entrained in the ejecta cloud (Retallack 1996).

Bioassay

The pH of modern lakes fluctuates daily and seasonally, and it is convenient to assess acidification by bioassay, that is to say from the presence or absence of species of fish or invertebrates of known pH tolerance (Howells 1995). In the case of the Cretaceous-Tertiary boundary, such bioassay would be crude and done at high taxonomic levels (fig. 2.7), because pH tolerances of individual extinct species are not known. The patterns of survival of freshwater mollusks, fish, and amphibians in Montana are particularly instructive (table 2.3). Aquatic mollusks suffered severe extinctions, with termination of most lineages of shelled bivalves and gastropods (Hartman 1998). Amphibians and fish were less affected, but suffered moderate extinction (table 2.3; Archibald and Bryant 1990; Weil 1994), especially if the diverse Bug Creek Anthills assemblage is disregarded as a fauna largely of earliest Paleocene fossils, with some admixed latest Cretaceous fossils (Lofgren 1995). Shelled mollusks would have been excluded by pH less than 5.5, but greater losses of fish and amphibians would have been expected at pH less than 4 (Weil 1994; Howells 1995). This crude bioassay thus limits terminal Cretaceous freshwater acidification in Montana to between pH 4 and 5.5. This result is in agreement with the idea advanced above, that acid buffering by calcareous smectitic soils of Montana was sufficient to prevent pH depression below 4, and to keep podzolization and heavy metal toxicity from Montana during the Cretaceous-Tertiary transition.

Yet other bioassays have been attempted for the ocean by d'Hondt et al. (1994) who noted that surface dwelling ammonites and coccolithophores suffered severe extinction, but other bottom-dwelling mollusks, as well as radiolarians and acid-sensitive dinoflagellates, survived (table 2.3). From this pattern of extinction the whole ocean could

TABLE 2.3
End-Cretaceous extinctions of fossil species from the upper Hell Creek Formation of Montana and adjacent states

Taxon	Number of Late Cretaceous Species	Percent of Species Extinct at K/T Boundary
Plant leaves (ferns, horsetails, conifers, angiosperms)	40	79
Pteridophtye spores (ferns, horsetails)	36	25
Gymnosperm pollen (seed ferns, cycadeoids, conifers)	14	36
Angiosperm pollen (flowering plants)	149	51
Bivalvia (freshwater clams)	28	100
Insecta (damage types to leaves)	51	27
Elasmobranchi (sharks and rays)	5	100
Chondrostei and Holostei (gars and sturgeons)	7	29
Teleostei (bony fish)	8	50
Amphibia (frogs and salamanders)	8	0
Choristodera (champsosaurs)	1	0
Crocodylia (alligators)	5	40
Testudines (turtles)	17	12
Squamata (lizards, snakes)	10	70
Pterosauria (flying reptiles)	1	100
Ornithischia (ceratopsians, ankylosaurs, duckbills)	10	100
Saurischia (tyrannosaurs and other theropods)	9	100
Aves (birds)	20	55
Multituberculata (multituberculate mammals)	10	50
Eutheria (condylarths and other early mammals)	6	0
Metatheria (marsupials)	11	91

Sources: Brodkorb (1963), Estes (1964), Estes and Berberian (1970), Nichols et al. (1990), Nicols and Fleming (1990), Johnson and Hickey (1990), Archibald and Bryant (1990), Lofgren (1995), Archibald (1996), Hartman (1998), Labandeira et al. (2002).
Note that the concept of species varies from pollen to leaves, and from plants to animals.

not have been acidified to less than pH 7.6, though more substantial acidification could have occurred in shallow and isolated parts of the sea. Whole ocean pH depression to 7.6 could still accommodate a total acid load of 1×10^{16} moles, such is the mixing and buffering power of the ocean (d'Hondt et al. 1994).

A different kind of bioassay for carbonic acid has recently emerged from studies of *Ginkgo* leaf cuticles across the Cretaceous-Tertiary boundary (Retallack 2001). With postindustrial atmospheric pollution, the leaves of vascular plants have reduced the density of their stomates, because this allows intake of comparable amounts of carbon dioxide while curbing water transpiration in a warmer atmosphere. The observed decline since 1888 in stomatal index, which is density normalized to changing cell size, together with greenhouse experiments (Beerling et al. 1998), has been used to estimate atmospheric carbon dioxide from stomatal index of fossil *Ginkgo* leaves, with the following results: earliest Paleocene atmospheric carbon dioxide of 4,511+/<330 ppmV, much higher than Late Cretaceous (1,194+/<505 ppmV) and later Paleocene (1,350+/<350 ppmV) concentrations of this greenhouse and acidic gas (Retallack 2001). This transient spike of carbon dioxide would have come from biomass decay and burning (Ivaney and Salawich 1993), but a methane-dissociation component cannot be ruled out (as for the Aptian isotopic excursion documented by Jahren et al. 2001). The greenhouse high of carbon dioxide would have lasted no longer than 50,000 years, judging from the duration of the associated carbon isotopic anomaly (Arens and Jahren 2000). Carbon dioxide would have been rained out as carbonic acid, but the effects of this on soils are difficult to appraise, because comparably high concentrations of carbonic acid are created by soil respiration even under modern conditions (Brook et al. 1983).

Acid Rain and Selective Extinction

The pedoassays, chemoassays and bioassays presented above indicate that acid rain was a consequence of the latest Cretaceous asteroid impact in Yucatan, and put quantitative limits on its magnitude (tables 2.1, 2.2). The differential extinction of plants, insects, mammals, birds, and dinosaurs in Montana and adjacent states can be reassessed from this perspective. Also of interest is how consistent patterns and circumstances of extinctions are with other kill mechanisms, such as darkness from clouds of ejecta for days to months (Zahnle 1990), impact winter for 1–2 months (Wolfe 1991), wildfires for weeks to months (Wolbach et al. 1988; Tinus and Roddy 1990), and postapocalyptic greenhouse (Wolfe 1990; Landis et al. 1996). The duration of the postapocalyptic greenhouse has been estimated as 0.5–1 million years from paleobotanical data (Wolfe 1990), but its reflection in the carbon isotopic excursion in Montana is only about 50,000 years (Arens and Jahren 2000). Direct kill by vaporization, tsunami, and burial by debris within the blast zone of the Caribbean was probably not significant in Montana, 3,330 km north of ground zero in Yucatan (Kring 1995).

Terrestrial Plants

Acidic trauma may explain the transition in Montana from eutrophic angiosperm-dominated semievergreen forests, to a fern-dominated recovery flora, and then to oligotrophic conifer-dominated swampland and woodlands of deciduous dicots (Wolfe and Upchurch 1987; Johnson and Hickey 1990). Acidic browning, then burning and decay of green plants, may partly explain the marked lightening in $\Delta^{13}C_{org}$ across the Cretaceous-Tertiary boundary (Ivany and Salawich 1993), but excursions of as much as −2.8 ‰ $\Delta^{13}C_{org}$ in non-marine facies of Montana (Arens and Jahren 2000) approach levels better explained by methane hydrate dissociation (Jahren et al. 2001). Extinction levels are much higher among angiosperm pollen than among conifer pollen and fern spores (Nichols and Fleming 1990). Both acid rain and the noxious nitrous and sulfurous precursor gases would have browned broad evergreen leaves, thus selecting for deciduous plants, as seen among both dicots and conifers (Wolfe 1987). Acid rain leaching into soil would have created less fertile substrates (such as the Ultisol of the Sikahk paleosol of Retallack 1994), thus selecting for oligotrophic taxa.Other extinction mechanisms are not as well supported by the fossil plant record. Although it could be argued that impact winter would have similar effects of extinguishing evergreen rather than deciduous angiosperms (Wolfe 1987), survival of palm pollen (*Arecipites* of Nichols et al. 1990) and leaves (Wolfe 1991) is evidence against fatal chilling. Furthermore, evergreen plants did not re-emerge from refugia during the Early Paleocene, as they did after full glacial chillings of the Pleistocene (Wolfe 1987), which modeling reveals was comparable to impact chilling from massive aerosol loadings (Covey et al. 1990). Damage to leaf cuticles at the Cretaceous-Tertiary boundary at Teapot Dome, Wyoming, has been claimed as evidence of transient freezing in spring, when water lily and lotus were in bloom (Wolfe 1991). However, comparable lesions on anticlinal walls of plant cells also can be created by acid pollution of living leaves (Adams et al. 1984). I have observed similar deformation with my own overenthusiastic preparation of fossil plant cuticles. Wildfire is an unlikely cause of the plant extinctions, even though charcoal is common in the paleosols (Retallack 1994), because less flammable evergreen dicots were less prone to extinction than resinous conifers (Johnson and Hickey 1990). Recolonization after fire could explain the earliest Paleocene abundance of fern spores and, in the Raton Basin, of fern leaves like those of *Stenochlaena,* a well-known volcanic recolonizer today (Wolfe and Upchurch 1987). Nor is darkening of the sky with dust apparent from the pattern of plant extinction, because this alone would etiolate evergreen dicots, rather than kill and brown their leaves. Warming during a multimillenial postapocalyptic greenhouse is indicated by ecophysiological interpretation of fossil leaves (Wolfe 1990), but not from their taxonomic composition. Compared with those of the Cretaceous, Paleocene fossil plants have decidedly more temperate climatic modern affinities, presumably as an indirect legacy of boundary events (Wolfe 1987).

Terrestrial Invertebrates

Few insects are known from the Cretaceous-Tertiary boundary beds of Montana and adjacent states, but much can be learned from insect damage of abundant fossil leaves (Labandeira et al. 2002). Serpentine leaf mines like those of gracillariid moths, slot-hole feeding like that of Curculionid beetles, and pit feeding like that of Cerambycid beetles in Late Cretaceous leaves are specific to particular kinds of leaves, but earliest Paleocene insect damage lacks specificity for particular host plants and is less diverse (table 2.3). Host-specific insects become extinct with their plant hosts, but generalist plant-feeding insects turn to dead plant material when all other sources of food become unavailable. The record of insect trace fossils thus supports the contention of Sheehan and Hansen (1986) that feeding on dead plant material (detritivory) was critical to survival across the Cretaceous-Tertiary boundary.

Earthworms (Oligochaeta) are also detritivorous and are represented by ellipsoidal fecal pellets in petrographic thin sections of both Late Cretaceous and Early Paleocene paleosols (Retallack 1994). Burrowing bugs (Cydnidae) also are partly detritivorous, and may be represented by backfilled burrows in earliest Paleocene sediments at Pyramid Butte, North Dakota (Johnson 1989). Detritivorous soil invertebrates would be least affected by impact winter, postapocalyptic greenhouse, wildfires, or acid rain, with both their food and environment buffered from surficial chemical and physical assault.

Aquatic Invertebrates

Non-marine snails and clams suffered severe extinction in North America (Hartman 1998). Extinction among mollusks was less severe in the ocean (Jablonski 1996), which was a much larger and better-buffered system than streams and lakes of Montana. Nevertheless, many calcareous shelled organisms such as ammonites were terminated, and others such as coccolithophores and foraminifera suffered heavy extinction (d'Hondt et al. 1996; Marshall and Ward 1996). Analysis of different lineages whose feeding behavior can be inferred from living relatives reveals that the keys to survival across the Cretaceous-Tertiary boundary were starvation resistance (for example by low basal metabolic rate), sessile habit, detritivory, and non-planktotrophic larvae (Jablonski 1996). Among phytoplankton, those with cysts and spores suffered less severe extinction than those without resting stages (Kitchell et al. 1986). These patterns of extinction are consistent with carbon isotopic and other evidence for a catastrophic crash in primary productivity in the ocean (Arthur et al. 1987; Rhodes and Thayer 1991; Gallagher 1991). The primary production loss in the ocean was mainly unicellular phytoplankton rather than higher plants affected on land, and it is likely that non-marine phytoplankton production was also affected.

Aquatic Vertebrates

Most aquatic animals such as turtles, frogs, salamanders, and champsosaurs fared relatively well across the Cretaceous-Tertiary boundary (Sheehan and Fastovsky 1992). Dead and decaying plant detritus is at the base of the food chain in many aquatic ecosystems, supporting amphibians, small fish, and turtles, which in turn support crocodiles and large fish (Webster 1983). Environmental acidification would be expected to generate widespread leaf browning and fall, clogging streams and lakes with plant detritus. Survival of large aquatic ectotherms such as champsosaurs and crocodiles is evidence that chilling due to impact winter was either not severe or very limited in duration (Wolfe 1991).

Sharks and rays are rare and suffer heavy terminal-Cretaceous extinction in non-marine facies of Montana (table 2.3), in contrast to marine facies where such fossils are common and do not show profound overturn (Kordikova et al. 2001). A few sharks and rays may have colonized fresh or estuarine waters during the Late Cretaceous (Archibald 1996). Their local extinction in Montana may have been due to change toward less marine-influenced facies, and shifting marine-biogeographic connections from Tethyan to Boreal from Cretaceous to Paleocene (Archibald et al. 1993).

Non-Dinosaurian Reptiles

Most Montanan lizards and snakes were extinguished at the Cretaceous-Tertiary boundary. Many species found in the upper Hell Creek Formation, including boas and monitor lizards, were probably terrestrial insectivores and carnivores. One species was perhaps a shell crusher (Archibald 1996). A large (3-m-long) monitor lizard (*Palaeosaniwa*) was probably an ambush carnivore. With limited ties to aquatic ecosystems, poor burrowing ability, and naked skin unprotected by fur or feathers, these creatures would have fared poorly in acidified soils.

Only one pterosaur, an azhdarchid comparable to *Quetzalcoatlus*, is known from latest Cretaceous rocks of Montana and adjacent states (Estes 1964; Archibald 1996). This was probably a scavenger of large dinosaur carcasses, and became extinct with the dinosaurs (Wellnhofer 1991). Perhaps piscivorous, insectivorous, and detritivorous niches of earlier Mesozoic pterosaurs had already been usurped by birds during the Late Cretaceous.

Mammals

Extinction was heavy among mammals at the Cretaceous-Tertiary boundary, though not as severe as indicated by the figures for species-level extinction given here (table 2.3), because eutherians, marsupials, and multituberculates all survived into the Paleocene and founded a dramatic evolutionary radiation (Sloan et al. 1986). Eutherians and deltatherians all had sharp-cusped teeth and were probably insectivorous (Kielan-Jaworowska et al. 1979a,b). Late Cretaceous multituberculates and marsupials included insectivores, omnivores, shell crushers, and herbivore-frugivores, but only insectivorous and omnivorous forms survived into the Paleocene (Clemens 1979; Clemens and Kielan-Jaworowska 1979; Krause 1984; Archibald 1996). Mollusks, fresh leaves, and fruit were probably scarce at the Cretaceous-Tertiary boundary, when simplified food chains were based on dead plant matter, with insects and other invertebrates that feed on such detritus (Sheehan and Hansen 1986). Although clear fossorial adaptations are not yet known from Cretaceous mammals (Kielan-Jaworowska et al. 1979a; Novacek 1996), all were small. They may have survived transient cold, dark, and heat within hollow logs, caves, or burrows.

Dinosaurs

Large carnivorous and herbivorous dinosaurs were terminated at the Cretaceous-Tertiary boundary instantaneously within limits of resolution (10^1–10^3 years) of the record (Sheehan et al. 1991, 2000; Pearson et al. 1999). Redeposition of bone in paleochannels, lack of bone preservation in non-calcareous paleosols, and statistical artifacts of range extensions all potentially compromise this conclusion (Fassett et al. 2002), but have been addressed at length (Retallack 1994; Lofgren 1995; Marshall 1998). An especially promising, and still little exploited, line of evidence is fossil footprints, which are clear evidence of living dinosaurs in place. Tracks of large duckbill dinosaurs have been found on the last sandstone bed suitable for their preservation, only 37 cm below the Cretaceous-Tertiary boundary beds in Colorado (Lockley 1991), again effectively at the boundary within temporal resolution for this fluvial sequence.

Latest Cretaceous duckbill and ceratopsian dinosaur herbivores had specialized dental batteries for processing large quantities of fresh leaves. Specializations for carnivory in latest Cretaceous theropod dinosaurs include sickle claws and fearsome arrays of sharply crested and serrated teeth, well suited for killing and slicing large quantities of fresh meat (Paul 1988a). None of these herbivorous or carnivorous specializations had any parallel among Late Cretaceous or Early Paleocene mammals or birds. Large dinosaurian herbivore extinction may have followed leaf browning and fall from acidification. Large carnivores would follow herbivores into extinction, unable to sustain themselves on small insectivorous and piscivorous mammals and birds.

Other extinction mechanisms are less appealing. Dinosaurs were no strangers to wildfires, either in the earlier Mesozoic (Harris 1957, 1981) or in the latest Cretaceous of Montana (Retallack 1994, 1997). Dinosaur bones and eggs in Indian paleosols between the first few basalts of the Deccan Traps are found only as high as the iridium anomaly (Courtillot et al. 1990; Bhandari et al. 1996; Sahni et al. 1994; Chatterjee 1997), and so support the idea of extinction related to impact rather than due to local volcanic eruption. Climatic extremes are unlikely to have extinguished dinosaurs either, because they lived at both high and low latitudes (Paul 1988b; Clemens and Nelms 1993; Vickers-Rich and Rich 1993; Hammer and Hickerson 1994). Wildfires, volcanic eruptions, impact winter, transient dust clouds, and post-apocalyptic greenhouse conditions are unlikely to have terminated such a successful, diverse, and widespread clade of ecologically dominant organisms.

Birds

Latest Cretaceous and earliest Paleocene birds remain poorly known. Eight species of birds including two loon-like divers (Gaviiformes), a flamingo-like wader (Ciconiiformes), four shorebirds (transitional Charadriformes), and a toothed bird (Ichthyornithiformes) have been described from the Late Cretaceous Lance Formation of Wyoming (Brodkorb 1963; Estes 1964). A latest Cretaceous enantiornithiform (*Avisaurus*) is known from the upper Hell Creek Formation of Montana (Chiappe 1992; Archibald 1996), along with many undescribed species (Stidham, 2002). In North America there were at least 20 species of latest Cretaceous birds, but only seven sure and two dubious species of earliest Paleocene birds (Stidham pers. comm. 2002). The earliest Paleocene, Bug Creek Anthills assemblage includes bones of at least one species referable to the Late Cretaceous genus *Cimolopteryx* (Estes and Berberian 1970), which was an avocet-sized shorebird (transitional Charadriformes) with some similarities to avocets (Recurvirostridae), thick-knees (Burhinidae), and pratincoles (Glareolidae: Brodkorb 1963). Both molecular biological (Cooper and Penny 1997) and paleontological evidence (Chiappe 1995) indicate that crown-group birds (Neornithes) survived, whereas archaic hesperornithiformes and enantiornithines went extinct (Stidham 2002). The small-bodied shorebirds that are known to have survived were probably piscivores or scavengers of aquatic ecosystems, and may have been sustained through the crisis by food chains based on plant detritus (Sheehan and Hansen 1986).

Stratigraphic and taxonomic considerations compromise other potential evidence of terminal Cretaceous bird survival. Late Cretaceous loons (Gaviiformes) are known from Chile and Antarctica (Chatterjee 1997), and flightless ground birds (*Patagopteryx*) from Patagonia (Chiappe 1996), but exactly when they went extinct is unknown. Similarly, a Late Cretaceous giant ground bird (*Gargantuavis*) from southern France is unrelated to Paleogene giant ground birds, and thus is unknown from the Tertiary (Buffetaut 2002). The late Paleocene shorebird *Dakotornis* from nearby North Dakota (Erickson 1975), and the adaptive radiation of birds documented from Eocene lacustrine deposits (Chatterjee 1997) are too geologically young to be relevant to Cretaceous–Tertiary survival. Some Late Cretaceous records, such as parrots (Stidham 1998), have been questioned (Dyke et al. 1999). Well-preserved shorebirds from New Jersey (Olson & Parris 1987; Gallagher 2002) may be Paleocene rather than Cretaceous (Benton 1999), although this is also debated (Dyke et al. 1999).

Acid Rain and Quality of Fossil Record

Much of what we know about Cretaceous-Tertiary extinctions on land comes from long-term studies of the badlands of eastern Montana, where the total extinction of dinosaurs and massive extinctions within many other groups of organisms offer a grim picture (fig. 2.8). Ironically, the calcareous and smectitic floodplains of Montana offered

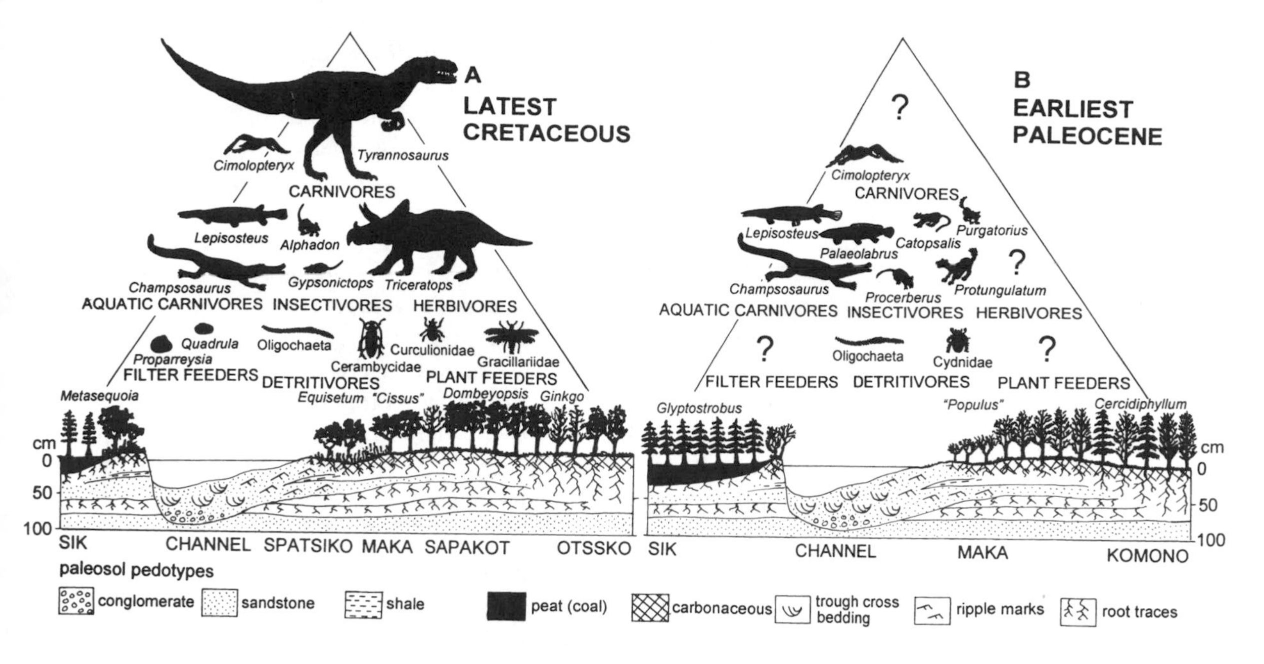

Figure 2.8. Changes in trophic structure across the Cretaceous-Tertiary boundary in non-marine communities of Montana, U.S. (Data from Brodkorb 1963; Estes and Berberian 1970; Lofgren 1995; Hartman 1998; and Labandeira et al. 2002)

considerable buffering of acid rain, and may have been one of the best places to weather the acid storm. In addition, this sequence of little-oxidized paleosols and paleochannels with calcareous cements and smectite clays was also one of few sequences favorable to preserving a record of both bones and plant debris (Retallack 1998). Acid rain may have compromised our fossil record of bones from the last meter or so in Bug Creek, but geochemical variability indicates little influence deeper than that (Retallack 1994). Paleosol sequences of Montana and adjacent states were thus uncommon preservational environments both for creatures weathering an acid crisis and for their fossils.

Other promising Cretaceous-Tertiary fossiliferous successions of calcareous paleosols have been discovered in southeastern China (Stets et al. 1996), peninsular India (Sahni et al. 1994; Bhandari et al. 1996), and southern France (Cousin et al. 1994). However, these red, highly oxidized paleosols are unlikely to yield a good record of fossil plants (Retallack 1998). These areas also may have been important refugia and preservational environments for animals with shells, bone, teeth, and shelled eggs, because their soils supplied abundant calcium and phosphorus for calcite and apatite, and also buffered these mineralized skeletal parts from acid rain. In contrast, areas of granitic and siliceous soils, like currently acidified regions of Scandinavia and northeastern North America (Howells 1995), would have lost more lives and fossils to acidification at the Cretaceous-Tertiary boundary. Acid rain may explain why end-Cretaceous fossil bones and shells were not preserved in many sequences.

Acid rain offers selectivity of mortality and extinction that may account for the complex pattern of extinctions in Montana and adjacent states, where the biota of the Cretaceous-Tertiary boundary is best understood. Even in this area there remains much to be learned about patterns and mechanisms of extinction from insect-damaged leaves, fossil footprints, and the isotopic chemostratigraphy of carbon, sulfur, and nitrogen. There is also promise for extending such studies to other parts of the world. Although the hypothesis of impact-induced acid rain as a proximate cause of dinosaur demise and bird success remains unfalsified by existing data, it is clearly a hypothesis deserving and capable of further testing. Simply put, the standing hypothesis is that the fatal component of end-Cretaceous events for dinosaurs and other creatures was browning of leaves and other transient destruction of primary productivity by impact-induced acid rain.

Acknowledgments. G. Leahy, P. Sheehan, G. Goles, D. Fastovsky, J. D. MacDougall, H. D. Holland, W. Holser, K. Caldeira, L. Derry, and R. Bakker offered useful discussion. I also thank D. Eberth and J. Wright for comments and editorial assistance.

References

Adams, C. M., N. G. Dengler, and J. Hutchinson. 1984. Acid rain effects on foliar histology of *Artemisia tilesi. Canadian Journal of Botany* 62: 463–474.

Alvarez, L. W., W. Alvarez, F. Asaro, and H. V. Michel. 1980. Extraterrestrial cause for the Cretaceous-Tertiary boundary extinction. *Science* 208: 1095–1108.

Alvarez, W., P. Claeys, and S. W. Kieffer. 1995. Emplacement of Cretaceous–Tertiary boundary shocked quartz from Chicxulub Crater. *Science* 269: 930–935.

Archibald, J. D. 1996. *Dinosaur extinction and the end of an era: what the fossils say.* New York: Columbia University Press.

Archibald, J. D., and L. J. Bryant. 1990. Differential Cretaceous/Tertiary extinctions of nonmarine vertebrates: evidence from northeastern Montana. *Geological Society of America* Special Paper 247: 549–562.

Archibald, J. D., P. M. Sheehan, and D. E. Fastovsky. 1993. Major extinction of land-dwelling vertebrates at the Cretaceous-Tertiary boundary, eastern Montana: comment and reply. *Geology* 21: 90–93.

Arens, N. C., and A. H. Jahren. 2000. Carbon isotopic excursion in atmospheric CO_2 at the Cretaceous-Tertiary boundary: evidence from terrestrial sediments. *Palaios* 15: 314–322.

Arthur, M. A., J. C. Zachos, and D. S. Jones. 1987. Primary productivity and the Cretaceous/Tertiary boundary event in the oceans. *Cretaceous Research* 8: 43–54.

Bakker, R. T. 1986. *The dinosaur heresies.* New York: William Morrow.

Beerling, D. J., J. C. McElwain, and C. P. Osborne. 1998. Stomatal response of the "living fossil" *Ginkgo biloba* L. to change in atmospheric CO_2 concentrations. *Journal of Experimental Botany* 49: 1603–1607.

Bell, R. E. 1965. Geology and stratigraphy of the Fort Peck fossil field, northwest McCone County, Montana. Master's thesis, University of Minnesota.

Benton, M. J. 1999. Early origins of modern birds and mammals: molecules vs. morphology. *BioEssays* 21: 1043–1051.

Bhandari, N., P. N. Shukla, Z. G. Ghevariya, and S. M. Sundaram. 1996. K/T boundary layer in Deccan Intertrappeans at Anjar, Kutch. *Geological Society of America* Special Paper 307: 417–427.

Bohor, B. F. 1990. Shocked quartz and more: impact signatures in Cretaceous/Tertiary boundary clays. *Geological Society of America* Special Paper 247: 335–342.

Boslough, M. B., E. P. Chael, T. B. Trucano, D. A. Crawford, and D. L. Campbell. 1996. Axial focusing of impact energy in the Earth's interior: a possible link to flood basalts and hotspots. *Geological Society of America* Special Paper 307: 441–550.

Brett, R. 1992. The Cretaceous-Tertiary extinction: a lethal mechanism involving anhydrite target rocks. *Geochimica et Cosmochimica Acta* 56: 3603–3606.

Brodkorb, P. 1963. Birds from the Upper Cretaceous of Wyoming. *13th Ornithological Conference Proceedings,* pp. 55–70.

Brook, G. A., M. E. Folkoff, and E. O. Box. 1983. A world model of soil carbon dioxide. *Earth Surface Processes and Landforms* 8: 79–88.

Bryant, L. J. 1989. Non-dinosaurian lower vertebrates across the Cretaceous-Tertiary boundary in northeastern Montana. *University of California Publications in Geological Sciences* 134: 1–107.

Buffetaut, E. 2002. Giant ground birds at the Cretaceous-Tertiary boundary. In C. Koeberl and K. G. MacLeod (eds.), *Catastrophic events and mass extinctions: impacts and beyond. Geological Society of America* Special Paper 356: 303–306.

Burnham, D. A., K. L. Derstler, P. J. Currie, R. J. Bakker, Z.-H. Zhou, and J. H. Ostrom. 2000. Remarkable new birdlike dinosaur (Theropoda, Maniraptora) from the Upper Cretaceous of Montana. *University of Kansas Paleontological Contributions* 13: 1–14.

Caldeira, K. G., and M. R. Rampino. 1990. Deccan volcanism, greenhouse warming, and the Cretaceous/Tertiary boundary. *Geological Society of America* Special Paper 247: 117–123.

Carlisle, D. B., and D. R. Braman. 1991. Nanometre-size diamonds in the Cretaceous/Tertiary boundary clay of Alberta. *Nature* 352: 708–709.

Chatterjee, S. 1997. *The rise of birds: 225 million years of evolution.* Johns Hopkins University Press, Baltimore.

Chiappe, L. M. 1992. Enantiornithiforme (Aves) tarsometatarsi and the avian affinities of the Late Cretaceous Avisauridae. *Journal of Vertebrate Paleontology* 12: 344–350.

———. 1995. The first 85 million years of avian evolution. *Nature* 378: 349–355.

———. 1996. Late Cretaceous birds of southern South America: anatomy and systematics of Enantiornithes and *Patagopteryx deferriisi.* In G. Arratia (ed.), *Contributions of southern South America to vertebrate paleontology. Münchner Geowissenschaftliche Abhandlungen* 30: 203–244.

Claeys, P., W. Kiessling, and W. Alvarez. 2002. Distribution of Chicxulub ejecta at the Cretaceous-Tertiary boundary. In C. Koeberl and K. G. MacLeod (eds.), *Catastrophic events and mass extinctions: impacts and beyond. Geological Society of America* Special Paper 356: 55–68.

Clemens, W. A. 1979. Marsupalia. In J. A. Lillegraven, Z. Kielan-Jaworowska, and W. A. Clemens (eds.), *Mesozoic mammals: the first two thirds of mammalian history,* pp. 193–220. Berkeley: University of California Press.

Clemens, W. A., and Z. Kielan-Jaworowska. 1979. Multituberculata. In J. A. Lillegraven, Z. Kielan-Jaworowska, and W. A. Clemens (eds.), *Mesozoic mammals: the first two thirds of mammalian history,* pp.99–149. Berkeley: University of California Press.

Clemens, W. A., and L. G. Nelms. 1993. Paleoecological implications of

Alaskan terrestrial vertebrate fauna in latest Cretaceous time at high paleolatitudes. *Geology* 21: 503–506.

Cooper, A., and D. Penny. 1997. Mass survival of birds across the Cretaceous–Tertiary boundary. *Science* 275: 1109–1113.

Courtillot, J., D. Vandamme, J. Besse, J. J. Jaeger, and M. Javoy. 1990. Deccan volcanism at the Cretaceous/Tertiary boundary: data and inferences. *Geological Society of America* Special Paper 247: 401–409.

Cousin, R., G. Breton, R. Fournier, and J.-P. Watté. 1994. Dinosaur egg-laying and nesting in France. In K. Carpenter, K. F. Hirsch, and J. R. Horner (eds.), *Dinosaur eggs and babies,* pp. 57–74. Cambridge: Cambridge University Press.

Covey, C., S. J. Ghan, J. J. Walton, and P. R. Weissman. 1990. Global environmental effects of impact-generated aerosols: results from a general circulation model. *Geological Society of America* Special Paper 247: 263–270.

Cronan, C. S. 1985. Chemical weathering and solution chemistry in acid forest soils: differential influence of soil type, biotic processes and H^+ deposition. In J. I. Drever (ed.), *The chemistry of weathering,* pp. 175–195. Dordrecht: D. Riedel.

Davenport, S. A., T. J. Wdowiak, D. D. Jones, and P. Wdowiak. 1990. Chondritic metal toxicity as a seed stock kill mechanism in impact-caused mass extinctions. *Geological Society of America* Special Paper 247: 71–76.

Demchuck, T. D., and D. A. Nelson-Glatiotis. 1993. The identification and significance of kaolinite-rich volcanic ash horizons (tonsteins) in the Ardley Coal Zone, Wabamon, Alberta, Canada. *Bulletin of Canadian Petroleum Geology* 41: 464–469.

d'Hondt, S., M. E. Q. Pilson, H. Sigurdsson, A. K. Hanson, and S. Carey. 1994. Surface water acidification and extinction at the Cretaceous-Tertiary boundary. *Geology* 22: 983–986.

d'Hondt, S., T. D. Herbert, J. King, and C. Gibson. 1996. Planktic foraminifera, asteroids and marine production: death and recovery at the Cretaceous-Tertiary boundary. *Geological Society of America* Special Paper 307: 303–318.

Duncan, R. A., and D. G. Pyle. 1988. Rapid eruption of the Deccan flood basalts at the Cretaceous-Tertiary boundary. *Nature* 333: 841–843.

Dyke, G. J., G. Mayr, and T. A. Stidham. 1999. Did parrots exist in the Cretaceous period? *Nature* 399: 317–318.

Erickson, B. R. 1975. *Dakotornis cooperi,* a new Paleocene bird from North Dakota. *Science Museum of Minnesota Scientific Publications* 3(1): 1–7.

Estes, R. 1964. Fossil vertebrates from the Late Cretaceous Lance Formation eastern Wyoming. *University of California Publications in Geological Science* 49: 1–180.

Estes, R., and P. Berberian. 1970. Paleoecology of a Late Cretaceous vertebrate community from Montana. *Breviora* 343: 1–35.

Fassett, J. E., R. A. Zielinski, and J. R. Budahn. 2002. Dinosaurs that did not die: evidence for Paleocene dinosaurs in the Ojo Alamo Sandstone, San Juan Basin, New Mexico. In C. Koeberl and K. G. MacLeod (eds.), *Catastrophic events and mass extinctions: impacts and beyond. Geological Society of America* Special Paper 356: 307–336.

Fastovsky, D. E., and K. McSweeney. 1987. Paleosols spanning the Cretaceous-Paleogene transition, eastern Montana and western North Dakota. *Geological Society of America Bulletin* 99: 66–77.

Fastovsky, D. E., K. McSweeney, and L. D. Norton. 1989. Pedogenic development at the Cretaceous-Tertiary boundary, Garfield County, Montana. *Journal of Sedimentary Petrology* 59: 758–767.

Fölster, H. 1985. Proton consumption rates in Holocene and present-day weathering of acid forest soils. In J. I. Drever (ed.), *The chemistry of weathering,* pp. 197–209. Dordrecht: D. Riedel.

Gallagher, W. B. 1991. Selective extinction and survival across the Cretaceous/Tertiary boundary in the northern Atlantic coastal plain. *Geology* 19: 967–970.

———. 2002. Faunal changes across the Cretaceous-Tertiary (K-T) boundary in the Atlantic coastal plain of New Jersey: restructuring the marine community after the K-T mass extinction event. In C. Koeberl and K. G. MacLeod (eds.), *Catastrophic events and mass extinctions: impacts and beyond. Geological Society of America* Special Paper 356: 291–301.

Gilmour, I., W. S. Wolbach, and F. Anders. 1990. Early environmental effects of the terminal Cretaceous impacts. *Geological Society of America* Special Paper 247: 383–390.

Hallam, A. 1987. End-Cretaceous mass extinction event: argument for terrestrial causation. *Science* 238: 1237–1242.

Hammer, W. R., and W. J. Hickerson. 1994. A crested theropod dinosaur from Antarctica. *Science* 264: 828–830.

Harris, T. M. 1957. A Rhaeto-Liassic flora in South Wales. *Royal Society of London* Proceedings B147: 289–308.

———. 1981. Burnt ferns in the English Wealden. *Proceedings of the Geologists Association* 92: 47–58.

Hartman, J. H. 1998. The biostratigraphy and paleontology of latest Cretaceous and Paleocene freshwater bivalves from the western Williston Basin, Montana, U.S.A. In P. A. Johnston and J. W. Haggart (eds.), *An eon of evolution: paleobiological studies honoring Norman D. Newell,* pp. 317–345. Calgary: University of Calgary Press.

Heymann, D., L. P. F. Chibante, R. B. Brooks, W. S. Wolbach, J. Smit, A. Korochantsev, M. A. Nazarov, and R. E. Smalley. 1996. Fullerenes of possible wildfire origin in Cretaceous-Tertiary boundary sediments. *Geological Society of America* Special Paper 307: 453–464.

Hildebrand, A. R., G. T. Penfield, D. A. Kring, D. Pilkington, A. Camargo, S. B. Jacobsen, and W. V. Boynton. 1991. Chicxulub Crater; a possible Cretaceous-Tertiary boundary impact crater on the Yucatan Peninsula. *Geology* 19: 867–871.

Hotton, C. L. 1988. *Palynology of the Cretaceous-Tertiary boundary in central Montana, U.S., and its implications for extraterrestrial impact.* Unpublished PhD thesis, University of California, Berkeley.

Howells, G. 1995. *Acid Rain and Acid Waters* (2d ed). Chichester: Ellis-Horwood.

Ivany, L. C., and P. J. Salawich. 1993. Carbon isotopic evidence for biomass burning at the K/T boundary. *Geology* 21: 487–490.

Izett, G. A. 1990. The Cretaceous/Tertiary boundary interval, Raton Basin, Colorado and New Mexico, and its content of shock-metamorphosed minerals: evidence relevant to the K/T boundary impact theory. *Geological Society of America* Special Paper 249, 100 pp.

Jablonski, D. 1996. Mass extinctions: persistent problems and new directions. *Geological Society of America* Special Paper 307: 1–9.

Jahren, A. H., N. C. Arens, G. Sarmiento, T. Guerro, and R. Amundson. 2001. Terrestrial record of methane hydrate dissociation in the Early Cretaceous. *Geology* 29: 159–162.

Johnson, K. R. 1989. *A high resolution megafloral biostratigraphy span-*

ning the Cretaceous-Tertiary boundary in the northern Great Plains. Unpublished PhD thesis, Yale University.
Johnson, K. R., and L. J. Hickey. 1990. Megafloral change across the Cretaceous/Tertiary boundary in the northern Great Plains and Rocky Mountains, U.S.A. *Geological Society of America* Special Paper 247: 433–444.
Kielan-Jaworowska, Z., T. M. Bown, and J. A. Lillegraven. 1979a. Eutheria. In J. A. Lillegraven, Z. Kielan-Jaworowska, and W. A. Clemens (eds.), *Mesozoic mammals: the first two thirds of mammalian history,* pp. 221–258. Berkeley: University of California Press.
Kielan-Jaworowska, Z., J. G. Eaton, and T. M. Bown. 1979b. Theria of Metatherian-Eutherian grade. In J. A. Lillegraven, Z. Kielan-Jaworowska, and W. A. Clemens (eds.), *Mesozoic mammals: the first two thirds of mammalian history,* pp. 183–191. Berkeley: University of California Press.
Kitchell, J. A., D. L. Clark, and A. M. Gombos. 1986. Biological selectivity of extinction: a link between background and mass extinction. *Palaios* 1: 504–511.
Kordikova, E. G., P. D. Polly, V. A. Alifanov, Z. Roček, G. F. Gunnell, and A. O. Averianov. 2001. Small vertebrates from the Late Cretaceous and Early Tertiary of the northeastern Aral Sea region, Kazakhstan. *Journal of Paleontology* 75: 390–400.
Krause, D. W. 1984. Mammalian evolution in the Paleocene: beginning of an era. In P. D. Gingerich and C. E. Badgley (eds.), *Mammals, notes for a short course,* pp. 87–109. University of Tennessee, Department of Geology, Studies in Geology 8.
Kring, D. A. 1995. The dimensions of the Chicxulub impact crater and impact melt sheet. *Journal of Geophysical Research* 100: 1679–1686.
Labandeira, C. C., K. R. Johnson, and P. Wilf. 2002. Impact of the terminal Cretaceous event on plant-insect associations. *Proceedings National Academy of Sciences U.S.* 99: 2061–2066.
Landis, G. P., J. K. Rigby, R. E. Sloan, R. Hengston, and L. W. Smee. 1996. Pelée hypothesis: ancient atmospheres and geologic-geochemical controls in evolution, survival and extinction. In N. MacLeod and G. Keller (eds.), *Cretaceous-Tertiary mass extinction: biotic and environmental change,* pp. 519–556. New York: W. W. Norton.
Lewis, J. S., G. H. Watkins, H. Hartman, and R. G. Prinn. 1982. Chemical consequences of major impact events on Earth. *Geological Society of America* Special Paper 190: 215–221.
Lockley, M. G. 1991. Tracking dinosaurs: a new look at an ancient world. New York: Cambridge University Press.
Lofgren, D. L. 1995. The Bug Creek problems and the Cretaceous-Tertiary transition in McGuire Creek, Montana. *University of California Publications in Geological Sciences* 140: 1–185.
Lündstrom, V. S., N. van Bremen, and D. Bain. 2000. The podzolization process: a review. *Geoderma* 94: 91–107.
MacDougall, J. D. 1988. Seawater strontium isotopes, acid rain and the Cretaceous-Tertiary boundary. *Science* 239: 485–487.
Marshall, C. F. 1998. Determining stratigraphic ranges. In S. K. Donovan and C. R. C. Paul (eds.), *Adequacy of the fossil record,* pp. 23–53. Chichester: John Wiley and Sons.
Marshall, C. F., and P. D. Ward. 1996. Sudden and gradual molluscan extinction in latest Cretaceous eastern European Tethys. *Science* 274: 1360–1363.

Martin, E. E., and J. D. MacDougall. 1991. Seawater Sr isotopes at the Cretaceous/Tertiary boundary. *Earth and Planetary Science* Letters 104: 166–180.

McHone, J. F., R. A. Nieman, L. F. Lewis, and A. M. Yates. 1989. Stishovite at the Cretaceous-Tertiary boundary, Raton, New Mexico. *Science* 243: 1182–1184.

McSweeney, K., and D. E. Fastovsky. 1987. Micromorphological and SEM analysis of Cretaceous-Paleogene petrosols from eastern Montana and western North Dakota. *Geoderma* 40: 49–63.

Morgan, J., M. Warner, and Chicxulub Working Group. 1997. Size and morphology of the Chicxulub impact crater. *Nature* 390: 472–476.

Nichols, D. J., and R. F. Fleming. 1990. Plant microfossil record of the terminal Cretaceous event in the western United States and Canada. *Geological Society of America* Special Paper 247: 445–455.

Nichols, D. J., R. F. Fleming, and N. D. Fredericksen. 1990. Palynological evidence of the terminal Cretaceous event on terrestrial floras in western North America. In E. G. Kauffman and O. R. Walliser (eds.), *Extinction events in Earth history,* pp. 351–364. Berlin: Springer.

Norris, R. D., and J. V. Firth. 2002. Mass wasting of Atlantic continental margins following the Chicxulub impact event. In C. Koeberl and K. G. MacLeod (eds.), *Catastrophic events and mass extinctions: impacts and beyond. Geological Society of America* Special Paper 356: 79–95.

Novacek, M. 1996. *Dinosaurs of the Flaming Cliffs.* New York: Doubleday.

Olson, S. L., and D. C. Parris. 1987. The Cretaceous birds of New Jersey. *Smithsonian Contributions to Paleobiology* 63: 1–22.

Orth, C. J., M. Attrep, and L. R. Quintana. 1990. Iridium abundance patterns across bio-event horizons in the fossil record. *Geological Society of America* Special Paper 247: 45–59.

Padian, K. 1998. When is a bird not a bird? *Nature* 393: 729–730.

Paul, G. S. 1988a. Predatory dinosaurs of the world. New York: Simon and Schuster.

Paul, G. S. 1988b. Physiological, migratorial, climatological, geophysical, survival and evolutionary implications of polar dinosaurs. *Journal of Paleontology* 62: 640–652.

Pearson, D. A., T. Schaefer, K. R. Johnson, and D. J. Nicols. 1999. Vertebrate biostratigraphy of the Hell Creek Formation in southwestern South Dakota. *Geological Society of America* Abstracts 31(7): A72.

Prinn, R. G. and B. Fegley. 1987. Bolide impacts, acid rain, and biospheric traumas at the Cretaceous-Tertiary boundary. *Earth and Planetary Science* Letters 83: 1–15.

Qiang, J., P. J. Currie, M. A. Norell, and S.-A. Ji. 1998. Two feathered dinosaurs from northeastern China. *Nature* 393: 753–761.

Retallack, G. J. 1994. A pedotype approach to latest Cretaceous and earliest Tertiary paleosols in eastern Montana. *Geological Society of America* Bulletin 106: 1377–1397.

Retallack, G. J. 1996. Acid trauma at the Cretaceous-Tertiary boundary in eastern Montana. *Geological Society of America Today* 6(5): 1–7.

Retallack, G. J. 1997. Dinosaurs and dirt. In D. Wolberg, E. Stump, and G. Rosenberg (eds.), *Dinofest International,* pp. 345–359. Philadelphia: The Academy of Natural Sciences.

Retallack, G. J. 1998. Fossil soils and completeness of the rock and fossil record. In S. K. Donovan and C. R. C. Paul (eds.), *The adequacy of the fossil record,* pp. 131–162. Chichester: John Wiley and Sons.

Retallack, G. J. 2001. A 300-million-year record of atmospheric carbon dioxide from fossil plant cuticles. *Nature* 411: 287–290.

Retallack, G. J., G. D. Leahy, and M. D. Spoon. 1987. Evidence from paleosols for ecosystem changes across the Cretaceous–Tertiary boundary in Montana. *Geology* 15: 1090–1093.

Rhodes, M. C., and C. W. Thayer. 1991. Mass extinctions: ecological selectivity and primary production. *Geology* 19: 877–880.

Rigby, J. K., Sr., and J. K. Rigby, Jr. 1990. Geology of the Sand Arroyo and Bug Creek Quadrangles, McCone County, Montana. *Brigham Young University Geology Studies* 36: 69–134.

Sahni, A., S. K. Tandon, A. Jolly, S. Bajpai, A. Sood, and S. Srinivasan. 1994. Upper Cretaceous dinosaur eggs and nesting sites from the Deccan volcano-sedimentary sequence of peninsular India. In K. Carpenter, K. F. Hirsch, and J. R Horner (eds.), *Dinosaur eggs and babies,* pp. 204–226. Cambridge: Cambridge University Press.

Sharpton, V. L., K. Burke, A. Camargo-Zanoguera, S. A. Hall, D. S. Lee, L. E. Marin, G. Suárez-Reynoso, J. M. Quizada-Muñeton, P. D. Spudis, and J. U. Fucugauchi. 1993. Chicxulub multiring impact basin: size and other characteristics derived from gravity analyses. *Science* 261: 1564–1567.

Sharpton, V. L., L. E. Marin, J. L. Carney, S. Lee, G. Ryder, B. C. Schuraytz, P. Sikora, and P. D. Spudis. 1996. A model of the Chicxulub impact basin based on evaluation of geophysical data, well logs and drill core samples. *Geological Society of America* Special Paper 307: 55–74.

Sheehan, P. M., and T. A. Hansen. 1986. Detritus feeding as a buffer to extinction at the end of the Cretaceous. *Geology* 14: 868–870.

Sheehan, P. M., D. E. Fastovsky, R. G. Hoffman, C. B. Berghaus, and D. L. Gabriel. 1991. Sudden extinction of the dinosaurs, latest Cretaceous, upper Great Plains, U.S. *Science* 254: 835–839.

Sheehan, P. M., and D. E. Fastovsky. 1992. Major extinctions of land dwelling vertebrates at the Cretaceous-Tertiary boundary, eastern Montana. *Geology* 20: 556–560.

Sheehan, P. M., P. J. Coorough, and D. E. Fastovsky. 1996. Biotic selectivity during the K/T and Late Ordovician extinction events. *Geological Society of America* Special Paper 307: 477–491.

Sheehan, P. M., D. E. Fastovsky, C. Barreto, and R. G. Hoffman. 2000. Dinosaur abundance was not declining in a "3 m gap" at the top of the Hell Creek Formation, Montana and North Dakota. *Geology* 28: 523–526.

Sigurdsson, H., S. d'Hondt, M. A. Arthur, T. J. Bralower, J. C. Zachos, M. van Fossen, and J. E. T. Channell. 1991. Glass from the Cretaceous/Tertiary boundary in Haiti. *Nature* 349: 482–487.

Sigurdsson, H., S. d'Hondt, and S. Carey. 1992. The impact of the Cretaceous/Tertiary bolide on evaporite terrane and generation of major sulfuric acid aerosol. *Earth and Planetary Science* Letters 109: 543–559.

Sloan, R. E., J. K. Rigby, L. Van Valen, and D. Gabriel. 1986. Gradual dinosaur extinction and simultaneous ungulate radiation in the Hell Creek Formation. *Science* 232: 629–633.

Smit, J., W. A. van der Kaars, and J. K. Rigby, Jr. 1987. Stratigraphic aspects of the Cretaceous-Tertiary boundary in the Bug Creek area of eastern Montana. *Memoires de la Société géologique de la France* 150: 53–73.

Smit, J., A. Montanari, N. M. H. Swinburne, W. Alvarez, A. R. Hildebrand,

S. V. Margolis, P. Claeys, W. Lowrie, and F. Asaro. 1992. Tektite-bearing, deep-water clastic unit at the Cretaceous-Tertiary boundary in northeastern Mexico. *Geology* 20: 99–103.

Staub, J. R., and A. D. Cohen. 1978. Kaolinite enrichment beneath coals: a modern analog, Snuggedy Swamp, South Carolina. *Journal of Sedimentary Petrology* 48: 203–210.

Stets, J., A. R. Ashraf, H. K. Erben, G. Hahn, U. Hambach, K. Krumsiek, J. Jhein, and P. Wurster 1996. The Cretaceous-Tertiary boundary in the Nanxiong Basin (continental facies, southeast China). In N. MacLeod and G. Keller (eds.), *The Cretaceous-Tertiary mass extinction: biotic and environmental effects,* pp. 349–371. New York: W. W. Norton.

Stidham, T. A. 1998. A lower jaw from a Cretaceous parrot. *Nature* 396: 29–30.

Stidham, T. A. 2002. Extinction and survival of birds (Aves) at the Cretaceous-Tertiary boundary: evidence from western North America. *Abstracts Geological Society of America* 34(6): 355.

Swisher, C. C., L. Dingus, and R. F. Butler. 1993. $^{40}Ar/^{39}Ar$ dating and magnetostratigraphic correlation of terrestrial Cretaceous-Paleogene boundary and Puercan mammal age, Hell Creek–Tullock Formation, eastern Montana. *Canadian Journal of Earth Sciences* 30: 1981–1996.

Tinus, R. W., and D. J. Roddy. 1990. Effects of global atmospheric perturbations on forest ecosystems in the Northern Temperate Zone: predictions of seasonal depressed-temperature kill mechanisms, biomass production, and wildfire soot mechanisms. *Geological Society of America* Special Paper 247: 77–86.

Vickers-Rich, P., and T. H. Rich. 1993. Wildlife of Gondwana. Chatswood, N.S.W., Australia: Reed.

Vonhof, H. B., and J. Smit. 1997. High resolution late Maastrichtian-early Danian oceanic$^{86}Sr/^{87}Sr$ record: implications for the Cretaceous-Tertiary boundary events. *Geology* 25: 347–350.

Webster, J. R. 1983. The role of benthic macroinvertebrates in detritus dynamics of streams: a computer simulation. *Ecological Monographs* 53: 383–404.

Weil, A. 1994. K/T survivorship as a test of acid rain hypothesis. *Geological Society of America* Abstracts 26: A335.

Wellnhofer, P. 1991. The illustrated encyclopedia of pterosaurs. New York: Crescent Book.

Williams, M. E. 1994. Catastrophic versus noncatastrophic extinction of the dinosaurs: testing, falsifiability, and the burden of proof. *Journal of Paleontology* 68: 183–190.

Wolbach, W. L, I. Gilmour, E. Anders, C. J. Orth, and R. R. Brooks. 1988. A global fire at the Cretaceous/Tertiary boundary. *Nature* 334: 665–669.

Wolfe, J. A. 1987. Late Cretaceous–Cenozoic history of deciduousness and the terminal Cretaceous event. *Paleobiology* 13: 215–226.

Wolfe, J. A. 1990. Paleobotanical evidence for a marked temperature increase following the Cretaceous/Tertiary boundary. *Nature* 343: 152–154.

Wolfe, J. A. 1991. Palaeobotanical evidence for a June "impact winter" at the Cretaceous-Tertiary boundary. *Nature* 352: 420–423.

Wolfe, J. A., and G. R. Upchurch. 1987. Leaf assemblages across the Cretaceous-Tertiary boundary in the Raton Basin, New Mexico and

Colorado. *Proceedings National Academy of Sciences U.S.* 84: 5096–5100.
Yang, W., T. J. Ahrens, and G. Chen. 1996. Shock vaporization of anhydrite and calcite and its effect on global climate from the K/T impact crater at Chicxulub. *Lunar and Planetary Science Conference* Abstracts 27: 1473–1474.
Yoder, C. M., D. J. Siegel, C. R. Newton, D. E. Fastovsky, and P. M. Sheehan. 1995. Sulfur isotopic anomaly near terrestrial K/T boundary. *Geological Society of America* Abstracts 27: A349.
Zahnle, K. 1990. Atmospheric chemistry by large impacts. *Geological Society of America* Special Paper 247: 271–288.

Section II. Osteology and Ichnology

3. New Information on *Bambiraptor feinbergi* (Theropoda: Dromaeosauridae) from the Late Cretaceous of Montana

David A. Burnham

Abstract

Aspects of the osteology of *Bambiraptor feinbergi,* a velociraptorine dromaeosaurid from the Upper Cretaceous Two Medicine Formation of Montana, are described. The holotype consists of a nearly complete skull and skeleton of an immature animal found in association with at least two other individuals of the same species, one of which is larger. Barely a meter in total length, the holotype probably weighed only two kilograms. As in most sub-adults, the orbits and braincase seem disproportionately large when compared with those of most dromaeosaurids. An endocast suggests that *Bambiraptor* had one of the largest dinosaurian brains known. The scapula and coracoid are unfused, the scapula has a pronounced acromion for contact with the furcula, and the glenoid is oriented posterolaterally. The coracoid articulates with a relatively large sternal plate. The arm-to-leg-length ratio (0.69) is one of the highest known for any non-avian dinosaur. The pelvis is opisthopubic, and the pubis has a well-developed pubic boot. The functionally didactylous foot supported a large, strongly curved raptorial claw, like that of other dromaeosaurids.

Introduction

In recent years, new discoveries have elucidated the evolutionary relationships between dinosaurs and birds. Much of this research centers on the Dromaeosauridae, a family of lightly built, agile, carnivorous maniraptorans that is believed by most workers to be closely related to the ancestors of birds. Some controversy remains, owing to a lack of unequivocal interpretation of morphological characters found in the two groups (Martin 1991; Martin and Feduccia 1998; Feduccia 1996; Ruben et al. 1997). However, many of the arguments surrounding this debate have focused more on systematic methodology (Gauthier 1986; Sereno 1999; Norell et al. 2001; Xu et al. 2002) and less on functional aspects of the skeleton. The purpose of this paper is to describe *Bambiraptor feinbergi* and previously unknown aspects of dromaeosaur anatomy, as well as to provide new insight into dromaeosaurid functional morphology and bird origins.

Bambiraptor feinbergi was briefly reported as a small, birdlike, predatory dinosaur (Burnham et al. 2000). While this fossil is geologically too young to be the progenitor of birds, analysis of the specimen reveals a sequence of character acquisitions that may have culminated in the earliest members of Aves. It also shows that fundamental avian features existed within dromaeosaurids prior to the origin of birds. Additionally, the holotype is well preserved, and an assessment of its functional morphological adaptations indicates it was a highly developed, birdlike predator with an advanced brain and well-coordinated skeletal system.

The holotype of *Bambiraptor* represents a small, sub-adult theropod dinosaur less than one meter long and weighing approximately two kilograms in life (fig. 3.1). *Bambiraptor* is assigned to the Maniraptora (Gauthier 1986) on the basis of having a forelimb that is more than 75 percent as long as the presacral vertebral column, a hand which is longer than the foot, a posteriorly bowed ulna, a semilunate carpal, and a thin, bowed third metacarpal. It conforms to the typical dromaeosaurid design with a retroverted pubis, a large, retractable pedal ungual on digit II, and a tail modified with bony extensions of the prezygapophyses and chevrons. It also has relatively long, slender limbs, a shoulder girdle with a laterally facing glenoid, a furcula, and large sternals. It can be identified as a velociraptorine dromaeosaurid because the anterior tooth denticles are significantly smaller than the denticles on the posterior carina (Currie et al. 1990). The skeleton is important because it is reasonably complete, well preserved, and represents a life stage that is not well represented in other small theropods. *Bambiraptor feinbergi* provides new insights into anatomy, functional morphology, and life habits of dromaeosaurid theropods.

One of the most influential descriptions of any theropod is the revolutionary monograph on *Deinonychus* (Ostrom 1969). This fossil material was crucial to understanding the evolutionary changes necessary for determining the probable ancestry of *Archaeopteryx* (Ostrom 1976) and other birds. It precipitated the dinosaur-bird debate with the

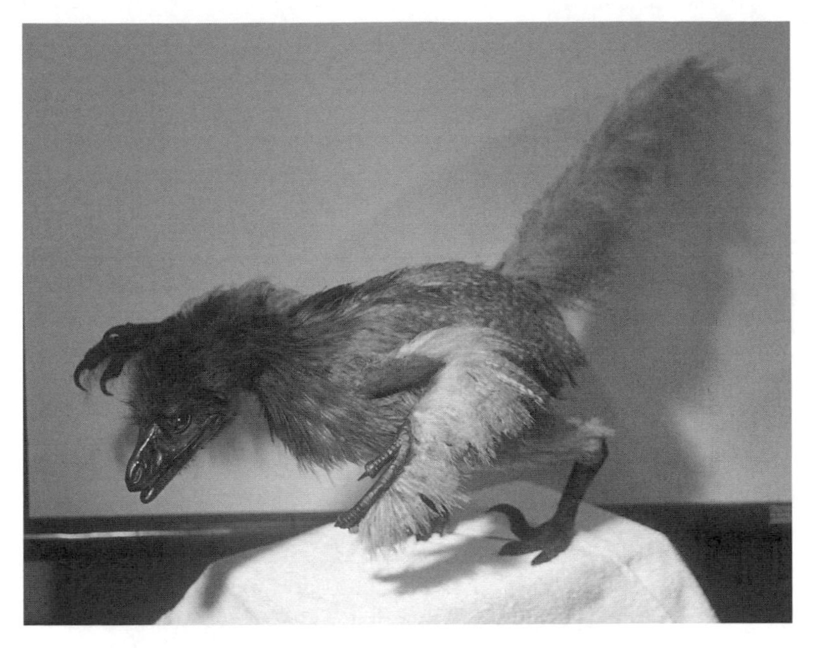

Figure 3.1. Life reconstruction of Bambiraptor feinbergi. (Sculpture ©2003 Brian Cooley)

discovery of a folding wrist mechanism (involving a semilunate carpal) in conjunction with a shoulder girdle that was an evolutionary precursor to the modern avian condition. Ostrom's description of dromaeosaurid osteology also enhanced our understanding of the killing mechanism of the foot, the rod-stiffened tail, and the overall *bauplan* for dromaeosaurids. Although specimens of *Deinonychus* are well preserved, some crucial skeletal elements, including cranial bones, remain unknown. Additionally, *Deinonychus* is larger and apparently less derived (Ostrom 1969) than the geologically younger *Bambiraptor.*

Well-preserved discoveries in Mongolia led workers to focus on *Velociraptor* (Osborn 1924; Sues 1977; Paul 1988). The discovery of a specimen with a furcula (Norell et al. 1997) fulfilled an important criterion required by Heilmann (1927) to be present in an avian ancestor. Recently described material from Mongolia includes articulated skulls (Barsbold and Osmólska 1999) and postcrania (Norell and Makovicky 1997, 1999).

The Yixian deposits (Upper Jurassic or Lower Cretaceous) of China have produced non-avian theropods that include some of the most primitive and smallest representatives of the Dromaeosauridae. Featherlike integumentary structures are preserved on nearly all of the dromaeosaurids from these localities. The dromaeosaurid *Sinornithosaurus* (Xu et al. 2000) is based on a semi-articulated, incomplete skeleton with a furcula and sternal plates. Another specimen of *Sinornithosaurus* has long feather plumes near the tail and hindlimb (Norell 2001). Closely related is a juvenile dromaeosaurid (Ji et al. 2001; Norell 2001). Less known is *Microraptor* (Xu et al. 2000), which is the smallest dromaeosaurid known.

Only partial dromaeosaurid skeletons are known from Canada. *Dromaeosaurus* Matthew and Brown 1922 is based on a specimen with a fairly complete skull and associated foot. The skull (Colbert and Russell 1969; Currie 1995) may be more primitive than other members of the group. *Saurornitholestes* Sues 1978 was initially established on less than 10 percent of a skeleton, which lacks many of the diagnostic features for this group. More recently collected specimens of *Saurornitholestes* in the collections of the Museum of the Rockies (MOR 660) and the Royal Tyrrell Museum of Palaeontology (TMP 88.128.1) will better define this taxon when described.

Other fossils have been assigned to the Dromaeosauridae, but remain poorly known because they are so incomplete. Nonetheless, these specimens help establish the geographic and temporal ranges of this family. They include the larger-bodied forms *Utahraptor* (Kirkland et al. 1993) from North America and *Achillobator* (Perle et al. 1999) from Asia. Additional occurrences include partial skeletons of *Adasaurus* (Barsbold 1983) and *Hulsanpes* (Osmólska 1982) from Mongolia, *Pyroraptor* (Allain and Taquet 2000) from France, and possibly *Unenlagia* (Novas 1998) from Argentina. The Cretaceous of Madagascar has produced *Rahonavis ostromi,* which was described as a bird (Forster et al. 1998) but shares with small maniraptorans the sickle claw on the foot and some other features.

Geology, Taphonomy, and Preservation

Bambiraptor was recovered from the Two Medicine Formation, which crops out along the flanks of the Rocky Mountains in northwestern Montana. Especially at Egg Mountain, the formation is famous for its dinosaur nesting grounds (Horner and Gorman 1988). This rock unit is approximately 600 meters thick (Lorenz 1981) and comprised of fluvial sediments deposited 83 to 74 million years ago (Rogers 1997) adjacent to the Cretaceous interior seaway. The sediments were deposited in a series of westward-dipping beds. The holotype of *Bambiraptor* was found north of the Willow Creek anticline in a non-marine, gray-green mudstone. A thin layer of ankerite surrounded the bones.

The stratigraphy of the Two Medicine formation is well documented, though no precise stratigraphic data was collected with the holotype of *Bambiraptor.* It is estimated that the site is 360 meters (± 50m) above the base of the Virgelle Sandstone/Two Medicine contact (D. Trexler pers. comm. 1999). Its association with a *Maiasaura* bone bed supports this stratigraphic interval because this hadrosaur is restricted to a narrow zone within the Two Medicine Formation (J. Horner pers. comm. 2001).

The *Bambiraptor* specimens, which include the holotype (AMNH 001) and isolated adult bones (AMNH 002–036), were collected from a single locality. A small outcrop is exposed along the northern edge of Blackleaf Creek (S 18, T 26 N, R 7 W) about 11 miles north of Bynum, Montana on the Jones (Tee Six, Inc.) Ranch. The site has been quarried for many years as a bone bed composed mostly of isolated hadrosaur bones with some partially articulated skeletons (Burnham et al. 1997).

Large theropods have also been found, including the articulated skull and partial skeleton of the tyrannosaurid *Gorgosaurus,* now in the Children's Museum of Indianapolis, along with a *Maiasaura* skeleton from the same site. Most of the fossils from this quarry have not yet been adequately studied, but initial observations show an interesting sample representing the Two Medicine fauna (Horner et al. 2001; Trexler 2001). Fish, amphibian, and non-dinosaurian reptiles are not known from the quarry although these fossils are reported from MOR sites thought to be part of the same bone bed (Horner et al. 2001). The only other materials collected are isolated theropod teeth and different types of eggshell fragments. It has not been determined if the dinosaur skeletons occur at different horizons than the isolated bones or whether it is a mixed assemblage of bones and skeletons in one interval. Because some of the material consists of portions of articulated skeletons of different dinosaurs interspersed with many isolated bones, the question remains whether a single event concentrated this material. Until the entire site can be studied and documented in more detail, the data herein must be considered preliminary. However, the occurrence of the holotype of *Bambiraptor* within a small outcrop with other well-preserved skeletons may represent a single catastrophic event.

As determined from the quarry maps (Burnham et al. 2000), the holotype of *Bambiraptor feinbergi* was found near a large hadrosaur skull. It was partially articulated, and was spread out over an area of less than one square meter. The degree of disarticulation shows that the skeleton was disturbed before burial (Weigelt 1989). Major portions of the right side of the skeleton were crushed, and there was considerable disturbance between the skull and the limbs. The nearly complete skull and lower jaws formed a collapsed mass of closely associated bones. The thin cranial bones are well preserved with intact delicate processes. The left side of the muzzle had been separated, twisted, and displaced. The lower jaws were joined at the symphysis and remained in articulation with the quadrates. Remarkably, these areas remained intact after the muzzle and dentaries were deflected onto the rest of the skull. This kind of disarticulation strongly suggests these elements were held together by soft tissues (muscles and integument) until shortly after death. Subsequent to this damage, teeth floated out of the jaws, the podials disassociated, and the skeleton separated into units. Loose teeth, some with roots, were found in the matrix surrounding the skull. The disassociation of the skeleton may have been accomplished by flowing water.

The axial skeleton was preserved as closely associated and articulated vertebral segments, although some were disarticulated in the neck and chest region. The anterior cervicals were in position behind the skull, although a fragment of the braincase (proximal portion of the exoccipital) was found under the cervical centra. The positions of the mid-to-posterior cervicals and anterior dorsals were not clearly recorded. The posterior dorsals and sacrals were articulated, but a small gap separated them from the first four caudals, which were preserved articulated in an upward curve. Most caudal vertebrae were held together by the bony rods of their prezygapophyses and hemal arches, and all but the most distal portion of the tail was recovered. At midpoint, the tail was upturned and slightly twisted, and curved anterodorsally almost 180°.

The appendicular elements were arranged on either side of the axial skeleton close to their positions in life. The scapulae lay in their respective positions, but were separated from the paired sternals. The right coracoid was crushed and partly folded near the right sternal although the glenoid articulation was never found. The nearly complete left coracoid was in close association with the left sternal. Unfortunately, the furcula was not in articulation and was found near the pelvis. Ribs and gastralia lay strewn about the sternal plates although a series of posterior ribs lay in articulation with the dorsal vertebrae. The arms, carpus, and manus were laid out in loose association. The pelvis lay collapsed on its right side, and a single, loose dorsal centrum was found lying under the ventral side. An anterior chevron was found between the ischia, which were in contact distally. The hindlimbs lay close to the pelvis, but neither femur was in the acetabulum. The tibia, fibula, and metatarsals were associated, whereas the pedal elements were in disarray with some missing bones. The metatarsals on the right side were still articulated.

Preservation of the bones was excellent due, in part, to spar calcite filling in the hollow spaces. The bones are black in color (similar to other Two Medicine fossil bones), and their surfaces show foramina and minute details of texture. Crushing was minimal and most bones are three-dimensional.

Sub-Adult Features of the Holotype

Growth series are relatively rare in the fossil record, but various workers have used bone fusion, delayed ossification of some elements, tooth counts, relative dimensions, and histology to determine ages at death of fossil tetrapods. These estimates are rarely accurate, because trends vary among dinosaurs (Varricchio 1997) and are influenced by many different factors. For example, the large heads and eyes of juvenile archosaurs become relatively smaller as the animals grow, but even mature modern birds have large skulls.

The extremes of the size range known for *Bambiraptor* specimens are close, and allometric trends cannot be determined without reference to related animals. Reid (1993) did histological work on the velociraptorine *Saurornitholestes,* and this kind of work may ultimately produce a method to estimate the age of dromaeosaurids. Carpenter and Smith (2001) believe femur length is more reliable for estimating age, especially when multiple specimens are available. Such is the case in *Bambiraptor,* in which the femora of three individuals were recovered from the same bone bed. The femur of the holotype of *Bambiraptor* is 69 percent of the length of the largest velociraptorine femur from the same site, which is presumably a more mature individual of the same species. Comparison of lengths between humeri suggests the holotype is 70 percent grown, and between the tibiae shows the holotype tibia is 74 percent that of the longest tibia. At least one other dromaeosaurid femur is known from the Two Medicine Formation (MOR 660). It lacks a femur, but the tibia of the holotype of *Bambiraptor* is 67 percent of the length of this tibia, and the humerus is only 63 percent of the length. Without cranial material, it is difficult to know if MOR 660 is *Bambiraptor, Saurornitholestes,* or a new type of dromaeosaurid.

The bones of the braincase are separate in the holotype of *Bambiraptor,* which is a clear indication of immaturity. Incomplete fusion, evident in the neural arches by the presence of visible "zigzag" sutures between posterior dorsal neural arches and centra, is another clue suggesting the sub-adult nature of the holotype at the time of death. Although this specimen is not mature, the presence of sternal plates and fusion of some skeletal elements show that it was not a hatchling either. At this time, it cannot be determined exactly how old the holotype was at the time of death. Its small size, along with associated characters such as relatively large orbits and brain, may be at least partially attributable to immaturity.

Materials and Methods

Most of the original bones of the holotype of *Bambiraptor* were molded and cast. A variety of silicone molding materials were used because of their capacity to record surface details (down to a microscopic level), to maintain dimensionality (very low shrinkage), and to release easily from delicate fossil bones. Casts were then poured using urethane plastic (Pro Cast 10) that also has low shrinkage and retains fine detail. Two sets of casts representing the holotype were produced: a research set of unaltered elements, and a working set restored and straightened to assemble a skeleton of the animal. The skull was assembled using casts of the individual elements. Missing portions (supraoccipital, right premaxilla) were sculpted. The nasals were restored posteriorly, although there is some uncertainty as to their total length and their contact with the frontals because of postmortem damage. The dentaries and posterior regions of the jaws were cast as found, but restoration was necessary near the intramandibular joints. The size and shape of the mandibular fenestra were not preserved. The resulting cast of the skull and jaws was straightened, missing teeth were added, and re-molded. Little sculpting or restoration was necessary for the post-cranial skeleton, although some bones (sacrum, tail, some podials) were straightened or partially restored (some vertebrae, right ilium, right coracoid, tips of manual unguals). Missing paired elements were reproduced as mirror images of their counterparts from the opposite side (phalanges, unguals). Rib shafts were sculpted based on information from *Velociraptor* and *Saurornitholestes* (MOR 660). Casts of the appendicular elements were articulated to help in determining their ranges of motion.

The cranial elements of *Bambiraptor feinbergi* were so well preserved that casts of the braincase were easily articulated and an endocast (fig. 3.2A) of silicone rubber rendered.

Institutional abbreviations: AMNH, American Museum of Natural History, New York; IVPP, Institute of Vertebrate Paleontology and Paleoanthropology, Beijing; MOR, Museum of the Rockies, Bozeman; NGMC, National Geological Museum of China, Beijing; TMP, Royal Tyrrell Museum of Palaeontology, Drumheller.

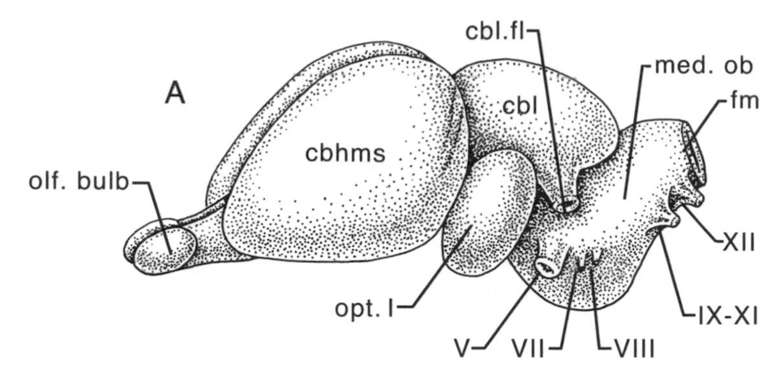

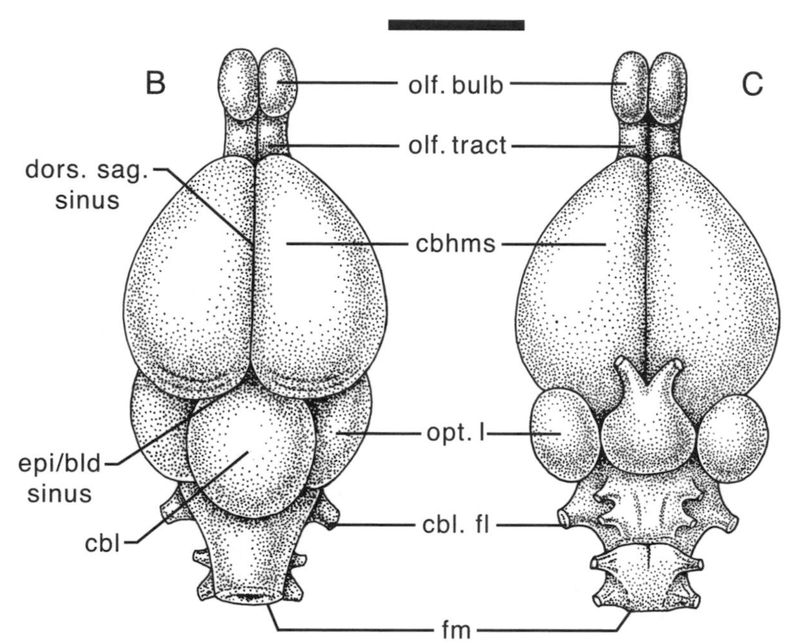

Figure 3.2. Reconstructed endocranial cast Bambiraptor feinbergi *(KUVP 129737) in left lateral view (a), with cranial nerves visible in Roman numerals, V—trigeminal, VII—facial, VIII—cochlear, IX–XI—vagus, glossopharyngeal, accessory spinal nerves, XII—hypoglossal (missing: II, III, IV, VI); dorsal view (b), and ventral view (c). Abbreviations: olfactory bulb (olf. bulb), olfactory tract (olf. tract), cerebral hemispheres (cbhms), cerebellum (cbl), optic lobe (opt. l), cerebellar flocculi (cbl. fl), dorsal sagittal sinus (dors. sag. sinus), epiphysis/blood sinus (epi/bld sinus), medulla oblongata (med. ob), foramen magnum (fm). Scale bar = 1 cm.*

Systematic Paleontology

Dinosauria Owen, 1842
Theropoda Marsh, 1881
Maniraptora Gauthier, 1986
Dromaeosauridae Matthew and Brown, 1922
Velociraptorinae Barsbold, 1983
Bambiraptor feinbergi Burnham, Derstler, Currie, Bakker, Zhou, and Ostrom, 2000.
Holotype: American Museum of Natural History AMNH 001, virtually complete skull and postcranium.
Horizon: Two Medicine Formation (Upper Cretaceous)
Locality and age: Teton County, Montana.

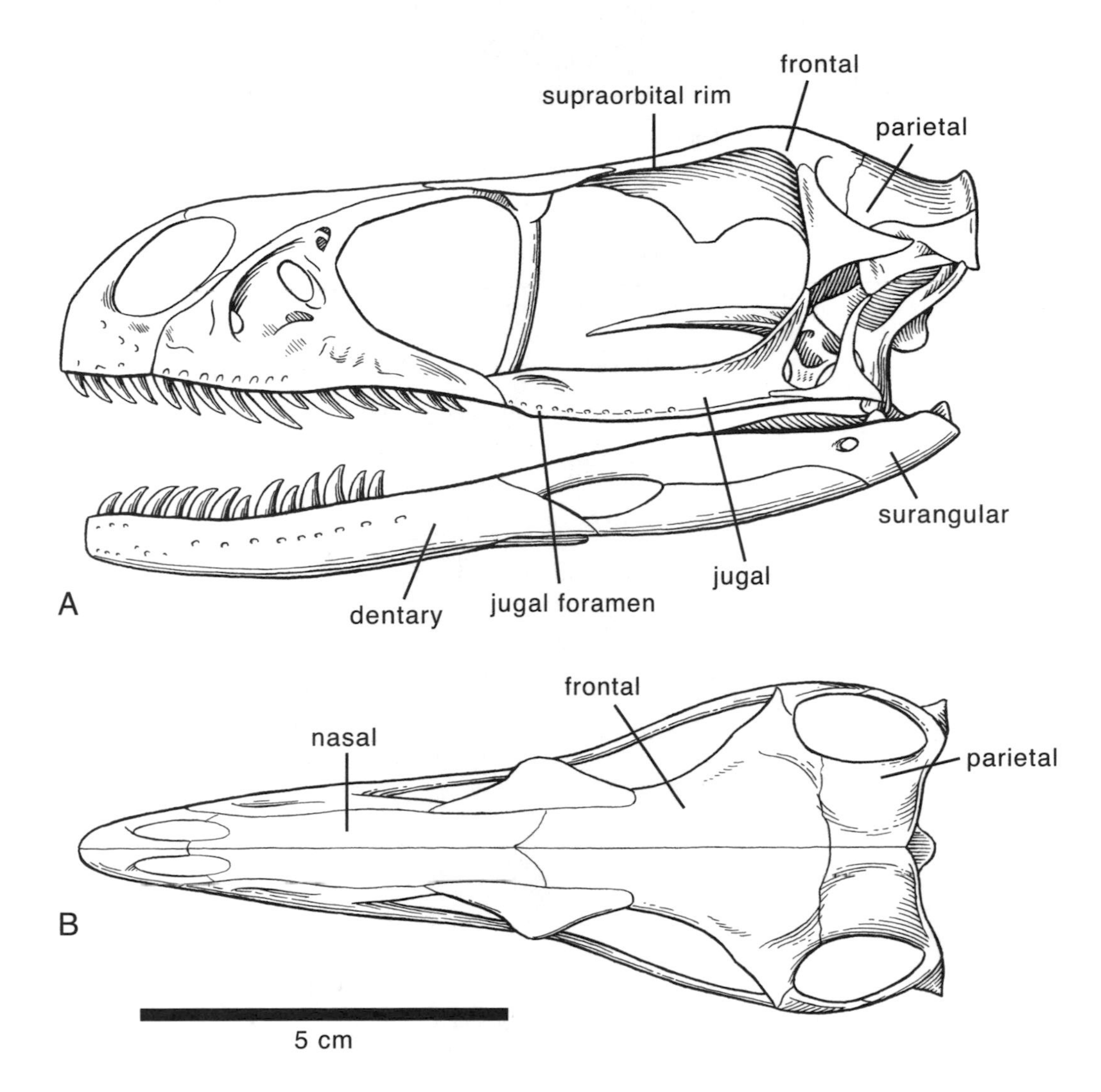

Figure 3.3. Reconstructed skull of Bambiraptor feinbergi *in left lateral (a) and dorsal (b) views. Scale bar = 5 cm.*

Description

The holotype of *Bambiraptor feinbergi* consists of a nearly complete skull and skeleton. The skull (table 3.1) measures 125 mm from the tip of the snout to the occipital condyle, and 55 mm maximum width across the posterior cranium. It has large, almost rectangular orbits (fig. 3.3). The snout is narrow (fig. 3.3B), allowing for a degree of forward, possibly stereoscopic, vision. The antorbital region comprises only 55 percent of the maximum skull length, which is relatively shorter than that of *Velociraptor.* The oval shape of the nares and premaxilla give the snout a small anterior bump in lateral view similar to, but less distinct than, that of *Velociraptor.* The temporal region of the skull is relatively short. The lightly built skeleton is less than 0.5 m tall and is about 1 meter long from the tip of snout to the end of the tail. The tail itself is 350 mm long.

TABLE 3.1
Skull measurements of *Bambiraptor feinbergi* (AMNH 001)

Element	Measured Measurement (in mm)
Maximum length of skull (paraoccipital process–tip of snout	127
Maximum width (across postorbitals)	60
Length of snout (rostral margin of orbit–tip of snout)	70
Maximum depth (skull roof–quadratic condyle)	53
Width of snout (in front of lacrimals)	25
Maxillary tooth row length	43
Upper tooth row length	57
Orbit height	35
Orbit length	36
Lower jaw length	122
Dentary tooth row length	47

Skull

The left premaxilla was found in close association with the left maxilla. No complete tooth is preserved in the four alveoli, but a tooth crown was found less than a centimeter away. The premaxilla is otherwise complete (fig. 3.4) and has a length to height index (Kirkland et al. 1993) of 150, which is closest to *Velociraptor* (164) among dromaeosaurids. The length of the tooth row is 15 mm. The superior (nasal) process is almost parallel to the inferior (maxillary) process, but is longer, more slender, and tapers to a point. The nasal process is straight and is directed posterodorsally at 45°. The maxillary process is stouter, and is concave ventrally for its contact with the maxilla.

There is an isolated crown of a premaxillary tooth, but there are roots within the alveoli of the left premaxilla. The crown has seven serrations per millimeter along the posterior keel, whereas the anterior carina lacks denticles.

The left maxilla (fig. 3.5) is well preserved, but the right maxilla is in two pieces. The triangular maxilla is relatively tall and foreshortened compared to that of *Velociraptor.* Anteriorly, the maxilla is bluntly squared-off where it contacts the premaxilla. The area anterior to the antorbital fenestra has at least two subsidiary fenestrae (fig. 3.5). The teeth are recurved and laterally compressed. The tooth count is at least nine, based on stereo x-ray examination of the alveoli (fig. 3.6) of the left maxilla, but there could have been as many as twelve if more alveoli

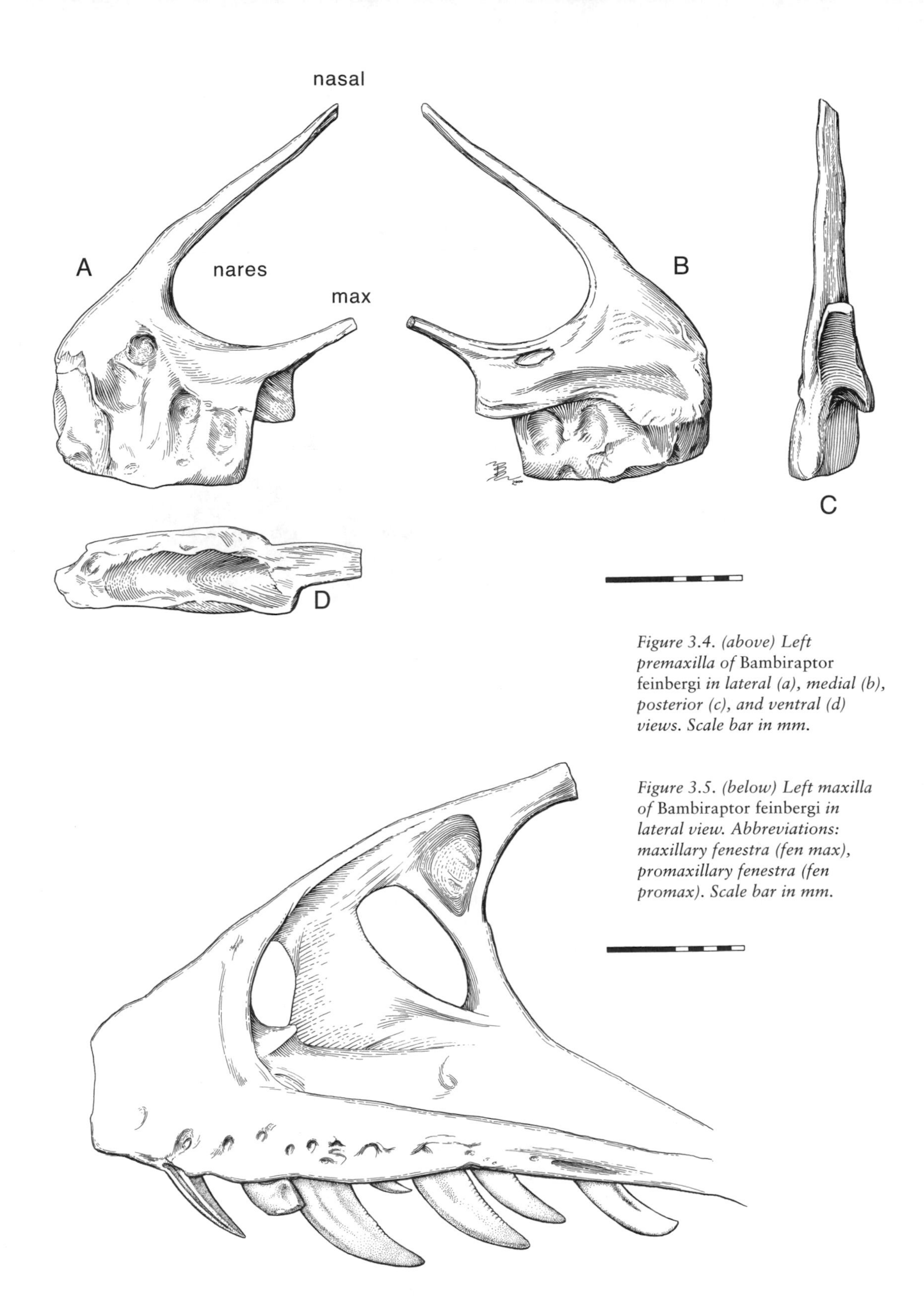

Figure 3.4. (above) Left premaxilla of Bambiraptor feinbergi *in lateral (a), medial (b), posterior (c), and ventral (d) views. Scale bar in mm.*

Figure 3.5. (below) Left maxilla of Bambiraptor feinbergi *in lateral view. Abbreviations: maxillary fenestra (fen max), promaxillary fenestra (fen promax). Scale bar in mm.*

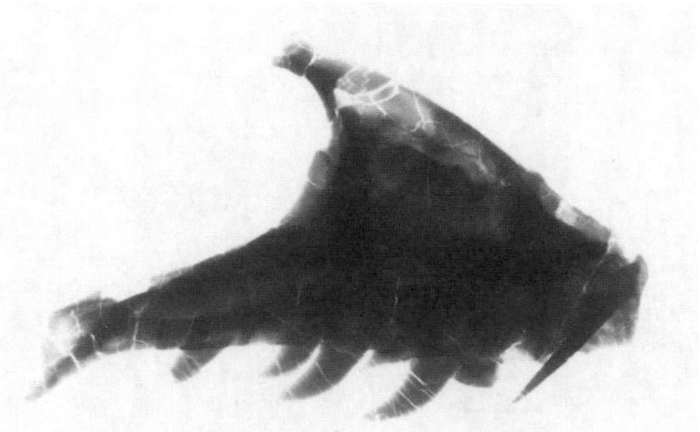

Figure 3.6. Stereo x-rays of left maxilla of Bambiraptor feinbergi *lateral view.*

were added as the animal grew. The largest teeth are positioned mid-length in positions 4, 5, and 6. Serrations are larger on the posterior carina than the anterior, and are sometimes completely absent anteriorly. The interdental plates seem to be separate from each other.

Both nasals suffered postmortem damage when they were separated from the rest of the skull. This long, thin bone bifurcated anteriorly to contact the premaxilla posterodorsal to the external naris.

The left frontal is nearly complete and measures 45 mm along the midline and 22 mm wide just behind the orbits. The anteriorly tapering, triangular shape resembles that of *Velociraptor, Saurornitholestes,* and *Archaeopteryx* more than *Dromaeosaurus,* which is squared off anteriorly (Currie 1995). The relatively longer orbital rim of the holotype of *Bambiraptor* with its raised lateral margin (Burnham et al. 2000) distinguishes it from *Saurornitholestes,* although this may be just an allometric growth feature associated with the relatively large orbit of the juvenile. Exposed dorsally along the anterolateral edge of the *Bambiraptor* frontal is an articular surface for the lacrimal. The suture with the parietal is thickened and grooved, forming a stout, immobile contact between the two bones. Brain morphology can be seen on the ventral surface of the frontals, with distinct depressions for the olfactory lobes and the cerebrum. Small convolutions reflect undulations in the tissues covering the brain. These attest to the tight fit of the brain and associated tissue to the skull roof.

The parietals were found in close association with the frontals and

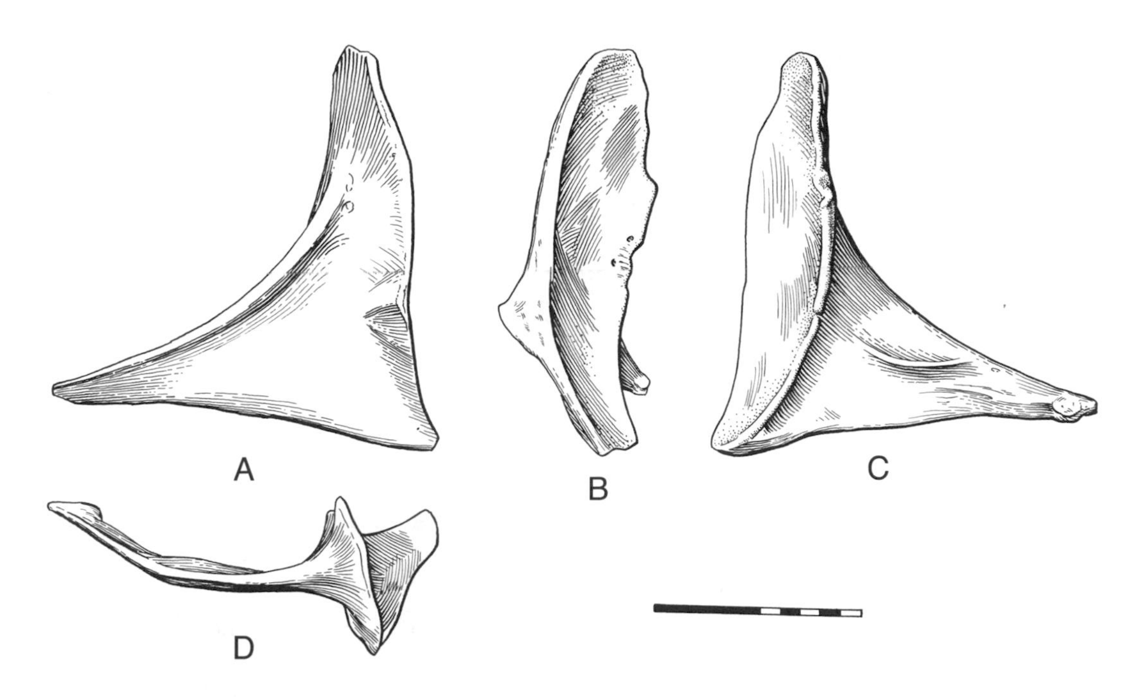

Figure 3.7. (above) Right postorbital of Bambiraptor feinbergi *in lateral (a), anterior (b), medial (c), and posterior (d) views. Scale bar in mm.*

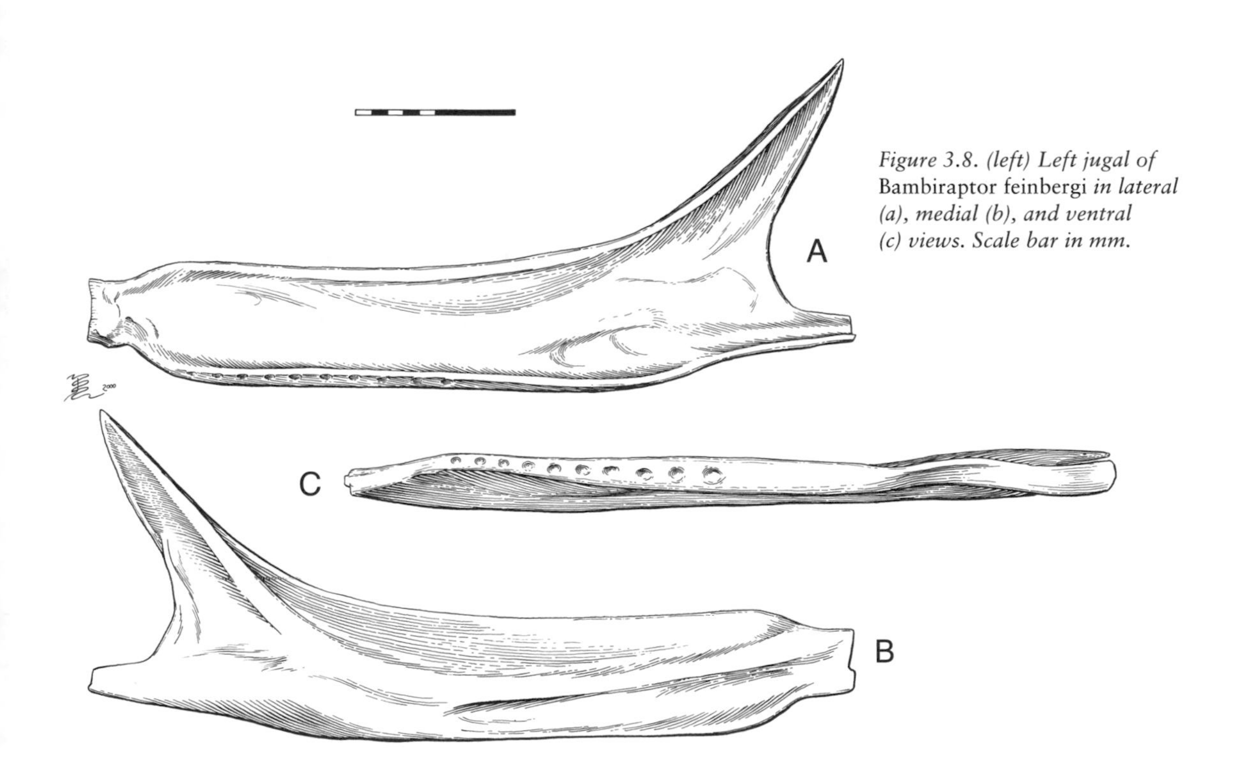

Figure 3.8. (left) Left jugal of Bambiraptor feinbergi *in lateral (a), medial (b), and ventral (c) views. Scale bar in mm.*

laterosphenoids as separate, unfused right and left elements. A suture also separates the parietals in *Sinornithosaurus,* but presumably represents immaturity, because all mature dromaeosaurids have fused parietals. In *Bambiraptor* the parasagittal crest bifurcates behind the parietal-frontal contact. The nuchal crest across the back of the parietals curves laterally downward to form a process that inserts between the squamosal and exoccipital. There is no evidence of a paraparietal process in the holotype as reported for *Sinornithosaurus* (Xu et al. 1999). The ventral surface of paired parietals has a large depression for the cerebellum.

The postorbital is almost triangular in outline (fig. 3.7), like that of *Sinornithosaurus* (Xu et al. 2000). In *Velociraptor* and *Dromaeosaurus,* the postorbital has better-defined processes and is triradiate rather than triangular. *Deinonychus* displays an intermediate condition. A thickened flange forms the back of the orbit and descends to partially overlap the jugal. The posterior process fits into a slot in the squamosal.

The parallel dorsal and ventral edges of the jugal (fig. 3.8) are laterally keeled. There is a row of eight tiny foramina on the ventral surface of each jugal (fig. 3.8c). This feature has not been reported in dromaeosaurids, but was also observed on the MOR specimen of *Deinonychus.* The bone is similar in overall shape to those of *Velociraptor* and *Dromaeosaurus,* but is unlike the jugal of *Deinonychus* in which the suborbital bar expands posteriorly in lateral view (Ostrom 1969).

The lacrimals are T-shaped bones in lateral aspect (fig. 3.3). The anterior nasal process is shorter than the frontal process. The upright preorbital process is channeled, giving it an I-beam appearance in cross section. The shaft of the lacrimal curves medially, clearing the line of sight for the eye. Ventrally, the shaft of the lacrimal flares out into a small boot-shaped contact with the jugal. The dorsal portion of the lacrimal is triangular, tapers anteriorly and posteriorly, and overlaps the frontal. In dorsal view, this bone has a lateral boss that is also found on *Velociraptor.* It has been suggested that the lacrimal is a compound element fused with the prefrontal as in *Deinonychus* (Witmer and Maxwell 1996; Currie and Dong 2001). However, there is no evidence for this in *Bambiraptor,* a sub-adult specimen. The lacrimal in *Velociraptor* is relatively longer anteroposteriorly, but is otherwise very similar to that of *Bambiraptor.*

The squamosal is a rectangular bone with a posterolaterally projecting process that contacts exoccipital and supraoccipital. Anteriorly there is a triangular slot for the postorbital. The contact with the quadratojugal appears loose as preserved.

Both quadratojugals are well preserved. Each is a delicate, triradiate bone (fig. 3.9) with an inverted T-shape (Paul 1988). It is similar to that of *Velociraptor,* but is more lightly built than the quadratojugals of *Deinonychus* (Ostrom 1969) and *Dromaeosaurus* (Currie 1995). The squamosal process is an ascending, curved, thin rod that is anteroposteriorly constricted dorsally where it inserts between the squamosal and quadrate. The curvature has not been reported in other known dromaeosaurid skulls (fig. 3.9c). The posteroventral quadrate process

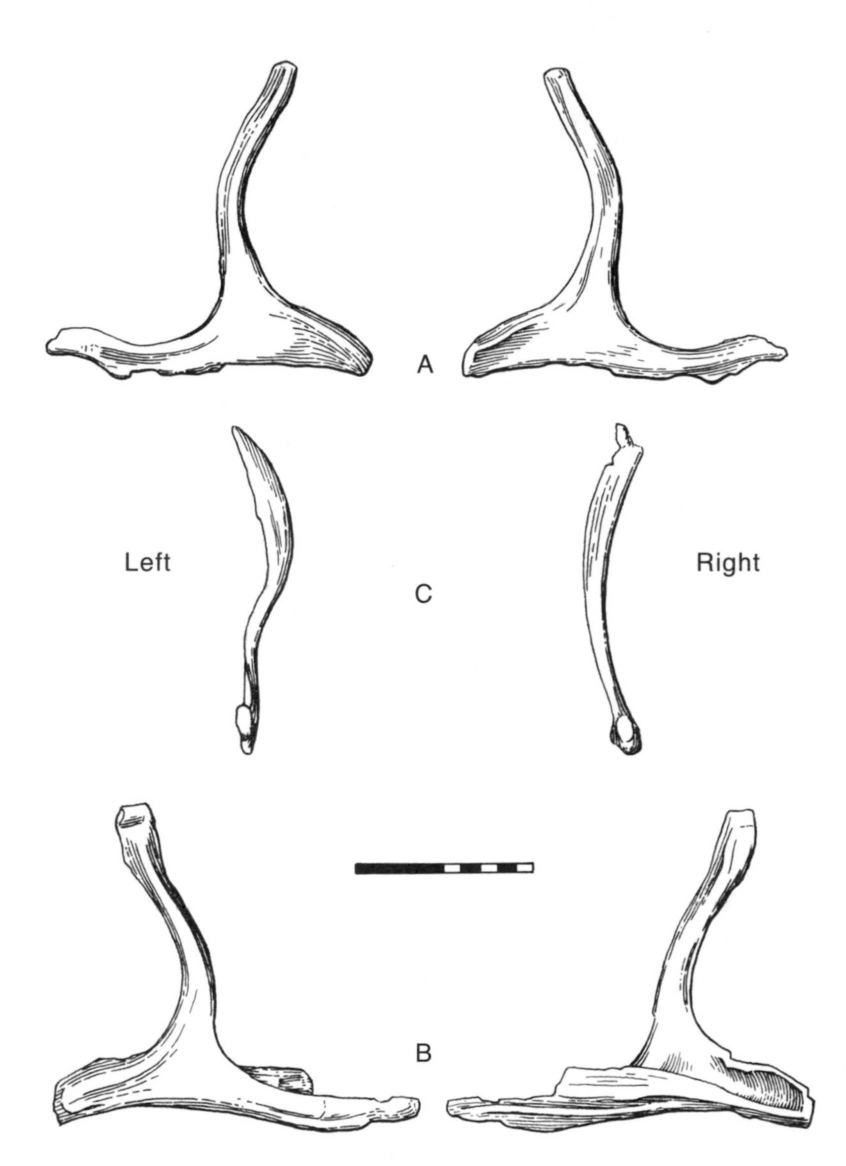

Figure 3.9. Left and right quadratojugals of Bambiraptor feinbergi *in lateral (a), medial (b), and anterior (c) views. Scale bar in mm.*

is short and stout as in all dromaeosaurids (Barsbold and Osmólska 1999) and attaches to the lateral condyle of the quadrate. The anteriorly projecting jugal process overlaps the lateral surface of the jugal as in all theropods.

Right and left quadrates were recovered in articulation with the lower jaws, quadratojugals, and squamosals. As in *Velociraptor, Dromaeosaurus,* and *Deinonychus,* there is a single-headed otic process (fig. 3.10). There is no evidence of any pneumatic foramina. The medial and lateral condyles for the mandibular articulation are separated by a shallow sulcus.

The palate consists of very thin elements that lie inside the crushed skull. A fissure in the matrix extended alongside the palate and caused some damage to these bones. The pterygoid and palatine are apparent-

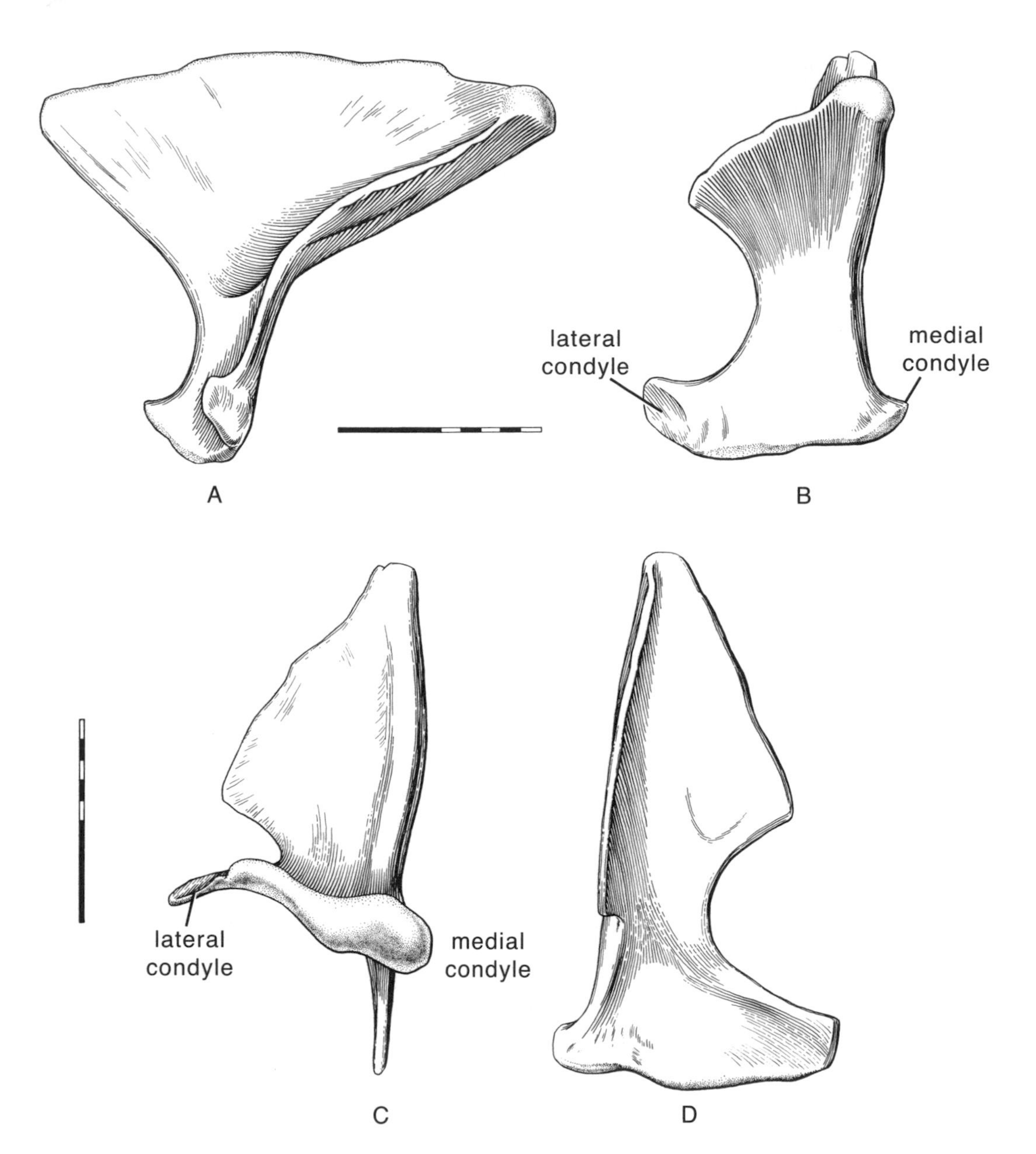

Figure 3.10. Left quadrate of Bambiraptor feinbergi *in lateral (a), posterior (b), dorsal (c), and anterior (d) views. Scale bar in mm.*

ly comparable with those of *Deinonychus* (Ostrom 1969). Both ectopterygoids were found with the skull, and are distinctively like those of most other theropods in design, with robust upwardly curved processes that meet the jugal.

Remnants of a small portion of the sclerotic ring were found in the left orbit. These bones indicate an approximate diameter of 15 mm for the eyeball.

With the exception of the supraoccipital, all elements of the braincase were found in close association. The prootic is notched for the exit of cranial nerve V, the front margin of which was formed by the

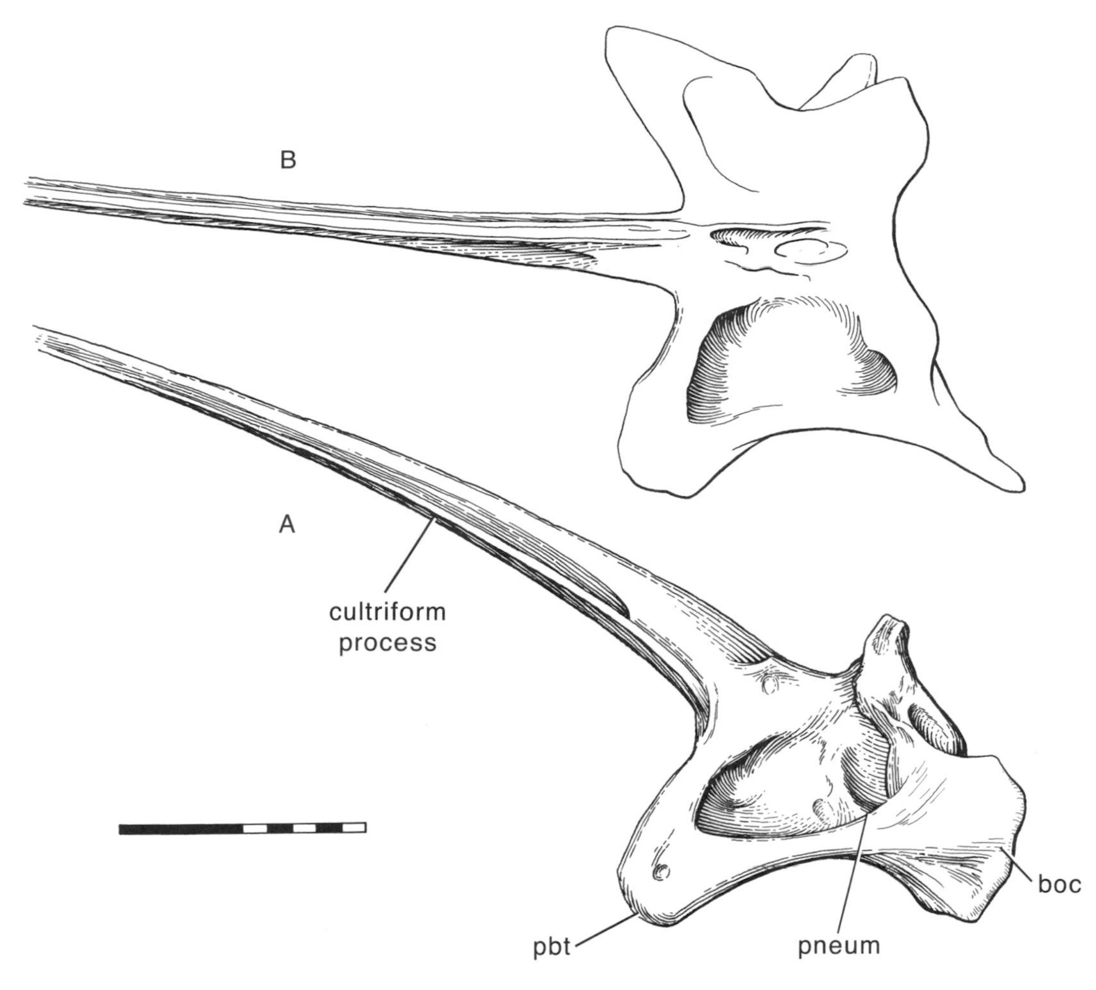

Figure 3.11. Parasphenoid of Bambiraptor feinbergi *in lateral (a) and ventral (b) views. Scale bar in mm.*

laterosphenoid. The basisphenoid-parasphenoid complex (parabasisphenoid) is pneumatic, with deep pockets along its lateral and ventral surfaces (fig. 3.11). The anterior tip of the elongate cultriform process is not preserved, and is probably missing a few millimeters. The basioccipital (fig. 3.12), which formed most of the occipital condyle, participated in the floor of the foramen magnum. The occipital condyle is only one-third the diameter of the foramen magnum. As in *Velociraptor,* the articular surface of the condyle is well rounded, and the basitubera flare out posteriorly (fig. 3.12a). The basitubera are separated by a cleft, and diverge ventrolaterally, unlike the condition in *Velociraptor* and *Dromaeosaurus* in which they are parallel. Pneumatic recesses penetrate the basioccipital-basisphenoid suture. The paroccipital process (exoccipital plus opisthotic) projects posterolaterally as in *Deinonychus* (Brinkman et al. 1998), but contrasts with the posteroventrally oriented process of the London *Archaeopteryx*.

An elongate, slender stapes, found along the side of the exoccipital, is broken lengthwise and is poorly preserved.

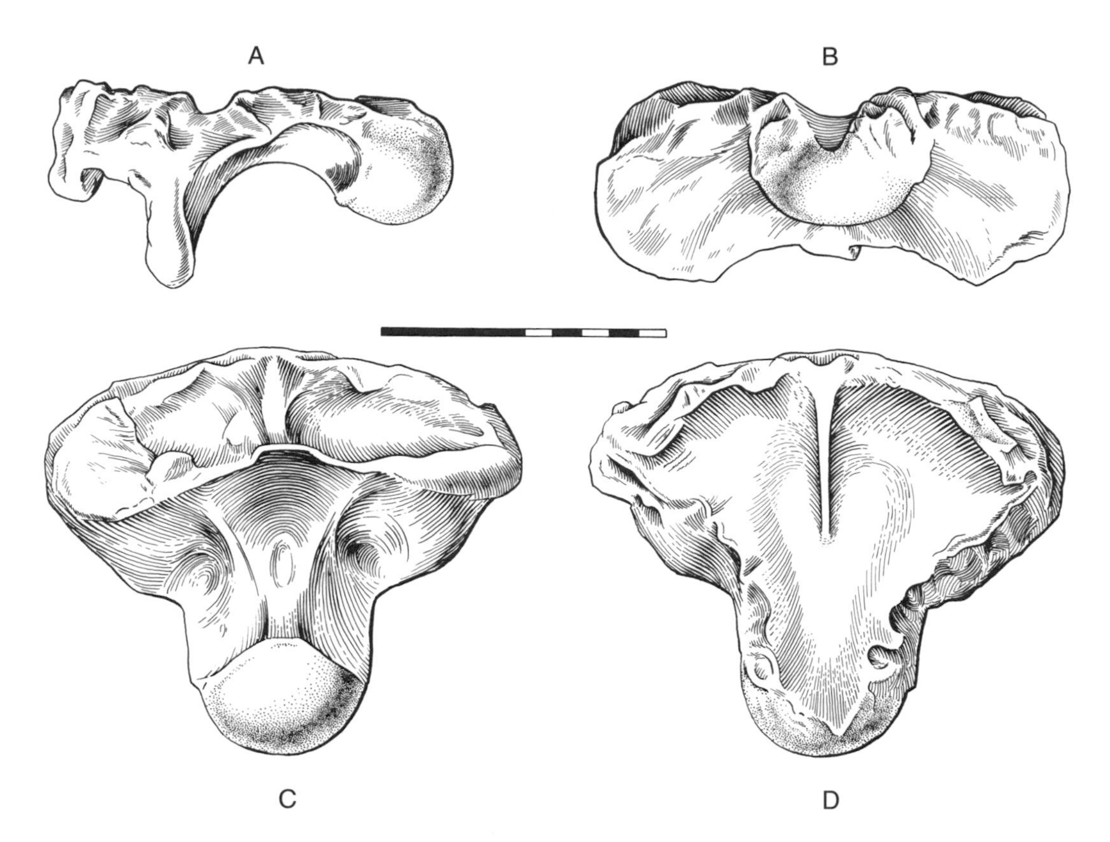

Figure 3.12. Basioccipital of Bambiraptor feinbergi *in lateral (a), posterior (b), ventral (c), and dorsal (d) views. Scale bar in mm.*

The dentaries were recovered in a separate block of matrix from the posterior mandibular bones, and were apparently displaced before burial. The dentaries lack their posterior margins, but include complete alveolar margins (fig. 3.13). In dorsal view (fig. 3.13c) the anterior ends of the dentaries curve toward the midline to meet in a symphysis, which is a small, flat, roughened area. The left dentary has 12 tooth positions. The anterior sockets are empty, but most of positions 6 through 12 have teeth in situ. Along the lateral surface, there is a row of foramina that are larger anteriorly, as well as a lower parallel row of smaller foramina. There are fused interdental plates as reported by Currie (1987) for other dromaeosaurids. The dentary tooth crowns show some wear. They fit the basic velociraptorine denticle pattern of having approximately seven denticles per mm on the posterior carina, but lack denticles on the anterior keel (Currie et al. 1990).

The left splenial was found separated from the dentary and other jaw elements, whereas the right splenial was recovered near the right jaw articulation. The splenial is similar to that of *Deinonychus* (Ostrom 1969) in lateral and medial views. Both surangulars were recovered but are missing their anterior portions. The posterior surangular foramen is a small opening positioned anterior to the jaw articulation. The angular is flat and fan-shaped where it overlaps the surangular as a thin sheet. Anteriorly, the dorsal margin is clearly evident and forms the margin of the external mandibular fossa. The external mandibular fenestra was

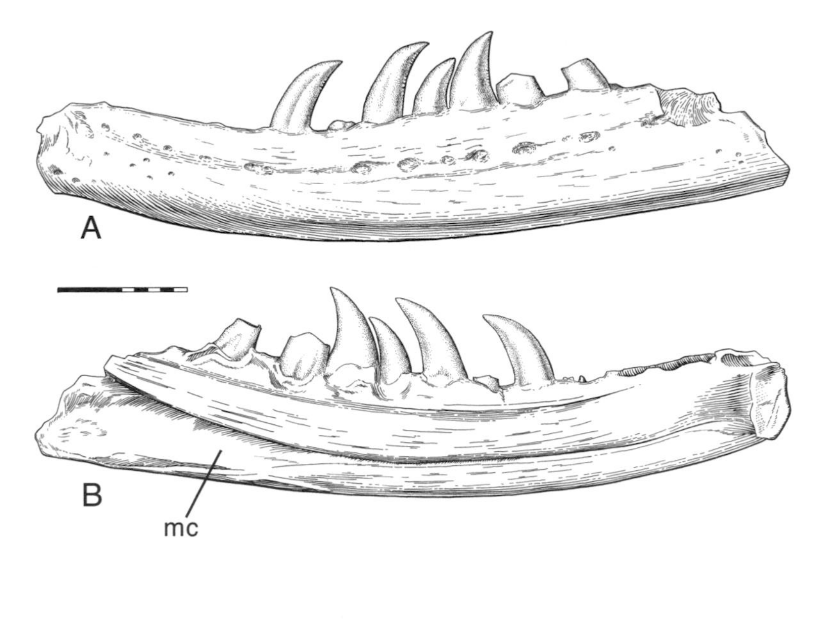

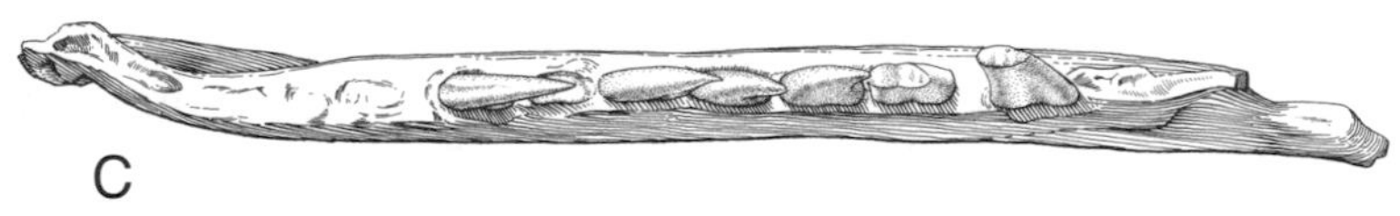

Figure 3.13. Left dentary of Bambiraptor feinbergi *in lateral (a), medial (b), and dorsal (c) views. Abbreviation: Meckelian canal (mc). Scale bar in mm.*

probably large, as described for *Deinonychus* (Ostrom 1969) and *Velociraptor* (Barsbold and Osmólska 1999). A small, flat, triangular bone found on the medial surface of the dorsal process of the right jugal is tentatively identified as a coronoid. The posterior end of the prearticular is exposed along the medial side of the surangular, but is covered anteriorly by portions of the palate. Right and left articulars were found in place on both surangulars and are unfused. Each is a robust bone with a prominent downwardly curved retroarticular process. A posterior buttress is present, as in *Dromaeosaurus* (Colbert and Russell 1969) and *Deinonychus* (Ostrom 1969).

The slightly expanded posterior end of the thin, rod-like right ceratohyal was in close association with the posterior end of the surangular. The bone is 45 mm long and 1mm in diameter, and is similar to the hyoid figured for *Sinornithosaurus* (Xu and Wu 2001).

Loose teeth recovered from the matrix surrounding the skull range from isolated crowns to perfectly preserved teeth with roots. Like all velociraptorine dromaeosaurids, the denticles are larger on the posterior carina than they are on the anterior ridge.

Paleoneurology

The only previously described endocast of a small, non-avian theropod is that of *Troodon*, which has a high degree of encephalization (Russell 1969; Currie 1985). The endocast taken from the holotype of

Bambiraptor is detailed enough (fig. 3.2) to show vascular imprints on the ventral surface of the skull roof. Because this suggests that the brain of the living animal occupied the entire braincase cavity, the endocast is a good indicator of how large the brain actually was (Jerison 1973). The three main areas of the brain (Cobb and Edinger 1962) are evident in the endocast. Whereas the optic lobe is relatively pronounced, the paired olfactory tracts are relatively small in comparison with *Troodon* and tyrannosaurids. Cranial nerves II, III, IV, and VI cannot be seen on the endocast, but V and VII–XII are represented.

The brain lies mostly behind the orbits, with the olfactory structures extending anteriorly toward the snout. The olfactory bulbs lie above and slightly anterior to the orbit (approximately at the frontal-nasal suture). The optic lobes are readily distinguished in the midbrain region and reside in a ventrolateral position, as in birds. There is some flexure in the brain (fig. 3.2a), but is not as pronounced as in modern birds.

The endocast, measured from the olfactory lobe to the foramen magnum, is 55.2 mm long; its maximum height is 31.3 mm; and at its widest point across the anterior portion of the cerebellum it is 27.5 mm. The endocast of the holotype of *Bambiraptor* displaced 14 cm^3 of water. If we assume that the brain had a specific gravity of 0.9, then the original brain weighed an estimated 12.6 g. Body mass is estimated to range from 1.86 to 2.24 kg (estimated using the circumference of the femur shaft in the formula developed by Anderson et al. 1985). This yields an REQ (Reptilian Encephalization Quotient, Jerison 1973) of 12.5 to 13.8 and a BEQ (Bird Encephalization Quotient, Jerison 1973) of 1.2 to 1.4, figures well above values found not only in dinosaurs, but birds as well (Wharton 2001). This estimate may even be conservative because the endocast does not preserve the portion of the brain bordered by the supraoccipital. However, the estimated EQs may be somewhat inflated because of the immaturity of the holotype (brain size shows negative allometry during growth in all tetrapods). Nevertheless, the estimate is high enough to suggest that the relative brain size of *Bambiraptor* was as large or larger than that of *Troodon* (Currie and Zhao 1993) and other coelurosaurs, which puts it in the lower part of the range of modern birds. Additionally, the posterior enlargement of the brain is also atypical of other coelurosaurs and is more birdlike.

Comparison of spinal cord data shows similarities in the cervicodorsal region between the holotype *Bambiraptor* and those of other dromaeosaurids. This cross-sectional area was described by Giffin (1990) as especially enlarged in this region. Spinal cord anatomy has been described for *Deinonychus* and *Saurornitholestes* by Giffin (1990). Other lines of evidence, referred to as the spinal quotient (SQ), indicate the biggest difference in brain size occurs between very young individuals and adults (Giffin 1990). SQ measure versus basal skull length for alligators showed that it was higher only for hatchlings, but medium-sized and large individuals were "remarkably constant" (Giffin 1990).

Axial Skeleton

Nearly the entire vertebral column (table 3.2) was recovered for the holotype of *Bambiraptor feinbergi*. The anterior cervicals (up to C-

TABLE 3.2
Vertebral measurements* for AMNH 001

Vertebra position	Maximum length	Posterior width	Width at transverse process	Maximum height of neural canal	Maximum width of neural canal
C-1 (atlas)	—	—	—	—	—
C-2 (axis)	13	5	16.9	5	6.1
C-3	13.5	11.2	24a†	5.8	7.9
C-4	13.5	10.8	22.4	5.7	7
C-5	16	10	—	7	7.7
C-6	14.5	7.4	—	6.7	5.8
C-7	15a	9.4	—	7	7
C-8	12	11.3	—	—	7.4
C-9	13	—	—	—	—
C-10	—	—	—	—	—
D-1	11	9.7	34a	8	8.4
D-2	12	12.7	29.5a	7.2	7.5
D-3	12	—	—	—	—
D-4	11.8	—	—	—	—
D-5	11	—	—	—	—
D-6	13	—	—	—	—
D-7	10	—	—	—	—
D-8	10.5	—	—	—	—
D-9	9.5	—	—	—	—
D-10	10	—	—	—	—
D-11	10.5	—	—	—	—
D-12	10.8	—	—	—	—
D-13	10.5	—	—	—	—
S-1	11	—	—	6	7
S-2	15	—	—	?	?
S-3	15	—	—	?	?
S-4	13	—	—	?	?
S-5	11.5	31	—	5.7	6
CA-1	9.1	8.6	25a	3.6	6.1
CA-2	12.6	8.4	28a	—	—
CA-3	13.1	7.3	28.7a	—	—
CA-4	14.5	7.9	29a	3.5a	4a
5	15.6	—	—	—	—
6	16.1	—	—	—	—
7	17.9	—	—	—	—
8	20.3	—	—	—	—
9	21.6	—	—	—	—
10	24.2	—	—	—	—
11	24.6	—	—	—	—
12	26	—	—	—	—
13	27a	—	—	—	—
14	27	—	—	—	—
15	26a	—	—	—	—
16	24a	—	—	—	—
17	23	—	—	—	—
18	22	—	—	—	—
19	?	—	—	—	—
20	?	—	—	—	—
21	?	—	—	—	—
22	?	—	—	—	—

*All measurements are in mm.
†a denotes approximation.

4) were loosely articulated, after which the fifth cervical to the sixth dorsal vertebrae were disarticulated. Behind the position of the sternal plates, the vertebral column is continuous almost to the end of the tail.

The cervical vertebrae strongly resemble those described for *Deinonychus* (Ostrom 1969) in having large divergent zygapophyses and relatively short centra. The centra are angled differentially along the cervical column to form an S-shaped neck. Each pneumatic centrum has lateral pleurocoels, and pneumatopores penetrate the neural arch below the diapophysis. The relatively large neural canal increases in diameter posteriorly until the cervicodorsal transition. At this point, the canal is larger in diameter than it is in any of the dorsal vertebrae, indicating a prominent brachial plexus and extensive innervation of the forelimbs (Giffin 1995).

The disarticulated atlas neural arch, centrum, intercentrum, and odontoid were found closely associated in the anterior cervical region. The axis was recovered from the matrix near the back of the skull behind the atlas elements. It closely resembles that of *Deinonychus*. However, in ventral view the centrum tapers posteriorly and does not have a distinct keel. The axis includes parapophyses and diapophyses for cervical ribs, even though no ribs were recovered.

The complete cervical series is presumed to number ten, although this cannot be ascertained because some vertebrae were badly damaged when collected. Cervical 7 is the worst-preserved vertebra in the specimen and cannot be reconstructed. Cervical centra are wider than tall and have somewhat heterocoelous articular surfaces. The laterally positioned pleurocoels increase in size posteriorly, and are subdivided by struts in the ninth and tenth cervical centra. Fragmentary anterior cervical ribs were found crushed onto the lateral surfaces of the centra.

Some anterior dorsals are crushed, and are missing spines and transverse processes. Neural arches on some dorsals have distinct, interdigitating sutures with the centra, although these elements are not completely fused. The anterior dorsals have large neural canals and prominent ventral keels. Pleurocoels are present on all dorsal centra, with the most anterior ones having multiple openings on each side. The parapophyses are cupped, circular facets anteroventral to the transverse processes. They diminish in size posteriorly and move up in position along the neural arch. The tall neural spines of the posterior dorsals are rectangular in shape and are about twice the height of the centra. The centra of the posterior dorsals are rounded in cross section and are best described as amphiplatyan, although they are slightly platycoelus as in *Deinonychus* (Ostrom 1969). These vertebrae have centra that are constricted at mid-length, and transverse processes that arch posterodorsally.

The sacrum, composed of five vertebrae, was crushed between the ilia. This resulted in the loss of most of the sacral ribs and posterior neural spines, even though the centra are well preserved. The sacrum described for *Velociraptor* (Norell and Makovicky 1997) is similar in morphology to that of *Bambiraptor*. The third, fourth, and last sacrals have fused ribs. There are foramina on the lateral surfaces of the centra just under the transverse sacral ribs as reported for *Saurornitholestes*

TABLE 3.3
Pectoral girdle measurements

Element	Maximum Length (in mm)	Maximum Width (in mm)
Right sternal plate	63	29
Left sternal plate	67	29
Right scapula	85	—
Left scapula	83*	—
Right coracoid	19*	28*
Left coracoid	22	33

*Approximation.

(Norell and Makovicky 1997) but in contrast with *Velociraptor*, which lacks pleurocoels in the fifth sacral. The ventral surface of the third sacral has a deep sulcus.

As with most non-avian coelurosaurs, the caudals increase in length until the middle of the series. The first few caudal centra have small foramina on their lateral surfaces, but there are no pneumatopores. The articular surfaces of the pre- and postzygapophyses are offset 45° to the neural spine. Transverse processes protrude from low on the arch and are directed lateroventrally. The elongate extensions of the prezygapophyses of more distal vertebrae extend anteriorly onto the third and fourth caudals.

Elongate anterior zygapophyses and chevrons stiffened the distal part of the tail as described in detail for *Deinonychus* (Ostrom 1969). However, there is a degree of flexibility of this system that was not apparent initially. The stiffening rods in the holotype of *Bambiraptor* and several *Saurornitholestes* specimens from Alberta (Currie pers. comm. 2000) bent enough to allow the distal part of the tail to curve gently dorsally. In short, the stiffening rods restricted the mobility of the distal part of the tail, but did not stop it entirely from bending.

A single proximal hemal arch was discovered in the matrix between the articulated ischia. The chevron is tall and thin in comparison with the anteroposteriorly elongate ones associated with more posterior caudals.

Appendicular Skeleton

Proximally, the scapula (fig. 3.14) is robust and almost triangular in cross section (table 3.3). There was no fusion with the coracoid, and the sutural contact is smooth. Most of the glenoid articular surface is smoothly concave, but the convex edges form a lip or buttress anterodorsally. There is a prominent anteromedially directed acromion process with a roughened surface, presumably for the attachment of the furcula (Norell and Makovicky 1999). Distally, the elongate, strap-like scapular blade becomes mediolaterally thinner. It curves gently dorsoposteriorly to conform to the rib cage. Compared with *Deinonychus* (Ostrom 1969) and recently described dromaeosaurids (Norell and Makovicky 1999), the scapular blade of *Bambiraptor* is more gracile and tapers distally. The acromion in *Bambiraptor* is relatively longer

Figure 3.14. *(right) Right scapula of* Bambiraptor feinbergi *in lateral (a), dorsal (b), and anterior (c) views. Scale bar in mm.*

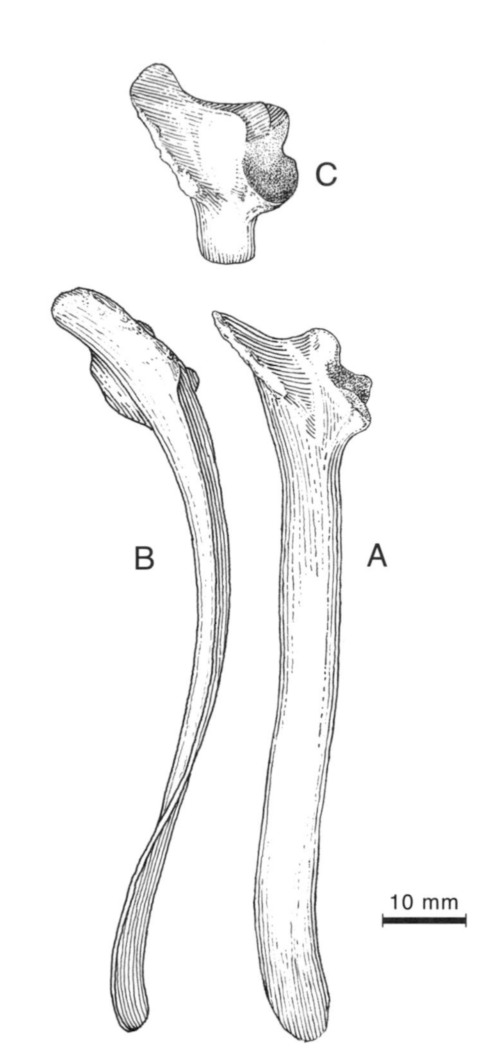

Figure 3.15. *(below) Left coracoid of* Bambiraptor feinbergi *in anterior (a), posterior (b), dorsal (c), and ventral (d) views. Scale bar = 5 mm.*

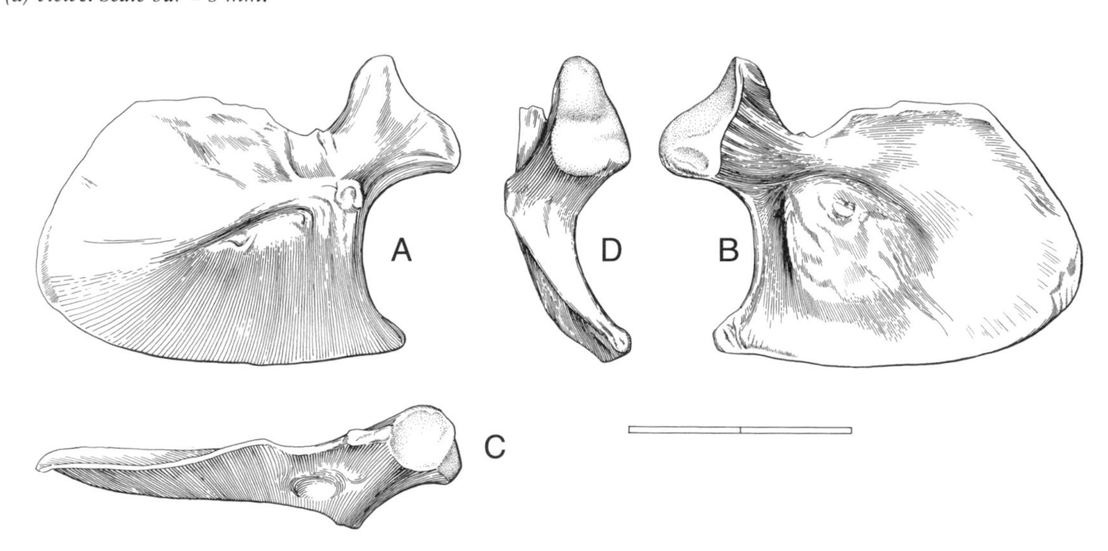

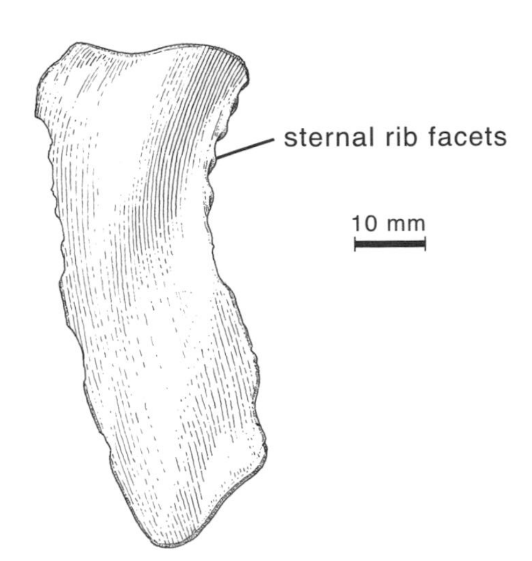

Figure 3.16. Left sternal of Bambiraptor feinbergi *in ventral view. Scale bar = 1 cm.*

and wraps around the front of the chest (fig. 3.14c). In *Sinornithosaurus* (Xu et al. 1999) the elongate scapula is also strap-like and has a large, forwardly directed acromion.

The left coracoid is complete (fig. 3.15), but the right one was crushed, and lacks the sutural contact for the scapula. Overall, the coracoid is similar in shape to the larger, quadrangular ones described for other dromaeosaurids by Norell and Makovicky (1999). However, it is longer anteroposteriorly than tall dorsoventrally, the glenoid seems to be supported on a prominent strut-like neck, and there is no coracoid foramen. The absence of the latter can be attributed to the presence of a deep notch in the bone, and may represent an immature state. The recovery of a more mature *Bambiraptor* coracoid will be necessary to determine whether this is an autapomorphy of the genus. The biceps or coracoid tubercle is positioned anteroventral to the glenoid.

Both sternal plates are well preserved and three-dimensional, although the right one is crushed at the coracoid-sternal articulation. When found, the two plates were touching on the midline posteriorly, but were slightly displaced anteriorly. Each elongate sternal is thin and sub-rectangular (fig. 3.16), and is similar in shape to the dromaeosaurid sternals described by Norell and Makovicky (1997, 1999). There is a transverse groove for the coracoid along the thick anterior margin. The lateral margins are scalloped with facets for the attachment of four, possibly five, sternal ribs. The anteroventral surface of each sternal plate is shallowly concave, probably for muscle attachment.

The furcula (fig. 3.17) is a well-preserved bone shaped like a flared "U" or boomerang. It has a flattened cross section with a grooved dorsal surface. Within this channel, a nutrient foramen can be found on each ramus of the radiale (fig. 3.17b), about 10 mm from the apex. The angle between the rami is approximately 80°. The distal ends of the rami taper and have striated attachment surfaces, which are especially prominent on the ventral side. Near the midline of the furcula, the bone thickens on the dorsal surface (fig. 3.17a), although there is no hint of

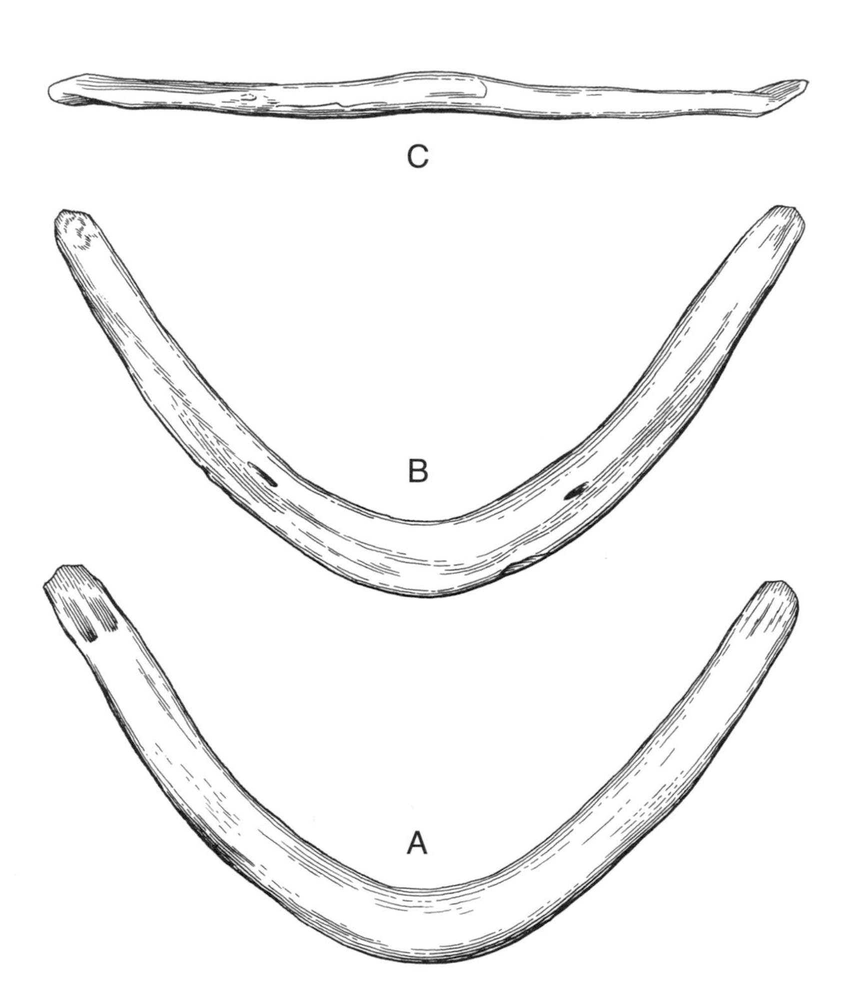

Figure 3.17. Furcula of Bambiraptor feinbergi *in anterior (a), posterior (c), and ventral (c) views. Scale bar = 5 cm.*

a hypocleidium. This bone is in sharp contrast to the robust, V-shaped furcula of *Velociraptor* (Norell et al. 1997), and is closer in appearance to that of *Archaeopteryx,* which is also U-shaped and flattened, has a low clavicular angle, and lacks a hypocleidium.

The head of the humerus (fig. 3.18) is strongly convex, has a smooth articular surface, and is larger than that of *Deinonychus* (Ostrom 1969). The shaft is relatively long and slender, and supports a well-developed pectoral crest. Distally, the humerus expands into radial (larger) and ulnar condyles separated by a shallow groove. The distal condyles are separated by a depression in the holotype. The slightly concave proximal end of the ulna is triangular in section (fig. 3.19e). The olecranon forms a strong ulnar ridge on the exterior surface similar to *Velociraptor* (Norell and Makovicky 1999). The shaft is bowed and flares into a thin, wide distal articular surface (fig. 3.19f). This condyle turns slightly medially and forms a flange that seems to be absent in *Deinonychus* (Ostrom 1969). The radius (fig. 3.20) is a

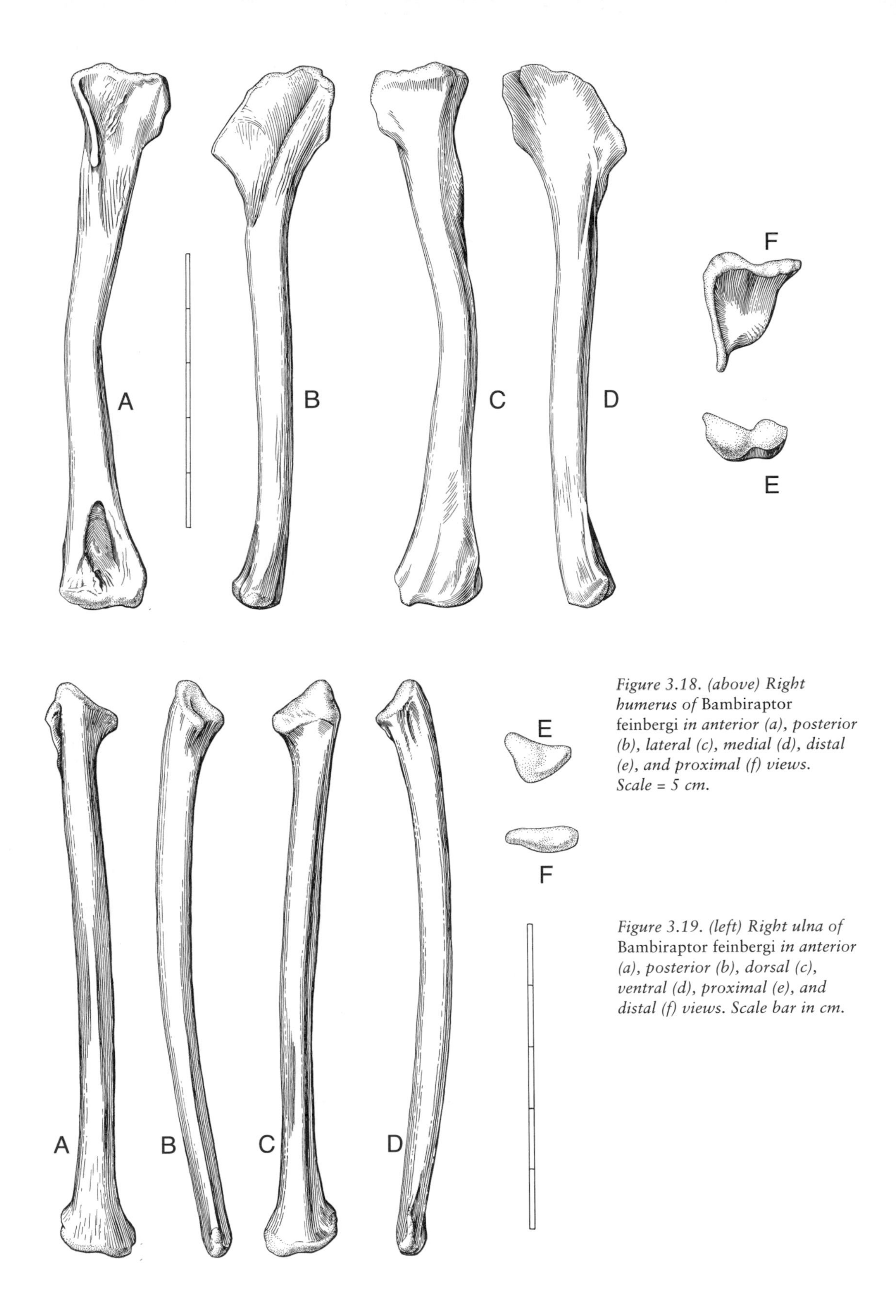

Figure 3.18. (above) Right humerus of Bambiraptor feinbergi *in anterior (a), posterior (b), lateral (c), medial (d), distal (e), and proximal (f) views. Scale = 5 cm.*

Figure 3.19. (left) Right ulna of Bambiraptor feinbergi *in anterior (a), posterior (b), dorsal (c), ventral (d), proximal (e), and distal (f) views. Scale bar in cm.*

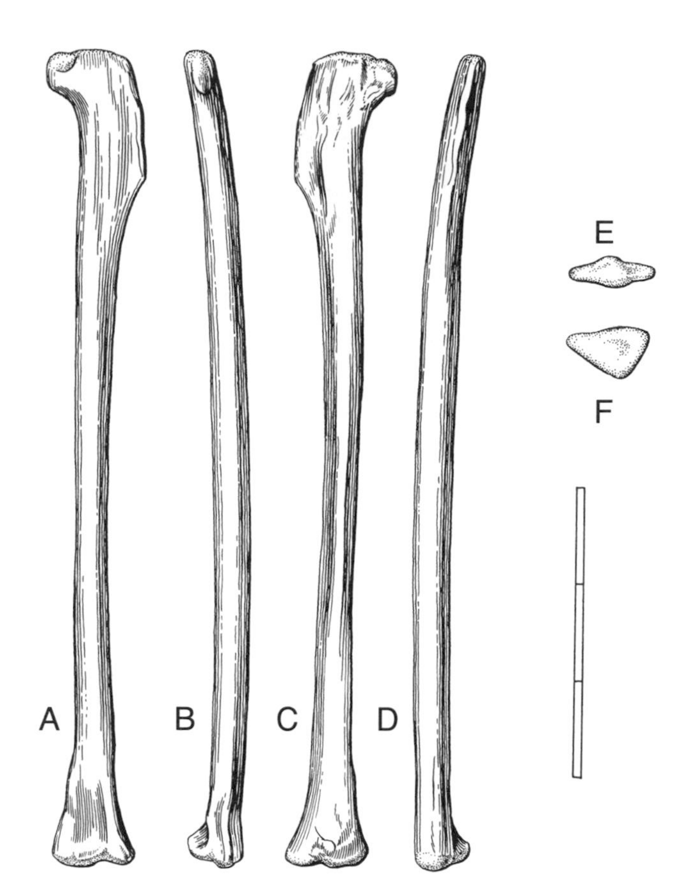

Figure 3.20. Right radius of Bambiraptor feinbergi *in anterior (a), posterior (b), lateral (c), medial (d), distal (e), and proximal (f) views. Scale bar in cm.*

slender, thin-shafted bone that is circular in cross section. A flattened, striated area on the medial surface of the proximal end fits between the proximal tubercles of the ulna.

The hands bear long, curved claws although the manual digits, except for the first digit, are relatively inflexible compared with the pedal digits. The wrist contains two carpal bones (a semilunate bone and radiale) that allow for a folding back of the hand as well as a slight degree of lateral flexure.

Both semilunate bones (radiale of Ostrom 1969; fused distal carpals 1 and 2 of Chure 2001 and others) were recovered with the holotype, and another was found with the adult *Bambiraptor* (AMNH 002). The semilunate has a saddle-shaped proximal articular surface (carpal trochlea) that is notched (fig. 3.21) and articulates mostly with the ulna. Distally, the semilunate caps a portion of metacarpal I and the entire proximal surface of metacarpal II, as in most maniraptoriforms. The second, smaller carpal bone found in the holotype is the radiale (fig. 3.22). Both the right and left radiale were found near the semilunate carpals. This small, ovoid bone slid along the carpal trochlea of the semilunate.

Metacarpal I is short and robust, and has a ginglymoid distal articulation as in *Deinonychus* and *Velociraptor* (fig. 3.23). This bone contacts Metacarpal II for most of its length through a relatively flat

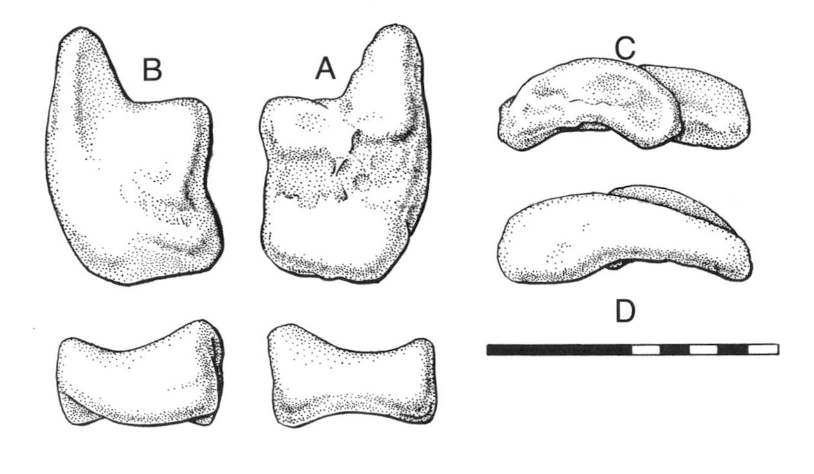

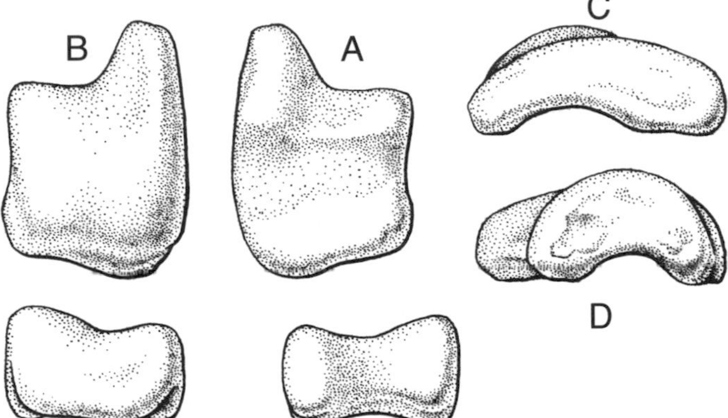

Figure 3.21. Right and left semilunate bones of Bambiraptor feinbergi *in proximal (a), distal (b), dorsal (c), and ventral (d) views. Scale bar in mm.*

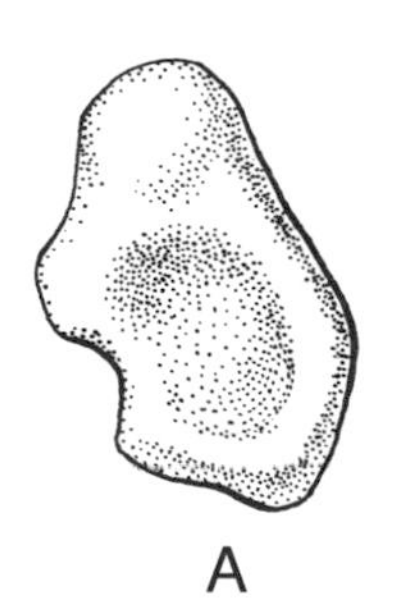
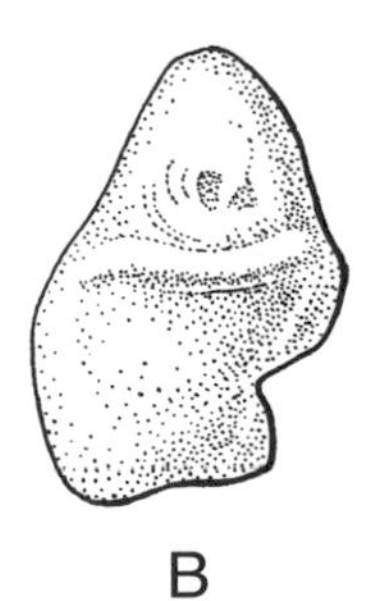
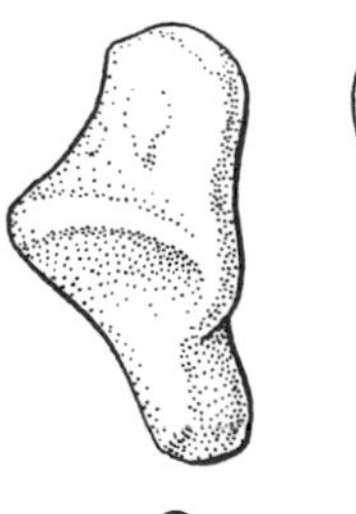
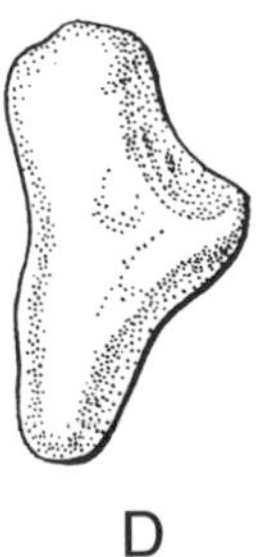

Figure 3.22. Right radiale of Bambiraptor feinbergi *in proximal (a), distal (b), dorsal (c), and ventral (d) views. Scale bar in mm.*

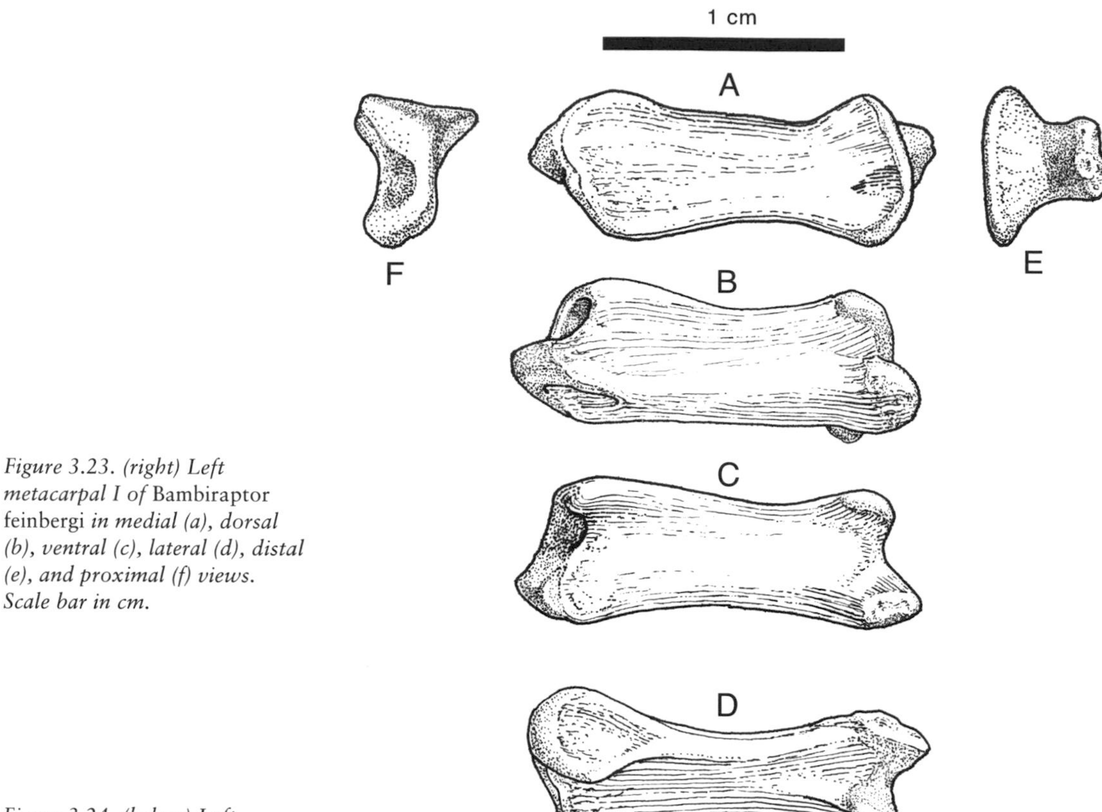

Figure 3.23. (right) Left metacarpal I of Bambiraptor feinbergi *in medial (a), dorsal (b), ventral (c), lateral (d), distal (e), and proximal (f) views. Scale bar in cm.*

Figure 3.24. (below) Left metacarpal II of Bambiraptor feinbergi *in medial (a), lateral (b), ventral (c), dorsal (d), distal (e), and proximal (f) views. Scale bar in cm.*

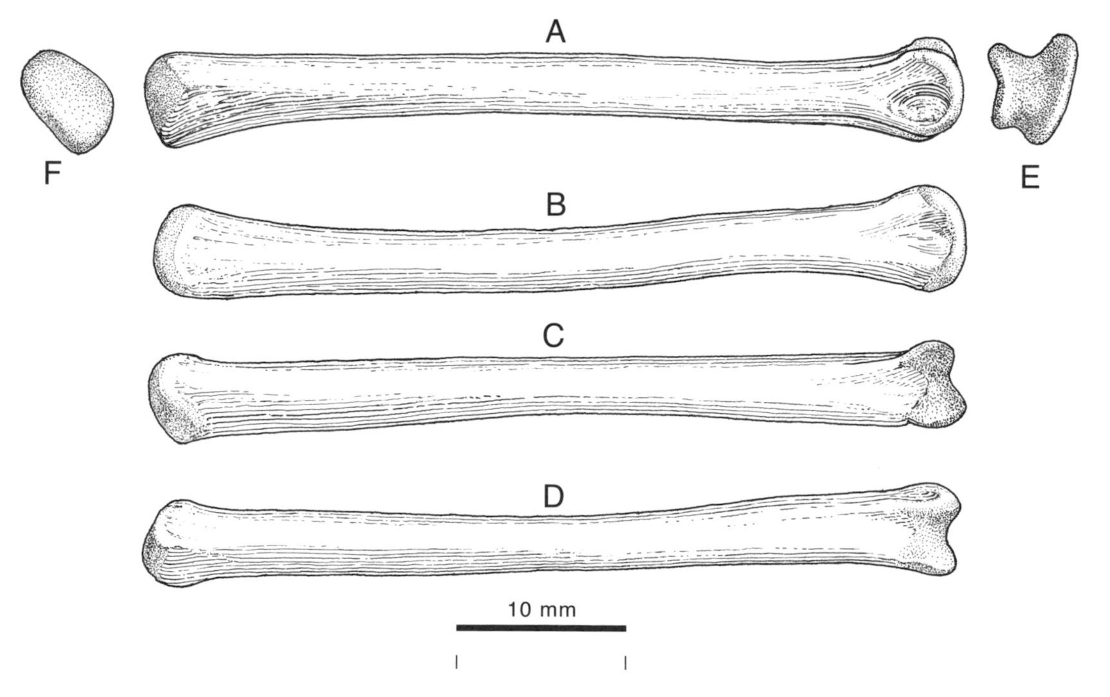

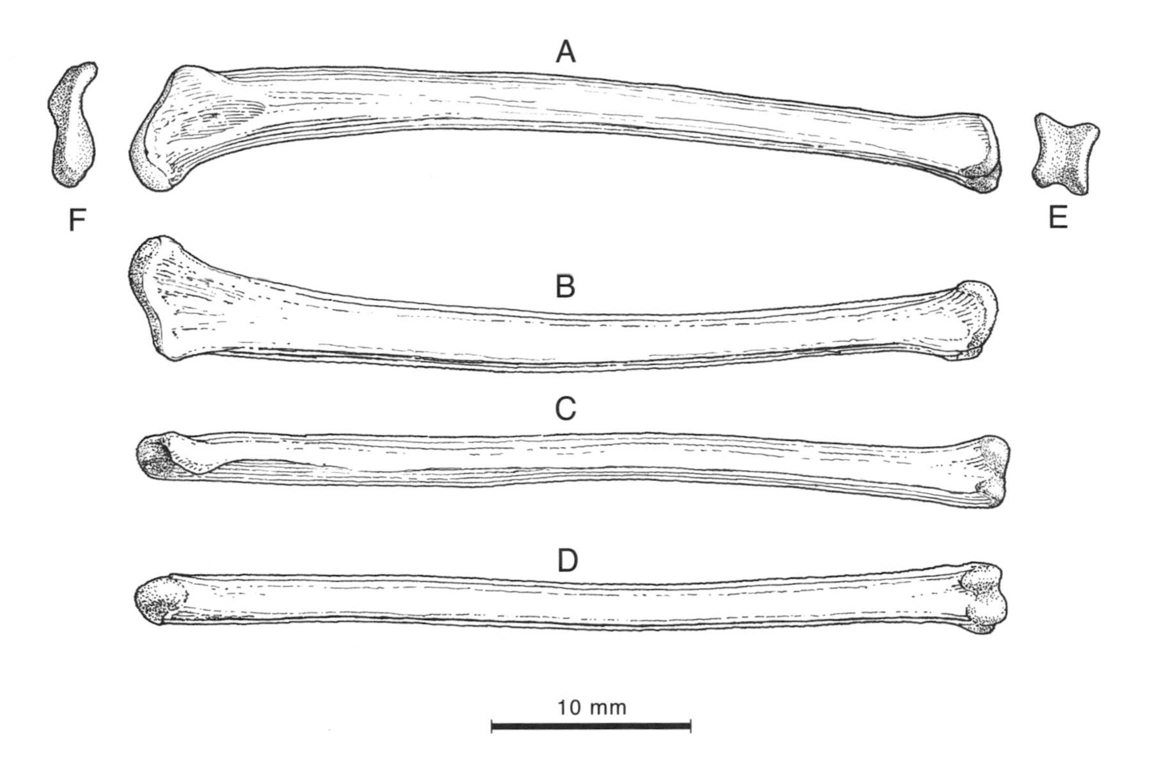

Figure 3.25. Left metacarpal III of Bambiraptor feinbergi *in medial (a), lateral (b), ventral (c), dorsal (d), distal (e), and proximal (f) views. Scale bar in cm.*

proximolateral surface. Metacarpal II (fig. 3.24), the longest of the three, is circular in cross section. Metacarpal III is the thinnest and is slightly bowed (fig. 3.25).

The three digits are long and gracile. The second is the longest, the third is nearly as long, and the first is the shortest. The phalangeal formula is 2-3-4. Phalanx I-1 is the longest and most robust phalanx of the hand, followed by II-1. Digit I has the largest claw, which has a greater range of flexion/extension than the other two fingers. When the hand closed, the phalanx and claw rotated toward the central axis of the hand. However, it crossed palmar to the other digits, and was not capable of grasping. The articular surfaces of the phalanges of the second digit suggest that this finger was relatively stiff and inflexible. Manual phalanx III-1 is thin and relatively short, but is almost four times the length of III-2. The longest phalanx of the third finger is III-3. The interphalangeal articulations cause this digit to move toward digit II during flexion. The manual unguals are laterally compressed, have prominent flexor tubercles, and are more strongly curved than the unguals of the pes. Digit III is the most gracile finger, and is directed inward when flexed. Other than being from a smaller animal, the manual bones of the *Bambiraptor* holotype are essentially the same as those of *Velociraptor* (Norell and Makovicky 1999).

Pelvic Girdle and Hindlimb

The pelvis of the holotype of *Bambiraptor* is more or less intact. Dromaeosaurid pelves are known for their high degree of pubic retro-

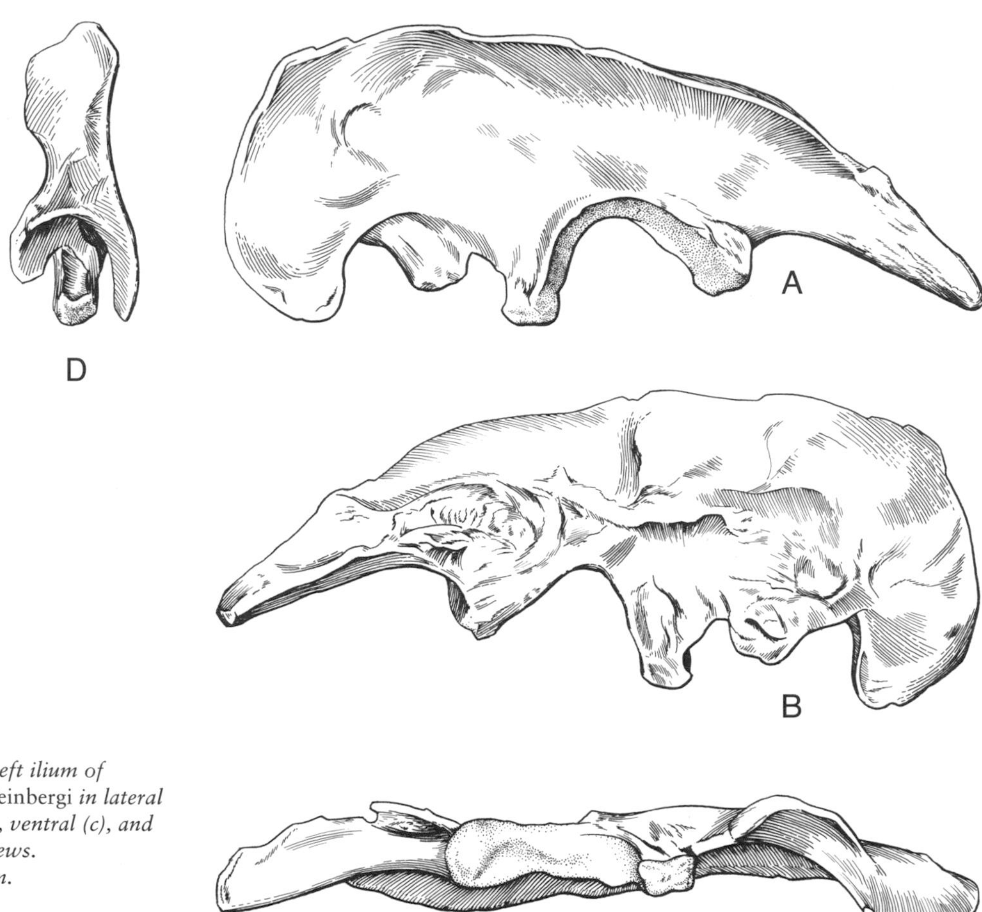

Figure 3.26. Left ilium of Bambiraptor feinbergi *in lateral (a), medial (b), ventral (c), and anterior (d) views. Scale bar in cm.*

version (Norell and Makovicky 1997), and *Bambiraptor* as a dromaeosaurid shares this feature. Additionally, the ischia are shorter than the pubis as in most maniraptoran coelurosaurs (Rasskin-Gutman 1997).

The postacetabular region of the ilium (fig. 3.26) is shorter than the preacetabular blade, a condition opposite that of the dromaeosaurid described by Norell and Makovicky (1997). The dorsal margin of the ilium is gently curved, whereas that of the *Velociraptor* is relatively straight. As a consequence, the distal end of the ilium of *Bambiraptor* is more strongly tapered. The pubic peduncle is tall, as is typical of *Deinonychus* and *Velociraptor* (Norell and Makovicky 1997). The ischial peduncle is sub-triangular and more pronounced than in other dromaeosaurids.

The pubis (fig. 3.27) was oriented posteroventrally as in other dromaeosaurids. It has a sub-triangular ischial peduncle. Proximally, the shaft is oval in section, but in the distal half of the bone extends medially into a pubic apron. There is a large pubic boot, which has no anterior component and tapers posterodorsally to end in a blunt tip.

The ischium (fig. 3.28) is similar to those described for *Deino-*

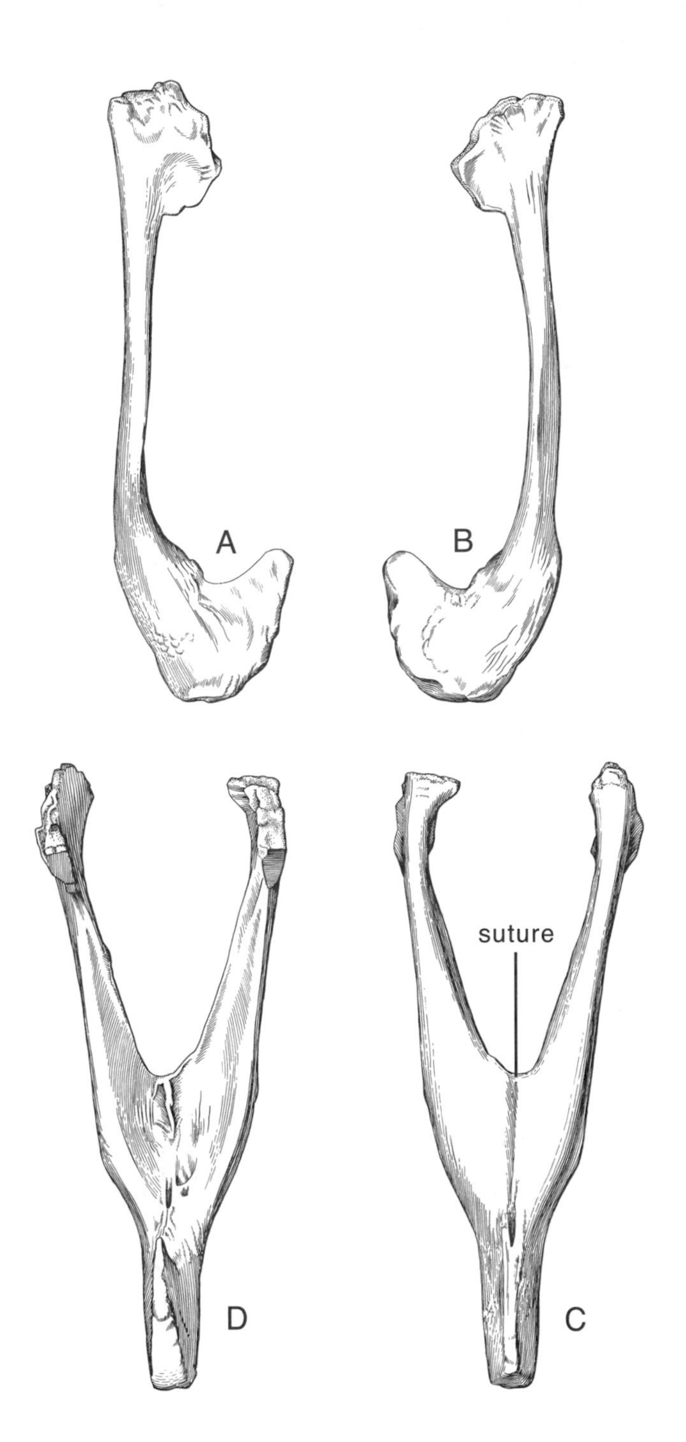

Figure 3.27. Right and left pubis of Bambiraptor feinbergi *in left lateral (a), right lateral (b), anterior (c), and posterior (d) views.*

nychus and other dromaeosaurids (Barsbold 1983; Norell and Makovicky 1997). This ischium is only half the length of the pubis as in most maniraptorans. Proximally, the pubic process is longer and narrower than the relatively short, stout iliac process. There is a low posterodorsal process near the proximal end of the flattened shaft of the ischium. This is similar in position to the more pronounced postero-

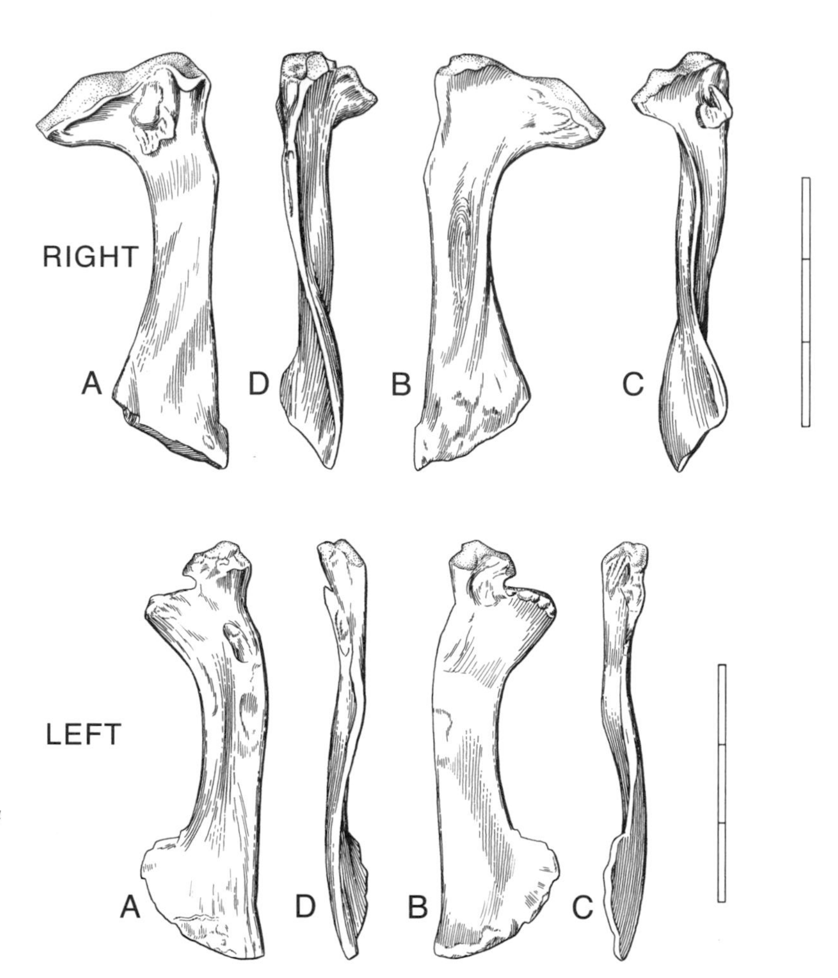

Figure 3.28. Right and left ischia of Bambiraptor feinbergi *in lateral (a), medial (b), anterior (c), and posterior (d) views. Scale bar in cm.*

dorsal processes of *Rahonavis* (Forster et al. 1998), *Microraptor, Sinornithosaurus, Unenlagia,* and *Archaeopteryx* (Xu et al. 2000). The distal ends of the ischia contact each other but are not fused. The relatively small obturator process is positioned at the end of the bone, which is presumably a juvenile trait.

There is a distinctive twist to the femoral head, which is also found in some other non-avian theropods (*Troodon*) and birds (*Archaeopteryx, Enantiornis*). The shaft of the femur (fig. 3.29) is strongly bowed as in many small coelurosaurs and enantiornithine birds (L. D. Martin pers. comm.). There is a prominent posterior trochanter (Ostrom 1969) near the proximal end of the shaft. A slight rugosity on the shaft may represent the insertion of the M. caudifemoralis brevis. Posterodistally, the popliteal fossa separates the medial and lateral condyles.

The tibia (fig. 3.30) is longer than the femur. At the level of the fibular crest, the shaft of the tibia is sub-triangular in cross section, and is penetrated by a nutrient foramen as in other theropods. The rod-like shaft of the fibula has a diameter of less than 1 mm. The distal end of the fibula overlaps the anterolateral margin of the astragalus where it

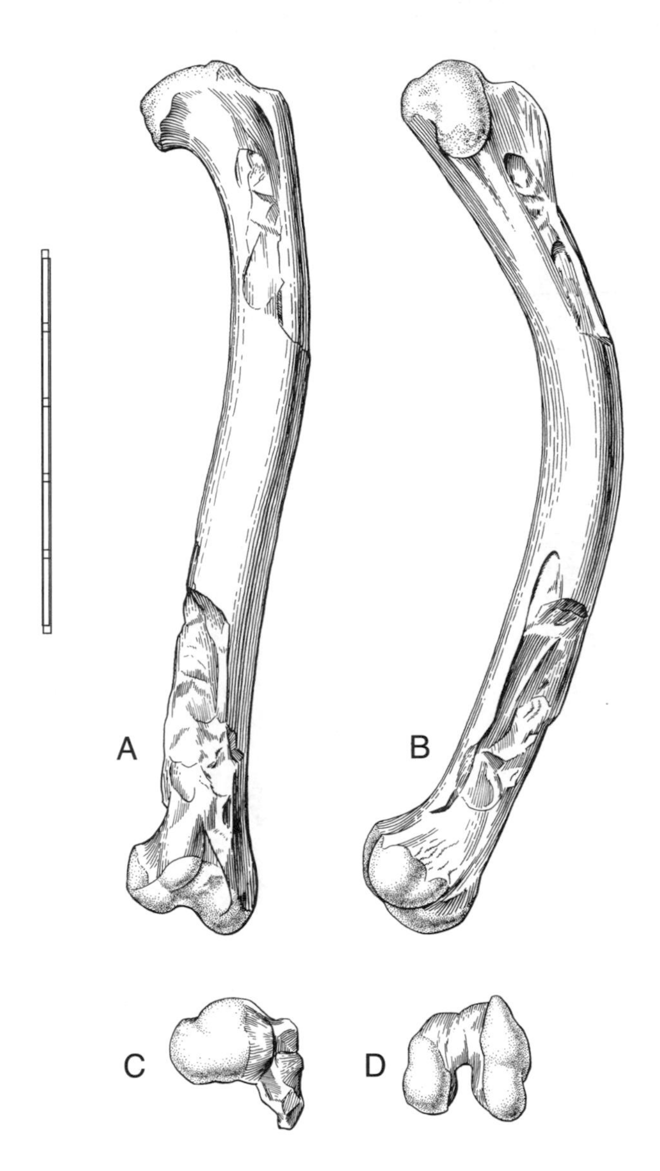

Figure 3.29. Left femur of Bambiraptor feinbergi *in lateral (a), medial (b), dorsal (c), and ventral (d) views. Scale bar in cm.*

contacts the proximal end of the calcaneum. The disc-shaped calcaneum is small relative to the astragalus as it is in all other dromaeosaurids. There is no sign of fusion between the two bones. The ascending pro-cess of the astragalus is 30 mm high. It tapers dorsally to a point offset toward the lateral edge of the tibia.

One right (III) and two left (III and IV) distal tarsals were found in association with the metatarsals of the holotype. They are similar to the same elements in *Deinonychus* (Ostrom 1969), *Velociraptor* (Norell and Makovicky 1997), and most other theropods that have these bones preserved.

Metatarsal I is a short, proximally tapering bone. The distal end has a single collateral ligament pit, and ends in a nearly ginglymoid articular surface. Metatarsals II, III, and IV were the weight-bearing

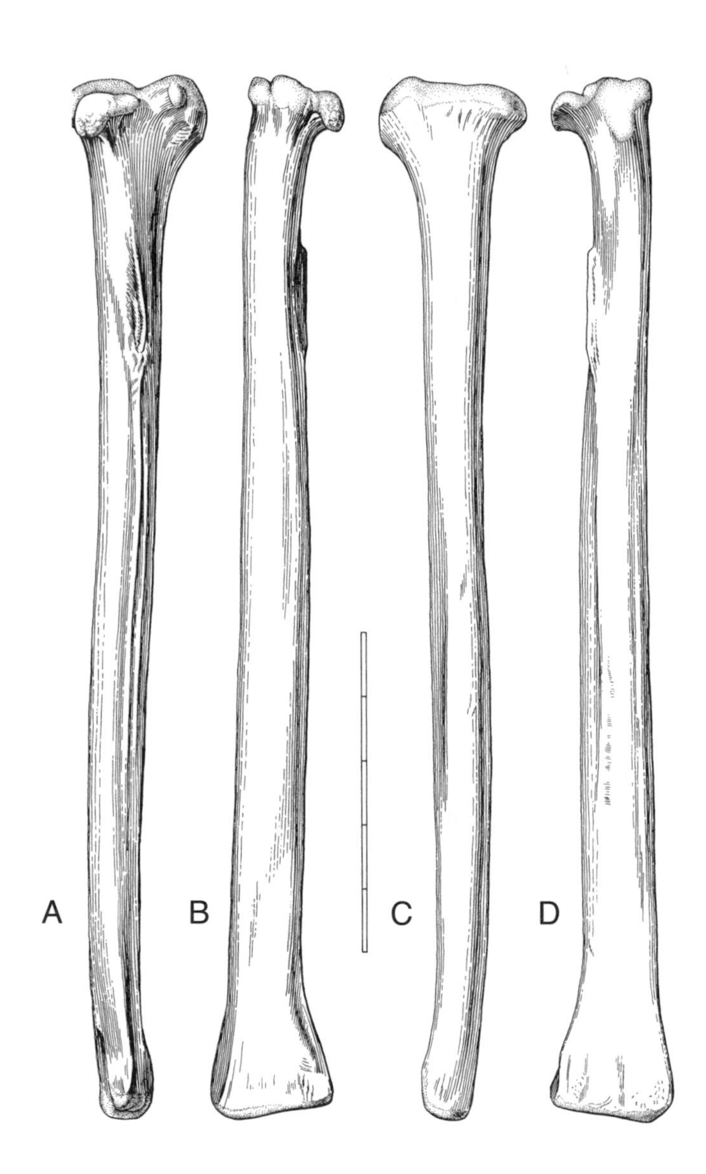

Figure 3.30. Right tibia of Bambiraptor feinbergi *in lateral (a), medial (b), anterior (c), and posterior (d) views. Scale bar in cm.*

portion of the foot. Metatarsal II is shorter than either the third or fourth metatarsals, and is less robust. Metatarsal III is mediolaterally flattened in cross section, and the proximal end is squeezed between its neighbors. The distal end has a relatively large semicircular articular surface, and characteristic of dromaeosaurids it would have permitted a wider range of parasagittal motion in the third digit than was possible in most other theropods. The rounded, distal articular surface of Metatarsal IV also would have permitted wide excursion of the associated toe. It is not as prominent as the distal end of the third metatarsal.

None of the phalanges were articulated, but were closely associated. Like other dromaeosaurids, the foot of *Bambiraptor* has characteristic phalanges (fig. 3.31) and the retractable, raptorial claw on the second digit. The other digits have significantly smaller and less strongly curved unguals (figs. 3.32, 3.33). The raptorial claw is supported by

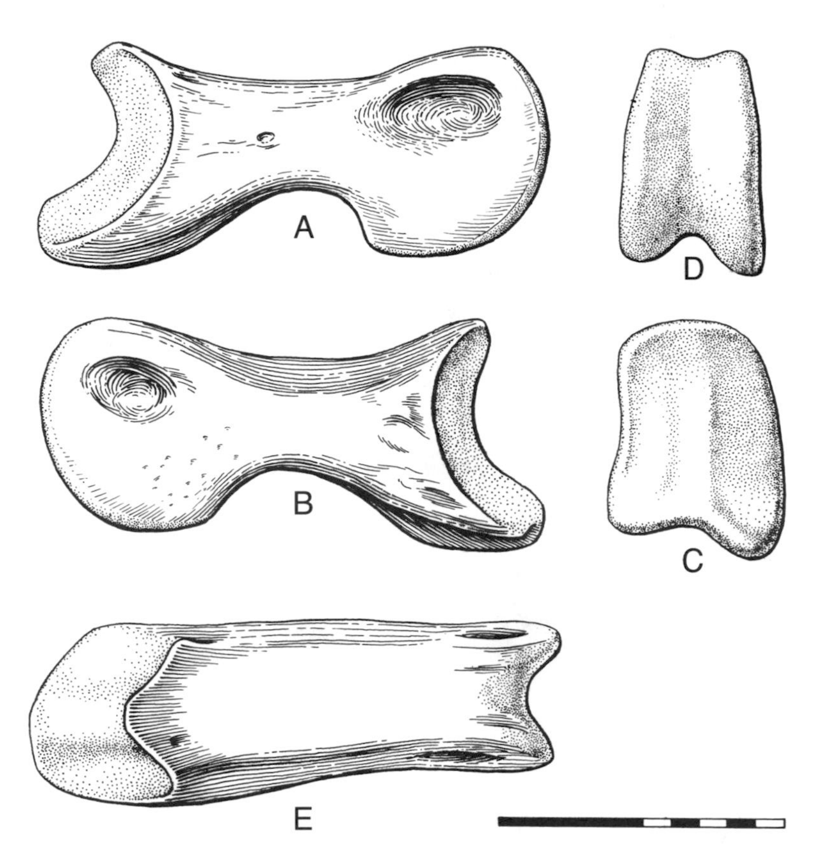

Figure 3.31. (left) Right pedal phalanx II-2 of Bambiraptor feinbergi *in lateral (a), medial (b), distal (c), proximal (d), and ventral views (e). Scale bar in mm.*

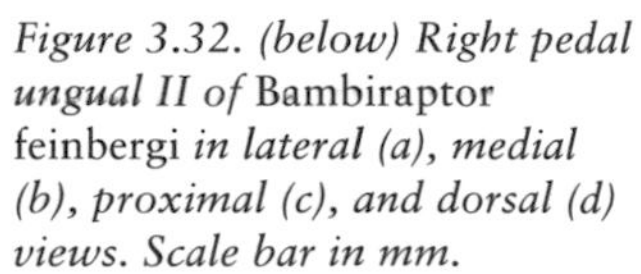

Figure 3.32. (below) Right pedal ungual II of Bambiraptor feinbergi *in lateral (a), medial (b), proximal (c), and dorsal (d) views. Scale bar in mm.*

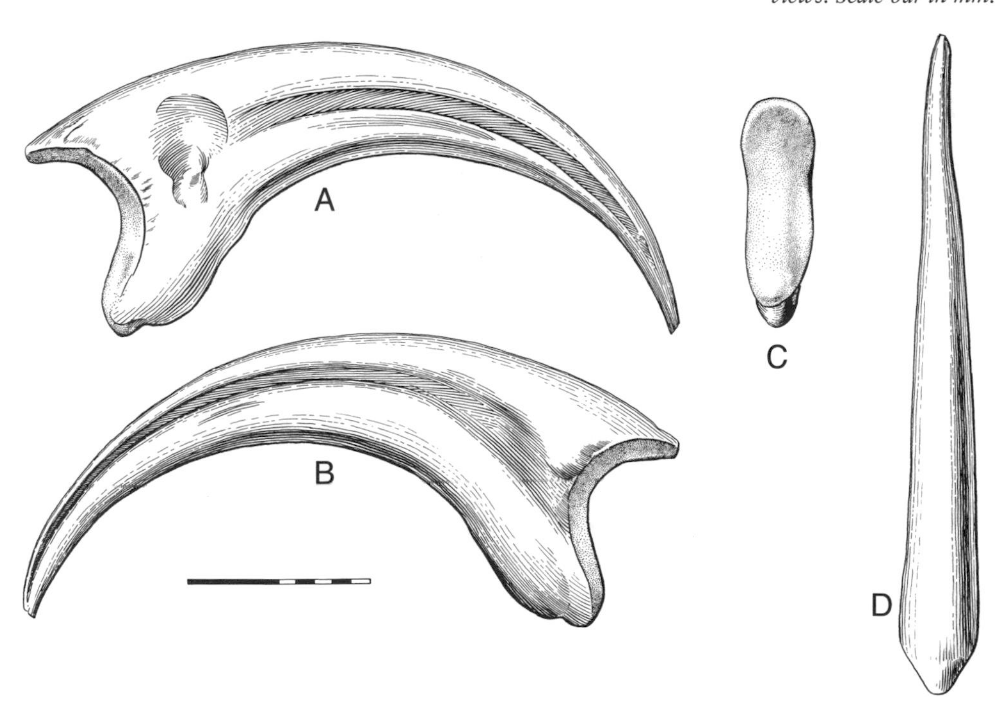

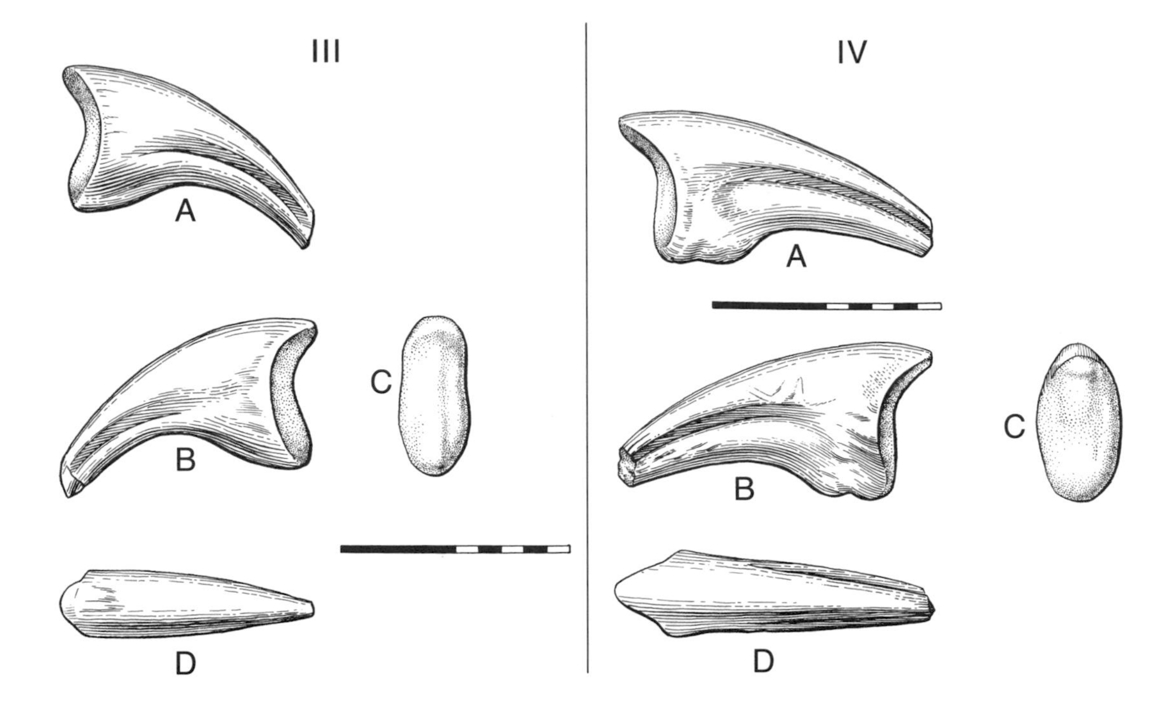

Figure 3.33. Right pedal unguals III and IV of Bambiraptor feinbergi *in lateral (a), medial (b), proximal (c), and dorsal (d) views. Scale bar in mm.*

two robust, specialized phalanges. In spite of the strength and size of the second pedal digit, it is shorter than the third and fourth toes, but has more contact with the metatarsal. Consequently, this toe probably rarely touched the ground. It is quite possible the foot acted as a didactyl unit as described by Ostrom (1969), although no convincing trackway evidence of "two-toed" dinosaurs has been published to date (J. O. Farlow pers. comm. 1999). The proximal and distal articulations of the second phalanx of the second digit (fig. 3.31) gave the raptorial claw a very specific range of motion. The sickle-shaped claw is laterally compressed, highly recurved, and ends in a sharp tip (fig. 3.32). Pedal III-1 is the longest phalanx of the foot. Similar to *Rahonavis,* the penultimate phalanges are longer than the antepenultimate ones (table 3.4). The unguals of the third and fourth digits are shorter and less strongly curved than that of the second digit (fig. 3.33).

Discussion

In *Bambiraptor,* the enlarged cerebellum suggests agility and higher intelligence than its contemporaries had (Jerison 1973; Bock 1985). Large optic lobes, combined with possibly overlapping fields of vision, probably indicate good vision (Allman 1999), although the small olfactory bulbs suggest its sense of smell was less acute than in tyrannosaurids, *Troodon,* and other theropods. The relatively large brain, overlapping fields of vision, small size, and elongate front limbs might indicate that *Bambiraptor* was arboreal. The complex environment encountered by a tree-dwelling animal may account for the evolution of

TABLE 3.4
Measurements of appendicular skeletal elements of *Bambiraptor*

Forelimb	**Left**	**Right**
Humerus	105*	100
Ulna	95	93
Radius	8585	
Manus	**Left**	**Right**
Metacarpal I	16.8	16.8
M-I-1	32.5	32.3
M-I-2 ungual	31+	20+
Metacarpal II	47.8	46.5
M-II-1	21.1	21.4
M-II-2	35	35
M-II-3 ungual	—	43
Metacarpal III	44.9	43.8
M-III-1	15.5	16.5
M-III-2	6.3	5.5
M-III-3	23.5	18+
M-III-4 ungual	—	—
Pelvis	**Left**	**Right**
Ilium	86	78†
Ischium	53	50
Pubis	103	103
Hindlimb	**Left**	**Right**
Femur	118	118
Tibia-Astragalus-Calcaneum	167	170
Fibula	170†	—
Pes	**Left**	**Right**
Metatarsal I	18.3	11+
P-I-1	—	12.1
Claw	10.6†	14.0
Metatarsal II	70	67.5
P- II-1	14.3	14.0
P- II-2	14.6	14.2
Ungual	46.0	tip only
Metatarsal III	77	81
P-III-1	29.6	27.8
P-III-2	17.6	15.4
P-III-3	—	16.9
Claw	—	24
Metatarsal IV	70	74
P-IV-1	23.6	23.4
P-IV-2	16.6	17.8
P-IV-3	11.6	12.3
P-IV-4	12.3	12.9
Claw	—	22
Metatarsal V	43+	33.2

*Maximum lengths are in millimeters unless otherwise noted.
All measurements of phalanges are "inter-condyle" lengths.
†Approximate measurement; e.g., distal tip of P-I ungual;
distal end of mt V. The right manus is missing M-II claw and M-III claw; the left manus is missing M-III claw. The left pes is missing the following bones: P-I-1, P-III-3, P-III ungual, P-IV ungual; the right pes is missing the proximal portion of the sickle claw P-II-3.

Figure 3.34. Life restoration of Bambiraptor feinbergi *in Two Medicine Formation time scenario.*

a large brain (Bock 1985). An alternative hypothesis is that brain size increased because it was hunting complex prey items (Radinsky 1974). This may have included lizards and mammals, which have been found in the gut region of *Sinosauropteryx* (Chen et al. 1998). Giffin (1990) also shows that *Coelurus* (3.04), *Deinonychus* (2.63), and *Allosaurus* (2.66) all had large SQs (high neural supply), implying manipulative ability. These values are higher than those of even modern birds that are active fliers (for example the SQ of mallard ducks is 2.33).

In the shoulder girdle, the glenoid is oriented laterally, which is similar to the conditions in some other non-avian maniraptorans (including *Unenlagia* and *Deinonychus*) and *Archaeopteryx*. It allowed the long arms a range of motion that was only restricted anteriorly. The scapula has an acromion that projects forward and medially, serving as the primary platform for the attachment of the furcula. The sternal plates are much longer than the coracoids, and form a large, flat ventral surface for the attachment of the pectoral musculature. Each has facets for five pairs of sternal ribs.

The humerus moves through a limited range of anterior motion and a wider range of dorsal-ventral movement. It can be folded back against the body, but cannot be brought forward much beyond a vertical plane passing through the glenoid. Furthermore, throughout its range of motion, the deltopectoral crest of the humerus is positioned anterodorsal to the glenoid. The pectoral girdle of *Bambiraptor* does have a sizable origin for the M. supracoracoideus. The humerus has the appropriate posture to allow the range of motion (flexure and longitudinal rotation) found in the flight stroke of modern birds and a potential insertion for the M. supracoracoideus.

In the manus, it is clear that the second and third digits worked in concert most of the time as a functional unit. If they had been separate, the weak construction of digit III would have made it vulnerable to breakage. Not only is the third digit constructed of slender phalanges, but the articular surfaces also forced the digit to fold against the middle digit. When the manus is considered as an operating unit, the first digit is the most robust and has the largest claw, but it was probably not capable of opposing the other fingers.

Bambiraptor is well adapted to a cursorial existence. This is indicated by the similarity in hindlimb proportions between *Bambiraptor* and modern running birds (Coombs 1978).

Acknowledgments. My sincere thanks to Michael and Ann Feinberg, who helped save *Bambiraptor* for science, and to Dr. Martin Shugar, for his enthusiastic support of this project. I am also grateful for the inspiration provided by discussions with my colleagues Larry D. Martin on functional morphology and Kraig L. Derstler for reconstruction of the animal, especially the endocast. I thank John Chorn for many helpful comments while editing this work, and for providing x-rays and stereophotographs used in the study of *Bambiraptor.* Additional sources of support and information include Robert T. Bakker, Philip J. Currie, John H. Ostrom, Deborah Wharton (EQ calculations), and Desui Maio. Chris Collins, Jack Horner, Mark Norell, David Varricchio, and Charles Zidar graciously provided access to specimens under their care. Much gratitude is owed to Tom Swearingen for his work on the life model of *Bambiraptor.* Additional help in this task was enthusiastically supplied by Peggy Williams, Katrina Gobetz, David Buzzanco, Donald Burnham, and Chip Robelen. Artistic talents devoted to this project include Pat Redman, Mary Tanner, Julie Martinez, Chris Sheil, and Gael Summer. The paper was greatly improved through reviews done by Philip Currie, Peter Makovicky, and Joanna Wright.

References

Allain, R., and P. Taquet. 2000. A new genus of Dromaeosauridae (Dinosauria, Theropoda) from the Upper Cretaceous of France. *Journal of Vertebrate Paleontology* 20: 404–407.

Allman, J. M. 1999. *Evolving Brains.* New York: Scientific American Library.

Anderson, J. F., A. Hall-Martin, and D. A. Russell. 1985. Long-bone circumference and weight in mammals, birds and dinosaurs. *Journal of Zoology, London* (A) 207: 53–61.

Barsbold, R. 1983. Carnivorous dinosaurs from the Cretaceous of Mongolia. *Joint Soviet-Mongolian Paleontological Expedition, Transactions* 19: 5–120.

Barsbold, R., and H. Osmólska. 1999. The skull of *Velociraptor* (Theropoda) from the Late Cretaceous of Mongolia. *Acta Palaeontologica Polonica* 44: 189–219.
Bock, W. J. 1985. The arboreal theory for the origin of birds. In M. Hecht, J. Ostrom, G. Viohl and P. Wellnhofer (eds.), *The beginnings of birds,* pp. 199–208. Eichstätt: Freunde des Jura-Museums.
Brinkman, D., R. Cifelli, and N. Czaplewski. 1998. First occurrence of *Deinonychus antirrhopus* (Dinosauria: Theropoda) from the Antlers Formation (Lower Cretaceous: Aptian-Albian) of Oklahoma. *Oklahoma Geological Survey Bulletin* 146: 1–27.
Burnham, D., K. L. Derstler, and C. Linster. 1997. A new specimen of *Velociraptor* (Dinosauria: Theropoda) from the Two Medicine Formation of Montana. In D. L. Wolberg, E. Stump, and G. D. Rosenberg (eds.), *Dinofest International,* Proceedings, pp. 73–75. Philadelphia: Academy of Natural Sciences.
Burnham, D. A., K. L. Derstler, P. J. Currie, R. T. Bakker, Z. Zhou, and J. H. Ostrom. 2000. Remarkable new birdlike dinosaur (Theropoda: Maniraptora) from the Upper Cretaceous of Montana. *The University of Kansas Paleontological Institute, Paleontological Contributions (n.s.)* 13: 1–14.
Carpenter, K. 1999. *Eggs, nests, and baby dinosaurs.* Bloomington: Indiana University Press.
Carpenter, K., and M. Smith. 2001. Forelimb osteology of and biomechanics of *Tyrannosaurus rex.* In D. H. Tanke, and K. Carpenter (eds.), *Mesozoic vertebrate life: new research inspired by the paleontology of Philip J. Currie,* pp. 90–116. Bloomington: Indiana University Press.
Chen P. J., Z. Dong, and S. N. Zheng. 1998. An exceptionally well-preserved theropod dinosaur from the Yixian Formation of China. *Nature* 391: 147–152.
Chure, D. 2001. The wrist of *Allosaurus.* In J. Gauthier and J. H. Ostrom (eds.), *New perspectives on the origin and evolution of birds:* Proceedings of the International Symposium in honor of John H. Ostrom, February 13–14, 1999, New Haven, Conn., pp. 283–300. New Haven, Conn.: Peabody Museum of Natural History, Yale University.
Cobb, S., and T. Edinger. 1962. The brain of the Emu (*Dromaeus novaehollandiae,* Lath): I. Gross anatomy of the brain and pineal body. *Brevioria* 170: 1–18.
Colbert, E. H., and D. A. Russell. 1969. The small Cretaceous dinosaur *Dromaeosaurus. American Museum Novitates* 2380: 1–49.
Coombs, W. P., Jr. 1978. Theoretical aspects of cursorial adaptations in dinosaurs. *Quarterly Review of Biology* 53: 393–418.
Currie, P. J. 1985. Cranial anatomy of *Stenonychosaurus inequalis* (Saurischia, Theropoda) and its bearing on the origin of birds. *Canadian Journal of Earth Sciences* 22: 1643–1658.
———. 1987. Bird-like characteristics of the jaws and teeth of troodontid theropods (Dinosauria, Saurischia). *Journal of Vertebrate Paleontology* 7: 72–81.
———. 1995. New information on the anatomy and relationships of *Dromaeosaurus albertensis* (Dinosauria, Theropoda). *Journal of Vertebrate Paleontology* 15: 576–591.
Currie, P. J., K. Rigby, Jr., and R. E. Sloan. 1990. Theropod teeth from the Judith River Formation of southern Alberta, Canada. In K. Carpenter and P. J. Currie (eds.), *Dinosaur systematics: approaches and perspectives,* pp. 107–125. Cambridge: Cambridge University Press.

Currie, P. J., and Zhao X.-J. 1993. A new troodontid (Dinosauria, Theropoda) braincase from the Judith River Formation (Campanian) of Alberta. *Canadian Journal of Earth Sciences* 30: 2231–2247.
Currie, P. J., and Dong Z.-M. 2001. New information on Cretaceous troodontids (Dinosauria, Theropoda) from the People's Republic of China. *Canadian Journal of Earth Sciences* 38: 1753–1766.
Feduccia, A. 1996. *The origin and evolution of birds.* New Haven: Yale University Press.
Forster, C., S. S. Sampson, L. M. Chiappe, and D. Krause. 1998. The theropod ancestry of birds: new evidence from the Late Cretaceous of Madagascar. *Science* 279: 1915–1919.
Gauthier, J. 1986. Saurischian monophyly and the origin of birds. *California Academy of Science Memoirs* 8: 1–55.
Giffin, E. B. 1990. Gross spinal anatomy and limb use in living and fossil reptiles. *Paleobiology* 16: 448–458.
Heilmann, G. 1927. *The origin of birds.* New York: D. Appleton.
Horner, J. R., and J. Gorman. 1988. *Digging dinosaurs.* New York: Workman Publishing.
Horner, J. R., J. Schmitt, F. Jackson, and R. Hanna. 2001. Bones and rocks of the Upper Cretaceous Two Medicine–Judith River clastic wedge complex, Montana. *Museum of the Rockies Occasional Paper* 3: 1–13.
Hwang, S. H., M. A. Norell, and K.-Q. Gao 2001. New information on Jehol theropods. *Journal of Vertebrate Paleontology* 21: 64A.
Jerison, H. J. 1973. *Evolution of the brain and intelligence.* New York: Academic Press.
Ji Q., M. A. Norell, Gao K.-Q., Ji S.-A., and Ren D. 2001. The distribution of integumentary structures in a feathered dinosaur. *Nature* 410: 1084–1088.
Kirkland, J. I., D. Burge, and R. Gaston. 1993. A large dromaeosaur (Theropoda) from the Lower Cretaceous of eastern Utah. *Hunteria* 2(10): 1–16.
Lorenz, L. 1981. Sedimentary and tectonic history of the Two Medicine Formation, Late Cretaceous (Campanian), northwestern Montana. PhD dissertation, Princeton University.
Martin, L. 1991. Mesozoic birds and the origin of birds. In H.-P. Shultze and L. Trueb (eds.), *Origins of the higher groups of tetrapods: controversy and consensus,* pp. 485–540. Ithaca: Cornell University Press.
Martin, L., and A. Feduccia. 1998. Theropod-bird link reconsidered. *Nature* 391: 754.
Matthew, W., and B. Brown. 1922. The family Deinodontidae, with notice of a new genus from the Cretaceous of Alberta. *Bulletin of the American Museum of Natural History* 46: 367–385.
Naples, V. L., and L. D. Martin. 1998. Cenozoic brain evolution as a result of global cooling. TERQUA Symposium, University of Kansas, March 9–10, 1998, p. 4.
Norell, M. A. 2001. The proof is in the plumage. *Natural History* 7/01—8/01: 58–63.
Norell, M. A., J. M. Clark, D. Demberelyin, R. Barsbold, L. M. Chiappe, A. Davidson, M. McKenna, A. Perle, and M. Novacek. 1994. A theropod dinosaur embryo and the affinities of the Flaming Cliffs dinosaur eggs. *Science* 266: 779–782.
Norell, M. A., J. M. Clark, L. M. Chiappe, and D. Dashzeveg. 1997. A *Velociraptor* wishbone. *Nature* 389: 447.

Norell, M. A., and P. J. Makovicky. 1997. Important features of the dromaeosaur skeleton: Information from a new specimen. *American Museum Novitates* 3215: 1–28.

———. 1999. Important features of the dromaeosaurid skeleton II: Information from newly collected specimens of *Velociraptor mongoliensis. American Museum Novitates* 3282: 1–45.

Norell, M. A., P. J. Makovicky, and P. J. Currie. 2001. The beaks of ostrich dinosaurs. *Nature* 412: 873–874.

Novas, F. E. 1998. *Megaraptor namunhuaiquii,* gen. et sp. nov., a large-clawed, Late Cretaceous theropod from Patagonia. *Journal of Vertebrate Paleontology* 18: 4–9.

Osborn, H. F. 1924. Three new Theropoda, *Protoceratops* zone, Central Mongolia. *American Museum Novitates* 144: 1–12.

Osmólska, H. 1982. *Hulsanpes perlei* n.g., n. sp. (Deinonychosauria, Saurischia, Dinosauria) from the Upper Cretaceous Barun Goyot Formation of Mongolia. *Neues Jahrbuch für Geologie und Paläontologie, Monatshefte* 1982: 440–448.

Ostrom, J. H. 1969. Osteology of *Deinonychus antirrhopus,* an unusual theropod from the Lower Cretaceous of Montana. *Peabody Museum of Natural History, Bulletin* 30: 1–165.

———. 1976. *Archaeopteryx* and the origin of birds. *Biological Journal of the Linnean Society* 8: 91–182.

———. 1990. Dromaeosauridae. In P. Dodson, D. Weishampel, and H. Osmólska (eds.), *The Dinosauria,* pp. 269–279. Berkeley: University of California Press.

Paul, G. S. 1988. *Predatory dinosaurs of the world.* New York: Simon and Schuster.

Perle, A., M. A. Norell, and J. M. Clark. 1999. A new maniraptoran theropod—*Achillobator giganticus* (Dromaeosauridae)—from the Upper Cretaceous of Burkhant, Mongolia. National University of Mongolia, Geology and Mineralogy, 107 pp.

Radinsky, L. 1974. Fossil evidence of anthropoid brain evolution. *American Journal of Physical Anthropology* 41: 15–27.

Rasskin-Gutman, D. 1997. Pelvis, comparative anatomy. In P. J. Currie and K. Padian (eds.), *Encyclopedia of Dinosaurs,* pp. 536–540. San Diego: Academic Press.

Reid, R. E. 1993. Apparent zonation and slowed late growth in a small cretaceous theropod. *Modern Geology* 18: 391–406.

Rogers, R. 1997. Two Medicine Formation. In P. J. Currie and K. Padian (eds.), *Encyclopedia of dinosaurs,* pp. 760–765. San Diego: Academic Press.

Ruben, J., T. Jones, N. Geist, and W. Hillenius. 1997. Lung structure and ventilation in theropod dinosaurs and birds. *Science* 278: 1267–1270.

Russell, D. A. 1969. A new specimen of *Stenonychosaurus* from the Oldman Formation (Cretaceous) of Alberta. *Canadian Journal of Earth Sciences* 6: 595–612.

Sereno, P. 1999. The evolution of dinosaurs. *Science* 284: 2137–2147.

Sues, H.-D. 1977. The skull of *Velociraptor mongoliensis,* a small Cretaceous theropod dinosaur from Mongolia. *Paläontologische Zeitschrift* 51: 173–184.

———. 1978. A new small theropod dinosaur from the Judith River Formation (Campanian) of Alberta, Canada. *Zoological Journal of the Linnean Society* 62: 381–400.

Trexler, D. 2001. Two Medicine Formation, Montana: geology and fauna.

In D. H. Tanke and K. Carpenter (eds.), *Mesozoic vertebrate life: new research inspired by the paleontology of Philip J. Currie,* pp. 298–309. Bloomington: Indiana University Press.

Varricchio, D. 1997. Growth and embryology. In P. J. Currie and K. Padian (eds.), *Encyclopedia of dinosaurs,* pp. 282–288. San Diego: Academic Press.

Weigelt, J. 1989. *Recent vertebrate carcasses and their paleobiological implications* (translation of 1927 German edition). Chicago: University of Chicago Press.

Wharton, D. S. 2001. The evolution of the avian brain. *Journal of Vertebrate Paleontology* 21: 113A.

Witmer, L. M., and D. Maxwell. 1996. The skull of *Deinonychus* (Dinosauria: Theropoda): new insights and implications. *Journal of Vertebrate Paleontology* 16: 73A.

Xu X., Wang X-L., and Wu X.-C. 1999. A dromaeosaurid dinosaur with a filamentous integument from the Yixian Formation of China. *Nature* 401: 262–266.

Xu X., Z. Zhou, and X. Wang. 2000. The smallest non-avian theropod dinosaur. *Nature* 408: 705–708.

Xu X., and Wu X.-C. 2001. Cranial morphology of *Sinornithosaurus millenii* Xu et al. 1999 (Dinosauria: Theropoda: Dromaeosauridae) from the Yixian Formation of Liaoning, China. *Canadian Journal of Earth Sciences* 38: 1739–1752.

Xu X., M. A. Norell, Wang X.-L., P. J. Makovicky, and Wu X. 2002. A basal troodontid from the Early Cretaceous of China. *Nature* 415: 780–784.

4. A New Dromaeosaurid from the Horseshoe Canyon Formation (Upper Cretaceous) of Alberta, Canada

Philip J. Currie and David J. Varricchio

Abstract

The discovery of a new dromaeosaurid in the Horseshoe Canyon Formation (uppermost Campanian–lowermost Maastrichtian, Upper Cretaceous) increases the known diversity of this interesting group of theropods, considered by many as the closest non-avian theropod relatives of *Archaeopteryx* and other more derived birds. The new animal, known from a partial skull, is relatively small. It differs from the contemporary *Bambiraptor, Saurornitholestes,* and *Velociraptor* in having a short, deep face. The teeth are more strongly inclined toward the throat than they are in most other dromaeosaurids, and are all almost the same size. Phylogenetic analysis suggests that the new dromaeosaurid may represent an independent lineage having origins back in the early Cretaceous.

Introduction

Dromaeosaurids are an important group of theropods that have been strongly implicated as being the closest known relatives of birds. *Dromaeosaurus* (Matthew and Brown 1922), *Velociraptor* (Osborn 1924), and *Deinonychus* (Ostrom 1969) have formed the core of our understanding of the Dromaeosauridae. Although little material is

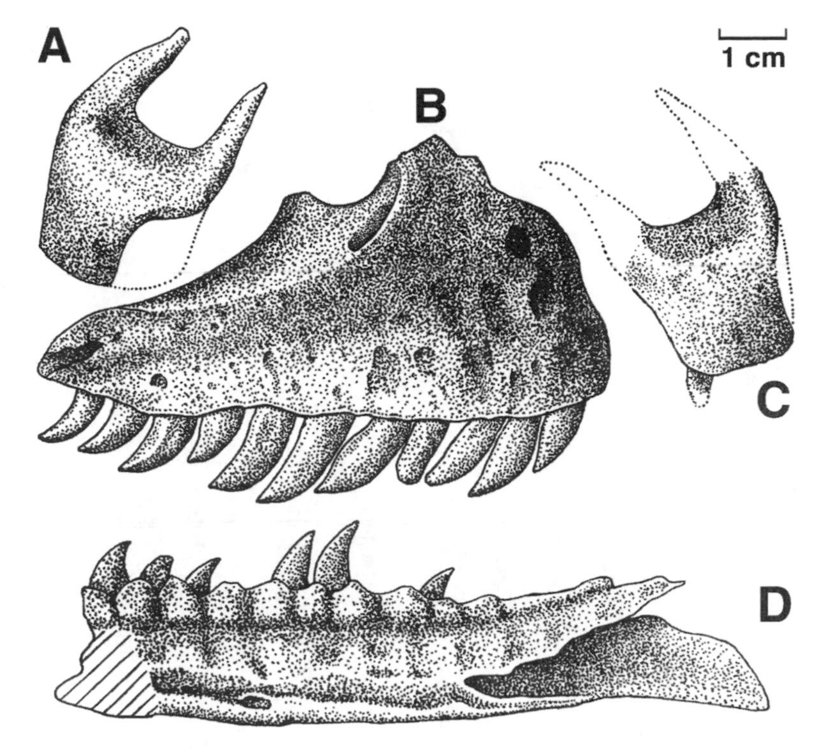

Figure 4.1. Atrociraptor marshalli *(TMP 95.166.1, holotype). A, left premaxilla, lateral view; B, right maxilla, lateral aspect; C, right premaxilla, lateral aspect; D, left dentary, medial view.*

known for *Dromaeosaurus* other than the holotype (Currie 1995), a great deal of new information is available for *Deinonychus* (Maxwell and Ostrom 1995; Brinkman et al. 1998) and *Velociraptor* (Barsbold and Osmólska 1999; Norell and Makovicky 1997, 1999; Norell et al. 1997). Two skeletons of velociraptorines have been collected by the Museum of the Rockies and the Royal Tyrrell Museum of Palaeontology (Varricchio and Currie 1991), and are being described by the authors of this paper. The known diversity of the group has been increased by the description of many new Cretaceous forms, including the giants *Utahraptor* (Kirkland et al. 1993) and *Achillobator* (Perle et al. 1999), the diminutive *Bambiraptor* (Burnham et al. 2000), and the feathered *Sinornithosaurus* (Xu et al. 1999). Other dromaeosaurid species, such as *Adasaurus* (Barsbold 1983) and *Hulsanpes perlei* (Osmólska 1982), are distinctive but incompletely known. A suspected dromaeosaur from Japan (Azuma and Currie 1995) has turned out to be a carnosaur (Azuma and Currie 2000). *Megaraptor* was also compared to dromaeosaurids because of its sickle-like claw (Novas 1998; Calvo et al. 2002), although it was always clear that it is not related.

In 1995, a partial skull of a dromaeosaurid was discovered close to the Royal Tyrrell Museum of Palaeontology in beds of the Horseshoe Canyon Formation. Fragments of jaws and teeth on the hillside led to the discovery of TMP 95.166.1 (fig. 4.1). The specimen was in a relatively hard, isolated block of sandstone capping softer, medium-grained sands. Preparation revealed the right maxilla (exposed in lateral view) and the medial surface of the right dentary. Although the

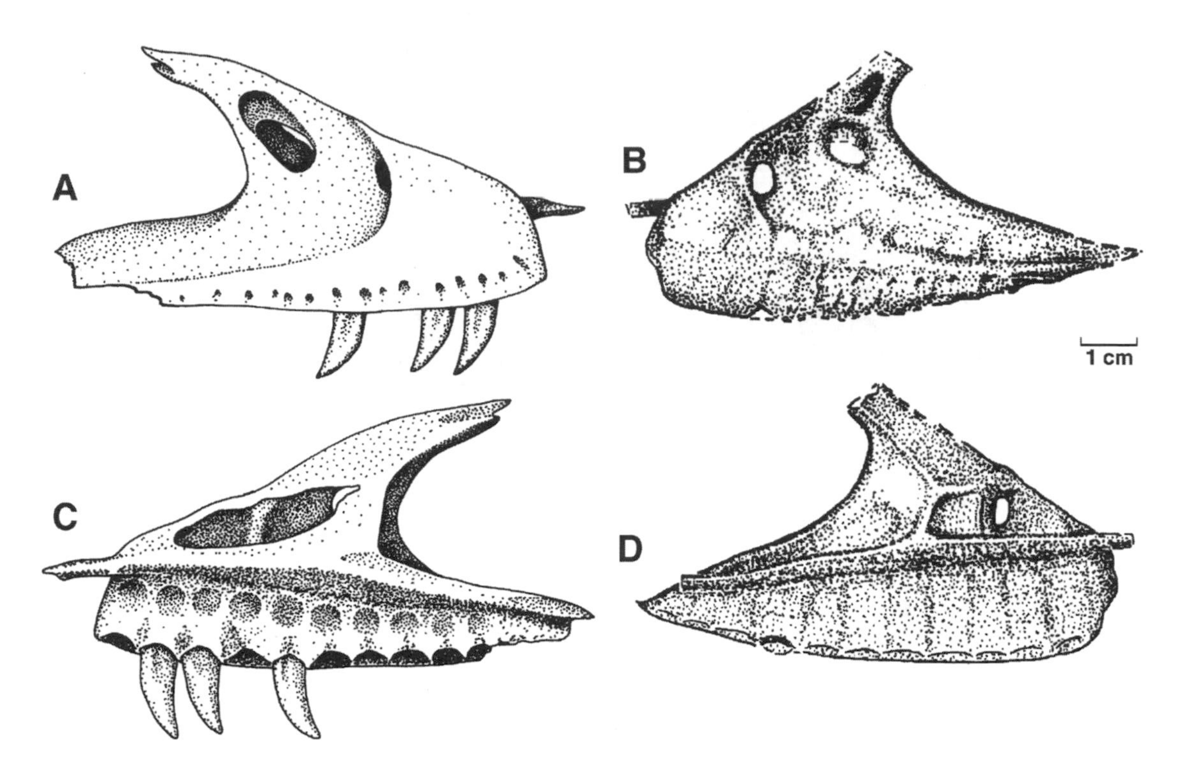

Figure 4.2. Maxillae of Saurornitholestes langstoni *(TMP 94.12.844) (A, C) and cf.* Bambiraptor feinbergi *(MOR 553S—7.30.91.274) (B, D) in lateral (A, B) and medial (C, D) views.*

teeth are clearly velociraptorine, the distinct osteology of the specimen immediately showed that it represents a new taxon.

In order to establish the identity of the new specimen, it is necessary to compare it with other dromaeosaurids known from the region. Most of the maxilla of *Dromaeosaurus albertensis* is known (Currie 1995), but this bone has not been described for *Saurornitholestes langstoni* (Sues 1978). Although the holotype of *Bambiraptor feinbergi* (Burnham et al. 2000) includes both maxillae, it represents a juvenile individual and introduces potential allometric complications into comparative analysis. Two velociraptorine maxillae recovered from Campanian beds of Alberta and Montana will therefore also be described in this paper. TMP 94.12.844 (figs. 4.2A, 4.2C) is an isolated right maxilla from the Dinosaur Park Formation (Campanian) of Dinosaur Provincial Park, Alberta. It lacks the posterior end of the jugal process, and only three of the teeth remain in position. The specimen can be identified as *Saurornitholestes langstoni,* the only known velociraptorine dromaeosaurid from the Park, on the basis of tooth denticulation, which is the same as that of the holotype. MOR 553S-7.30.91.274 (figs. 4.2B, 4.2D) is an isolated, left maxilla found in the Two Medicine Formation (Campanian) of Montana (South Quarry at Jack's Birthday Site near Cutbank, Montana). It is nearly complete, and lacks only a portion of the dorsal process for the lacrimal contact. Identification is more problematic in this case. In overall morphology, it closely matches *Bambiraptor feinbergi,* which is from the same formation 100 kilometers farther south. It is identified as cf. *Bambiraptor feinbergi* in this paper because of this similarity, and because it comes from the same formation. However, it is also morphologically similar to *Saurornitholestes,* the remains of which are found in a different but contemporaneous formation only 300 kilometers to the north.

Institutional Abbreviations: AMNH, American Museum of Natural History, New York, U.S.; CEU, College of Eastern Utah Prehistoric Museum, Price, Utah, U.S.; GIN, Institute of Geology, Ulaan Baatar, Mongolia; IVPP, Institute of Vertebrate Paleontology and Paleoanthropology, Beijing, China; MNUFR, Mongolia National University, Ulaan Baatar, Mongolia; TMP, Royal Tyrrell Museum of Palaeontology, Drumheller, Alberta, Canada; YPM, Yale Peabody Museum, New Haven, Connecticut, U.S.

Taxonomy

Dinosauria Owen, 1842
Saurischia Seeley, 1888
Theropoda Marsh, 1881
Dromaeosauridae Matthew and Brown, 1922
Velociraptorinae Barsbold, 1983
Atrociraptor marshalli, new genus, new species

Etymology: "Atroci" is a Latin word meaning savage, whereas "raptor" is Latin for robber. The species is named after Wayne Marshall of East Coulee, Alberta, who discovered the type specimen.

Holotype: TMP 95.166.1, a partial skull that includes premaxillae, the right maxilla, the right dentary, portions of the left dentary, teeth, and numerous bone fragments.

Locality and age: The holotype was recovered from strata about 5 m above the Daly Coal Seam #7 (Gibson 1977) in the Horseshoe Canyon Formation (upper Campanian or lower Maastrichtian, Upper Cretaceous) at UTM 12U 372,125 E, 5,708,055 N, which is about 5 km west of the Royal Tyrrell Museum of Palaeontology in Drumheller, Alberta.

Diagnosis: Small velociraptorine, dromaeosaurid theropod that differs from *Saurornitholestes* and *Velociraptor* in having a shorter, deeper face. Subnarial body of premaxilla is taller than its anteroposterior length as in *Deinonychus* and possibly *Dromaeosaurus.* Internarial and maxillary processes of premaxilla subparallel and oriented more dorsally than posteriorly. Larger maxillary fenestra than in any other velociraptorines. Maxillary fenestra is directly above the promaxillary fenestra, rather than well behind it as in all other dromaeosaurids. Maxillary teeth more strongly inclined toward the throat than in all other dromaeosaurids except *Bambiraptor* and *Deinonychus.* Maxillary dentition is essentially isodont.

Description of *Atrociraptor marshalli*

TMP 95.166.1 (fig. 4.1) consists of a pair of premaxillae, a right maxilla, two dentaries (only one of which is reasonably complete), and associated teeth and bone fragments from other parts of the same skull.

TABLE 4.1
Numbers of premaxillary, maxillary, and dentary tooth positions in dromaeosaurids

Species, Specimen No.*	Premax	Max	Dent
Achillobator, MNUFR 15	?	11	?
Atrociraptor, TMP 95.166.1	4	11	14†
Bambiraptor, AMNH 001	4	12	13
Bambiraptor, MOR 553S-7.30.91.274	-	12	-
Deinonychus, YPM 5232	4	15	16
Dromaeosaurus, AMNH 5356	4	9	11†
Dromaeosaurid, GIN 100/22	4	11	13
Saurornitholestes, TMP 88.121.39, TMP 94.12.844	4	12	15
Sinornithosaurus, IVPP V12811	4	11	12
Utahraptor, CEU 184v.400	4	?	?
Velociraptor, AMNH 6515	4	10	14
Velociraptor, GIN 100/25	4	11	?

*All counts taken directly from specimens.
† = estimate.

Although the premaxillae are free from matrix, the maxilla (exposed in lateral view) and the right dentary (lingual aspect exposed) have been left in the hard block of sandstone that they were found in.

There are four teeth in each relatively deep premaxilla, which is the same number in all other dromaeosaurids (table 4.1). Like *Deinonychus* (Kirkland et al. 1993), the subnarial body is taller than it is anteroposteriorly long (figs. 4.1A, 4.1C), whereas this relationship is the opposite in *Bambiraptor* (AMNH 001), *Saurornitholestes* (TMP 94.12.844), and *Velociraptor* (Barsbold and Osmólska 1999). In *Utahraptor,* the height is slightly greater than its anteroposterior length (Kirkland et al. 1993). As in other dromaeosaurids (Currie 1995; Barsbold and Osmólska 1999), there is an elongate subnarial extension that wedges between the nasal and maxilla. Because of the depth of the snout, however, the subparallel internarial and subnarial processes are oriented more dorsally than posteriorly, in contrast with other dromaeosaurids in which the reverse is true. The shallow lateral depression marking the anteroventral limit of the narial opening is nested between the bases of the internarial and subnarial processes (figs. 4.1A, 4.1C), whereas it extends more anteriorly in *Velociraptor* (Barsbold and Osmólska 1999). As in *Dromaeosaurus* (Currie 1995), there is no posteromedial maxillary process, but the anteromedial process of the maxilla contacts a smooth triangular facet on the posteromedial surface of the premaxilla.

The sizes of the right alveoli show that the second premaxillary tooth was the largest of the four as in other velociraptorines (Currie

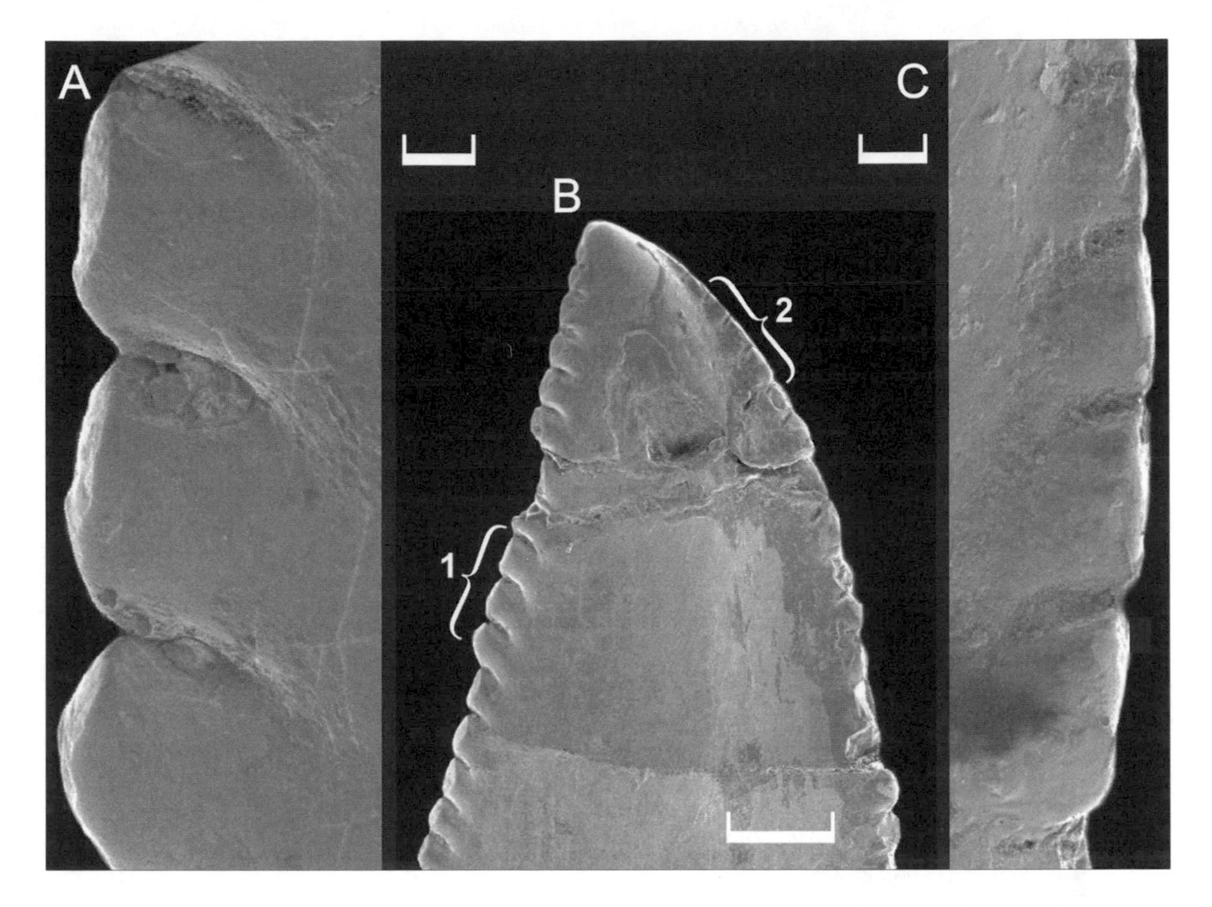

Figure 4.3. Scanning electron microscope photographs of isolated first right premaxillary tooth of Atrociraptor marshalli *(TMP 95.166.1). A, enlargement of the posterior denticles labeled "1" in B. C, enlargement of anterior serrations labeled "2" in B. Scale bars for A and C are 100 μm, and that for B is 1 mm.*

1995). The first premaxillary tooth of the right side is complete, as is the second premaxillary tooth of the left side. Both were found adjacent to the articulated premaxilla and maxilla, and had clearly fallen out of their sockets before burial and fossilization. In the right premaxilla, two functional premaxillary teeth remain in their sockets (second and fourth positions). The crown of the right third premaxillary tooth was broken and lost sometime after the specimen was exposed. The first alveolus of the left premaxilla still contains a germ tooth.

Like the teeth of *Saurornitholestes* (Currie et al. 1990), the anterior carina is on the posteromedial edge of the tooth, but is more anterior in position than the posterior carina. In cross section, the tooth is more J-shaped than D-shaped. The denticles on the premaxillary teeth have almost the same basal diameters on both anterior and posterior carinae (figs. 4.3, 4.4A; table 4.2), although the posterior denticles are taller. There are 2.3–3.0 denticles per millimeter (this figure is usually quoted as number of denticles per five millimeters in theropods with larger teeth, and multiplication gives a range of 11.5 to 15 denticles per five millimeters in *Atrociraptor*).

The maxilla (fig. 4.1B) is 92 mm long as preserved, but lacks part of the short, slender postalveolar process that articulates with the jugal. The maxillary tooth row is 85 mm long. The posterodorsal lacrimal-

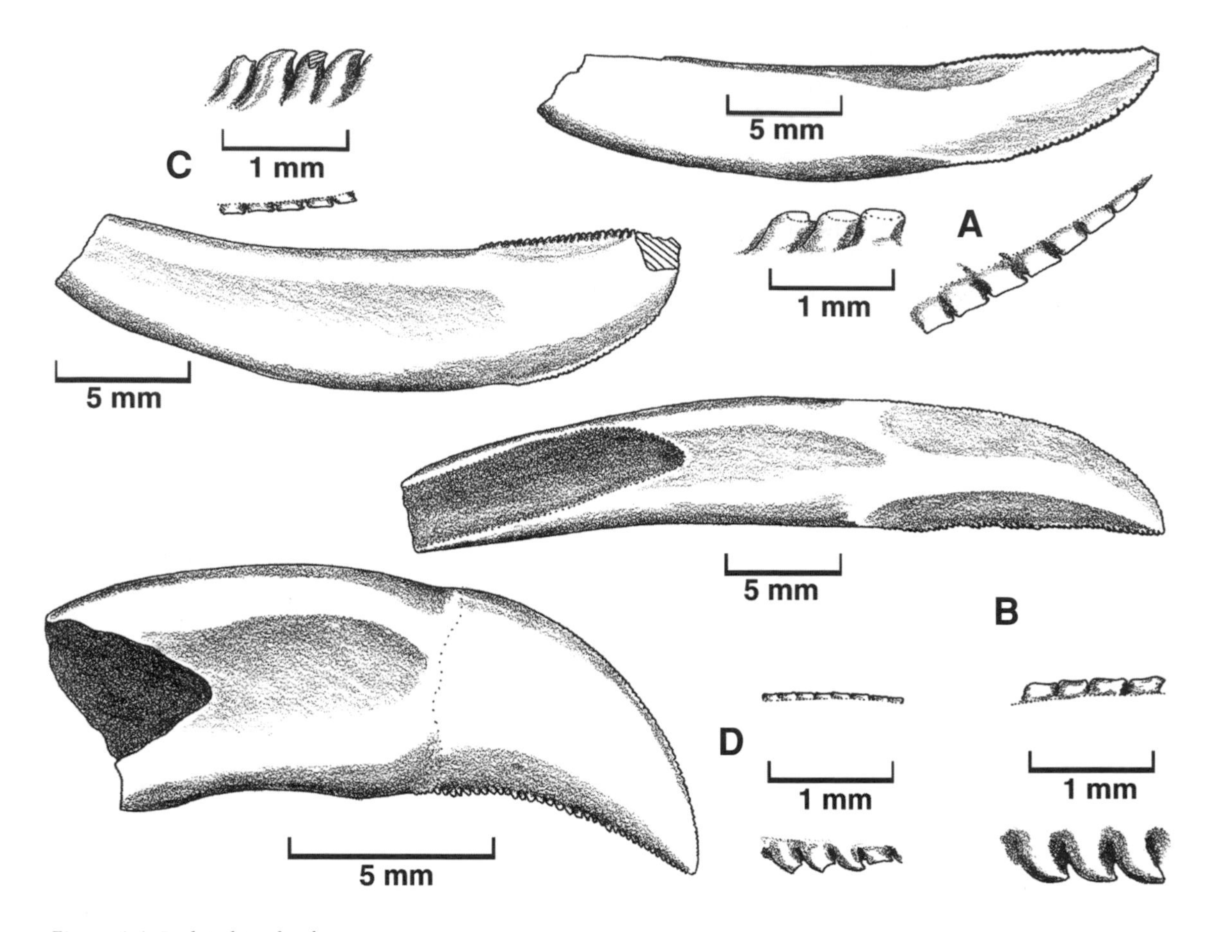

Figure 4.4. Isolated teeth of Atrociraptor marshalli *(TMP 95.166.1) with enlargements of anterior and posterior denticles. A, premaxillary tooth with enlargements of anterior and posterior denticles; B, maxillary tooth with enlargements of anterior and posterior serrations in the upper righthand corner; C, anterior dentary tooth with enlargements of posterior and anterior denticles; D, dentary tooth with anterior and posterior serrations enlarged above.*

nasal process was destroyed by erosion, and the maxilla as preserved is 45 mm high. The lower margin of a relatively large, round maxillary fenestra is evident. The promaxillary fenestra is tucked under the anterior margin of the antorbital fossa directly under the maxillary fenestra. In other velociraptorines where both features are known (fig. 4.5), the promaxillary fenestra is also at a lower level than the maxillary fenestra, but is also well anterior to it.

The roughly triangular maxilla is relatively deeper than other dromaeosaurid maxillae. The distance between the lower edge of the maxillary fenestra and the dentigerous margin is 32 mm, which when divided by the length of the maxillary tooth row gives a ratio of 0.38. The same ratio is 0.33 in *Bambiraptor* (MOR 553S-7.30.91.274) and 0.33 in *Deinonychus,* but both of these animals have relatively small maxillary fenestrae (figs. 4.5A, 4.5B, 4.5F). In other dromaeosaurids, the ratio is 0.31 in *Dromaeosaurus* (AMNH 5356), 0.28 in *Saurornitholestes* (TMP 94.12.844), and 0.19 in *Velociraptor* (GIN 100/25). Generally in dromaeosaurids, the height between the maxillary fenestra

TABLE 4.2
Teeth of *Atrociraptor* and *Saurornitholestes* (TMP 74.10.5, TMP 88.121.39)

	Tooth Position	TL	Crown	FABL	BW	ANT	POST
PM-1	L	xx	xx	4.5	xx	2.5	3
PM-2	R	xx	xx	6.5	4	xx	xx
PM-3?	L	26.5	10	5.0	3.5	2.7	2.3
PM-3	R	xx	xx	5.5	4	xx	xx
PM-4	R	xx	xx	5.5	4	4	3
Mx-1	R	xx	11.5	5.5	xx	xx	4
Mx-2	R	xx	13.5	6.3	xx	xx	4
Mx-3	R	xx	11.8	6.5	xx	xx	3.6
Mx-4	R	xx	11+	5.9	xx	xx	4
Mx-5	R	xx	13.0	6.1	xx	6	4
Mx-6	R	xx	15.3	6.6	xx	5	4
Mx-7	R	xx	11.9	5.5	xx	6	4
Mx-8	R	xx	7.8e	5.6	xx	7	4
Mx-9	R	xx	10.5	5.6	xx	5.5	4
Mx-10	R	xx	7.7	5.1	xx	xx	4.2
Mx-11	R	xx	7.0	4.4	xx	8	4.5
Mx-ant	L	32	12	5.5	3.5	3.5	3
D-1a	L	xx	xx	4.5	xx	xx	xx
D-2a	L	xx	xx	5.1	xx	xx	xx
D-3	L	xx	xx	5.2	3.0	xx	xx
D-4	L	xx	xx	4.5	2.6	5	4
D-4	R	xx	xx	4.9	xx	xx	xx
D-5	R	xx	xx	5.1	xx	xx	xx
D-6	R	xx	5.7e	3.2e	xx	6	5
D-7	R	xx	xx	xx	xx	xx	xx
D-8	R	xx	11.3	5.5	xx	5.2	4
D-9	R	xx	11.8	5.7	xx	xx	3.5
D-10	R	xx	xx	xx	xx	xx	xx
D-11	R	xx	xx	5.0	xx	xx	xx
D-ant	L	23+	7.5+	5.2	2.8	5	4
D-post	R	16	7	5.2	2.6	8	4
74.10.1	Max	xx	9.2	4.5	2.1	5	4
74.10.1	Dent	xx	8.9	3.9	xx	7	5
88.121.39	Dent	xx	9	5.1	2.3	6	4

a = anteroposterior alveolar length; ANT = lowest number of denticles per 1 mm along the anterior carina; BW = labial-lingual base width of crown; CROWN = height of the crown, measured from the tip to the proximal end of the posterior carina or to the edge of the enamel layer; e = erupting tooth; FABL = fore-aft base length, which is anteroposterior length of tooth at the base of the crown; POST = lowest number of denticles per 1 mm along the posterior carina; TL = total length of crown and root; xx, unknown; + = tip of tooth lost to wear.

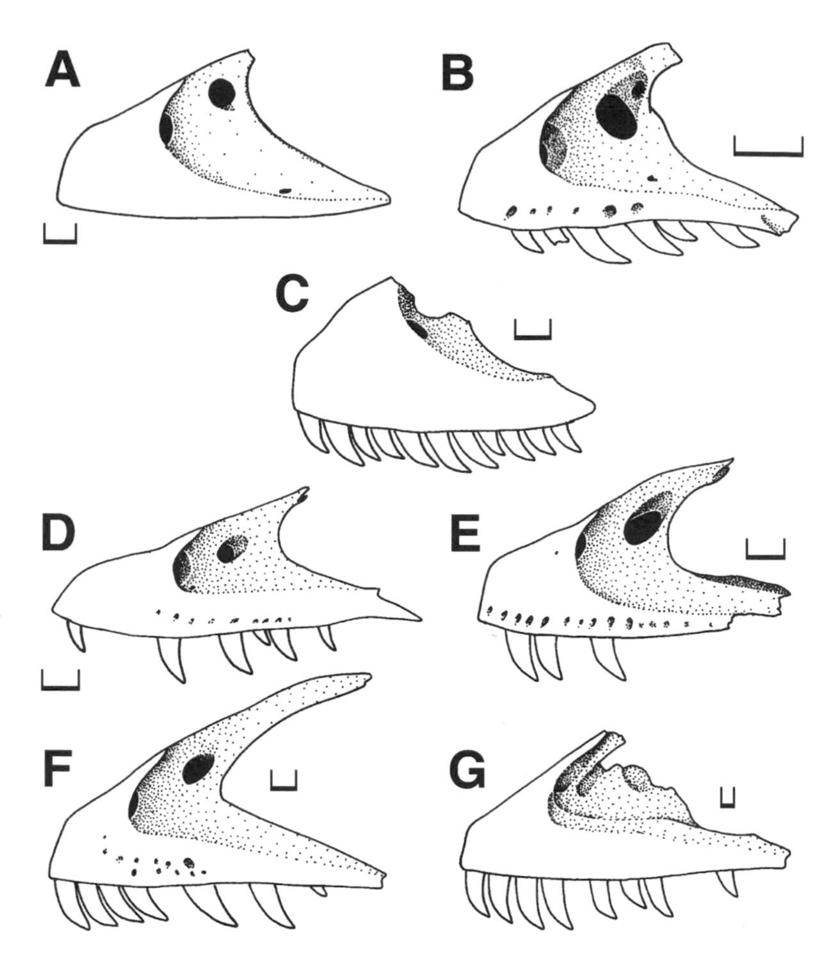

Figure 4.5. Left lateral views of dromaeosaurid maxillae:
A, cf. Bambiraptor feinbergi *(MOR 553S—7.30.91.274);*
B, Bambiraptor feinbergi *(AMNH 001);*
C, Atrociraptor marshalli *(TMP 95.166.1), reversed image of right maxilla;*
D, Velociraptor mongoliensis *(GIN 100/25);*
E, Saurornitholestes langstoni *(TMP 94.12.844), reversed image of right maxilla;*
F, Deinonychus antirrhopus *(YPM 5232);*
G, Achillobator giganticus *(MNUFR 15).*

and the dentigerous margin is less than twice (1.5 in *Bambiraptor,* 1.6 in *Deinonychus,* 1.6 in *Dromaeosaurus,* and 1.8 in *Saurornitholestes*) the height of the largest tooth, whereas in *Atrociraptor* it is more than twice (2.2) as high. If we assume that the teeth have the same relative heights in all of these animals, this would suggest that the short, deep appearance of the maxilla of *Atrociraptor* could be attributed to an increase in snout depth rather than to an abbreviation of the snout.

The anterior margin of the maxilla contacted the premaxilla in a tall butt joint that is notched at mid-height by a conspicuous subnarial foramen (fig. 4.1B). The maxilla is excluded from the narial border by the elongate, thin maxillary process of the premaxilla as in other dromaeosaurids. The margin of the antorbital fossa is restricted to the posterior 52 percent of the preserved length of the maxilla, whereas in *Velociraptor,* it occupies two-thirds of the total length (Barsbold and Osmólska 1999). Compared to most other dromaeosaurids, the antorbital fenestra was relatively small (fig. 4.7), the maxillary portion of it making up less than 43 percent of the maxillary length. The distance between the anterior margin of the antorbital fenestra and the anterior margin of the antorbital fossa is only 19 mm, which shows that the

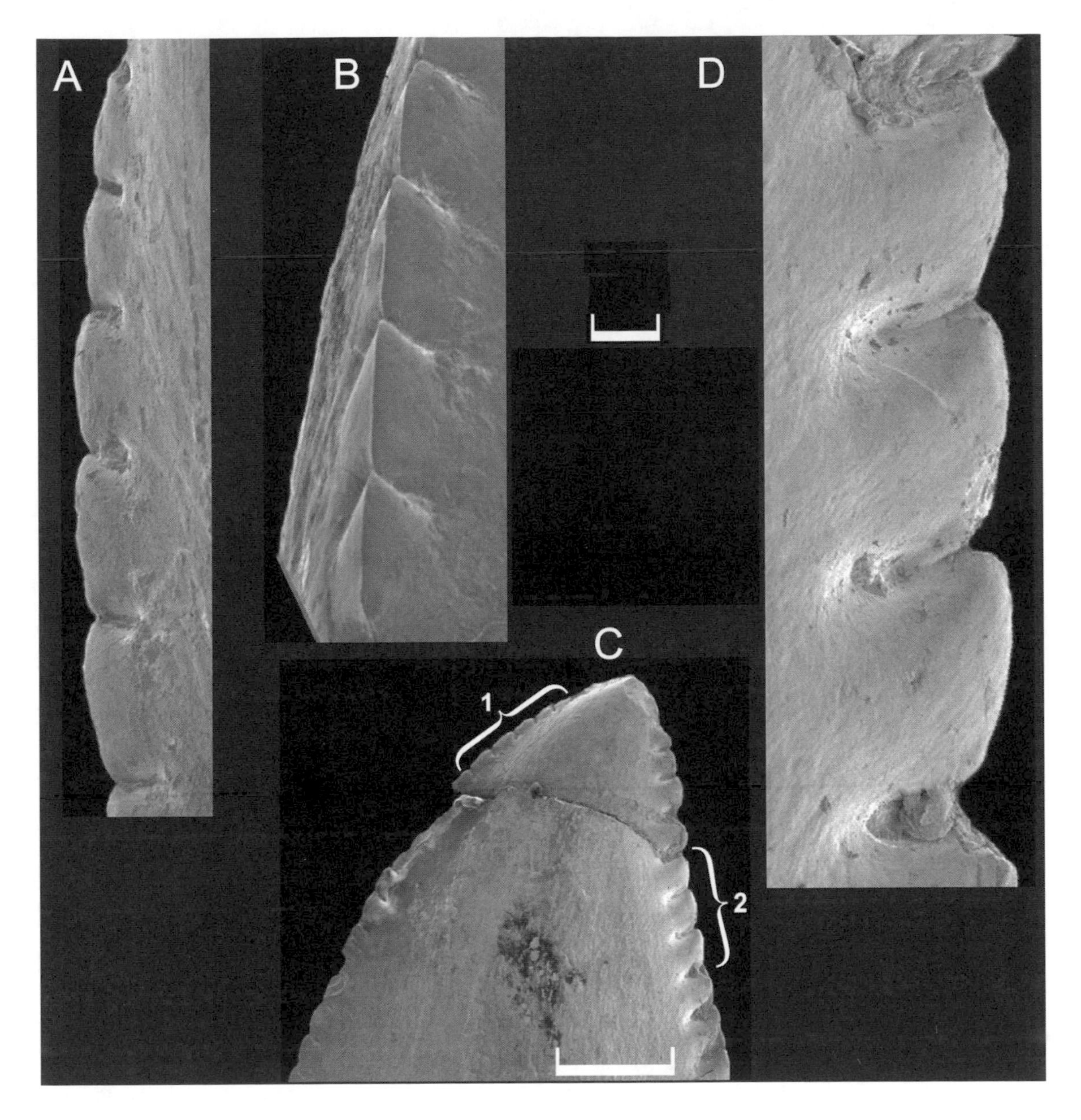

Figure 4.6. Scanning electron microscope photographs of isolated left maxillary tooth of Atrociraptor marshalli *(TMP 95.166.1). A, enlargement of the anterior denticles (5th to 9th from the tip) labeled "1" in C. B, enlargement of 7th to 10th anterior serrations from the tip of the tooth, mostly covered by region "1" in C. D, enlargement of posterior serrations labeled "2" in C. Scale bar for A, B and D is 100 μm, and that for C is 1 mm.*

fossa is also relatively smaller than other dromaeosaurids. The preserved ventral margin of the maxillary fenestra shows that this rounded opening was relatively large, with a diameter of about one centimeter.

There is a shallow posterior trough on the lateral surface at the back of the maxilla. Its anterior end is above the last maxillary tooth. Anterodorsal to this depression, the well-defined margin of the antorbital fossa slopes forward and upwards at a higher angle than in other dromaeosaurids. Most of the external surface of the maxilla is sculptured, and there is a row of neurovascular foramina just above the alveolar margin.

The ventral margin of the maxilla is strongly convex in lateral view. There are eleven closely packed teeth in their alveoli, with no gaps between the teeth. The number of tooth positions compares well with most other dromaeosaurids (table 4.1). All of the labiolingually narrow, bladelike maxillary teeth have a conspicuous posteroventral inclination (fig. 4.1B). The only dromaeosaurids that have a similar inclination are *Bambiraptor* (AMNH 001) and *Deinonychus* (YPM 5232). Because of the inclination of the teeth, the enamel at the base of the crown is also inclined. In contrast, the edge of the enamel is almost perpendicular to the longitudinal axis of the tooth in *Saurornitholestes, Dromaeosaurus,* and other genera where the teeth have more vertical orientations.

The maxillary dentition is almost isodont with no gaps left by shed teeth, which is unusual for a dromaeosaurid (fig. 4.5). The teeth vary relatively little in overall height, whereas in the anterior part of the tooth row of *Velociraptor* every second tooth is conspicuously longer than its neighbors (Barsbold and Osmólska 1999). The maxillary teeth (figs. 4.4B, 4.6; table 4.2) all have larger denticles (3–4.5 per mm) on the posterior carinae than they do on the anterior carinae (5–8 per mm). Posterior denticles have relatively straight, elongate shafts with distally hooked tips, and are much taller than the anterior denticles (figs. 4.4b, 4.6). As in other velociraptorines, and in contrast with *Dromaeosaurus* (Currie et al. 1990), the anterior and posterior carinae lie on the midlines of maxillary teeth. The maxillary teeth are closely comparable in terms of tooth shape, carina position, denticle size, and denticle shape with those of *Bambiraptor* (Burnham et al. 2000), *Deinonychus* (Ostrom 1969), *Saurornitholestes* (Currie et al. 1990), and *Velociraptor* (Barsbold and Osmólska 1999).

The dentary of *Atrociraptor* (fig. 4.1D) is comparable with those of other dromaeosaurids. Dorsal and ventral margins are almost parallel, although the height decreases somewhat toward the back of the dentigerous region. The external surface of the fragmentary left dentary has two rows of nutritive foramina. Below the external intramandibular process for the surangular (Currie and Zhao 1993), the posterior margin of the dentary slopes posteroventrally. There is no accommodation in that margin for the external mandibular fenestra, which suggests that this opening was small and low in position as in other dromaeosaurids. As in other velociraptorines, the dentary is thin labiolingually, the Meckelian canal is shallow, the dental shelf is narrow, and the interdental plates are fused to each other and to the margin of the dentary. The dental shelf splits posterior to the last alveolus to accept the anterior end of the surangular. The medial fork extends more posteriorly than the lateral one. The posteroventral edge of the dentary was excluded from the ventral margin of the jaw by a lateral extension of the splenial, which is a feature characteristic of dromaeosaurids and troodontids (Currie 1995). In addition to the shallow Meckelian groove, there is a shallow groove along the bases of the interdental plates for the dental artery. The exact number of tooth positions is unknown. Six teeth are positioned in ten alveoli in the right dentary, and six teeth and alveoli can be seen in the fragment of the left. The left

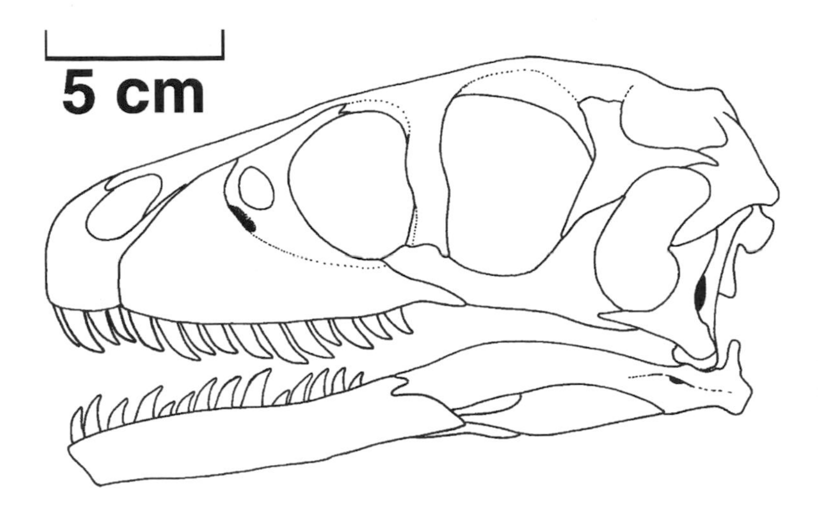

Figure 4.7. Reconstruction of the skull of Atrociraptor marshalli *with missing parts restored from* Dromaeosaurus *and* Velociraptor.

dentary fragment, which includes the symphysis, overlaps the right dentary, which lacks the anterior part. Based on the evidence from both dentaries, there would have been twelve or thirteen dentary teeth (fig. 4.7).

The dentary teeth seem to be generally smaller than the maxillary teeth (fig. 4.1), and are not as strongly inclined posteriorly. They are all labiolingually narrow, and are as bladelike as the maxillary teeth. The anterior denticles (5–8 per mm) are smaller (figs. 4.4C, 4.4D; table 4.2) and more numerous than the posterior denticles (3.5–5 per mm).

Two Isolated Velociraptorine Maxillae

The best maxilla of *Saurornitholestes langstoni* is TMP 94.12.844 (figs. 4.2A, 4.2C), which is 96.5 mm long as preserved, has a tooth row length of 82.5 mm, and has a maximum height of 47 mm. MOR 553S-7.30.91.274 (figs. 4.2B, 4.2D), identified as cf. *Bambiraptor feinbergi* in this paper, has a total length along the lateral margin of 91 mm, and the total preserved height is 46 mm. The two maxillae are similar enough to be described together, although differences will be noted.

Similar to other velociraptorine maxillae, the lateral surface above the tooth row is marked by irregular, short, and subvertical grooves terminating in ventrally opening neuro-vascular foramina. The margin that defines the anterior limit of the antorbital fenestra in each of the specimens forms a broadly open arc, much broader than in *Deinonychus* or *Velociraptor,* but apparently not as broad as in *Atrociraptor.* The margin of the antorbital fossa is well defined everywhere except above the sixth and seventh tooth sockets. Posteriorly it is at the posteroventral edge of the antorbital fenestra close to the alveolar margin, but anteriorly curves progressively more dorsally until it is vertical between the fourth and fifth alveoli in TMP 94.12.844, and between the third and fourth alveoli in MOR 553S-7.30.91.274. The fossa covers 70–75 percent of the total length of the maxilla. The distance between the front of the antorbital fenestra and the back of the maxilla is 43 mm (or 47 percent of the total length) in MOR 553S-7.30.91.274, but is only 38 percent of the total length in TMP 94.12.844. Relative to each other, the promaxillary and maxillary fenestrae are positioned much like those of *Bambiraptor, Deinonychus,* and *Velociraptor* (fig. 4.5). This region is incomplete in *Achillobator,* although the presence of two posterodorsally oriented channels suggests that the arrangement was similar. The subcircular promaxillary fenestra is tucked under the anterior margin of the antorbital fossa, and its lower edge is aligned with the bottom of the antorbital fenestra and the anteromedial process of the maxilla. The maxillary fenestra is positioned at a higher level in the base of the posterodorsal lacrimal-nasal process, slightly anterior to the margin of the antorbital fenestra. The maxillary fenestra is oval in TMP 94.12.844, and more rounded in MOR 553S-7.30.91.274. Like *Bambiraptor,* the maxillary fenestra is nested within a shallow depression, and opens anteroventrally into the more medial sinus system. In contrast, the maxillary fenestra of *Velociraptor* is relatively smaller, and is positioned well anterior to the antorbital fenestra.

The posterodorsal nasal-lacrimal process of the maxilla passes between the nasal bone and the antorbital fenestra, and bifurcates distally into dorsal and ventromedial prongs (figs. 4.2A, 4.2C) to embrace the anteroventral process of the lacrimal. The dorsal process wedges between the front of the lacrimal and the nasal, and the ventromedial fork is overlapped laterally by the lacrimal.

The ventral portion of the maxilla/premaxilla contact slopes anterodorsally and has a small triangular anterior projection that is medially concave.

On the lingual side, a medially directed horizontal ledge, roughly 10 mm wide, extends along the entire length of the maxilla. It angles slightly anterodorsally until it is 15 mm above the alveolar margin. Continuing forward from this ledge is a well-developed anteromedial process, the anterior end of which extends well anterior to the main premaxillary-maxillary contact in TMP 94.12.844 (figs. 4.2A, 4.2C), but is broken in MOR 553S-7.30.91.274. Its medial margin is grooved for contact with the vomer and the opposing maxilla. Sutures on the medial surface of the ledge show that the secondary palate extended posteriorly to at least the level of the maxillary fenestra. Dorsomedial to the last three alveoli, the inner surface of the ledge is also scarred for the palatine suture.

The sinus above the medial ledge is divided into chambers (figs. 4.2C, 4.2D) that connect with the antorbital, maxillary, and promaxillary fenestrae. A thin sheet of bone (postantral strut of Witmer 1997) extends dorsally from the medial ridge to the dorsomedial surface of the dorsoposterior nasal-lacrimal process of the maxilla. The sheet encloses a chamber (the maxillary antrum) medial to the maxillary fenestra, and ventrally forms the medial border of the passage between the sinus system and the antorbital fenestra. This sheet formed a partition that completely separated a posterior space opening into the antorbital fenestra, from the two anterior chambers and their associated fenestrae. Extending downward and then across the top of the medial ledge, a bar on the medial side of the maxilla weakly defines the anterior promaxillary recess, which connects to the promaxillary fenestra directly in front of the bar. The maxillary sinus system is basically the same as those of *Deinonychus* (Ostrom 1969), *Velociraptor* (Barsbold and Osmólska 1999), and other theropods (Witmer 1997).

As is typical of dromaeosaurids, the interdental plates are fused to the maxilla and each other in TMP 94.12.844 (fig. 4.2C) and MOR 553S-7.30.91.274 (fig. 4.2D). They can only be distinguished from the maxilla because of subtle textural differences in the surfaces, the interdental plates being more highly vascularized.

Twelve alveoli are present, although only three of the teeth remained in the sockets of TMP 94.12.844, and only the tip of the replacement tooth in the fourth alveolus was preserved in MOR 553S-7.30.91.274. In both specimens, the denticles of the anterior carina are slightly smaller in basal lengths than those of the posterior carina. The anterior carina lies wholly on the midline and shows none of the twisting of the carina onto the lingual side as in *Dromaeosaurus*.

Phylogenetic Analysis

A data matrix (Appendix 4.1) was assembled for the best-known dromaeosaurid genera, plus various outgroup taxa. The purpose of the analysis was not to determine the relationships of dromaeosaurids to other theropods or birds, but it was simply to see what could be learned about the position of *Atrociraptor* within the Dromaeosauridae. For this reason, the analysis was limited to cranial characters.

The phylogenetic analysis was performed using the beta version of PAUP 4.0 (Swofford 2001). The analysis included 42 characters, twelve of which could be coded for *Atrociraptor*. All of the characters were parsimony-informative. *Coelophysis* and *Allosaurus* were used as successively proximal outgroups, and Troodontidae were included in the analysis because of their purported relationship with dromaeosaurids. The Branch-and-Bound search method produced a single most parsimonious tree (tree length = 61, C.I. = 0.80, R.I. = 0.82, R.C. = 0.66) under an Acctran transformation. *Atrociraptor* sorted most strongly with *Deinonychus* (fig. 4.8), and secondarily with *Bambiraptor*.

Discussion

Atrociraptor can be identified as a dromaeosaurid, and distinguished from contemporary tyrannosaurids and troodontids by the collective evidence of its relatively small size, the sizes and positions of the antorbital and maxillary fenestrae, the presence of a subnarial-maxillary process on the premaxilla that extends posteriorly to wedge between the maxilla and nasal, the subparallel dorsal and ventral margins of the dentary, a labiolingually thin dentary, fusion of the interdental plates, and by its bladelike teeth. *Atrociraptor* is easily distinguishable from previously described dromaeosaurids by its short, deep snout (fig. 4.7).

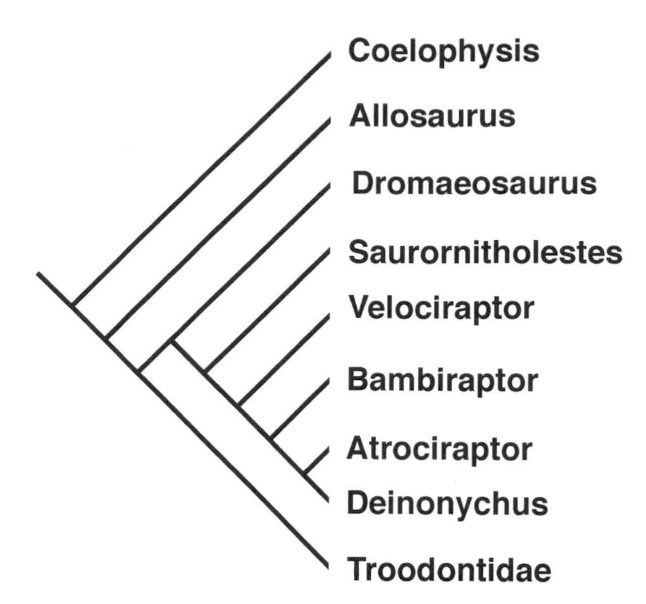

Figure 4.8. Most parsimonious tree generated in restricted analysis of phylogenetic relationships of Atrociraptor.

Atrociraptor can also be identified as a member of the Velociraptorinae because denticles on the anterior carinae are significantly smaller than posterior serrations in maxillary and most dentary teeth. Furthermore, the largest premaxillary tooth in *Atrociraptor* is in the second alveolus, as in other velociraptorines but not *Dromaeosaurus* (Currie 1995).

Atrociraptor is similar to *Deinonychus* in having maxillary teeth that are inclined sharply posteroventrally. The maxillary fenestra is relatively larger and more circular in *Atrociraptor.* The posterodorsal nasal-lacrimal process of the maxilla rises more steeply in *Atrociraptor* than it does in *Deinonychus,* but the process is relatively shorter and the anterior margin of the antorbital fenestra forms a more open curve. Whereas *Deinonychus* has 15 maxillary teeth, *Atrociraptor* has only 11.

Two other Early Cretaceous dromaeosaurids have been described, but both *Achillobator* (Perle et al. 1999) and *Utahraptor* (Kirkland et al. 1993) are fundamentally different than *Atrociraptor* because of allometric differences related to their much larger sizes.

Saurornitholestes is known from the middle Campanian beds of Dinosaur Provincial Park in southern Alberta (Sues 1978). Cranial material recovered with the type specimen includes a pair of frontals and teeth. TMP 94.12.844 can be assigned to *Saurornitholestes langstoni,* because it was found close (less than 5 kilometers) to the locality from which the holotype was recovered, comes from the same formation (Dinosaur Park Formation), and has teeth that are indistinguishable from the holotype. Other skull bones, including a premaxilla (TMP 86.36.117), frontals (Currie 1987b), several dentaries (Sues 1977; Currie 1987b), and hundreds of teeth (Currie et al. 1990), have been assigned to this genus on the basis of their recovery from the same region and formation. The subnarial body of the premaxilla of

Saurornitholestes is longer anteroposteriorly than it is high, thereby distinguishing it from *Atrociraptor.* The maxilla of *Saurornitholestes* is relatively longer and lower, the ventral rim of the antorbital fossa is almost horizontal for most of its length, the maxillary fenestra is higher in position but smaller, the alveolar margin is only shallowly convex, and the teeth are more heterodont. The dentary is similar in most respects, although the height does not increase anteriorly. The teeth are almost indistinguishable, although the enamel begins beneath the crown at almost the same level anteriorly and posteriorly. This character can be correlated with the differences in the angles that the teeth erupt from the jaws.

The diminutive holotype of *Bambiraptor* is based on an immature skeleton from Campanian strata of western Montana. Isolated bones from the same site show that larger individuals of this taxon would have been close to the same size as *Atrociraptor. Atrociraptor* can be distinguished from *Bambiraptor* (and other velociraptorines) by its deeper maxilla, by the more limited incursion of the antorbital fossa onto the maxilla anterior to the antorbital fenestra, by the sizes and relationships of the promaxillary and maxillary fenestrae, and by its isodont dentition.

An isolated maxilla from Montana MOR 553S-7.30.91.274 is morphologically similar to *Bambiraptor feinbergi* and comes from the same formation (Two Medicine Formation). For these reasons, it has been identified in this paper as cf. *Bambiraptor feinbergi.* Burnham et al. (2000) reported that *Bambiraptor* had only 10 maxillary teeth, which is significantly lower than the tooth count in MOR 553S-7.30.91.274. However, reexamination of the holotype of *Bambiraptor* (AMNH 0001) revealed that there are in fact 12 tooth positions (one tooth had broken postmortem through the posterior wall of its socket and now occupies two sockets, and two more posterior alveoli were difficult to see because of their tiny size and the fact that they were still filled with matrix). The teeth of *Bambiraptor* slope strongly backward like those of *Atrociraptor* and *Deinonychus,* whereas the bases of the maxillary teeth of *Saurornitholestes, Achillobator,* and *Velociraptor* are perpendicular to the jaw margin. Unfortunately, it is difficult to determine how strongly the teeth sloped in MOR 553S-7.30.91.274, which has lost all its functional teeth. There are other differences between *Bambiraptor* and *Saurornitholestes* in the sizes and shapes of the maxillary and promaxillary fenestrae, but these differences are not so great that they could not be accounted for by ontogenetic or individual variation. In short, it is not possible at this time to distinguish these genera on the basis of maxillae alone, and MOR 553S-7.30.91.274 could conceivably turn out to be *Saurornitholestes.* This conundrum is irrelevant to the diagnosis of *Atrociraptor* because MOR 553S-7.30.91.274 is different from the maxilla of *Atrociraptor* in the same ways that both *Saurornitholestes* and *Bambiraptor* are.

Velociraptor is closely related to *Saurornitholestes,* and the two are considered to be congeneric by some authors (Paul 1988a,b). It is not surprising then that *Velociraptor* shows the same differences as *Atrociraptor.*

Hulsanpes (Osmólska 1982) is a dromaeosaurid, but lacks cranial material and therefore cannot be compared with *Atrociraptor. Adasaurus* is known from several undescribed partial skulls and skeletons (GIN 100/20, 100/22, 100/23) from the Nemegt Formation at Büügiin Tsav (Barsbold 1983). As in *Atrociraptor,* there are 4 premaxillary, 11 maxillary, and 13 dentary teeth (GIN 100/23) in *Adasaurus*, although the teeth have anterior denticles that are the same size (20 denticles per 5 mm) as the posterior ones.

The phylogenetic analysis suggests that *Atrociraptor* is very closely related to *Deinonychus* from the early Cretaceous Cloverly Formation of Montana. The relationship may change, however, if more material is found. Perhaps most surprising about this analysis is that both taxa seem to be more derived than other dromaeosaurids, even though *Deinonychus* is one of the earliest known dromaeosaurids.

Acknowledgments. The authors are very grateful to Wayne Marshall, who discovered the holotype of *Atrociraptor marshalli.* The specimen was nicely prepared by Ken Kucher (Tyrrell Museum), and Brenda Middagh skillfully drew Figure 4.1. Figures 4.2C and 4.2D were drawn by Frankie Jackson, and the remainder by the first author. Rinchen Barsbold was kind enough to provide access to *Adasaurus* and *Velociraptor* specimens in Mongolia. TMP 94.12.844 was found by Don Brinkman, and prepared by Darren Tanke. Kevin Aulenback did some additional preparation and photographed the fossils in preparation for drawing. Darla Zelenitsky did the SEM and assembled Figures 4.3 and 4.6. The paper benefited from reviews by Dan Chure and Eric Snively. This study was supported by the Natural Sciences and Engineering Research Council of Canada (203091-98).

APPENDIX 4.1
Morphological characters used in this chapter.
0 represents the primitive state.

1. Articular, vertical columnar process on retroarticular process: 0, absent; 1, present (Currie 1995).
2. Basipterygoid process: 0, moderately long; 1, very short.
3. Braincase, endocranial cavity: 0, typical size; 1, enlarged, but temporal musculature extends onto frontals.
4. Braincase, trigeminal nerve, separation of ophthalmic branch: 0, no; 1, incipient; 2, complete (Bakker et al. 1988).
5. Dentary: 0, thick when compared to height, deep Meckelian groove; 1, thin and high with shallow MG and dental shelf (Currie 1995).
6. Dentary, lateral view: 0, tapers conspicuously anteriorly; 1, upper and ventral margins sub-parallel (Currie 1995).
7. Ectopterygoid, ventral recess: 0, absent; 1, present and comma-shaped; 2, present and sub-circular.
8. Exoccipital-opisthotic, paroccipital process: 0, no pneumatization; 1, pneumatized in proximal part.
9. Exoccipital-opisthotic, paroccipital process: 0, occipital surface of distal end oriented more posteriorly than dorsally; 1, conspicuous twist in the distal end oriented more dorsally than proximal end (Currie 1995).
10. External auditory meatus: 0, does not extend beyond level of inter-

temporal bar of postorbital and squamosal; 1, ventrolateral process of squamosal and lateral extension of paroccipital process beyond head of quadrate (Currie 1995).

11. Frontal: 0, anterior margin of supratemporal fossa straight or slightly sinuous; 1, sinusoidal with deep pit (Currie 1995).
12. Frontal, anterior part: 0, relatively broad and square with obtuse or W-shaped suture with nasals; 1, triangular, distinct acute angle.
13. Frontal, lacrimal-prefrontal contacts: 0, sutures on lateral, dorsal and/or ventral surfaces; 1, dorsal and ventral sutural surfaces connected by a vertical slot (Currie 1995).
14. Frontal, supratemporal fossa: 0, limited extension onto dorsal surfaces of frontal and postorbital; 1, covers most of frontal process of the postorbital and extends anteriorly onto dorsal surface of frontal (Currie 1995).
15. Interdental plates: 0, present and separate; 1, fused together; 2, absent (Currie 1987a).
16. Jugal: 0, does not participate in margin of antorbital fenestra; 1, participates in antorbital fenestra.
17. Jugal, pneumatic: 0, no; 1, yes.
18. Lacrimal shape in lateral view: 0, L-shaped; 1, T-shaped (Currie 1995).
19. Lacrimal, dorsal ramus: 0, dorsoventrally thick; 1, pinched and narrow; 2, absent.
20. Maxilla, anterior ramus size: 0, absent; 1, shorter anteroposteriorly than dorsoventrally; 2, as long or longer anteroposteriorly.
21. Maxilla, palatal shelf: 0, narrow; 1, wide and forms part of secondary bony palate (Makovicky and Sues 1998).
22. Maxilla: 0, no maxillary fenestra; 1, maxillary fenestra occupies less than half of the depressed area between the anterior margins of the antorbital fossa and the antorbital fenestra; 2, maxillary fenestra large and takes up most of the space between the anterior margins of the antorbital fossa and fenestra.
23. Orbit, length: 0, subequal to or longer than antorbital fenestra length; 1, shorter than antorbital fenestra length.
24. Orbit, margin: 0, smooth; 1, raised rim.
25. Palatine, recesses: 0, absent; 1, present.
26. Palatine, subsidiary fenestra between pterygoid and palatine: 0, absent; 1, present (Sues 1997).
27. Parietal, dorsal surface: 0, flat with ridge bordering supratemporal fossa; 1, parietals with sagittal crest (Russell and Dong 1993).
28. Prefrontal: 0, well-exposed dorsally; 1, reduced or absent.
29. Premaxilla, palatal shelf: 0, absent; 1, broad (Sues 1997).
30. Premaxilla, subnarial depth: 0, shallow; 1, higher than long.
31. Premaxilla, subnarial-maxillary process: 0, distal end separated from maxilla by nasal; 1, distal end separates nasal and maxilla; 2, no subnarial contact between premaxilla and nasal (Currie 1995).
32. Pterygoid flange: 0, includes major contribution from pterygoid; 1, is formed mostly by ectopterygoid.
33. Quadratojugal: 0, L-shaped; 1, Y- or T-shaped (Currie 1995).
34. Quadratojugal-Squamosal (qj-sq) contact: 0, tip of dorsal ramus of quadratojugal contacts tip of lateroventral ramus of squamosal; 1, dorsal ramus of qj does not contact squamosal; 2, broad contact between dorsal ramus of qj and lateroventral ramus of sq.
35. Splenial, forms notched anterior margin of internal mandibular fenestra: 0, absent; 1, present.

36. Splenial: 0, limited or no exposure of splenial on lateral surface of mandible; 1, conspicuous triangular process on external surface of mandible between dentary and angular (Currie 1995).
37. Surangular, horizontal shelf on lateral surface anterior and ventral to the jaw articulation; 0, absent or faint; 1, prominent and lateral; 2, prominent and pendant.
38. Teeth, maxillary, mandibular: 0, anterior and posterior denticles not significantly different in size; 1, anterior denticles, when present, significantly smaller than posterior denticles (Ostrom 1969).
39. Teeth, maxillary: 0, 13 to 15; 1, 11 or 12; 2, 8 to 10; 3, 16 or more; 4, none.
40. Teeth, maxillary: 0, almost perpendicular to jaw margin; 1, teeth inclined strongly posteroventrally (new).
41. Teeth, maxillary: 0, tooth height highly variable with gaps evident for replacement; 1, almost isodont dentition with no replacement gaps and with no more than a 30% difference in height between adjacent teeth (new).
42. Teeth, premaxillary tooth #1, size compared with crowns of premaxillary teeth 2 and 3: 0, slightly smaller or same size; 1, much smaller (Currie 1995).

APPENDIX 4.2

Data matrix used for phylogenetic analysis.

0 = primitive state; 1, 2, 3 = derived character states; ? = missing data.

Allosaurus	01020 01000 00000 01001 00100 00000 20021 02000 00
Atrociraptor	????? 1???? ????1 ????1 ??1?? ???11 1???? 1?111 1?
Bambiraptor	10101 1??11 11111 10111 1101? ?1?10 1110? 11111 01
Coelophysis	00000 01000 000?0 00000 00100 00000 20010 00000 0?
Deinonychus	1???1 11?11 1??11 11111 1101? 1?111 11101 11101 01
Dromaeosaurus	11101 12111 00111 10110 1?001 1?11? 11121 11020 00
Saurornitholestes	1?1?1 12??1 11111 1???2 1101? ???10 11121 11110 01
Troodontidae	00120 02100 01002 10100 12011 111?0 ??020 10030 10
Velociraptor	1?1?1 12111 11111 11112 1101? 111?0 11101 11110 01

References

Azuma, Y., and P. J. Currie. 1995. A new giant dromaeosaurid from Japan. *Journal of Vertebrate Paleontology* 15: 17A.

———. 2000. A new carnosaur (Dinosauria: Theropoda) from the Lower Cretaceous of Japan. *Canadian Journal of Earth Sciences* 37: 1735–1753.

Bakker, R. T., M. Williams, and P. J. Currie. 1988. *Nanotyrannus,* a new genus of pygmy tyrannosaur from the latest Cretaceous of Montana. *Hunteria* 1(5): 1–30.

Barsbold, R. 1983. Carnivorous dinosaurs from the Cretaceous of Mongolia. *Joint Soviet-Mongolian paleontological expedition, Transactions* 19: 5–120 (in Russian).

Barsbold, R., and H. Osmólska. 1999. The skull of *Velociraptor* (Theropoda) from the Late Cretaceous of Mongolia. *Acta Palaeontologica Polonica* 44: 189–219.

Brinkman, D. L., R. L. Cifelli, and N. J. Czaplewski. 1998. First occurrence of *Deinonychus antirrhopus* (Dinosauria: Theropoda) from the Ant-

lers Formation (Lower Cretaceous: Aptian-Albian) of Oklahoma. *Oklahoma Geological Survey, Bulletin* 146: 1–27.
Burnham, D. A., K. L. Derstler, P. J. Currie, R. T. Bakker, Z.-H. Zhou, and J. H. Ostrom. 2000. Remarkable new birdlike dinosaur (Theropoda: Maniraptora) from the Upper Cretaceous of Montana. *University of Kansas Paleontological Contributions* 13: 1–14.
Calvo, J. O., J. Porfiri, C. Veralli, and F. E. Novas. 2002. *Megaraptor namunhuaiquii (Novas 1998), a new light about its phylogenetic relationships.* Page 20 in Primer Congreso Latino Americano de Paleontología de Vertebrados, Resúmenes (Sociedad Paleontológica de Chile, Santiago, Chile).
Currie, P. J. 1987a. Bird-like characteristics of the jaws and teeth of troodontid theropods (Dinosauria, Saurischia). *Journal of Vertebrate Paleontology* 7: 72–81.
———. 1987b. Theropods of the Judith River Formation of Dinosaur Provincial Park. In P. J. Currie and E. H. Koster (eds.), *Fourth symposium on mesozoic terrestrial ecosystems,* pp. 52–60. Short Papers. Royal Tyrrell Museum of Palaeontology, Occasional Papers no. 3.
———. 1995. New information on the anatomy and relationships of *Dromaeosaurus albertensis* (Dinosauria: Theropoda). *Journal of Vertebrate Paleontology* 15: 576–591.
Currie, P. J., K. Rigby, Jr., and R. E. Sloan, 1990. Theropod teeth from the Judith River Formation of southern Alberta, Canada. In K. Carpenter and P. J. Currie (eds.), *Dinosaur systematics: approaches and perspectives,* pp. 107–125. New York: Cambridge University Press.
Currie, P. J., and Zhao X. J. 1993. A new carnosaur (Dinosauria, Theropoda) from the Jurassic of Xinjiang, People's Republic of China. *Canadian Journal of Earth Sciences* 30: 2037–2081.
Gibson, D. W. 1977. Upper Cretaceous and Tertiary coal-bearing strata in the Drumheller-Ardley region, Red Deer River Valley, Alberta. *Geological Survey of Canada,* Paper 76-35: 1–41.
Kirkland, J. I., D. Burge, and R. Gaston. 1993. A large dromaeosaur (Theropoda) from the Lower Cretaceous of eastern Utah. *Hunteria* 2(10): 1–16.
Makovicky, P. J., and H.-D. Sues. 1998. Anatomy and phylogenetic relationships of the theropod dinosaur *Microvenator celer* from the Lower Cretaceous of Montana. *American Museum Novitates* 3240: 1–27.
Matthew, W. D., and B. Brown. 1922. The family Deinodontidae, with notice of a new genus from the Cretaceous of Alberta. *American Museum of Natural History, Bulletin* 46: 367–385.
Maxwell, W. D., and J. H. Ostrom. 1995. Taphonomy and paleobiological implications of *Tenontosaurus-Deinonychus* associations. *Journal of Vertebrate Paleontology* 15: 707–712.
Norell, M. A., J. M. Clark, L. M. Chiappe, and D. Dashzeveg. 1997. A *Velociraptor* wishbone. *Nature* 389: 447.
Norell, M. A., and P. J. Makovicky. 1997. Important features of the dromaeosaur skeleton: information from a new specimen. *American Museum Novitates* 3215: 1–28.
———. 1999. Important features of the dromaeosaur skeleton II: information from newly collected specimens of *Velociraptor mongoliensis. American Museum Novitates* 3282: 1–45.
Novas, F. E. 1998. *Megaraptor namunhuaiquii,* gen. et sp. nov., a large-clawed, Late Cretaceous theropod from Patagonia. *Journal of Vertebrate Paleontology* 18: 4–9.

Osborn, H. F. 1924. Three new theropoda, *Protoceratops* Zone, central Mongolia. *American Museum of Natural History, Novitates* 144: 1–12.

Osmólska, H. 1982. *Hulsanpes perlei* n. gen., n. sp. (Deinonychosauria, Saurischia, Dinosauria) from the Upper Cretaceous Barun Goyot Formation of Mongolia. *Neues Jahrbuch für Geologie und Paläontologie,* Monatshefte 1982: 440–448.

Ostrom, J. H. 1969. Osteology of *Deinonychus antirrhopus,* an unusual theropod from the Lower Cretaceous of Montana. *Peabody Museum of Natural History, Bulletin* 30: 1–165.

Paul, G. S. 1988a. The small predatory dinosaurs of the mid-Mesozoic: the horned theropods of the Morrison and Great Oolite—*Ornitholestes* and *Proceratosaurus*—and the sickle-claw theropods of the Cloverly, Djadokhta and Judith River—*Deinonychus, Velociraptor* and *Saurornitholestes. Hunteria* 2(4): 1–9.

———. 1988b. *Predatory Dinosaurs of the World.* New York: Simon and Schuster.

Perle, A., M. A. Norell, and J. M. Clark. 1999. *A new maniraptoran theropod—Achillobator giganticus (Dromaeosauridae)—from the Upper Cretaceous of Burkhant, Mongolia.* Ulaan Baatar: National University of Mongolia, Geology and Mineralogy.

Russell, D. A., and Dong Z.-M. 1993. The affinities of a new theropod from the Alxa Desert, Inner Mongolia, People's Republic of China. *Canadian Journal of Earth Sciences* 30: 2107–2127.

Sues, H.-D. 1977. Dentaries of small theropods from the Judith River Formation (Campanian) of Alberta, Canada. *Canadian Journal of Earth Sciences* 14: 587–592.

———. 1978. A new small theropod dinosaur from the Judith River Formation (Campanian) of Alberta, Canada. *Zoological Journal of the Linnaean Society* 62: 381–400.

———. 1997. On *Chirostenotes,* an oviraptorosaur (Dinosauria: Theropoda) from the Upper Cretaceous of western North America. *Journal of Vertebrate Paleontology* 17: 698–716.

Swofford, D. L. 2001. PAUP: Phylogenetic analysis using parsimony (and other methods), version 4.0b6. Sinaeur Associates, Sunderland, Massachusetts.

Varricchio, D., and P. J. Currie, 1991. New theropod finds from the Two Medicine Formation (Campanian) of Montana. *Journal of Vertebrate Paleontology* 11: 59A.

Witmer, L. M. 1997. The evolution of the antorbital cavity of archosaurs: a study in soft-tissue reconstruction in the fossil record with an analysis of the function of pneumaticity. *Society of Vertebrate Paleontology Memoir* 3: 1–73.

Xu X., Wang X-L., and Wu X.-C. 1999. A dromaeosaurid dinosaur with a filamentous integument from the Yixian Formation of China. *Nature* 401: 262–266.

5. The Braincase of *Velociraptor*

MARK A. NORELL, PETER J. MAKOVICKY, AND JAMES M. CLARK

Abstract

Although dromaeosaurid braincases have been known from Asia for a long time, only a few detailed descriptions have ever been presented. Here we describe the braincase of *Velociraptor mongoliensis,* based on two newly collected specimens, and make comparisons with other dromaeosaurid taxa. The amount of variation (particularly in regard to the pattern and degree of pneumatization) that exists between these specimens is especially notable.

Introduction

Dromaeosaurid theropods have been considered to play an important role in the origin of avialans. Several phylogenetic analyses place them as the sister group of, or very closely related to, "living dinosaurs" (e.g., Gauthier 1986; Sereno 1997; Norell et al. 2001). Unfortunately, until recently dromaeosaurid theropods were poorly known, especially in regard to cranial material. The first two dromaeosaurids described, *Dromaeosaurus albertensis* and *Velociraptor mongoliensis,* are known from partial crania (Osborn 1924; Matthew and Brown 1922). *Deinonychus antirrhopus* has been well studied (Ostrom 1969a,b), but little braincase material has been recovered until recently. Brinkman et al. (1998) reported on a braincase of *Deinonychus,* but provided only a brief description of its structure. Aside from the excellent studies by Currie (1995) of *Dromaeosaurus* and by Barsbold and Osmólska (1999) of *Velociraptor,* the braincase of these animals has not been described in sufficient detail to garner character evidence, or to facilitate comparisons with other theropods.

Here we provide a comparative description of the braincase of *Velociraptor mongoliensis* that is based primarily on two specimens, GIN 100/976 and GIN 100/982, collected by field parties of the Mongolian Academy of Sciences and the American Museum of Natural History. This description supplements that provided by Barsbold and Osmólska (1999), and we have examined GIN 100/25 described by them. Comparisons with other dromaeosaurid taxa, including the recently described *Bambiraptor feinbergi* Burnham, Derstler, Currie, Bakker, Zhou, and Ostrom 2000, will be made.

Institutional abbreviations: GIN, Institute of Geology, Ulaan Baatar, Mongolia.

Anatomy of the Braincase

Figure 5.1. (opposite) Braincase of Velociraptor mongoliensis *(GIN 100/976). A, anterodorsal view; B, occipital view; C, posterodorsal view; D, ventral view; E, right lateral view; and F, left lateral view. Abbreviations: bot—basioccipital tuber; bsr—basisphenoid rostrum; dtr—dorsal tympanic recess; endo—opening for endolymphatic duct; ep—epiotic; fm—foramen magnum; fr—flocular recess; ms—metotic strut; mcv—opening for middle cerebral vein; oc—occipital condyle; or—otic recess; pop—paroccipital process; pr—prootic; ptr—posterior tympanic recess; so—supraoccipital; vvc—vertical vestibular canal; V, VI, and VII—cranial nerves 5, 6, and 7; X, XI, and XII—cranial nerves 10, 11, and 12.*

The braincase of GIN 100/976 is preserved as a single piece, broken and separated from the remainder of the skull (fig. 5.1). The braincase of GIN 100/982 (fig. 5.2) is missing the dorsal and left lateral regions; however, the floor and right wall of the braincase are exposed and were prepared. Both of these skulls represent adults, and all of the individual elements of the braincase are fused with no indication of the sutural boundaries between separate elements.

The posterior surface of the braincase in *Velociraptor mongoliensis* is a mosaic of four elements: the basioccipital ventrally, the exoccipital/opisthotics laterally, and the supraoccipital/epiotics dorsally. Together these bones surround the sub-circular foramen magnum. The fused supraoccipital and epiotics form an anterodorsally sloping surface dorsal to the foramen magnum. On either side of the midline, a shallow pit lies on this surface; presumably this is the demarcation between supraoccipital and epiotic elements. This epiotic region is slightly depressed relative to the supraoccipital component. Along the medial edge of this pit lies a slit-like posteroventrally directed canal for the middle cerebral vein, which passes through the braincase wall. Along the midline, there is a faint longitudinal ridge, and, anteriorly, the supraoccipital-epiotic plate sweeps dorsally.

The exoccipital/opisthotic forms the occipital surface lateral to the foramen magnum. This surface is nearly vertical relative to the sloping supraoccipital/epiotic surface as in *Bambiraptor.* On GIN 100/976 only the right paroccipital process is well preserved. On GIN 100/982 and AMNH 6515 the paroccipital processes are not preserved. On GIN 100/976, the paroccipital process is long and nearly vertical at its base adjacent to the foramen magnum. More laterally, the posterior surface of the paroccipital process twists slightly, giving it a slight anterodorsal-posteroventral orientation, as in *Dromaeosaurus* but not as in *Bambiraptor.* In posterior view, the paroccipital process tapers slightly lateral to its base and then expands and anteroposteriorly flattens distally. In *Deinonychus* (Brinkman et al. 1998), the paroccipital process appears to be angled ventrally. As in *Dromaeosaurus,* a shallow depression faces posteriorly, just lateral to the foramen magnum. It is separated

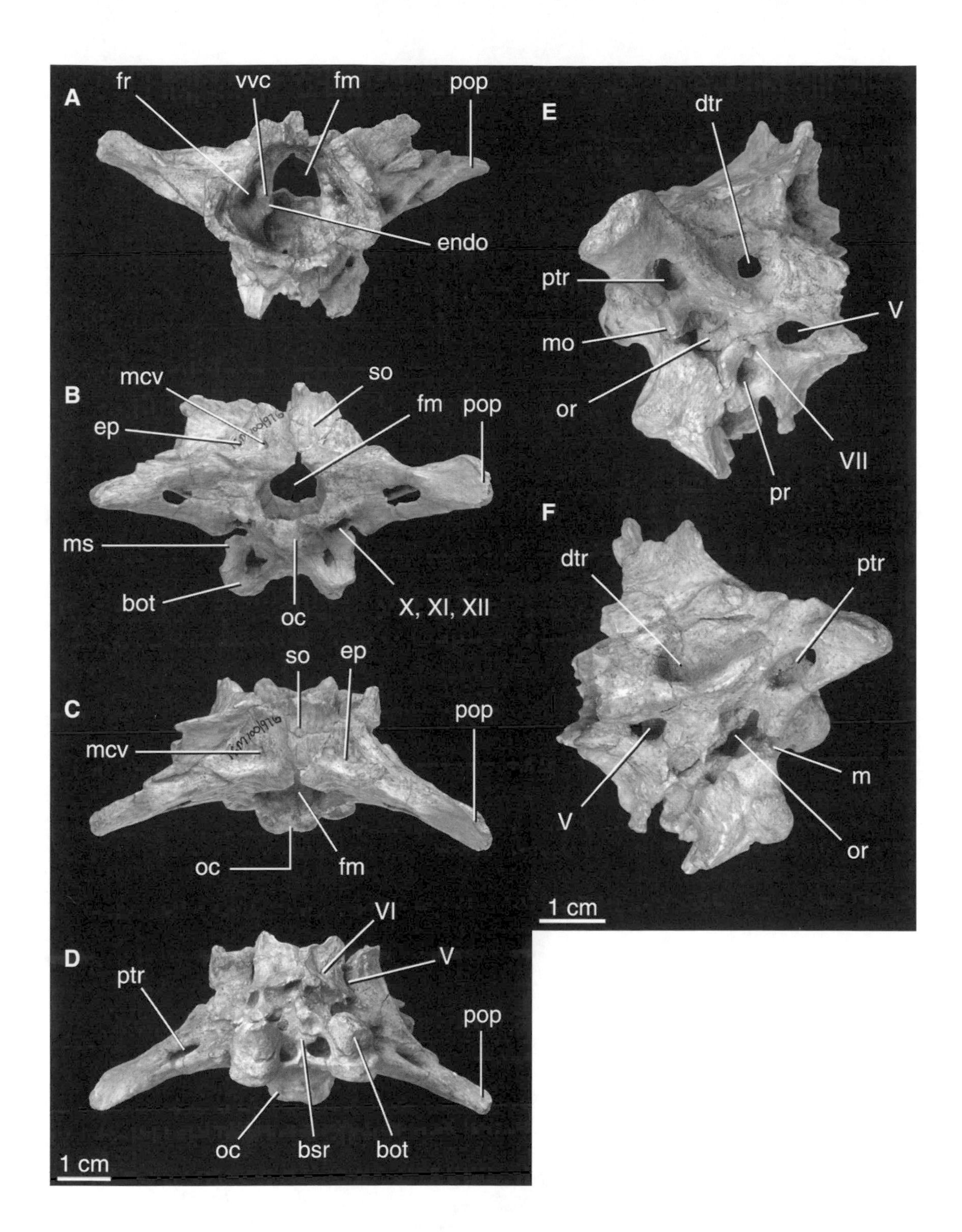
A
fr
vvc
fm
pop
endo
B
mcv
so
ep
fm
pop
ms
bot
oc
X, XI, XII
C
so
ep
pop
mcv
oc
fm
D
VI
V
ptr
pop
oc
bsr
bot
1 cm
E
dtr
ptr
V
mo
or
VII
pr
F
dtr
ptr
m
V
or
1 cm

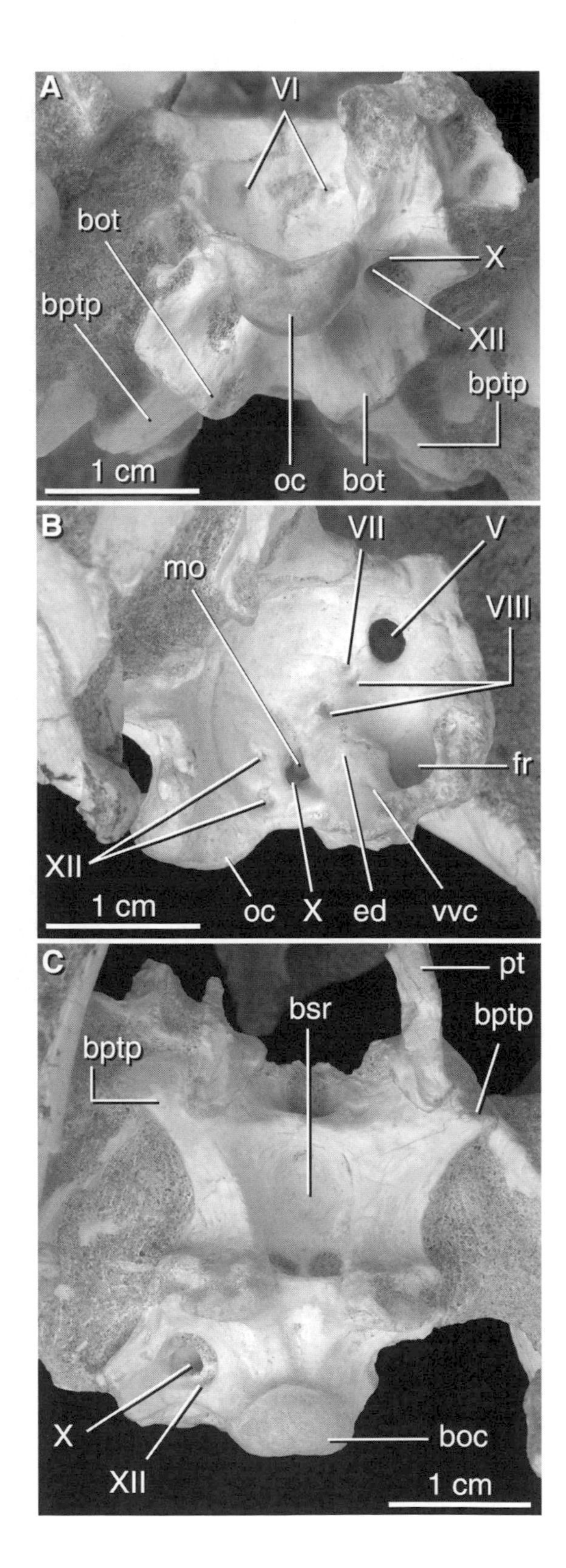

Figure 5.2. Braincase of Velociraptor mongoliensis *(GIN 100/982). A, occipital view; B, medial view (occiput at bottom); and C, ventral view. Abbreviations: bptp—basipterygoid process; bsr—basisphenoid recess; bot—basioccipital tuber; ed—opening for endolymphatic duct; fr—flocular recess; mo—metotic opening; oc—occipital condyle; pt—pterygoid; vvc—vertical vestibular canal; V, VII, and VIII—cranial nerves 5, 7, and 8; X, XII—cranial nerves 10 and 12.*

from the sloping supraoccipital surface by a small ridge that forms the angle between the dorsal and posterior surfaces of the braincase wall. This ridge continues laterally onto the paroccipital process for nearly two-thirds of its length. This depression and accompanying ridge are not present in *Bambiraptor,* but are seen in *Deinonychus* (Brinkman et al. 1998). The paroccipital process is perforated by a large foramen posteriorly on both the right and left sides of GIN 100/976. This foramen may represent erosion of a thin covering of bone over a pneumatic cavity in the paroccipital process. A foramen in this position was not reconstructed in GIN 100/25 (Barsbold and Osmólska 1999). However, examination of the specimen indicates the presence of a small opening in this region on both sides (evident in Barsbold and Osmólska 1999). Furthermore, the dorsolateral corners of both paroccipital foramina of GIN 100/976 have smooth edges that are confluent with the hollow internal cavity of the paroccipital process. This suggests that at least parts of the paroccipital openings are actual foramina and not erosional artifacts. This region of the paroccipital process of *Dromaeosaurus albertensis* is damaged, but the bone in this area is moderately thick and there is no evidence of an opening (Currie 1995). No opening is evident in *Deinonychus,* either (Brinkman et al. 1998). No indication of a posterior paroccipital foramen is apparent on the well-preserved paroccipital process of *Bambiraptor;* however, the bone is extremely thin in this area.

The paroccipital process differs greatly from that of troodontids (Norell et al. 2000; Makovicky and Norell in press) where the process is short, apneumatic, and pendant distally. Furthermore, as seen in *Byronosaurus,* the troodontid paroccipital process is oriented almost horizontally relative to the occipital surface of the braincase, as in the alvarezsaurids *Mononykus* (Perle et al. 1993) and *Shuvuuia* (Chiappe et al. 1996, 1998). Oviraptorids have long, pendant paroccipital processes, which appear to be pneumatic.

The exoccipital/opisthotic forms the dorsolateral corner of the elliptical occipital condyle. A ridge extends laterally from the exoccipital-opisthotic contribution to the condyle, and forms the roof to a large depression that opens ventrolaterally. Both branches of the hypoglossal (XII) and the vagus and accessorius (X and XI) cranial nerves were transmitted through the floor of this depression. The broad metotic strut, which extends from the paroccipital process onto the basal tubera, defines the ventrolateral extent of the posterior braincase wall, and forms the anterior wall of this depression. This large depression is not present in *Dromaeosaurus albertensis* or in *Bambiraptor feinbergi,* but is found in *Deinonychus* (Brinkman et al. 1998) and an undescribed taxon from Mongolia (Norell et al. in prep).

The occipital condyle is formed predominantly by the basioccipital. It is crescent-shaped with the dorsal surface concave and the ventral surface convex. The occipital condyle is unconstricted laterally and slightly constricted ventrally. The basioccipital tubera are short and subrectangular as in *Bambiraptor* and *Dromaeosaurus.* They project ventrolaterally, diverging from the sagittal midline as in *Bambiraptor.* A small, vertically oriented, oval depression lies on the surface of each tuber. Kurzanov (1976) interpreted these depressions as areas of attachment for the M. rectus capitis anterior in *Itemirus*. In GIN 100/976 the floor of this depression is eroded, exposing the pneumatic spaces inside the basioccipital basisphenoid. In *Dromaeosaurus*, the basal tubera are much less divergent and are nearly parallel in posterior view, and their ventral edge does not face laterally at all. Depressions are not developed on the posterior face of the tubera in *Dromaeosaurus*.

The ventral surface of the braincase is formed of the fused basioccipital and basisphenoid. The predominant feature is the presence of a large basisphenoidal recess in GIN 100/982, a common feature of theropods (Clark et al. 1994; Witmer 1990; Currie 1995). Although sutural boundaries are indistinct, this depression is bordered posteriorly by the basioccipital and laterally and anteriorly by the basisphenoid. The recess is shallower in GIN 100/982 than in *Dromaeosaurus albertensis*. In GIN 100/982, a thin web of bone connects the basal tuber forming the posterior wall of the basisphenoidal recess. Two anteriorly facing foramina perforate this posterior wall of the recess, as in GIN 100/24 (Barsbold and Osmólska 1999). In *Dromaeosaurus*, Currie (1995) reported the presence of similar foramina, which lead to blind pockets, in a slightly more lateral position; these are also present in *Bambiraptor*. However, some variation is observed in these features; as in GIN 100/976, the basal tubera are separate, due to the lack of a high posterior wall formed by the basisphenoid. In this specimen the pair of presumably pneumatic foramina that lie on the posterior wall of the basisphenoidal depression in *Dromaeosaurus albertensis* and in GIN 100/982 open instead into the flat, horizontal ventral surface of the braincase.

The remainder of the floor of the basisphenoidal recess is smooth and featureless, except for a small median foramen on the anterior wall that can be observed in GIN 100/982. Although most of this area is damaged in GIN 100/976, it also seems to possess the posterior border of this feature. The contents of this passage are unclear, but it appears also to be present in *Dromaeosaurus albertensis* and *Bambiraptor feinbergi* and it may communicate anteriorly with the subsellar depression (Witmer 1997) at the ventral base of the parabasisphenoid rostrum.

The basipterygoid processes lie anterior and lateral to the basisphenoidal depression. These are preserved on GIN 100/982 only. Unlike the condition in *Dromaeosaurus albertensis* where the basipterygoid processes lie at the same level as the basal tubera and are relatively long, they are short and project anteroventrally below this level in GIN 100/982. The basipterygoid processes taper and curve laterally at the distal end. Each has a large, round, and transverse articular surface for the pterygoid. In *Bambiraptor*, the basipterygoid processes project anteroventrally, but are not hooked. In *Dromaeosaurus* they also point anteroventrally and do not taper distally. As in other dromaeosaurids, and in contrast to troodontids, the basipterygoid processes are solid. The parabasisphenoid rostrum (cultriform process) is not preserved on either of the specimens described here.

The Lateral Wall of the Braincase

The lateral wall of the braincase is a complex region that is perforated or marked by several foramina, recesses, and pneumatic spaces. It is preserved and exposed only on GIN 100/976. Anteriorly, the large, single trigeminal cranial nerve V opening is circular and laterally directed. Anteromedial to the trigeminal opening lies a small foramen for VI (n. abducens) that is transmitted anteriorly through the floor of the braincase. Posteroventral to the posterior border of the trigeminal lies a small opening for the exit of the facial nerve (VII) from inside the braincase. Ventral to this foramen, and separated from it by a thin crista, is a large pneumatic space, the prootic recess (Witmer 1997). The prootic recess is bordered anteriorly by a weakly developed otosphenoidal crest, unlike the more developed crest of *Dromaeosaurus.* Posterior to the prootic recess lies the otic recess. The otic recess may connect with the basisphenoidal recess ventrally on GIN 100/976, although this region is somewhat fragmented. The opening for the internal carotid artery ventral to the otosphenoidal crest is not exposed on GIN 100/982 and this region is broken on GIN 100/976.

A large, deep dorsal tympanic recess excavates the lateral face of the opisthotic posterodorsal to the trigeminal foramen and dorsal to the middle ear. The border of the dorsal tympanic recess is circular, with a flat floor. The recess extends posterolaterally onto the anterior surface of the paroccipital process. This is unlike the condition seen in *Dromaeosaurus albertensis* where this region bears a shallow depression, as in *Archaeopteryx lithographica* (Walker 1985). Oviraptorids have a deep dorsal tympanic recess (Clark et al. 2002), although it is more complex than that of *Velociraptor.*

The interfenestral bar between the fenestra ovalis and the fenestra pseudorotunda is broken, and little structure is preserved within the otic recess. This region is much better preserved in an as yet undescribed dromaeosaurid from Ukhaa Tolgod, Mongolia. However, dorsal to the otic recess is the area in which the utriculus was housed, and it is proportionately larger than this area of the *Dromaeosaurus* braincase.

The Interior of the Braincase

In both GIN 100/982 and 100/976 the interior of the braincase is extremely well preserved. The presence of valecula on the interior surface of the braincase indicates that the brain was juxtaposed with the braincase over most of its outer surface, as in *Dromaeosaurus* (Currie 1995). The lateral wall of the braincase is formed by the exoccipital/opisthotic, prootic, and laterosphenoid, although no sutures can be detected between these elements (except for a part of the suture between the laterosphenoid and prootic, visible anterior to the trigeminal opening on the right side of GIN 100/982). Laterally, a single large trigeminal opening that transmitted V passes through the braincase wall, unlike the multiple openings of several theropods, such as *Troodon* and birds (Currie 1995). Medial to this, toward the midline on the braincase floor, is the anteriorly directed opening for VI. Dorsal to the floccular fossa, a groove comparable to that figured in *Dromaeosaurus* carried the middle cerebral vein (Currie 1995). The foramen leading into the canal for the vein is more posteriorly situated in GIN 100/976 than in *Dromaeosaurus,* and the canal for the middle cerebral vein through the supraoccipital/epiotic is much shorter in *Velociraptor.*

The laterosphenoid is well preserved only on the left side of GIN 100/982, where it is dislocated and incomplete dorsally. The anteromedial surface is strongly concave. The preserved portion does not include the openings for any nerves, other than the posteroventral border of the trigeminal opening. The capitate dorsal process extends laterally and is anteroposteriorly compressed. The ventral portion of the laterosphenoid is much thinner than the comparable portion preserved with the *Dromaeosaurus albertensis* holotype.

The floor of the braincase is best exposed in GIN 100/982, although it is also well preserved in GIN 100/976. The floor is formed by the basioccipital and the basisphenoid, and the suture between them is evident only as a transverse flexure between the metotic openings. The floor is deeply concave, and rises anteriorly to a nearly vertical sella turcica. In both GIN 100/982 and GIN 100/976 a small longitudinal ridge lies on the floor of the braincase, extending from just anterior to the base of the basioccipital condyle to the point where the floor of the braincase begins to dorsally rise.

Immediately posteroventral to the trigeminal opening is a small depression—the acoustic fossa—that transmits VII through the braincase wall, and the two branches of VIII (n. vestibulocochlearis) into the inner ear. The foramen for the vestibular branch of VIII lies almost directly dorsal to the exit for VII, and is removed from the more posterior exit for the cochlear branch of VIII. This configuration of nerve exits within the acoustic fossa is similar to that of *Dromaeosaurus* (Currie 1995), and differs from that of ornithomimids and *Troodon* (Makovicky and Norell 1998). Posterodorsal to the acoustic fossa is the deep floccular recess, which is surrounded by the inflated, vertical vestibular canal. This recess is oriented posteriorly and is so deep that it terminates in the paroccipital process. Posteroventrally, the vestibular canal is even more expanded into a pyramidal structure with a transverse ridge that represents the junction of the prootic and the exoccipital/opisthotic. At the apex of this prominence lies the opening for the tiny endolymphatic duct. On the dorsal edge of the vertical vestibular canal lies the endocranial opening for the middle cerebral vein. Just anterodorsal to the vertical vestibular canal lies the foramen for the anterior canal of the middle cerebral vein. This vein appears to have been within a shallow canal along the anterior margin of the vertical vestibular canal and exited the braincase through the canal piercing the supraoccipital-epiotic. The relative positions of the two canals and the groove for the middle cerebral vein appear to be rotated more posteriorly in *Velociraptor* than in *Dromaeosaurus,* such that the anterior canal is more dorsally positioned and the posterior canal is more posteriorly located in the former taxon.

The swelling around the vertical vestibular canal surrounding the floccular recess is much more pronounced in *Velociraptor mongoliensis* than it is in *Dromaeosaurus albertensis* and most other non-avialan theropods. The expansion of this area was cited as an avian feature (also present in *Protoavis*) by Chatterjee (1991).

Posterior to the floccular recess lies the metotic foramen, which transmitted cranial nerves IX–XI, and the internal jugular vein. It is obliquely oriented, and is broader posterodorsally and constricted ventrally. Two hypoglossal openings are visible ventral to the metotic fissure, the posterior one being the larger. All three foramina exit ventrally through the depression on the exoccipital adjacent to the occipital condyle.

Summary

The braincase of *Velociraptor mongoliensis* is superficially similar to that of *Dromaeosaurus albertensis* but differs in several significant aspects. In *Velociraptor*, the basal tubera are shallower, divergent, and bear depressions, presumably for the rectus capitis anterior muscles. Shallow divergent basal tubera are also present in *Bambiraptor*. The basipterygoid processes are short and extend ventrolaterally, rather than anteroventrally as in *Dromaeosaurus*. The basipterygoid processes are curved and tapering in *Velociraptor*, unlike the tab-like basipterygoid processes of *Dromaeosaurus*. As in *Deinonychus*, the hypoglossal and vagus foramina exit within a fossa on the occiput in *Velociraptor*. Such a fossa is not present in either *Dromaeosaurus* or *Bambiraptor*. Another feature shared by *Deinonychus* and *Velociraptor*, but not by *Dromaeosaurus*, is a well developed dorsal tympanic recess. The basisphenoidal recess is much more shallow in *Velociraptor* than in *Dromaeosaurus*, a condition also observed in *Bambiraptor*. A prootic recess is present anterior to the otic recess in *Velociraptor*, as in many other coelurosaurs, but not in *Dromaeosaurus*. The swelling around the vertical semicircular canal is more pronounced, and a foramen appears to be present on the occipital surface of the paroccipital process. The many differences between *Velociraptor* and *Dromaeosaurus* are surprising given the supposed close relationship of these two genera. Similarities between these two taxa (and *Itemirus* [Kurzanov 1976]) include the relative positions of the cranial nerve exits within the acoustic fossa, when compared to ornithomimids and troodontids (Makovicky and Norell 1998). Unfortunately the neurocranial anatomy of other dromaeosaurid taxa, such as *Deinonychus*, *Saurornitholestes*, and *Bambiraptor*, is too poorly known to allow for extensive comparisons. More detailed data on maniraptoran braincases are required to determine character polarities and potential derived braincase characters within the group.

Acknowledgments. We thank the Mongolian Academy of Sciences, especially D. Dashzeveg, the leader of the Mongolian side of the expedition, and our many friends with whom we have worked in Mongolia over the years. In New York we thank the members of our field crews over the years. Amy Davidson prepared GIN 100/982, and Michael Ellison illustrated this paper. Funding that made this work possible came from the National Geographic Society, the Epply Foundation,

IREX, the Philip McKenna Foundation, the Frick Laboratory Endowment, and the Department of Vertebrate Paleontology at the American Museum of Natural History.

References

Barsbold, R., and H. Osmólska. 1999. The skull of *Velociraptor* (Theropoda) from the Late Cretaceous of Mongolia. *Acta Palaeontologica Polonica* 44(2): 189–219.

Brinkman, D. L., R. L. Cifelli, and N. J. Czaplewski. 1998. First occurrence of *Deinonychus antirrhopus* (Dinosauria: Theropoda) from the Antlers Formation (Lower Cretaceous: Aptian—Albian) of Oklahoma. *Oklahoma Geological Survey Bulletin* 146, 27 pp.

Burnham, D. A., K. L. Derstler, P. J. Currie, R. T. Bakker, Z. Zhou, and J. H. Ostrom. 2000. Remarkable new birdlike dinosaur (Theropoda: Maniraptora) from the Upper Cretaceous of Montana. *University of Kansas Paleontological Contributions* 13: 1–14.

Chatterjee, S. 1991. Cranial anatomy and relationships of a new Triassic bird from Texas. *Philosophical Transactions of the Royal Society, London* (B) 309: 395–460.

Chiappe, L., M. A. Norell, and J. M. Clark. 1996. Phylogenetic position of *Mononykus* (Aves: Alvarezsauridae) from the Late Cretaceous of the Gobi Desert. *Memoirs of the Queensland Museum* 39(3): 557–582.

———. 1998. The skull of a relative of the stem-group bird *Mononykus*. *Nature* 392: 275–278.

Clark, J. M., A. Perle, and M. Norell. 1994. The skull of *Erlicosaurus andrewsi,* a Late Cretaceous "segnosaur" (Theropoda: Therizinosauridae) from Mongolia. *American Museum Novitates* 3113: 1–39.

Clark, J. M., M. A. Norell, and R. Barsbold. 2001. Two new oviraptorids (Theropoda: Oviraptorosauria) late Cretaceous Djadokhta Formation, Ukhaa Tolgod, Mongolia. *Journal of Vertebrate Paleontology* 21: 209–213.

Clark, J. M., M. A. Norell, and T. Rowe. 2002. Cranial anatomy of *Citipati osmolskae* (Theropoda, Oviraptorosauria), and a reinterpretation of the *Oviraptor philoceratops* holotype. *American Museum Novitates* 3364: 1–24.

Currie, P. J. 1995. New information on the anatomy and the relationships of *Dromaeosaurus albertensis* (Dinosauria: Theropoda). *Journal of Vertebrate Paleontology* 15(3): 576–591.

Gauthier, J. A. 1986. Saurischian monophyly and the origin of birds. *Memoirs of the California Academy of Sciences* 8: 1–55.

Kurzanov, S. M. 1976. Braincase structure of the carnosaur *Itemirus* gen. nov. and some problems of the cranial anatomy of dinosaurs. *Paleontological Journal* 10: 361–369.

Makovicky, P. J., and M. A. Norell. 1998. A partial ornithomimid braincase from Ukhaa Tolgod (Upper Cretaceous, Mongolia). *American Museum Novitates* 3247: 1–16.

———. Submitted for publication. Troodontidae. In D. Weishampel et al. (eds.), *The Dinosauria* (2d ed.). Berkeley: University of California Press.

Matthew, W. D., and B. Brown. 1922. The family Deinodontidae, with notice of a new genus from the Cretaceous of Alberta. *Bulletin of the American Museum of Natural History* 46: 367–385.

Norell, M. A., P. J. Makovicky, and J. M. Clark. 2000. A new troodontid from Ukhaa Tolgod, Late Cretaceous, Mongolia. *Journal of Vertebrate Paleontology* 20: 7–11.

Norell, M. A., J. M. Clark, and P. J. Makovicky. 2001. Phylogenetic relationships among coelurosaurian theropods. In J. Gauthier and L. F. Gall (eds.), *New perspectives on the origin and early evolution of birds,* pp. 49–68. New Haven: Peabody Museum of Natural History, Yale University.

Osborn, H. F. 1924. Three new Theropoda, *Protoceratops* zone, central Mongolia. *American Museum Novitates* 144: 1–12.

Ostrom, J. H. 1969a. Osteology of *Deinonychus antirrhopus,* an unusual theropod from the Lower Cretaceous of Montana. *Peabody Museum of Natural History, Bulletin* 30: 1–165.

———. 1969b. A new theropod dinosaur from the Lower Cretaceous of Montana. *Postilla* 128: 1–17.

Perle, A., M. A. Norell, L. M. Chiappe, and J. M. Clark. 1993. Flightless bird from the Cretaceous of Mongolia. *Nature* 362: 623–626.

Sereno, P. 1997. The origin and evolution of dinosaurs. *Annual Review of Earth and Planetary Sciences* 25: 435–489.

Walker, A. D. 1985. The braincase of *Archaeopteryx*. In M. K. Hecht, J. H. Ostrom, G. Viohl, and P. Wellnhofer (eds.), *The beginnings of birds.* Proceedings of the International *Archaeopteryx* Conference, pp. 123–134. Eichstätt: Freunde des Jura-Museums.

Witmer, L. 1990. The craniofacial air sac system of Mesozoic birds (Aves). *Zoological Journal of the Linnean Society* 100: 327–378.

———. 1997. The evolution of the antorbital cavity of archosaurs: a study in soft-tissue reconstruction in the fossil record with an analysis of the function of pneumaticity. *Journal of Vertebrate Paleontology, Memoir* 3: 1–73.

6. A Theropod (Dromaeosauridae, Dinosauria) Sternal Plate from the Dinosaur Park Formation (Campanian, Upper Cretaceous) of Alberta, Canada

Stephen J. Godfrey and Philip J. Currie

Abstract

A right sternal plate of a small velociraptorine (Dromaeosauridae) theropod, recovered from the Dinosaur Park Formation (Campanian, Upper Cretaceous) of Alberta, Canada, exhibits several birdlike features. The length of the sternum is equal to or greater than the width of the paired sternal plates, a wide and well-developed coracoidal sulcus is present along nearly all its cranial margin, and there is a clustering of at least three costal processes on the craniolateral half of the element for the reception of three sternal ribs. Primitive features of the sternum include its paired structure with no development of a sternal keel.

Introduction

Fossilized vertebrate remains from Dinosaur Provincial Park are numerous and remarkably diverse taxonomically. Approximately 130 species, of which 38 are dinosaurs, have been described from the Dinosaur Park Formation (Mid-Campanian, Upper Cretaceous) of the

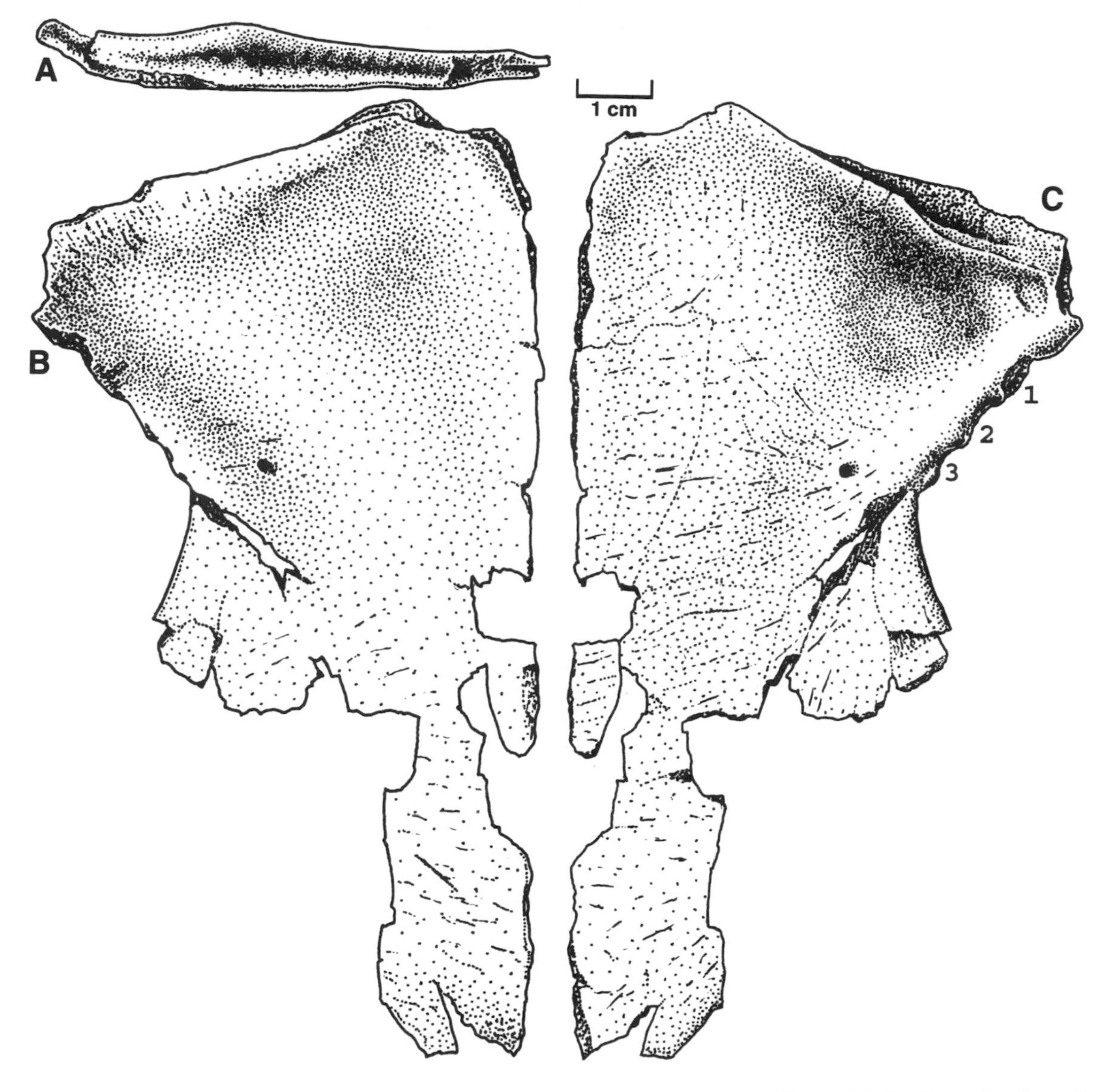

Figure 6.1. Specimen drawings of a right sternal plate (TMP 92.36.333). A, view of the craniolateral end, principally the coracoidal sulcus; B, ventral view; and C, dorsal (visceral) view. Scale bar equals 1 cm. #1–#3 mark the sites for the attachment of three sternal ribs.

Judith River Group within the park (Eberth et al. 2001). Two dromaeosaurids—*Dromaeosaurus* and *Saurornitholestes*—have been collected from these beds, although no complete skeletons have ever been found. During the 1992 field season, most of an isolated right sternal plate (fig. 6.1) was collected from Bone Bed 47 (UTM [WGS 84] 12U 0,463,976; 5,620,718) within the park. Although ornithischia sterna are relatively common, theropod sternals, especially those of small theropods, are exceedingly uncommon. In spite of its not being complete, the excellent preservational quality and rarity of this element prompted this note.

Institutional Abbreviations: GIN, Mongolian Institute of Geology, Ulaan Baatar, Mongolia; TMP, Royal Tyrrell Museum of Palaeontology, Drumheller, Alberta, Canada.

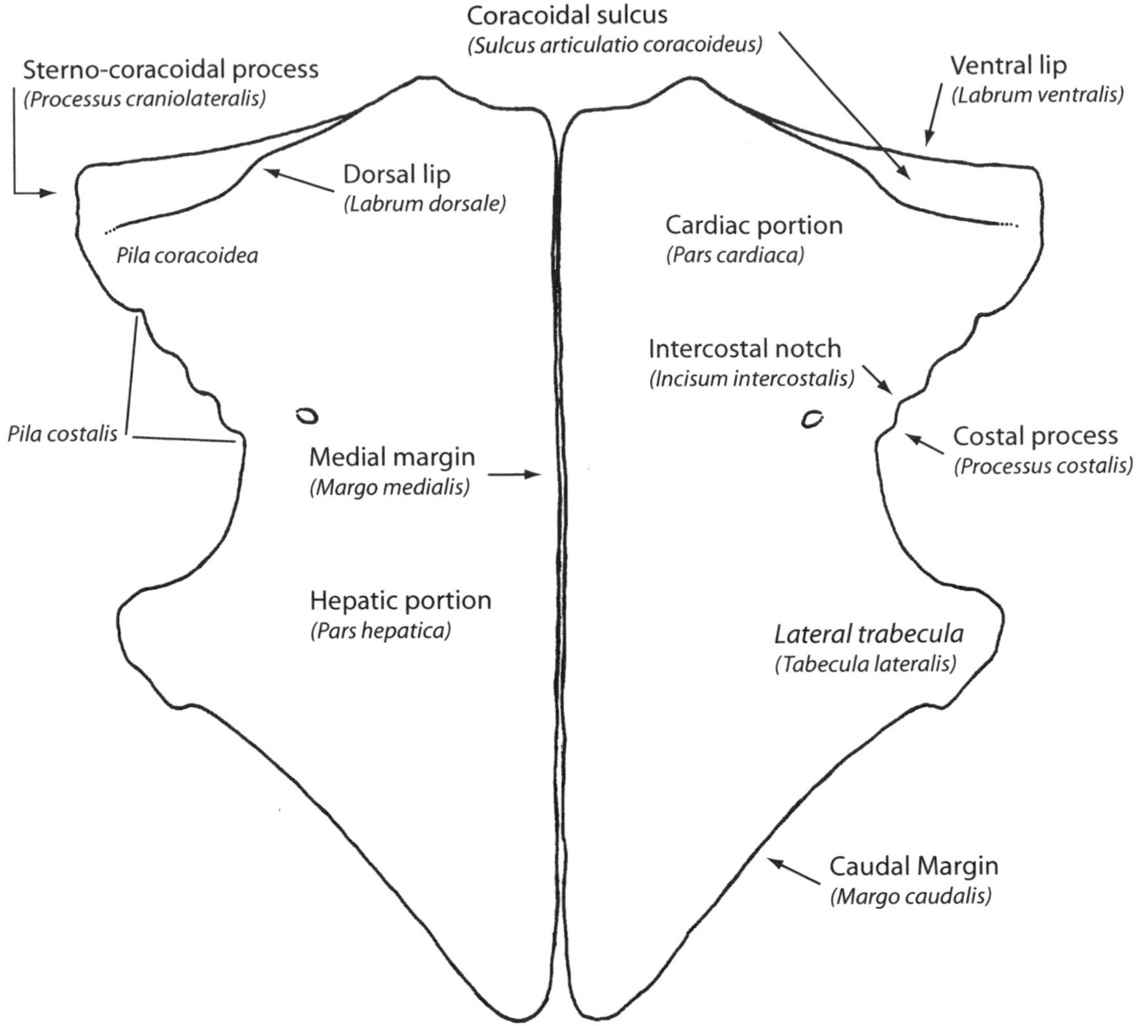

Figure 6.2. A restoration of the sternum based on TMP 92.36.333, drawn in dorsal (visceral) view. Both halves of the sternum are shown lying in the horizontal plane. In life, they were probably angled slightly dorsolaterally, resulting in a reduction in their width.

Description

The sandstone in which the bone was encased was removed mechanically and no restoration was performed (fig. 6.1). Although incompletely preserved, the bone in its present condition appears to have suffered little postmortem distortion. Approximately three-quarters of the element was preserved. It consists of an essentially flat, roughly triangular plate of bone conspicuously indented along its lateral margin. The sternal plate is 127 mm long, 66.5 mm wide, and 9 mm deep through the *pila coracoidea*. (Latin terminology, as applied to avian sterna, is taken from Baumel et al. 1979, and Baumel and Witmer 1993.)

The nearly straight medial margin of TMP 93.36.333 indicates that the two sternal plates abutted throughout most, if not all, of their lengths. At the midpoint in its length, the bone is only 2 mm thick along its medial margin. Both cranial and caudal ends of the medial margin thicken to 3.5 mm.

The cranial end of the bone is dorsoventrally thickened to accommodate the coracoidal sulcus (figs. 6.1A, 6.2). Except for a small damaged segment mediocranially, this sulcus occupies the entire cranial end of TMP 92.36.333. The bone within the sulcus is pitted and pustulate, reminiscent of cartilage-covered bone. In dorsal view, the dorsal lip of the coracoidal sulcus obscures the ventral lip medially. Laterally, however, the ventral lip is clearly seen as the dorsal lip curves dorsally to form the *pila coracoidea.* The sterno-coracoidal process is incompletely preserved in TMP 92.36.333. As one passes caudolaterally along the coracoidal sulcus, the bone curves dorsally in the area of the sterno-coracoidal process.

Immediately caudal to the *pila coracoidea* the visceral surface of the bone is marked by a conspicuous circular concavity. Medial to this concavity, the cardiac portion of the sternum is slightly convex. In dorsal view, the bone medial to the cardiac portion is depressed slightly before thickening dorsoventrally along the medial margin of the bone. Within the cardiac portion of the sternal plate, the contours on the ventral surface of TMP 92.36.333 parallel those of the dorsal surface.

Immediately caudal to the sterno-coracoidal process, the dorsoventrally thickened bone (5.5 mm) curves caudomedially. This area, the *pila costalis,* presents on its lateral face three well-developed costal processes, each separated by an intercostal notch. We assume that each costal process articulated with one sternal rib. The centers of adjacent costal processes are separated by about 7 mm. The caudal-most process forms a smooth, roughly circular swelling. The other two processes appear to have been damaged, exposing the spongiosa within. From the top of the *pila costalis,* the bone dips forward into the aforementioned circular depression within the cardiac portion of the sternum. A small foramen pierces TMP 92.36.333 medial to the caudal-most costal process. On the ventral surface, the bone also thins medially, forming a shallow trough that extends from the sterno-coracoidal process back toward the hepatic region of the sternum. The bone also thins from 5.5 mm to 2 mm. Behind the costal processes, the bone curves abruptly caudolaterally. Were the bone complete in this area, it would likely have formed the lateral trabecula as seen in *Velociraptor* (Barsbold 1983) and birds (Baumel and Witmer 1993).

Although the caudal half of the sternal plate is incomplete, short segments of both the medial and caudal margins are intact, and were helpful in producing the restoration (fig. 6.2). The bone of the hepatic portion of the sternum is only 1.5 to 2 mm thick.

Discussion

Of the known sterna of small theropod dinosaurs, TMP 92.36.333 most closely resembles those of *Bambiraptor feinbergi* (Burnham et al. 2000) and *Velociraptor mongoliensis* (Barsbold 1983; Norell and Makovicky 1997, 1999). Close examination of the sternals of *Velociraptor mongoliensis* (GIN 100/25, 100/985, and a specimen with the field number GIN 940728) confirms the presence of all the anatomical features displayed by TMP 92.36.333. The length of the sternum in each of TMP 92.36.333 and *Velociraptor* is equal to or greater than the combined width of the paired sternal plates. The sterna of oviraptorosaurs (Barsbold 1983; Clark et al. 1999) and *Gorgosaurus* (Lambe 1917; but see Brochu 2002), on the other hand, are shorter craniocaudally and wider laterally than are those of *Bambiraptor, Velociraptor,* or TMP 92.36.333. This would lead us to suggest that short and wide sterna represent the primitive pattern from which the long dromaeosaur sternum is derived. Elongation of the sternum may be the only feature that renders the dromaeosaurid sternum more avian than that of oviraptorids. The most conspicuous non-avialan features of TMP 92.36.333 include its paired structure with no development of a sternal carina. That there are paired sternal plates in an immature Mongolian dromaeosaur, GIN 100/985 (Norell and Makovicky 1997), leaves open the possibility that TMP 92.36.333 was also derived from a juvenile.

Acknowledgments. We would like to thank Darren Tanke (Royal Tyrrell Museum) for collecting and preparing this element. Furthermore, we wish to extend our gratitude to Dr. Mark Norell and one anonymous reviewer whose comments in review improved the final version.

References

Barsbold, R. 1983. [Carnivorous dinosaurs from the Cretaceous of Mongolia.] [Joint Soviet-Mongolian Paleontological Expedition, Transactions] 19: 5–120 [in Russian].

Baumel, J. J., A. S. King, A. M. Lucas, J. E. Breazile, and H. E. Evans (eds.). 1979. *Nomina anatomica avium.* London: Academic Press.

Baumel, J. J., and L. M. Witmer. 1993. Osteologia. In J. J. Baumel, A. S. King, J. E. Breazile, H. E. Evans, and J. C. Vanden Berge (eds.), *Handbook of avian anatomy: nomina anatomica avium* (2d ed.), pp. 45–132. Cambridge, Mass.: Publications of the Nuttall Ornithological Club.

Brochu, C. A. 2002. Osteology of *Tyrannosaurus rex:* Insights from a nearly complete skeleton and high-resolution computed tomographic analysis of the skull. *Journal of Vertebrate Paleontology Memoir* 7: 1–138.

Burnham, D. A., K. L. Derstler, P. J. Currie, R. T. Bakker, Z-H. Zhou, and J. H. Ostrom. 2000. Remarkable new birdlike dinosaur (Theropoda: Maniraptora) from the Upper Cretaceous of Montana. *University of Kansas Paleontological Contributions* 13: 1–14.

Clark, J. M., M. A. Norell, and L. M. Chiappe. 1999. An oviraptorid skeleton from the Late Cretaceous of Ukhaa Tolgod, Mongolia, preserved in an avian-like brooding position over an oviraptorid nest. *American Museum Novitates* 3265: 1–36.

Eberth, D. A., P. J. Currie, D. B. Brinkman, M. J. Ryan, D. R. Braman, J. D. Gardner, V. D. Lam, D. N. Spivak, and A. G. Neuman. 2001. Alberta's dinosaurs and other fossil vertebrates: Judith River and Edmonton Groups (Campanian-Maastrichtian). In C. L. Hill (ed.), *Guidebook for the field trips: Society of Vertebrate Paleontology 61st Annual Meeting, Mesozoic and Cenozoic Paleontology in the Western*

Plains and Rocky Mountains, Museum of the Rockies, Occasional Paper 3: 49–75.

Lambe, L. M. 1917. The Cretaceous theropodous dinosaur *Gorgosaurus. Geological Survey of Canada,* Memoir 100: 1–84.

Norell, M. A., and P. J. Makovicky. 1997. Important features of the dromaeosaur skeleton: information from a new specimen. *American Museum Novitates* 3215: 1–28.

———. 1999. Important features of the dromaeosaur skeleton II: information from newly collected specimens of *Velociraptor mongoliensis. American Museum Novitates* 3282: 1–45.

7. Avian Traits in the Ilium of *Unenlagia comahuensis* (Maniraptora, Avialae)

Fernando E. Novas

Abstract

The pelvic girdle of *Unenlagia comahuensis* (Río Neuquén Formation, Turonian, NW Patagonia) seems to be intermediate in many anatomical respects between that of dromaeosaurid maniraptorans and that of early birds (e.g., *Rahonavis ostromi, Archaeopteryx lithographica, Confuciusornis sanctus*). This paper remarks on some details of the *Unenlagia comahuensis* ilium that concern the acquisition of avian characteristics.

In the ilia of *Saurornitholestes langstoni* (MOR 660), *Deinonychus antirrhopus* (AMNH 3015, MCZ 4371), and *Velociraptor mongoliensis* (GIN 100/985), a precursor of the avian processus supratrochantericus is expressed as a faint transverse enlargement of the posterodorsal iliac margin. *Unenlagia comahuensis,* instead, is more derived toward the avian condition, because this process is more prominent and triangular. Also, it is associated with a ridge that runs posterodorsally along the lateral surface of the ilium above the acetabulum.

The rounded preacetabular blade of *Unenlagia comahuensis* resembles that of early birds in extending anteriorly, in contrast with less derived coelurosaurs in which the cranial margin of the ilium is straight (e.g., oviraptorosaurs, ornithomimids) or notched (e.g., *Tyrannosaurus rex,* most dromaeosaurids).

Contrary to recent claims suggesting that dromaeosaurids are more derived than *Unenlagia comahuensis* in the loss of a medial antiliac shelf for M. cuppedicus, it must be said that this crest is present and prominent in *Deinonychus antirrhopus* AMNH 3015, *Saurornitholestes langstoni* (MOR 660), *Velociraptor mongoliensis* (GIN 100/985), and *Rahonavis ostromi* (UA 5686).

The morphological features of the ilium strengthen the hypothesis that *Unenlagia comahuensis* is closer to birds than to dromaeosaurids.

Introduction

Recent discoveries of almost complete dromaeosaurid skeletons (*Bambiraptor feinbergi, Sinornithosaurus millenii, Velociraptor mongoliensis*) have led to a consistent increase in knowledge of maniraptoran anatomy. For this reason, Laurasian dromaeosaurids became a prominent source of information for comparisons with early birds, and play a central role in interpretations on avian origins (Norell and Makovicky 1997, 1999; Xu et al. 1999; Burnham et al. 2000).

However, information from other taxa represented by *less* complete specimens should not be overlooked. One example is *Unenlagia comahuensis,* from the Río Neuquén Formation (presumably Turonian; Novas 1997) of NW Patagonia. This dinosaur was originally described as a derived maniraptoran theropod more closely related to birds than dromaeosaurids (Novas and Puerta 1997; Novas 1999). Albeit incomplete, the holotype specimen of *Unenlagia comahuensis* is highly informative for revealing the sequence of evolutionary novelties leading to birds.

Forster et al. (1998) interpreted *Unenlagia comahuensis* either as the sister group of birds, or as a member of a basal avian subgroup that also included *Archaeopteryx lithographica* and *Rahonavis ostromi.* Norell and Makovicky (1999) were less inclined to consider *Unenlagia comahuensis* more closely related to birds than dromaeosaurids, pointing out that dromaeosaurids like *Velociraptor mongoliensis* share derived features with avians (for example, a reduced antiliac shelf) that are not seen in the Patagonian taxon. Moreover, in regards to the phylogenetic relationships of *Unenlagia comahuensis,* Norell and Makovicky (1999) expressed the opinion that "its status as a phylogenetically intermediate link between non-avialan dinosaurs and avialans needs to be examined within a larger phylogenetic context, one that includes *Velociraptor mongoliensis* and related taxa."

Following that suggestion, this chapter describes one of the most birdlike bones of *Unenlagia comahuensis,* the ilium, with the aim of clarifying the evolutionary transformations in the dinosaur-bird transition.

Institutional Abbreviations: AMNH, American Museum of Natural History, New York, U.S.; BMNH, British Museum of Natural History, London, England, U.K.; CM, Carnegie Museum, Pittsburgh, Pennsylvania, U.S.; GIN, Geological Institute, Mongolian Academy of Sciences, Ulaan Baatar, Mongolia; MCF PVPH, Museo Municipal "Carmen Funes," Plaza Huincul, Neuquén, Argentina; MCZ, Museum of Comparative Zoology, Cambridge, Massachusetts, U.S.; MOR, Museum of the Rockies, Montana, U.S.; UA, University of Antananarivo, Madagascar; USNM, United States National Museum, Washington, D.C., U.S.; YPM, Yale Peabody Museum, New Haven, Connecticut, U.S.

Materials and Methods

The following systematic and nomenclatural statements form the framework for this study: Tyrannosauridae is included within Coelurosauria (Bakker et al. 1988; Novas 1991, 1992; Holtz 1994, 1995). Alvarezsaurids are not interpreted as birds, but are considered as the sister taxon of a clade composed of therizinosauroids, oviraptorosaurs, troodontids, dromaeosaurs, *Unenlagia comahuensis,* and birds (Novas and Pol 2002). Dromaeosauridae is assumed to be a monophyletic group (defying this hypothesis exceeds the scope of the present paper). Following the original definition by Gauthier (1986), the node-based group Maniraptora is composed of the common ancestor of Aves, dromaeosaurs and all its descendants (contra Holtz 1994, 1995; and Padian et al. 1999). I adopt recommendations by Padian et al. (1999) in applying the term Aves to the group formed by *Archaeopteryx lithographica,* extant birds, and their common ancestor. Following this, the stem-group name Avialae (originally coined by Gauthier 1986), is reserved for a more inclusive group that covers Aves plus all these maniraptorans that are more closely related to birds than to Dromaeosauridae. Thus, *Unenlagia comahuensis* enters within Avialae, if the hypothesis of sister group relationships with birds (Novas and Puerta 1997; Novas 1999; Forster et al. 1998) proves to be correct.

Specimens examined during this study were: *Albertosaurus libratus* (AMNH 5458, 5664); *Archaeopteryx lithographica* (London and Berlin specimens); *Archaeornithomimus asiaticus* (AMNH 6566, 6567, 6570); *Bambiraptor feinbergi* (AMNH 001); *Chirostenotes pergracilis* (BMNH R 10903); *Deinonychus antirrhopus* (AMNH 3015, MCZ 4371, YPM 5205, 5206, 5236); *Ornithomimus velox* (USNM 2164, AMNH 5421, 5339); *Rahonavis ostromi* (UA 8656); *Saurornitholestes langstoni* (MOR 660); *Tyrannosaurus rex* (CM 9380 and AMNH 5017); *Unenlagia comahuensis* (MCF PVPH 78); *Velociraptor mongoliensis* (GIN 100/985).

The Ilium of *Unenlagia comahuensis*

The ilium of *Unenlagia comahuensis* (fig. 7.1) is a low and long bone, trapezoidal in lateral view, with the preacetabular wing dorsoventrally deeper than the postacetabular one. The preacetabular blade is lobe-shaped and cranially expanded, while the postacetabular blade is short and sharply acuminate. The dorsal margin of the ilium is anteroposteriorly convex, with a strong posterodorsal inflexion that marks the cranial end of the postacetabular wing (Hutchinson 2001). In contrast, the dorsal margin of the postacetabular blade is concave in side view.

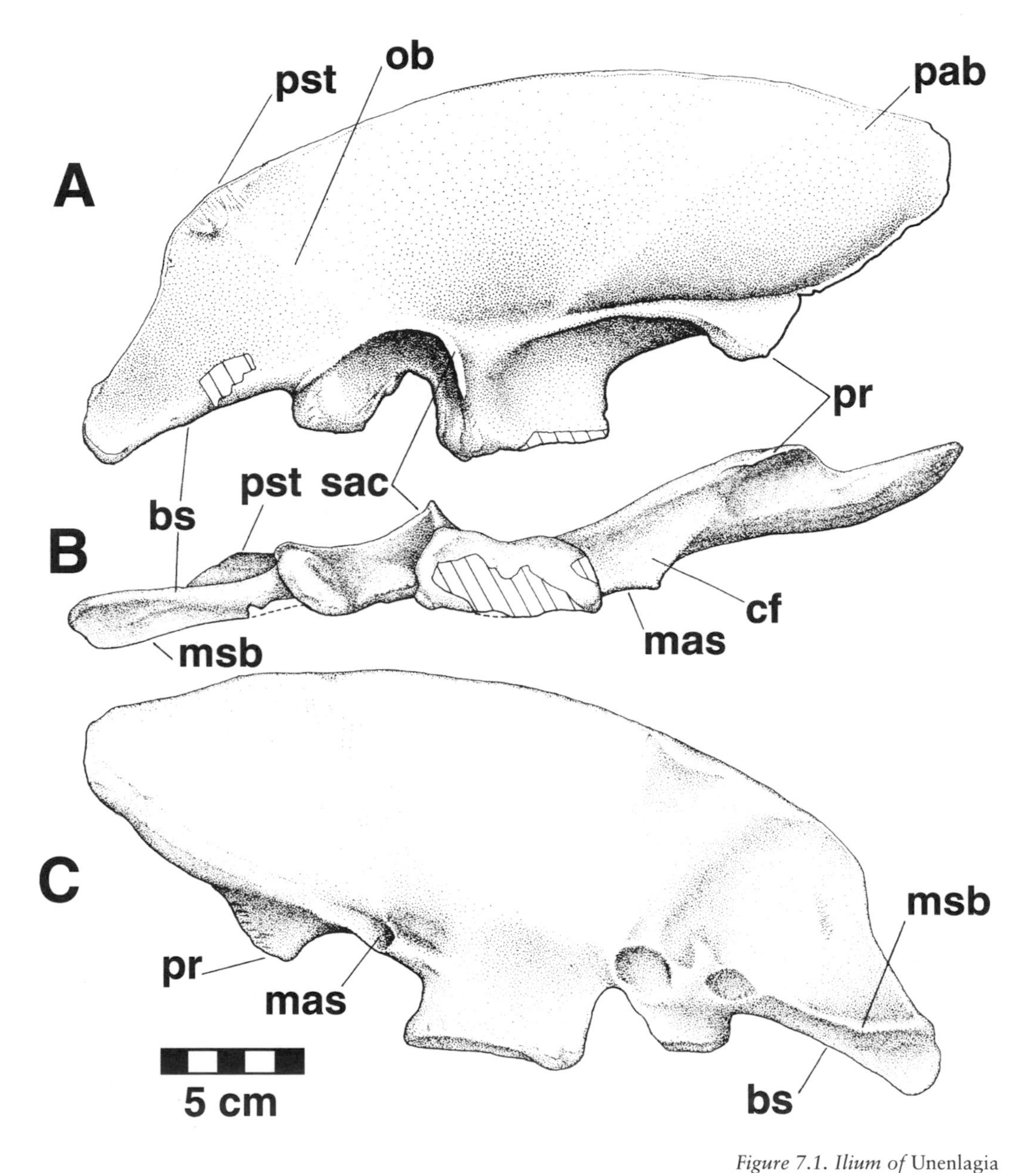

Figure 7.1. Ilium of Unenlagia comahuensis *(MCF PVPH 78). A, lateral; B, ventral; and C, medial views. Abbreviations: bs, brevis shelf; cf, fossa for M. cuppedicus; mas, medial antiliac shelf; msb, medial shelf of brevis fossa; ob, oblique ridge; pab, preacetabular blade; pr, "pointed process" of preacetabular blade; pst, processus supratrochantericus; sac, supracetabular crest. Scale bar is 5 cm.*

The external surface of the ilium is almost smooth, except for a notable posterodorsal triangular prominence, decorated by well-marked grooves. An oblique ridge extends between this prominence and the dorsal margin of the acetabulum, and presumably separated areas for muscle origin.

A deeply excavated and transversely wide cuppedicus fossa, which is C-shaped in transverse section, occupies the ventral edge of the preacetabular ala. The fossa is craniocaudally elongate and is laterally and medially bounded by two strong horizontal ridges (the medial

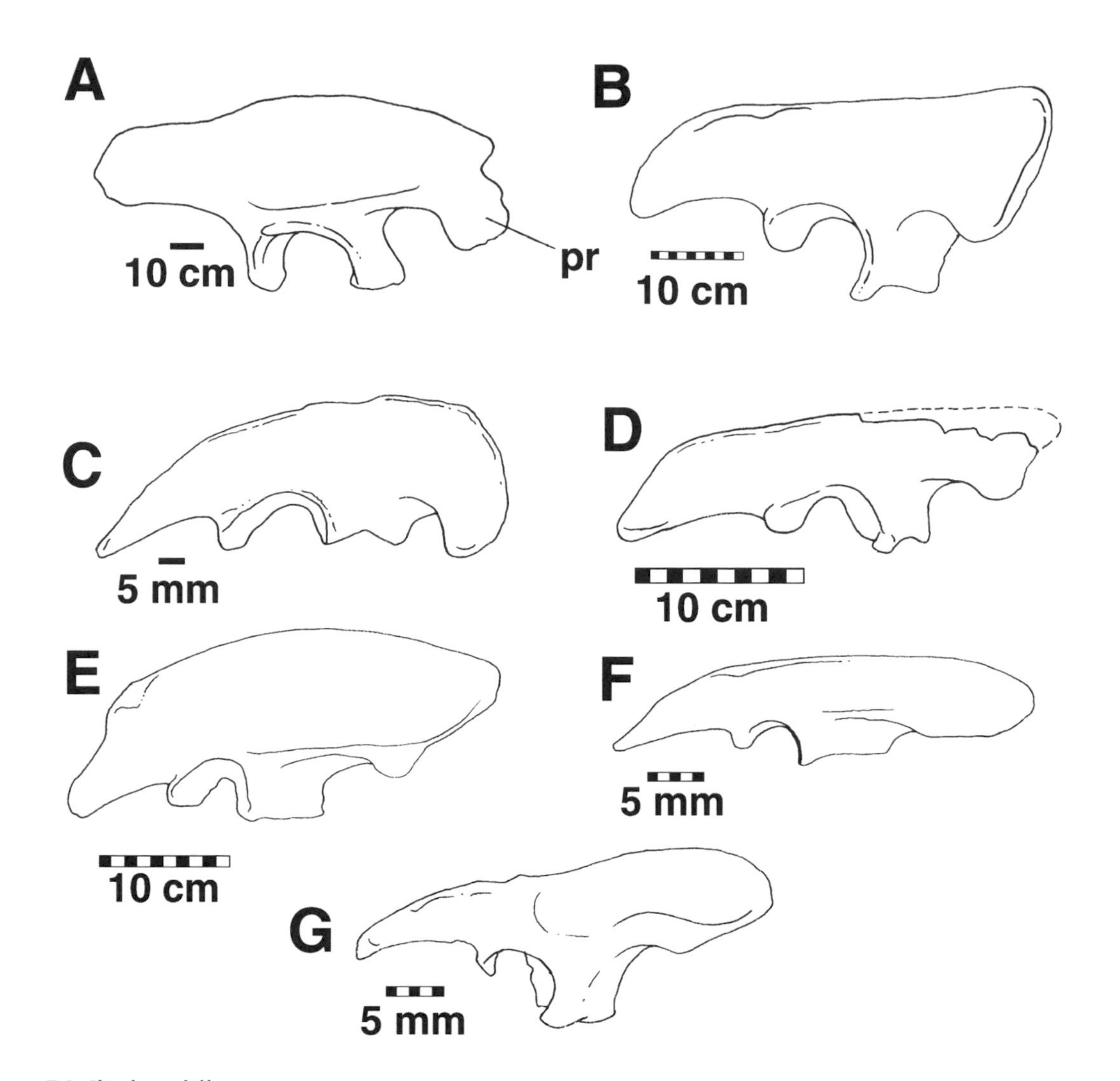

Figure 7.2. Ilia from different coelurosaurian taxa in right lateral view.
A, Tyrannosaurus rex;
B, Deinonychus antirrhopus;
C, Bambiraptor feinbergi;
D, Velociraptor mongoliensis;
E, Unenlagia comahuensis;
F, Rahonavis ostromi; *and*
G, Archaeopteryx lithographica.
Abbreviations:
pr, "pointed process" of preacetabular blade.
Scale is 10 cm, except C, F, and G, in which scale bar is 5 mm.

ridge, or medial antiliac shelf, in fig. 7.1). In agreement with other authors (Norell and Makovicky 1999), it is hypothesized here that the M. cuppedicus not only originated from the ventral margin of the preacetabular ala, but also from most of the lateral surface of the pubic peduncle, which is deeply excavated and is cranially contiguous with the cuppedicus fossa.

The supracetabular crest of the ilium forms a strong craniodorsal ridge. The pubic peduncle is craniocaudally expanded, and projects more ventrally than the ischiac peduncle. The ischiac peduncle is anterolaterally to medioposteriorly compressed and transversely expanded. The antitrochanter is represented by a slightly raised, craniolaterally faced surface, which lacks well-defined perimeters.

The brevis fossa is transversely narrow and craniocaudally short. It is bounded laterally by the brevis shelf, which merges with the ischiac peduncle, and medially by the medial shelf.

Character Analysis

Among recent papers on avian origins in which *Unenlagia comahuensis* was considered (Burnham et al. 2000; Chiappe et al. 1999; Forster et al. 1998; Norell and Makovicky 1997, 1999), those by Norell and Makovicky have included extensive comments on the anatomy of the Patagonian taxon. Their observations on the ilium of the Patagonian taxon were particularly addressed to the morphology of the preacetabular wing and the shape of the fossa for insertion of the cuppedicus muscle.

In regard to the ilium of *Velociraptor mongoliensis,* Norell and Makovicky (1997) comment that the preacetabular wing forms in its posteroventral corner a "small pointed process" (fig. 7.2) that is "proportionately much smaller than the anteroventral hook of the ornithomimid or tyrannosaurid ilium." On the basis of that observation, Norell and Makovicky (1997) conclude that "the dromaeosaur condition approaches that of Avialae [sensu Gauthier 1986] where there is no trace of an anteroventral process." The pointed process of the preacetabular ilium was reduced in avian evolution, as it is seen in basal birds (e.g., *Rahonavis ostromi, Archaeopteryx lithographica,* Enantiornithes, and *Patagopteryx deferrarisii*), but the dromaeosaur condition does not approach birds more than *Unenlagia comahuensis.* The development of the "pointed process" looks proportionally the same in both *Velociraptor mongoliensis* and *Unenlagia comahuensis.* To quantify the degree of development of such a "pointed process" (fig. 7.2), its depth was compared with the distance taken from the tip of this pendant process to the mid-point of the arc formed by the supracetabular rim. The proportions obtained show that the "pointed process" is strongly developed in *Allosaurus fragilis, Struthiomimus altus,* and *Tyrannosaurus rex,* while it is considerably less developed in *Unenlagia comahuensis, Velociraptor mongoliensis, Archaeopteryx lithographica,* and *Rahonavis ostromi.* An intermediate group is made up of the dromaeosaurids *Deinonychus antirrhopus, Saurornitholestes langstoni,* and *Bambiraptor feinbergi* (fig. 7.2). This demonstrates that the "pendant process" is not uniformly reduced toward the avian condition within Dromaeosauridae.

In addition, Norell and Makovicky (1997, 1999) listed some characters that purportedly place *Unenlagia comahuensis* outside a group formed by dromaeosaurids and avialans. For example, these authors state: "Another derived condition of the ilium shared by Avialae and dromaeosaurids is the absence of a medial antiliac shelf for the origin of the M. cuppedicus muscle. . . . Such a shelf is primitive for tetanurans . . . and is present in other advanced, small theropods such as ornithomimids, troodontids, *Microvenator celer,* and *Unenlagia comahuensis*" (Norell and Makovicky 1997). They conclude that the absence of a medial shelf along the antiliac blade in dromaeosaurids is an advanced character shared with birds, but is not present in other maniraptoran lineages. Because this shelf is present in *Unenlagia co-*

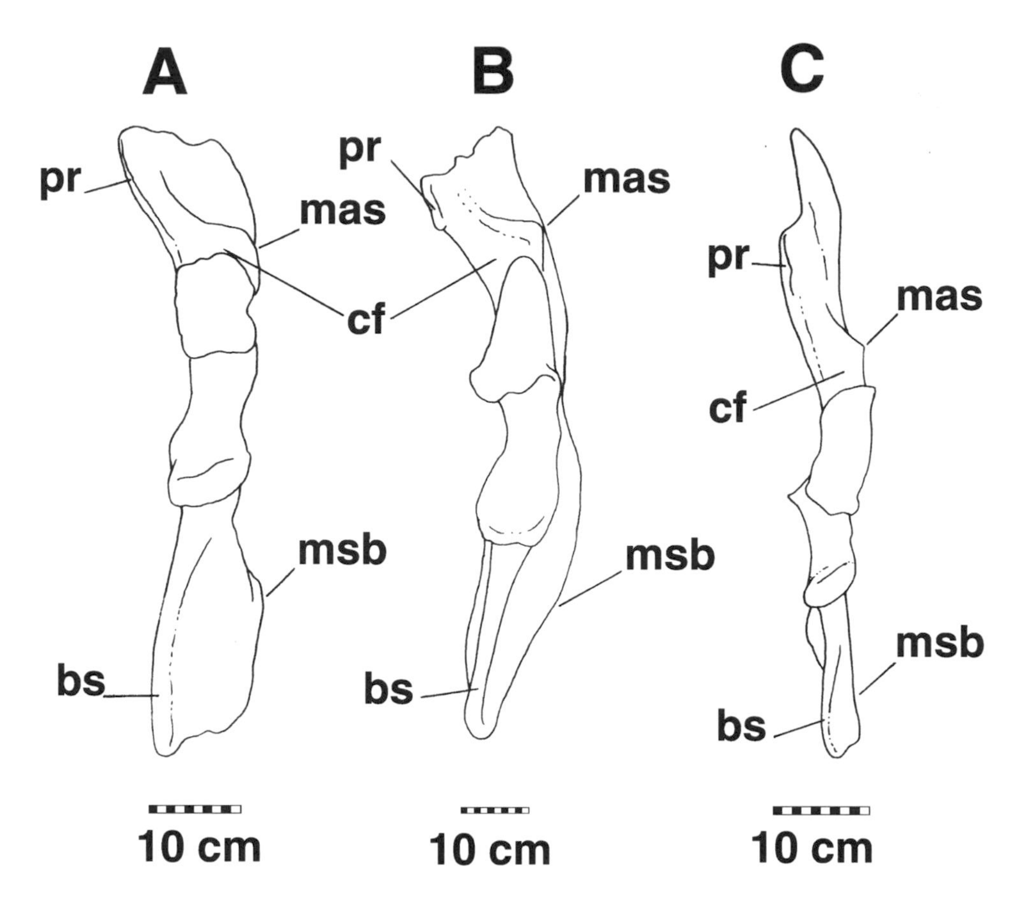

Figure 7.3. Right ilia from coelurosaurian theropods in ventral view.
A, Tyrannosaurus rex;
B, Deinonychus antirrhopus; *and*
C, Unenlagia comahuensis *(MCF PVPH 78).*
Abbreviations:
bs, brevis shelf;
cf, fossa for M. cuppedicus;
mas, medial antiliac shelf; msb, medial shelf of brevis fossa;
pr, "pointed process" of preacetabular blade.
Scale bar is 10 cm. (A, redrawn from Osborn 1917; B, from Novas 1996)

mahuensis (Novas and Puerta 1997), they conclude that in this respect *Unenlagia* is more primitive than dromaeosaurids. However, in a more recent paper, Norell and Makovicky (1999) interpret the cuppedicus fossa on the antiliac shelf of *Unenlagia comahuensis* as an artifact, "because the right ilium of the specimen shows a clear fracture and displacement of the ventral preacetabular margin in this region. The left ilium is more similar to that of *Velociraptor mongoliensis* and *Archaeopteryx lithographica* in that it has a laterally deflected ventral margin on the antilium and an expansion of this area onto the pubic peduncle."

Examination of the specimen clearly shows that the cuppedicus fossa on the antiliac shelf of *Unenlagia comahuensis* is not an artifact due to postmortem distortion. A cuppedicus fossa is present on both left and right ilia of the type specimen (MCF PVPH 78). Although it is true that the right ilium has a fracture on the preacetabular wing, it does not affect the morphology of the cuppedicus fossa. In both ilia, it is laterally and medially bound by strong longitudinal and sub-parallel ridges. Therefore, not only the left ilium of *Unenlagia comahuensis*, but also the right one resemble those of *Velociraptor mongoliensis* and *Archaeopteryx lithographica*. In summary, the right ilium of *Unenlagia comahuensis* shows, as does the left ilium, a laterally deflected ventral margin on the antilium with an expansion of this area onto the pubic peduncle. These are features that occur in many other coelurosaurian theropods, including dromaeosaurids.

Norell and Makovicky (1997) emphasized the presence of a primitive feature—a medial antiliac shelf (fig. 7.3)—in the ilium of *Unenlagia comahuensis,* which is purportedly absent in *Velociraptor mongoliensis,* other dromaeosaurids, and early birds. There are, however, some points of contention with regard to this character. First, it is not correct that dromaeosaurids lack a medial antiliac shelf for M. cuppedicus. On the contrary, the medial shelf is present (and is prominent) in AMNH 3015 (*Deinonychus antirrhopus,* Novas 1996) and MOR 660 (*Saurornitholestes langstoni*). In these taxa, as in *Unenlagia comahuensis,* the medial shelf of the cuppedicus fossa forms a strong, transversely rounded ridge, which is medially inflected when observed in ventral aspect (fig. 7.3). The morphology described is not unique to *Unenlagia comahuensis* and dromaeosaurids, but is also present in more basal coelurosaurs such as *Tyrannosaurus rex* (CM 9380 and AMNH 5017), *Albertosaurus libratus* (AMNH 5458, 5664), *Ornithomimus velox* (USNM 2164, AMNH 5421, 5339), *Archaeornithomimus asiaticus* (AMNH 6576), and *Chirostenotes pergracilis* (BMNH R 10903), as well as in basal birds like *Rahonavis ostromi* (UA 8656). Furthermore, both ilia of at least one specimen of *Velociraptor mongoliensis* (GIN 100/985) have medial crests homologous with those of other coelurosaurs. This condition also seems to be present in another specimen (GIN 100/986) of the same taxon (Norell and Makovicky 1999). *Velociraptor mongoliensis* apparently differs from other coelurosaurs in that the medial shelf of the cuppedicus fossa describes a rounded or continuous convexity in ventral view, instead of the sharp inflection seen in, for example, *Deinonychus antirrhopus* and *Unenlagia comahuensis.*

A second aspect to be considered concerns the condition of the cuppedicus fossa in basal birds. A deeply excavated fossa for the M. cuppedicus seems not to be present in *Archaeopteryx lithographica* (fig. 7.5) and Enantiornithes (Walker 1981). However, this fossa is inferred from the excavated lateral surface of the pubic peduncle and the ventral margin of the preacetabular wing, which is set off laterally from the rest of the ilium. Unfortunately, the presence or absence of the condition of the medial antiliac shelf is difficult to determine in early avians because of their two-dimensional preservation. For example, in most (if not all) of the available specimens of *Archaeopteryx lithographica,* the cranial portions of ilia are unexposed in both medial or ventral aspects, or are lost. The possible exception is the Eichstätt specimen (Wellnhofer 1974), but it is not possible to check for the presence of a cuppedicus fossa. Our knowledge of the cuppedicus fossa is no better for basal birds more derived than *Archaeopteryx lithographica.* The ilium is broken in *Iberomesornis romerali* (Sanz et al. 1988), and the medial side of the ilium was neither described nor illustrated for *Confuciusornis sanctus* (Chiappe et al. 1999) and Enantiornithes (Walker 1981).

However, a medial antiliac crest is present, at least, in *Rahonavis ostromi,* which closely resembles *Unenlagia comahuensis* (fig. 7.4).

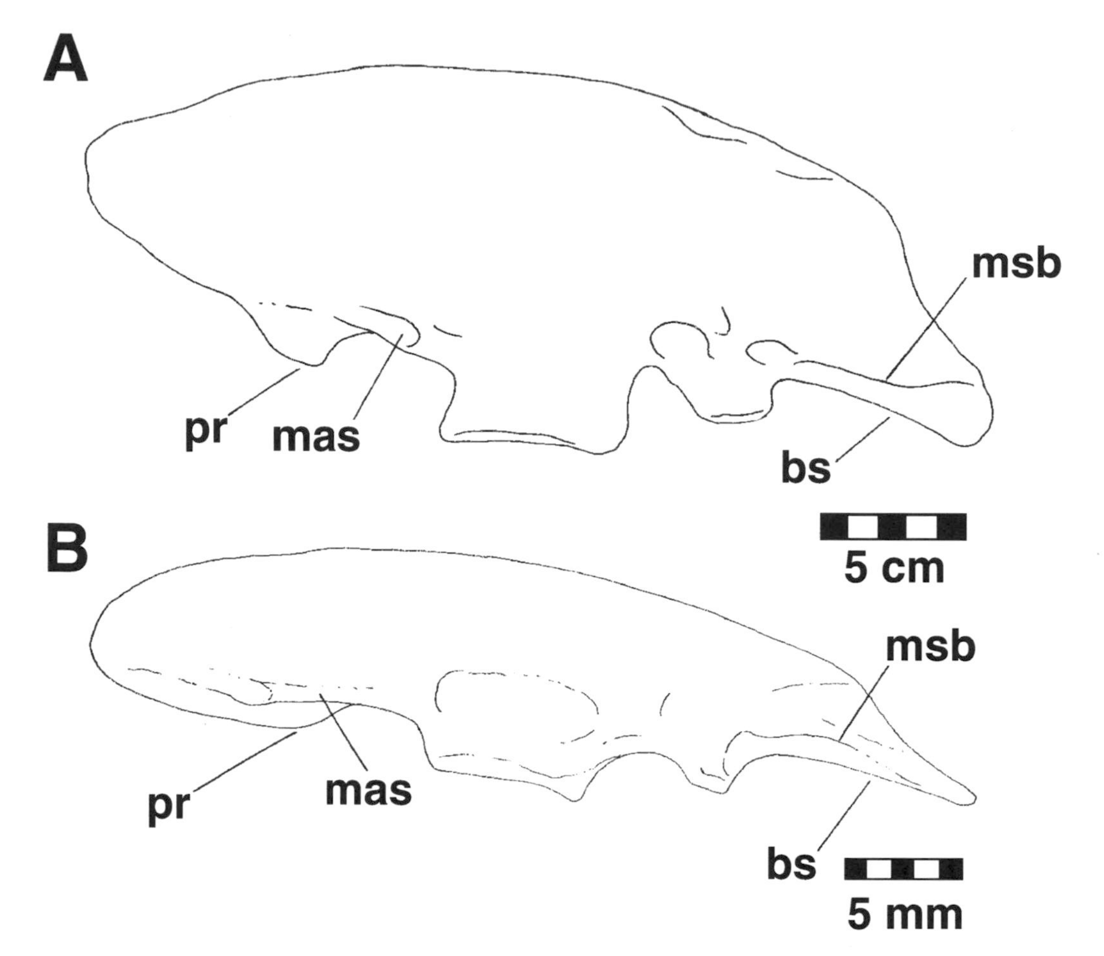

Figure 7.4. Right ilia of avialian theropods in medial view.
A, Unenlagia comahuensis *(MCF PVPH 78); and*
B, Rahonavis ostromi *().*
Abbreviations:
bs, brevis shelf;
mas, medial antiliac shelf;
msb, medial shelf of brevis fossa;
pr, "pointed process" of preacetabular blade.
Scale bar is 5 cm for A, and 5 mm for B. (Modified from Forster et al. 1998)

There is evidence that the fossa for the M. cuppedicus is lacking in more derived birds. Chiappe (1996) interpreted the loss of cuppedicus fossa as a synapomorphy of the clade formed by *Patagopteryx deferrarisii* plus Ornithurae, and Novas (1996, 1997) noted the absence of this fossa also in the non-avian alvarezsaurid maniraptorans (Novas 1996, 1997; Novas and Pol 2002).

In summary, the ilia of *Velociraptor mongoliensis* and all other dromaeosaurids known to date do not have characters that are more derived toward the avian condition than *Unenlagia comahuensis.*

Unenlagia comahuensis shares with birds many apomorphic features that are absent in dromaeosaurids, and which are enumerated and discussed below.

1) *Lobe-shaped preacetabular wing.* The lobe-shaped contour of the preacetabular blade of *Unenlagia comahuensis* results from cranial expansion of the blade, rostral to the anteroventral corner (the "pendant process" of Norell and Makovicky 1997) of the preacetabular ala, as well as the rounded contour of the cranial edge of ilium. This derived condition is shared by *Unenlagia comahuensis*

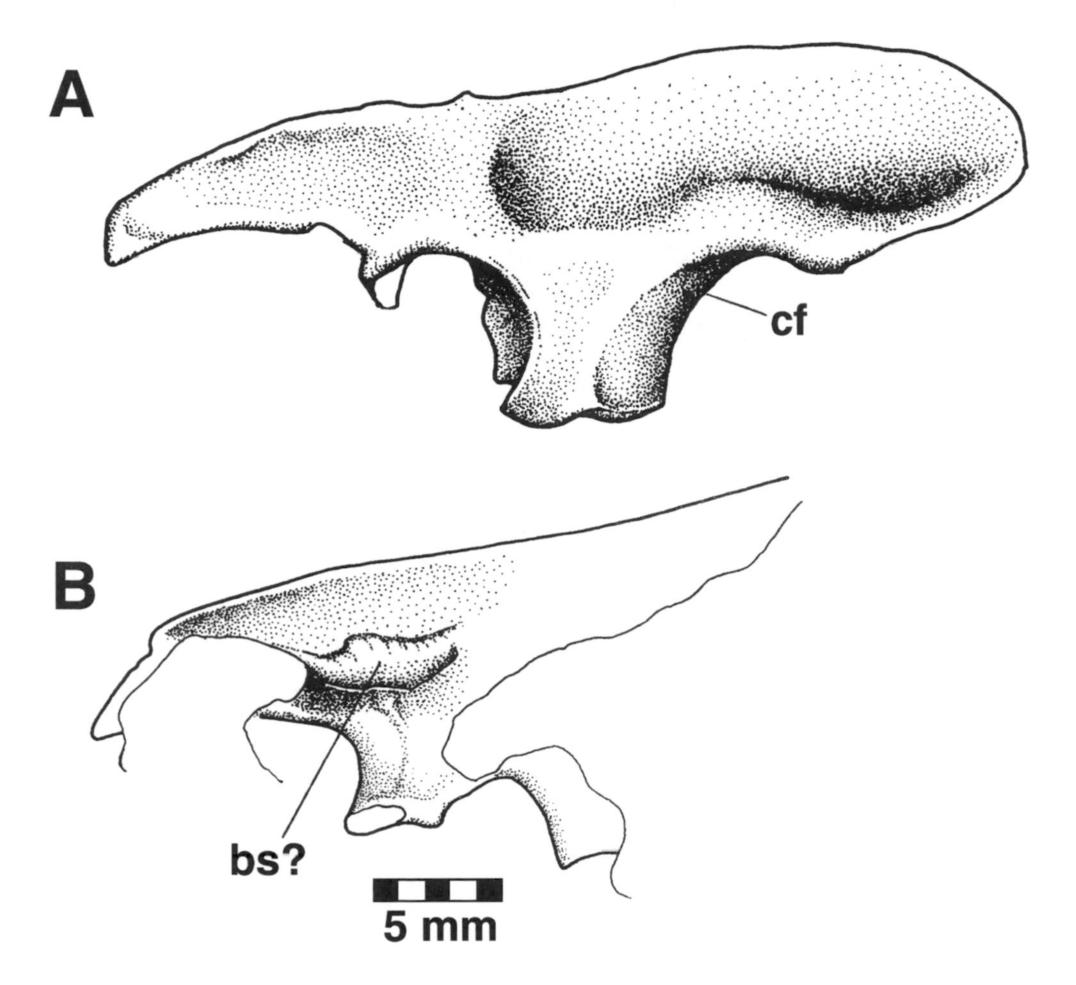

Figure 7.5. Ilia of Archaeopteryx lithographica *(London specimen). A, right ilium in lateral aspect, and B, and left one in medial view.*
Abbreviations:
bs?, brevis shelf?;
cf, fossa for M. cuppedicus.
Scale bar is 5 mm.

and early birds (e.g., *Rahonavis ostromi, Archaeopteryx lithographica, Confuciusornis sanctus, Patagopteryx deferrarisii*). In contrast, less derived coelurosaurs have a preacetabular wing that is not cranially projected with respect to the anteroventral corner, and the cranial margin of ilium is straight (oviraptorosaurs, ornithomimids), notched (*Tyrannosaurus rex;* Osborn 1917), or has a craniodorsal projection separated by a notch from the pendant process as in *Deinonychus antirrhopus* (Ostrom 1969, 1976), *Achillobator giganticus* (Perle et al. 1999) and *Saurornitholestes langstoni* (MOR 660). In *Bambiraptor feinbergi* (Burnham et al. 2000) the anterior margin of the ilium is nearly convex, but it is not cranially projected with respect to the anteroventral corner (pendant process). Unfortunately, the majority of ilia of the Mongolian specimens illustrated by Norell and Makovicky (1997, 1999) do not preserve the complete cranial edge of the ilium.

2) *Presence of processus supratrochantericus.* Located on the posterodorsal corner of the ilium of *Unenlagia comahuensis* is a conspicuous triangular prominence that is associated with an oblique ridge running from the prominence toward the acetabulum (fig.

7.1). The morphology and position of these structures suggest that they are homologous with the processus supratrochantericus and associated oblique ridge of living birds, which divide areas of origin of the M. iliotibialis from the M. iliofemoralis (Baumel and Witmer 1993). In the dromaeosaurids *Saurornitholestes langstoni* (MOR 660), *Deinonychus antirrhopus* (AMNH 3015, MCZ 4371), and *Velociraptor mongoliensis* (Makovicky and Norell 1997, 1999), there exists a precursor of the processus supratrochantericus, expressed as a transverse thickening of the posterodorsal portion of the iliac margin. Norell and Makovicky (1999) note that a poorly defined vertical ridge is present in GIN 100/982 above the acetabulum to the anterior and posterior muscle fossa, but point out that it is absent on other specimens. Moreover, in the tiny dromaeosaur *Bambiraptor feinbergi* a supratrochanteric process is absent (Burnham et al. 2000; AMNH 001). In comparison with these dromaeosaurid specimens, the processus supratrochantericus of *Unenlagia comahuensis* is better defined (anteroposteriorly restricted), triangular, and is clearly associated with the posterodorsally projected caudal ridge of the lateral iliac surface. This set of features in *Unenlagia comahuensis* is closer to the avian condition than any known dromaeosaurid, including *Velociraptor mongoliensis*. A similar supratrochanteric process is also seen in the following basal birds: *Archaeopteryx lithographica* (Berlin specimen), Enantiornithes (Walker 1981), *Rahonavis ostromi* (Forster et al. 1998), and *Confuciusornis sanctus* (Chiappe et al. 1999).

3) *Brevis fossa reduced.* In dromaeosaurids, *Unenlagia comahuensis,* and early avians (*Rahonavis ostromi, Archaeopteryx lithographica, Confuciusornis sanctus*), the postacetabular blade has a more basic shape than in ancestral non-avian Theropoda: it is transversely flat and its caudoventral margin is cranially continuous with the ischiac peduncle (fig. 7.2). Also in contrast to less derived theropods, the medial blade of the brevis fossa of *Unenlagia comahuensis, Deinonychus antirrhopus,* and basal birds rises in a more dorsal position with respect to the ventral margin of the postacetabular blade. Consequently, the brevis fossa of maniraptoran theropods is shallow, oriented almost medioventrally, and not exposed in lateral view.

The brevis fossa of *Unenlagia comahuensis* (fig. 7.1B) is fan-shaped in ventral aspect (with its caudal end transversely expanded), a condition that superficially resembles the shapes in basal coelurosaurs (tyrannosaurids, ornithomimids). This fan-shaped condition contrasts with those of *Deinonychus antirrhopus* (Novas 1996) and *Rahonavis ostromi* (UA 8656) in which the brevis fossa forms an elliptical contour in ventral view. However, the brevis fossa of *Unenlagia comahuensis* is minute, in sharp contrast with that in other coelurosaurs, including dromaeosaurids, in which the fossa is transversely wider and craniocaudally longer. Notably, the medial horizontal ridge of the brevis fossa of *Unenlagia comahuensis* is restricted caudal to the ischiac peduncle, as in *Rahonavis ostromi* (fig. 7.4). However, this contrasts with *Deino-*

nychus antirrhopus (AMNH 3015) in which the ridge is not only more prominent (medially projected), but also reaches a more cranial position, ending at the level of the center of the acetabulum. Norell and Makovicky (1997) state that the postacetabular wing forms a wide and flat brevis shelf in a number of undescribed specimens of *Velociraptor mongoliensis,* including GIN 100/981 and GIN 100/982. In short, the reduction observed in the brevis fossa of *Unenlagia comahuensis* is interpreted as a derived feature, shared with early avians, that is absent in other theropods, including dromaeosaurids.

Although reduced, the brevis fossa of *Unenlagia comahuensis* is excavated and has robust delimiting borders. In contrast, the area corresponding with the brevis fossa is represented by an almost flat surface in early avians such as *Rahonavis ostromi* (Forster et al. 1998). The brevis fossa is absent in *Confuciusornis sanctus* (Chiappe et al. 1999). However, the situation is perplexing in *Archaeopteryx lithographica,* because the medial side of the left ilium of the London specimen (fig. 7.5) has a prominent ridge that resembles the condition described above for *Deinonychus antirrhopus.*

Aside from the derived features that the ilium of *Unenlagia comahuensis* shares with Aves (preacetabular wing lobe-shaped, processus supratrochantericus present, and brevis fossa reduced), there still remain some other characteristics which merit brief discussion.

a) *Postacetabular blade dorsally concave.* The dorsal margin of the postacetabular blade of *Unenlagia comahuensis* is concave. Only *Rahonavis ostromi,* among maniraptoran theropods, shares with the Patagonian taxon a similar condition (figs. 7.2 and 7.4). In the remaining coelurosaurs known, the dorsal margin of the postacetabular wing of ilium is convex (Burnham et al. 2000; Norell and Makovicky 1997, 1999; Ji et al. 1998; Zhou et al. 2000; Novas 1996) or straight (Walker 1981; Chiappe et al. 1999). However, the dorsal margin of the ilium of *Archaeopteryx lithographica* varies from convex (Wellnhofer 1974) to concave (Wellnhofer 1993).

b) *Acetabulum medially constricted.* The medial closure of the acetabular aperture has been cited as an avian feature by Martin (1983, 1991). Novas and Puerta (1997) agreed with this opinion, but showed that this condition is also present in *Unenlagia comahuensis.* In this taxon, as in *Archaeopteryx lithographica* and later birds, the medial acetabular opening is medially constricted compared to dromaeosaurids and less derived theropods, suggesting that this feature may constitute an avialan synapomorphy. However, the acetabulum in *Rahonavis ostromi* (Forster et al. 1998) is wide open, with no indication of medial closure. This demonstrates that the distribution of this interesting feature is variable and must be considered in the context of a wider phylogenetic analysis.

c) *Pubic peduncle posteroventrally oriented.* Paul (1984) cited presence of a posteroventrally oriented pubic peduncle as a derived feature uniting *Deinonychus antirrhopus* with birds. A posteroventrally oriented pubic peduncle is present, for example, in *Archaeopteryx lithographica* (Wellnhofer 1993), *Confuciusornis sanctus* (Chiappe et al. 1999), and in enantiornithine birds (Walker 1981). *Unenlagia comahuensis* has a pubic peduncle that is anteroventrally oriented, a condition that may be interpreted as evidence for placing the Patagonian taxon outside of a clade formed by Dromaeosauridae plus Aves. However, *Deinonychus antirrhopus* seems to be unique among dromaeosaurids in the posteroventral orientation of the pubic peduncle. In *Saurornitholestes langstoni* (MOR 660), *Adasaurus mongoliensis* (Barsbold 1983), and *Velociraptor mongoliensis* (Norell and Makovicky 1997, 1999), the pubic peduncle is anteroventrally oriented. This orientation may also be expressed as the angle formed between the cranial margin of the pubic peduncle and the imaginary line uniting the distal tips of the ischiac pedicle and pendant process. Moreover, *Rahonavis ostromi* also has this latter condition (Forster et al. 1998).

d) *Supracetabular crest prominent.* The supracetabular crest of *Unenlagia comahuensis* is prominent and projects slightly laterally. The crest is almost restricted to the cranial portion of the acetabular aperture, and is continuous anteriorly with the lateral border of the pubic peduncle. In *Deinonychus antirrhopus* (MCZ 4371, AMNH 3015) and tyrannosaurids (Osborn 1917), the supracetabular crest is represented by a slight buttress over the cranial half of the acetabular aperture that is continuous with the lateral margin of the pubic peduncle. A supracetabular crest is absent in *Chirostenotes pergracilis* (BMNH R10903) and *Sinornithoides dongi* (IVPP V9612; contra Russell and Dong 1993b), and the same seems to be true for Oviraptorosauria and Therizinosauridae (Russell and Dong 1993a). In *Archaeopteryx lithographica* (Berlin specimen) the external margin of the acetabulum forms an almost continuous ridge from pubic to ischiac peduncles, and is straight in dorsal view. In this sense, the presence of a supracetabular crest, albeit reduced, in *Unenlagia comahuensis* may be interpreted as less derived than other maniraptorans.

Conclusions

Norell and Makovicky (1999) referred to *Unenlagia comahuensis* as extremely fragmentary, unusual, and enigmatic. It is true that *Unenlagia comahuensis* is far from complete (no more than twenty bones compose the holotype), but its anatomy is not so strange. The available bones exhibit a congruent set of unequivocal theropod, maniraptoran, and avialan characteristics (Novas and Puerta 1997). Maniraptoran synapomorphies present in *Unenlagia comahuensis* include a humerus with a proximodistally elongate internal tuberosity, a postacetabular blade that is acuminate, a ventrally directed pubis, an ischium that is nearly 50% of the pubic length, and a reduced or absent adductor fossa and associated anterodistal crista at the distal end of the femur. Moreover, its anatomy is intermediate between dromaeosaurids and *Archaeopteryx lithographica,* and most of its bones look strikingly similar to those of the Malagasy taxon *Rahonavis ostromi.* Having demonstrated elsewhere (Novas and Puerta 1997; Novas 1999; Forster et al. 1998) the maniraptoran and avialan pedigree of *Unenlagia comahuensis,* and given the lack of bizarre autapomorphies, it is probably misleading to refer to *Unenlagia comahuensis* as an unusual or enigmatic maniraptoran theropod.

Pelvic characteristics identified by Norell and Makovicky (1997, 1999) in support of a group formed by *Velociraptor mongoliensis* and birds, exclusive of *Unenlagia comahuensis,* are not as strong as presented. Furthermore, the ilium of *Unenlagia comahuensis* has a set of derived features shared with birds that are absent in dromaeosaurids and less derived theropods (such as the presence of a lobe-shaped preacetabular wing, a prominent supratrochanteric process, and a reduced brevis fossa). The avian traits of the ilium are in agreement with some other characteristics identified in the remaining bones of *Unenlagia comahuensis.* For example, the femur is elongate, has a reduced proximal head, and lacks a fourth trochanter. This set of features resembles *Archaeopteryx lithographica, Rahonavis ostromi,* and Enantiornithes, but sharply contrasts with most dromaeosaurids (except *Bambiraptor feinbergi*) in which the femur is proportionally robust, has a prominent articular head, and retains a faint prominence representing the fourth trochanter. The ischium of *Unenlagia comahuensis* has a prominent proximodorsal process that is shared with the dromaeosaurid *Microraptor zhaoianus* (Xu et al. 2000) and primitive birds such as *Archaeopteryx lithographica,* Enantiornithes, *Iberomesornis romerali,* and *Confuciusornis sanctus* (Novas and Puerta 1997; Forster et al. 1998). In particular, *Unenlagia comahuensis* shares with *Rahonavis ostromi* a craniocaudally concave dorsal margin of the postacetabular blade.

The available anatomical information still needs to be considered within a major phylogenetic context, based on character distribution in many coelurosaurian specimens. Also important, following suggestions by Makovicky (pers. comm.), is the identification of character distribution in each of the member species of Dromaeosauridae, with the aim of testing the monophyly of this coelurosaurian group. These methodological procedures may shed light on the contradictory character distribution of *Unenlagia,* which has dromaeosaurid features recognized in the dorsal vertebrae (Norell and Makovicky 1999) as well as some plesiomorphic features retained on the proximal femur (the presence of a deep cleft between the greater and lesser trochanters).

However, new evidence on ilium morphology supports the hypothesis depicting *Unenlagia comahuensis* as closer to birds than are dromaeosaurids (Novas and Puerta 1997; Novas 1999; Forster et al. 1998).

Acknowledgments. I wish to thank the organizers of the Florida

Symposium on Dinosaur-Bird Evolution for their kind invitation to participate in the meeting. M. Norell (AMNH) kindly allowed me to access the many specimens of *Velociraptor mongoliensis* and other Mongolian taxa collected by expeditions of the AMNH. A. Milner (BMNH) and W.-D. Heindrich (Humboldt Museum für Naturkunde) are thanked for access to the, respectively, London and Berlin specimens of *Archaeopteryx lithographica;* P. Currie for access to the theropod collections of the Royal Tyrrell Museum of Palaeontology, including casts of MOR 660 (*Saurornitholestes langstoni*); J. H. Ostrom and J. A. Gauthier for access to the specimens *Deinonychus antirrhopus* housed at the Yale Peabody Museum; C. Forster for allowing me to examine the skeleton of *Rahonavis ostromi;* R. Coria (MCF PVPH) for allowing me to borrow the holotype of *Unenlagia comahuensis* for many years; P. Makovicky (Field Museum of Natural History) for his clever observations and comments on an early version of this paper; and J. González for his skillful artwork. I would like to express my deep gratitude to the National Geographic Society (Washington), The Dinosaur Society (New York), and the Agencia Nacional de Promoción Científica y Tecnológica (Buenos Aires) for their valuable financial support.

References

Bakker, R. T., M. Williams, and P. J. Currie. 1988. *Nanotyrannus,* a new genus of pygmy tyrannosaur, from the latest Cretaceous of Montana. *Hunteria* 1(5): 1–30.

Barsbold, R. 1983. [Carnivorous dinosaurs from the Cretaceous of Mongolia.] *[Joint Soviet–Mongolian Paleontological Expedition, Transactions]* 19: 5–120 [in Russian].

Baumel, J. J., and L. M. Witmer. 1993. Osteologia. In J. J. Baumel, A. King, J. Breazile, H. Evans, and J. Vanden Berge (eds.), *Handbook of avian anatomy: Nomina anatomica avium,* pp. 45–132. Cambridge, Mass.: Publications of the Nuttall Ornithological Club.

Burnham, D. A., K. L. Derstler, P. J. Currie, R. T. Bakker, Z. Zhou, and J. H. Ostrom. 2000. Remarkable new birdlike dinosaur (Theropoda: Maniraptora) from the Upper Cretaceous of Montana. *The University of Kansas, Paleontological Contributions* 13: 1–14.

Chiappe, L. M. 1996. Late Cretaceous birds of southern South America: anatomy and systematics of Enantiornithes and *Patagopteryx deferrarisii.* In G. Arratia (ed.), *Contributions of southern South America to vertebrate paleontology. Münchner Geowissenschaftliche Abhandlungen* 30: 203–244.

Chiappe, L. M., Ji S.-A., Ji Q., and M. Norell. 1999. Anatomy and systematics of the Confuciusornithidae (Theropoda: Aves) from the Late Mesozoic of Northern China. *American Museum of Natural History, Bulletin* 242: 1–89.

Forster, C. A., S. D. Sampson, L. M. Chiappe, and D. W. Krause. 1998. The theropod ancestry of birds: new evidence from the Late Cretaceous of Madagascar. *Science* 279: 1915–1919.

Gauthier, J. A. 1986. Saurischian monophyly and the origin of birds. *California Academy of Sciences, Memoir* 8.

Holtz, T. R., Jr. 1994. The phylogenetic position of the Tyrannosauridae:

implications for theropod systematics. *Journal of Paleontology* 68: 1100–1117.
———. 1995. The arctometatarsalian pes, an unusual structure of the metatarsus of Cretaceous Theropoda (Dinosauria: Saurischia). *Journal of Vertebrate Paleontology* 14: 480–519.
Hutchinson, J. R. 2001. The evolution of pelvic osteology and soft tissues on the line to extant birds (Neornithes). *Zoological Journal of the Linnean Society* 131: 169–197.
Ji Q., P. J. Currie, M. A. Norell, and Ji S.-A. 1998. Two feathered dinosaurs from northeastern China. *Nature* 393: 753–761.
Martin, L. 1983. The origin and early radiation of birds. In A. Bush and G. Clark, Jr. (eds.), *Perspectives in ornithology,* pp. 291–338. New York: Cambridge University Press.
———. 1991. Mesozoic birds and the origin of birds. In H.-D. Schultze and L. Trueb (eds.), *Origin of the higher groups of tetrapods,* pp. 485–540. Ithaca: Comstock Publishing Associates.
Norell, M. A., and P. J. Makovicky. 1997. Important features of the dromaeosaur skeleton: information from a new specimen. *American Museum Novitates* 3215: 1–28.
———. 1999. Important features of the dromaeosaurid skeleton II: information from newly collected specimens of *Velociraptor mongoliensis. American Museum Novitates* 3282: 1–45.
Novas, F. E. 1991. Los Tyrannosauridae, gigantescos dinosaurios celurosaurios del Cretácico Tardío de Laurasia. *Ameghiniana* 28: 410.
———. 1992. La evolución de los dinosaurios carnívoros. In J. L. Sanz and A. Buscalioni (eds.), *Los dinosaurios y su entorno biótico,* pp. 125–163. Actas II Curso de Paleontología en Cuenca. Instituto "Juan de Valdés," Ayuntamiento de Cuenca, España.
———. 1996. Alvarezsauridae, Late Cretaceous maniraptorans from Patagonia and Mongolia. *Queensland Museum Memoirs* 39: 675–702.
———. 1997. Anatomy of *Patagonykus puertai* (Theropoda, Avialae, Alvarezsauridae), from the Late Cretaceous of Patagonia. *Journal of Vertebrate Paleontology* 17: 137–166.
———. 1999. *Dinosaur.* In McGraw-Hill 1999 Yearbook of Science & Technology, pp. 133–135. New York.
Novas, F. E., and P. F. Puerta. 1997. New evidence concerning avian origins from the Late Cretaceous of Patagonia. *Nature* 387: 390–392.
Novas, F. E., and D. Pol. 2002. *Mononykus* relationships reconsidered. In L. Chiappe and L. Witmer (eds.), *Above the heads of the dinosaurs,* pp. 121–125. Berkeley: University of California Press.
Osborn, H. F. 1917. Skeletal adaptations of *Ornitholestes, Struthiomimus, Tyrannosaurus. Bulletin of the American Museum of Natural History* 35: 733–771.
Ostrom, J. H. 1969. Osteology of *Deinonychus antirrhopus,* an unusual theropod from the Lower Cretaceous of Montana. *Peabody Museum of Natural History,* Bulletin 30: 1–165.
———. 1976. On new specimen of the Lower Cretaceous theropod dinosaur *Deinonychus antirrhopus. Breviora* 439: 1–21.
Padian, K., J. R. Hutchinson, and T. R. Holtz, Jr. 1999. Phylogenetic definitions and nomenclature of the major taxonomic categories of the carnivorous Dinosauria (Theropoda). *Journal of Vertebrate Paleontology* 19: 69–80.
Paul, G. 1984. The archosaurs: a phylogenetic study. In W.-E. Reif and

F. Westphal (eds.), *Third symposium on Mesozoic terrestrial ecosystems,* pp. 175–180. Tübingen: Attempto Verlag.
Perle, A., M. A. Norell, and J. M. Clark. 1999. A new maniraptoran theropod—*Achillobator giganticus* (Dromaeosauridae)—from the Upper Cretaceous of Burkhant, Mongolia. National University of Mongolia, Geology and Mineralogy. 107 pp.
Russell, D., and Z. Dong. 1993a. The affinities of a new theropod from the Alxa Desert, Inner Mongolia, People's Republic of China. *Canadian Journal of Earth Sciences* 30: 2107–2127.
———. 1993b. A nearly complete skeleton of a new troodontid dinosaur from the Early Cretaceous of the Ordos Basin, Inner Mongolia, People's Republic of China. *Canadian Journal of Earth Sciences* 30: 2163–2173.
Sanz, J. L., J. F. Bonaparte, and A. Lacasa Ruiz. 1988. Unusual Early Cretaceous birds from Spain. *Nature* 331: 433–435.
Walker, C. A. 1981. New subclass of birds from the Cretaceous of South America. *Nature* 292: 51–53.
Wellnhofer, P. 1974. Das funfte Skelettexemplar von *Archaeopteryx lithographica. Palaeontographica* A (147): 169–216.
———. 1993. Das siebte Exemplar von *Archaeopteryx lithographica* aus den Solnhofener Schichten. *Archaeopteryx* 11: 1–47.
Xu X., Wang X.-L., and Wu X.-C. 1999. A dromaeosaurid dinosaur with a filamentous integument from the Yixian Formation of China. *Nature* 401: 262–266.
Xu X., Zhou Z., and Wang X.-L. 2000. The smallest known non-avian theropod dinosaur. *Nature* 408: 705–708.
Zhou Z.-H., Wang X.-L., Zhang F.-C., and Xu X. 2000. Important features of *Caudipteryx*—evidence from two nearly complete new specimens. *Vertebrata PalAsiatica* 10: 241–254.

8. Bird-Like Features of Dinosaur Footprints

Joanna L. Wright

Abstract

The footprints of some dinosaurs are morphologically similar to those of birds; this has the potential to cause problems in trackmaker identification, given that the stratigraphic record of dinosaur and bird footprints overlaps by approximately 80 million years.

Theropod footprints are similar to those of birds in the number of phalangeal pad impressions and the visible claw marks. However, the footprints of most theropods are distinctly asymmetrical due to the relatively large difference in length between digits II and IV. The footprints of arctometatarsalian theropods are more symmetrical and may also show separate heel pad impressions formed by the soft tissues beneath metatarsal III.

Most ornithopod footprints do not show claw marks but a distinctive "heel" is often present. This is formed by a fleshy pad at the back of the foot. Ornithopod footprints are symmetrical because of the reduction in phalangeal length in digit IV. However, these footprints are rarely mistaken for those of birds because the digit impressions tend to be very wide and rounded.

Most fossil footprints of Mesozoic birds are those of wading or ground-dwelling birds. The footprints of wading birds are very distinctive with narrow digit impressions, sometimes evidence of webbing, or a hallux and very high divarification angles (greater than 110°). Dinosaur footprints rarely have a digit divarification of more than 90°. Bird footprints are symmetrical and often show a metatarsal pad.

Distinguishing between ground bird and dinosaur tracks is potentially most problematic in rocks of mid to Late Cretaceous age. By this time birds were diverse, and large ground-dwelling birds appeared. In addition, some arctometatarsalian dinosaurs had foot structures convergent with modern birds such as ratites.

Footprints reflect function more than systematics. The footprints of birds would have been similar to those of dinosaurs until bird feet became specialized for other functions such as perching or wading.

Introduction

The similarity of dinosaur footprints to those of birds was recognized from the first modern scientific studies in the early 1800s (Hitchcock 1836; Deane 1861; Beckles 1851). In fact, these early workers thought that they had found the traces of ancient birds and gave them names to match: *Ornithichnites* (Hitchcock 1836), *Ornithoidichnites* (Hitchcock 1841), and later *Grallator* (Hitchcock 1858). Beckles (1851, 1854) described three-toed footprints from the Purbeck and Wealden Groups of southern England, and used the term "ornithoidichnites" as a general description. Beckles (1851, 1854) at first thought that these tracks were of "ornithic" origin, but later realized that dinosaurs such as *Iguanodon* could have made such tracks (Beckles 1862). Beckles had a couple of advantages over Hitchcock in realizing the nature of the trackmakers. Not only did Beckles start his studies later than Hitchcock, when more specimens and more complete skeletons of dinosaurs had been discovered, but he was documenting footprints in strata that contained skeletal remains as well as footprints. The rocks of the Connecticut Valley, where Hitchcock's dinosaur footprints were found, are almost completely devoid of vertebrate remains. Beckles compared the appearance of the footprints to the structure of the foot of *Iguanodon* and concluded that *Iguanodon* could certainly have made such an impression, but, with commendable caution, he also noted that other dinosaurs probably had similar feet and that these footprints should not be referred exclusively to *Iguanodon* (Beckles 1862). Most subsequent writers conveniently ignored this caution (Mansell-Pleydell 1888, 1896; Calkin 1933; Delair 1960; Sarjeant 1974).

With the realization that most ornithoidichnites were of dinosaurian origin, the battle to ascribe them to their trackmakers began. In the U.K. the problem was simpler, and the answer even more so. Until 40 years ago all dinosaur footprints discovered there were found on the southeast coast, all the footprints were large and tridactyl, and they were all attributed to *Iguanodon*. It is now known that two large tridactyl footprint morphotypes are found in Early Cretaceous rocks of the U.K., iguanodontid and theropod. A third theropod track morphotype has been reported (Radley 1994a); it is considerably smaller with more slender digits. In addition sauropod, ankylosaur (nodosaur), and pterosaur footprints have been reported from the Purbeck and Wealden Groups (Ensom 1987, 1995; Radley 1994b; Wright et al. 1997, 1998).

In the United States, Lull (1904) was the first to redescribe Hitchcock's collection and attempt to ascribe the footprints to dinosaurian trackmakers. Lull's interpretation of the producers of the main dinosaurian ichnotaxa is, in general, very similar to our understanding today. Most of the footprints seem to have been made by theropods of varying sizes, with a few made by early ornithischians/ornithopods and with a question mark hanging over the possibility of prosauropod tracks.

More and more tracks are being discovered every year around the world. Some tracks have only, or initially, been recognized for what they are from an associated unusual feature. For instance, manus impressions may help decide between conflicting interpretations (Norman 1980; Wright 1999). However, in the absence of such clues, how can the affinities of tridactyl tracks be determined?

Institutional Abbreviations: CU–MWC, University of Colorado (Denver)/Museum of Western Colorado Joint Collection, Denver, Colo., U.S.; DMNH, Denver Museum of Natural History, Denver, Colo., U.S.; RSM, Royal Scottish Museum, Edinburgh, Scotland; UCM, University of Colorado Museum, Boulder.

Trackmaker Candidates

In general there are four main categories of producers of tridactyl tracks: non-arctometatarsalian theropods, arctometatarsalian theropods, ornithopods, and birds. (For the purposes of this chapter, for brevity, the use of the term "theropods" will be confined to non-avian and non-arctometatarsalian theropods except where otherwise stated.) The structure of the feet of these different groups may help to distinguish among them but they also have similarities to one another. Most members of these groups have a pedal phalangeal formula (of the digits functional in walking) of 3-4-5, with rare exceptions such as some moas and dromornithids, which have reduced the phalanges in digit IV to 4, and ostriches, which have lost digit II.

Theropods

Theropod feet are the most primitive foot structures under consideration. This type of foot structure is the nearest to that of the ancestral dinosaur, and in many theropods this structure was not greatly modified through time. Most theropod feet are functionally tridactyl; the hallux may make an impression but it is not a weight-bearing digit. Theropod feet are asymmetrical; the individual phalanges in digit IV are relatively long and therefore digit IV is considerably longer than digit II (fig. 8.1). This does not mean that digit IV had a greater role in weight-bearing in most theropods. On the contrary, digit IV is usually narrower than digit II and tends to leave a shallower impression (fig. 8.2). This indicates that theropods placed most of their weight on the inside and center of their feet.

The relatively longer phalanges of theropods tend to translate into discrete phalangeal pads in footprints. Many theropod footprints do not preserve phalangeal pads. Although in most cases this can be shown to be a result of poor preservation, it has been suggested that in at least some cases, with larger footprints, the absence of pad impressions reflects the anatomy of the trackmaker's foot (e.g., Moratalla 1993; Lockley et al. 1998). However, this has yet to be demonstrated

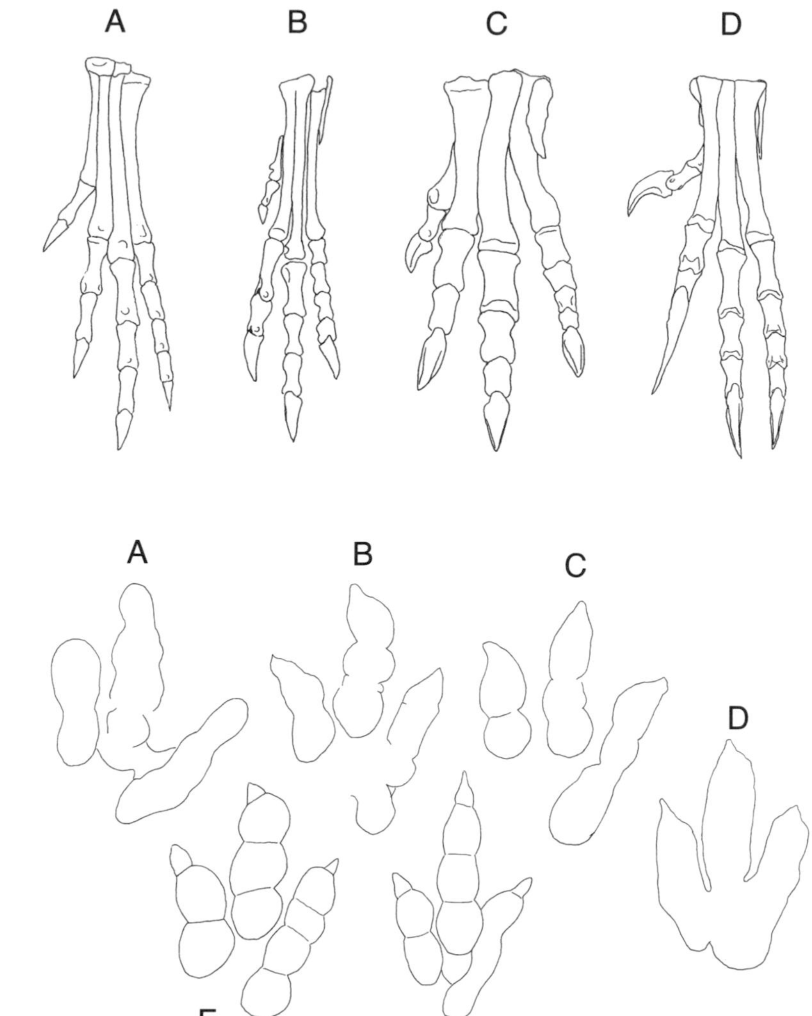

Figure 8.1. Theropod feet:
A, Coelophysis;
B, Megapnosaurus;
C, Allosaurus;
D, Deinonychus.
For ease of comparison, all drawn to the same approximate length, and all anterior views of the left pes. Note that Coelophysis *and* Deinonychus *have the most asymmetrical foot structures. (A, after Colbert 1989; B, after Rowe and Gauthier 1990; C, after Madsen 1993; D, after Ostrom 1969)*

Figure 8.2. Theropod footprints.
A, unnamed theropod track (GLAHM X1101, Hunterian Museum, Glasgow);
B, C, Megalosauripus *footprints;*
D, Eubrontes;
E, F, Grallator *footprints.*
All are drawn as apparent right foot impressions and to the same approximate length for ease of comparison. (A, after Wright 1996; B, C, after Lockley et al. 1996; D–F, specimens from the Hitchcock Collection, Pratt Museum, Amherst College)

convincingly, and preservational causes of this have not been ruled out. The creases between phalangeal pads tend to be single and at an oblique angle to the digital axis. Claw impressions, if present, may be either sharp or blunt, and may vary in shape from acicular to triangular. Larger theropod footprints tend to have blunter, more triangular claws, and smaller footprints tend to have sharper, more elongate claws (Wright 1996). Claw impressions in theropod footprints tend to proceed from the sides of the digit impressions; those on digits II and III extend from the medial side and those on digit IV from the lateral side. This is most pronounced in digit II. In addition the claw impressions on digits II and III tend to be oriented medially with respect to the digital axes and that on digit IV tends to be oriented laterally with respect to the digital axis.

In general, the impressions of digits II and IV extend a similar distance anteriorly, but the posterior end of the impression of digit IV forms the most posterior part of the footprint. The most posterior part of this digit impression may actually be made by the distal end of the fourth metatarsal (Baird 1957; Farlow and Chapman 1997).

Interdigital angles are not very useful in trying to distinguish between theropod and ornithopod tracks. The angle often varies a great deal within a single trackway; the variation may be due to a change in substrate consistency or the speed of the trackmaker, but often the cause is indiscernible. While the lowest values of interdigital angles are recorded from theropod tracks, with total divarification (between digits II and IV) as low as 15° (Wright 1996), the value can approach 90°, even though it is rarely more than 60–70°. Although the average digit divarification of a theropod trackway is often 10–20° less than that of an ornithopod trackway, the ranges overlap considerably (Wright 1996) and it cannot be considered a reliable discriminator. Larger tracks tend to show wider divarification angles than smaller tracks do, and the differences between divarification angles of theropod and ornithopod tracks decrease with increasing size (Thulborn 1990).

The axis of digit III in a theropod track tends to be parallel to the trackway axis, or slightly toed-outward. However, theropod tracks may be toed-inward, particularly if they are larger animals (Thulborn 1990; Wright 1996), which is another tendency convergent with large ornithopods.

Trends in theropod feet from primitive theropods to tetanurans include a length reduction in the phalanges of digit IV. Each phalanx tends to be reduced by a similar amount, unlike the extreme compression of phalanges IV-2, -3, and -4 in derived ornithopods.

Dromaeosaurids are considered to be one of the groups most closely related to birds, and yet their feet retained a relatively primitive theropod structure. Digit IV was not reduced in length at all, and in fact it increased in length slightly and became more robust (compare fig. 8.1A with 8.1D). This is probably related to the hyperextension of digit II and the large ungual on that toe, so that digits III and IV would have been the main weight-bearing digits. These modifications for even weight distribution on the outer two toes supports the hypothesis that dromaeosaurids bore very little weight on digit II. This also means that dromaeosaurids would have made very distinctive two-toed footprints, although no such footprints have yet been described. The two-toed footprints these dinosaurs would have made also means that their footprints are unlikely to be mistaken for those of birds.

Arctometatarsalians

Arctometatarsalian theropods have more symmetrical feet than less derived theropods (fig. 8.3). The arctometatarsalian condition mimics the tarsometatarsus of birds. The functions of these are thought to be similar—providing more rigidity for the foot, in arctometatarsalians for running (Wilson and Currie 1995), and in birds presumably for perching. Enantiornithine birds fuse the metatarsals, but it has been

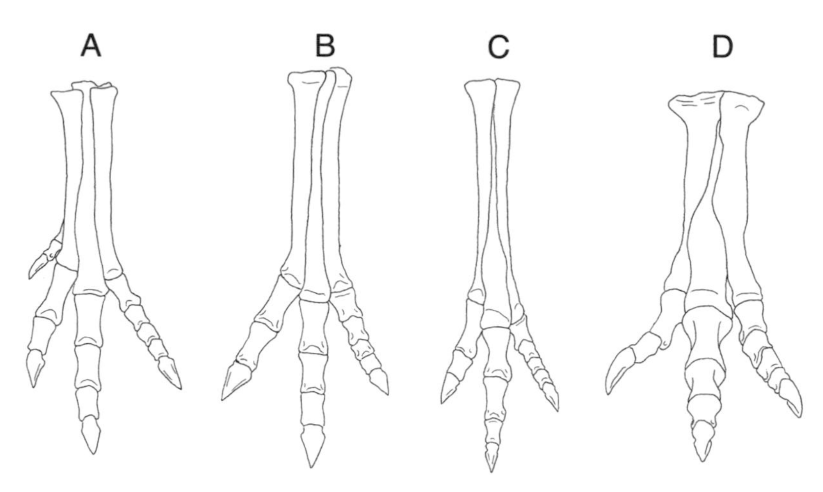

Figure 8.3. Arctometatarsalian feet. A, Garudimimus; *B,* Harpymimus; *C,* Struthiomimus; *D,* Tarbosaurus. *(A–C, after Barsbold and Osmólska 1990; D, after Molnar et al. 1990)*

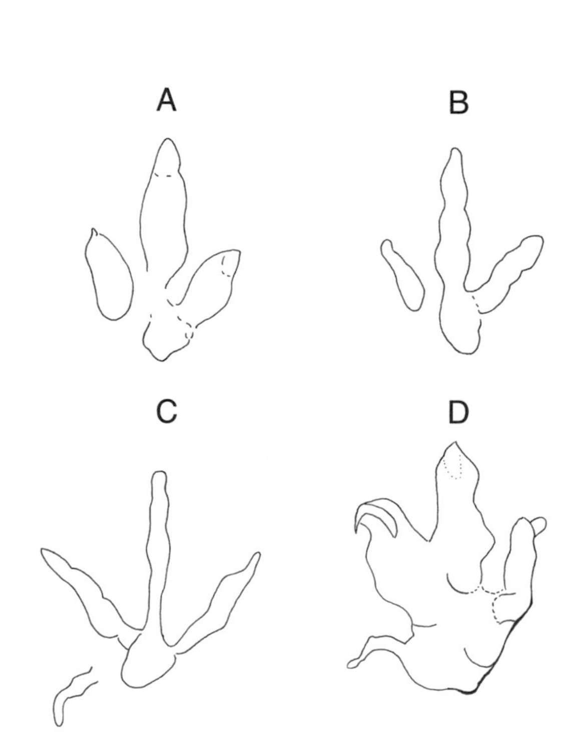

Figure 8.4. Possible arctometatarsalian footprints. A, Ornithomimipus angustus, *Upper Cretaceous Horseshoe Canyon Formation, Alberta, Canada; B, unnamed theropod track, Upper Cretaceous Lance Formation, Wyoming; C,* Saurexallopus, *Upper Cretaceous Harebell Formation, Wyoming, DMNH 5985; D,* Tyrannosauripus, *Upper Cretaceous Raton Formation, New Mexico.*

argued that this is an independent development separate from that of crown group birds (Feduccia 1996). Some dinosaurs (e.g., *Elmisaurus*) independently developed a form of tarsometarsus, where some of the tarsal elements and metarsals are fused (Hutchinson and Padian 1997).

For the purposes of this paper, the structure of the foot is more pertinent than the phylogenetic position and therefore arctometatarsalians will herein refer to dinosaurs and primitive birds having that condition, rather than to the Arctometatarsalia sensu stricto—Tyrannosauridae and Bullatosauria (Holtz 1996). Interestingly, *Megapnosaurus*, a small, relatively primitive theropod, also shows fusion of the proximal

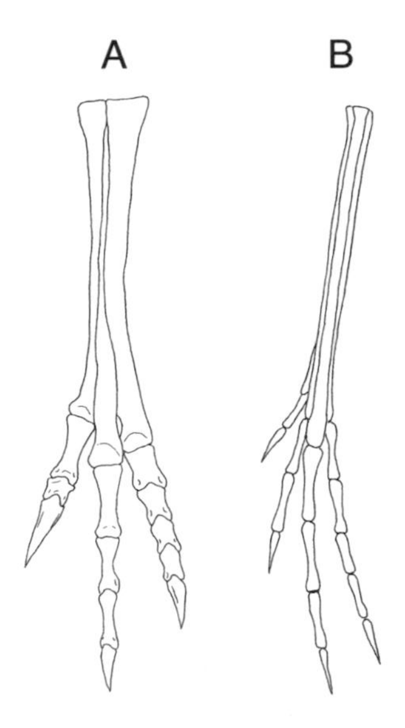

Figure 8.5. Left pedal skeleton of A, Borogovia, *a troodontid; B,* Archaeopteryx. *(A, after Osmólska and Barsbold 1990; B, after Feduccia 1996)*

ends of the metatarsals, although metatarsal III is not pinched (Raath 1969; Rowe 1989).

In addition to having a pinched third metatarsal, which increases the locomotory emphasis on the third digit, arctometatarsalians also tend to reduce the length of digit IV by reducing the lengths of individual phalanges. However, unlike advanced ornithopods (camptosaur grade and above) and some birds, which greatly reduce phalanges IV-2 to IV-4, arctometatarsalians reduce phalanges IV-1 to IV-4 fairly evenly. This even reduction is similar to that observed in some tetanurae (e.g., *Allosaurus*), and the primitive large ornithopod *Tenontosaurus. Megapnosaurus* has also reduced the phalanges of digit IV, which may or may not be related to the fusion of its metatarsals, but is in strong contrast to the structure of the foot of *Coelophysis* (figs. 8.1A, B).

The symmetrical nature of the arctometatarsalian foot should be reflected in the footprints that these dinosaurs would have made. Several different types of footprints have been attributed to arctometatarsalian dinosaurs (fig. 8.4) on the basis of similarity to the foot structure, including *Ornithomimipus* and *Tyrannosauripus.* Many footprints of arctometatarsalians also preserve an impression of a pad at the rear of the footprint. By analogy with bird footprints, this is thought to have been made by a pad under the metarsal bundle, mainly under metarsal III. The trackmaker of *Saurexallopus,* which also shows symmetrical foot structure and a metatarsal pad impression, has so far eluded identification because of the long hallux impression. Most arctometatarsalian dinosaurs have lost or reduced their hallux (fig. 8.3). However, *Tyrannosauripus* shows a relatively long hallux impression, probably because it is a very deep footprint that allowed the hallux to contact the ground; the hallux impression is much shallower than the rest of the footprint. *Saurexallopus* footprints, however, tend to be shallow.

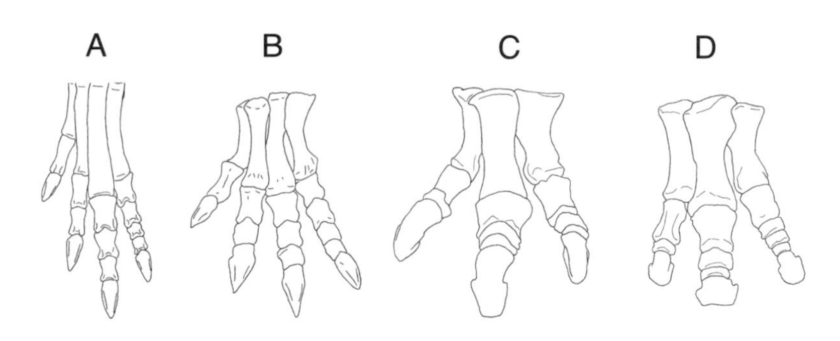

Figure 8.6. Ornithopod feet.
A, Hypsilophodon;
B, Tenontosaurus;
C, Iguanodon; *and*
D, Edmontosaurus.
(All after Norman 1985)

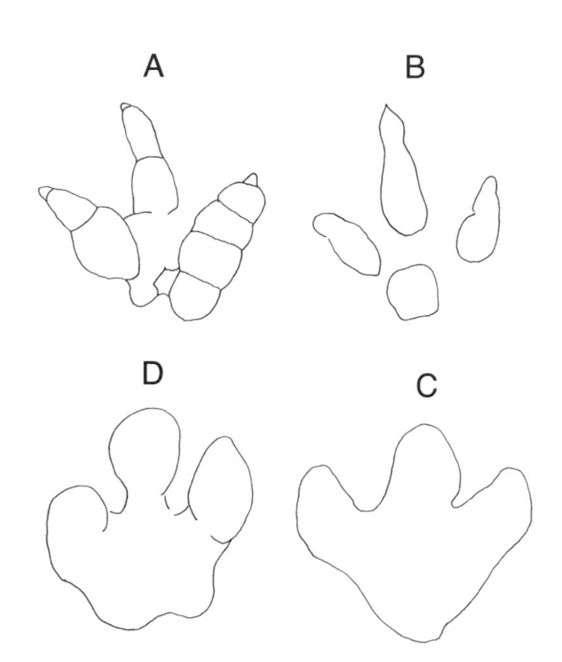

Figure 8.7. Footprints of ornithopods.
A, Anomoepus, *Lower Jurassic, Connecticut Valley, U.S.;*
B, Dinehichnus;
C, unnamed iguanodontid footprint from the Purbeck Limestone Group, southern England, now held in the Royal Scottish Museum (RSM 1974.16.1);
D, Caririchnium *from Mosquero Creek, Colorado.*
(B, after Lockley et al. 1996; C, after Wright 1996; D, after Lockley and Hunt 1995)

Troodontids have both a pinched third metatarsal and an enlarged claw on digit II whose phalanges indicate an ability to hyperextend. It seems likely that as with dromaeosaurids these dinosaurs put little or no weight on digit II and in consequence digit IV is more robust than digit III and is almost the same length. This is the opposite of the trend in most arctometatarsalians (compare fig. 8.5A with figs. 8.3A–D).

Ornithopods

Primitive ornithopods, such as *Heterodontosaurus* and *Hypsilophodon,* have asymmetrical feet in which digit IV is longer than digit II, although the asymmetry in hypsilophodontids is not as pronounced as in theropods (fig. 8.6). Larger primitive ornithopods such as *Tenontosaurus* have reduced the asymmetry in lateral digit lengths, and the trend toward reduction of the three distal phalanges on digit IV can already be seen in *Dryosaurus* and slightly more so in *Camptosaurus.*

More derived ornithopods—iguanodontids and hadrosaurids—

have almost symmetrical feet. The three most distal phalanges in digit IV are greatly reduced in length (fig. 8.6). However, unlike theropods, phalanx 2 of digit II is also strongly reduced. In *Iguanodon* the long ungual on digit II makes up for this reduced length, while in *Edmontosaurus* phalanx 1 of digit II is longer than the corresponding bone in digit IV. Thus, in general terms iguanodontids and hadrosaurs tend to reduce the lengths of phalanges from the middle of the toe relative to the first and last phalanges (Farlow and Chapman 1997). This strong reduction of phalanges 2–4, especially in larger animals, means that there was only a single pad beneath each of these three phalanges, and this is clearly seen in footprints (fig. 8.7C, D), especially those from the Wealden of southern England and the Late Cretaceous of North America (Currie et al. 1991). In footprints from both areas, details of skin impressions indicate that the presence of a single pad under each digit is not a preservational bias.

Iguanodontids and hadrosaurids also had very different foot posture from all the other groups discussed herein. Most groups of functionally tridactyl dinosaurs placed most of the length of their toes on the ground when standing. The distal ends of some metatarsals may, in some cases (especially arctometatarsalians), also have rested on the ground. In most modern birds the proximal ends of the lateral toes are raised off the ground and only the pad under the distal end of the bird tarsometatarsus contacts the ground (Farlow et al. 2000). Iguanodontids and hadrosaurids placed only the distal portion of their toes on the ground. The proximal ends of the toes and the distal ends of the metatarsals were raised off the ground and the back part of the foot was supported by a thick pad (Norman 1980) clearly visible in footprints (fig. 8.7).

Ornithopod metatarsals were tightly appressed proximally (Christiansen 1997), which gave their feet rigidity similar to that found in arctometatarsalians and birds, although each group has achieved it by different means. This increased rigidity, especially well developed in iguanodontids and hadrosaurids, may have helped to support the weight of these large animals, although it would undoubtedly have also been useful when they were moving at higher speeds.

Birds

Birds fuse the metatarsals into the tarsometatarsus. This is obviously the ultimate in metatarsal rigidity and may have developed as an adaptation for increased stability during perching. The tarsometatarsus is a structure unique to birds; the closest dinosaur relatives to birds and *Archaeopteryx* itself do not fuse the metatarsals or even pinch metatarsal III (figs. 8.1, 8.5, 8.8).

Bird feet have been adapted for a variety of functions (Feduccia 1996) and so it is difficult to generalize about the structure of all bird feet. However, many modern birds have symmetrical feet. The majority of fossil bird footprints described in the literature, especially from the Mesozoic, were made by wading birds (fig. 8.9). Ground birds might also be expected to have left many footprints although relatively few of

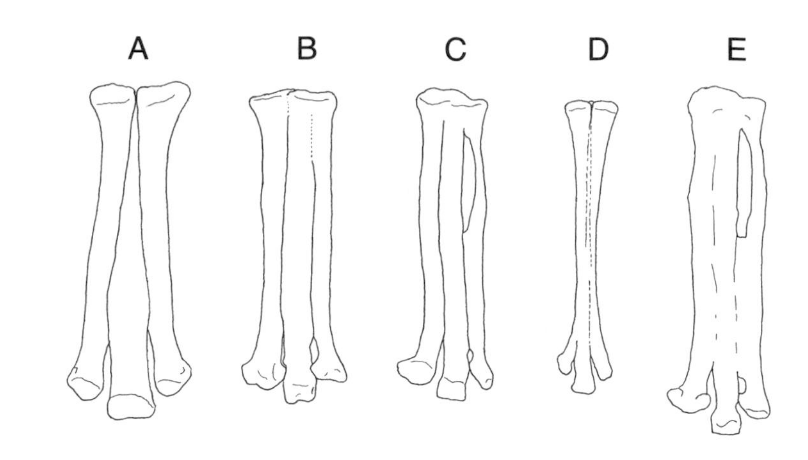

Figure 8.8. Bird tarsometarsi compared with the dinosaurian arctometatarsalian condition.
A, metatarsus of Ornithomimus velox *(arctometatarsalian dinosaur);*
B, tarsometarsus of Avisaurus archibaldi;
C, Avisaurus *sp.;*
D, Meleagris *(juvenile, turkey);*
E, Soroavisaurus australis. *(All after Feduccia 1996)*

these footprints have been described. Recent finds have demonstrated the existence of large ground birds in the Late Cretaceous (e.g., Buffetaut et al. 1995), but perhaps they did not frequent environments where their footprints would be preserved. Some large ground bird footprints, such as those of moas, are very dinosaurlike in morphology. The preponderance of wading bird tracks in the fossil record is not surprising as these types of birds spend a great deal of time walking around looking for food, and making tracks in environments where they are most likely to be preserved, such as shorelines and riverbanks (Lockley et al. 1992).

Wading birds have very distinctive feet. The function of the feet is to spread the weight of the bird over a larger area so that it is less likely to sink into the sediment. The toes themselves are usually narrow, and digit divarification is high, usually greater than 110°. Webbing around or between the toes may often be visible in footprints. They have a medially or posteriorly directed hallux, which may be very short. It is unlikely that the tracks of wading birds would be confused with those of dinosaurs.

Ground bird feet are more convergent with those of dinosaurs. The digit divarification tends to be about 90°, although in birds such as rheas it is much less (Farlow et al. 2000) and they often lack a hallux. The tracks of modern ground birds are distinct in morphology from those of dinosaurs. They are descended from flying ancestors, which had a modified bird foot structure. However, early ground birds may have retained a dinosaurian type of foot structure, and would therefore have made footprints identical to those of some non-avian dinosaurs because the structure of the foot and functions of the limb had not yet been significantly modified. Indeed moa tracks closely resemble some ornithopod tracks; they do not show discrete digital pads although they do have a slightly wider divarification angle and less rounded toes than dinosaur tracks (Farlow et al. 2000).

Archaeopteryx still had a dinosaurian foot structure (fig. 8.5); the metatarsals were not fused and the foot was asymmetrical, with digit IV much longer than digit II. It would have left dinosaurlike tracks.

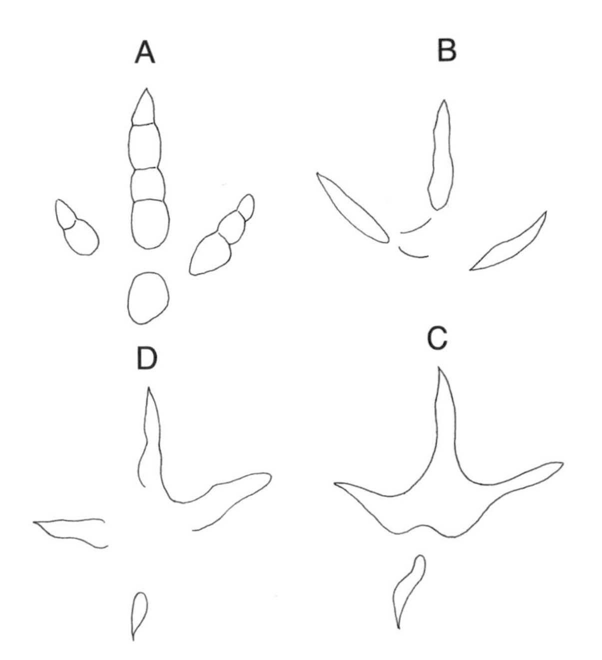

Figure 8.9. Bird footprints.
A, footprint of Rhea americana;
B, CU-MWC 230.9, Green River Formation, Utah;
C, UCM 17614, Lower Cretaceous Dakota Group, Colorado;
D, Upper Cretaceous Lance Formation.
(A, after Padian and Olsen 1989; C, after Lockley and Hunt 1995)

Discussion: Footprints, Feet, Phylogeny, and Function

Fossil footprints can be referred to likely trackmakers through an understanding of osteological features and soft tissue structures (Farlow 2001). Footprints are not merely an impression of the underside of the trackmaker's foot, but are a product of the kinetic interaction of the trackmaker's foot and the substrate at the time of track formation. Factors such as speed, stance, and gait influence the appearance of the footprint that eventually becomes a trace fossil. Perception of the quality of footprint preservation is therefore usually a reflection of how closely it seems to preserve the soft tissue morphology of the underside of the foot. Even under ideal circumstances, where an exact impression of the underside of the animal's foot may be preserved, there are still many uncertainties as to the correspondence between the soft tissues and the skeletal structure of the trackmaker's foot.

Footprints reflect hindlimb function. The footprints of large theropods are convergent with those of large ornithopods (Wright 1996; Farlow and Chapman 1997). Both sets of animals had to transport a large body on two legs and needed to keep their feet as close to the trackway midline as possible to provide greatest stability. This seems to have produced a toed-in trackway in both groups. In addition, both groups have relatively short feet compared to leg length (from the primitive state), and there is a tendency in both groups toward more symmetrical feet. Larger bipedal dinosaurs also tended to tighten the metatarsals into a single structure in order to better transport large weights.

Trackway patterns hold some clues to the nature of the trackmaker. It seems that, in general, theropods have relatively longer stride lengths that ornithopods or birds (Thulborn and Wade 1984). This difference is more marked in smaller trackmakers; larger trackways of all these groups converge on relatively shorter stride lengths.

It is a commonly held idea that dinosaur feet (perhaps especially theropod feet) are conservative, and that therefore footprints will not be particularly informative about the type of dinosaur that made them. However, dinosaur feet actually vary considerably within groups. For instance, *Coelophysis* and *Megapnosaurus* are closely related dinosaurs with very different feet—*Megapnosaurus* has a much more symmetrical foot (fig. 8.1) and, because of its fused metatarsals, it is also a much more rigid foot. These features presumably reflect differences in lifestyle or habitat of these two genera.

Feet and footprints do not show unidirectional trends, toward, for example, becoming more birdlike. Because they reflect function, changes in phylogenetic states may be hard to detect. For instance, primitive ground-dwelling birds such as *Mononykus* would have left dinosaurlike footprints because functionally it was very similar to a non-avian theropod. Derived theropods, like *Deinonychus,* are not arctometatarsalian nor have they greatly reduced the length of the phalanges of digit IV. This is probably related to the loss of digit II as a weight-bearing digit, which would mean that the two remaining digits would become similar in length for more stability. Even *Archaeopteryx* has a similar condition and would not have made birdlike but rather dinosaurlike footprints.

Summary and Future Work

Dinosaur feet are usually dismissed as being very conservative and of little use in differentiating between dinosaurs at almost any taxonomic level. However, while some aspects of dinosaur foot morphology, such as the phalangeal formula, may be conservative, dinosaur feet vary to a surprising degree within groups. The question is, can that variation be detected in footprints and are the feet of these groups well enough known?

Footprints reflect function more than systematics. As such they may not always allow determination of the first occurrences of certain taxa (such as birds), but they may allow determination of when birds became more distinctively "birdy" with respect to terrestrial locomotion. For instance, not only did birds adapt their feet to different functions, as diverse as perching and wading; they also underwent postcranial changes that altered their walking mechanisms. The loss of the heavy counterbalancing dinosaurian tail meant that birds had to compensate by using knee-driven, rather than hip-driven locomotion (Carrano 1997). This change may be detectable in trackways, by looking for alterations in the weight distribution of the feet and also possibly by changes in relative stride lengths. An understanding of the birdlike aspects of dinosaur feet may allow separation of some of these factors.

Plate 1

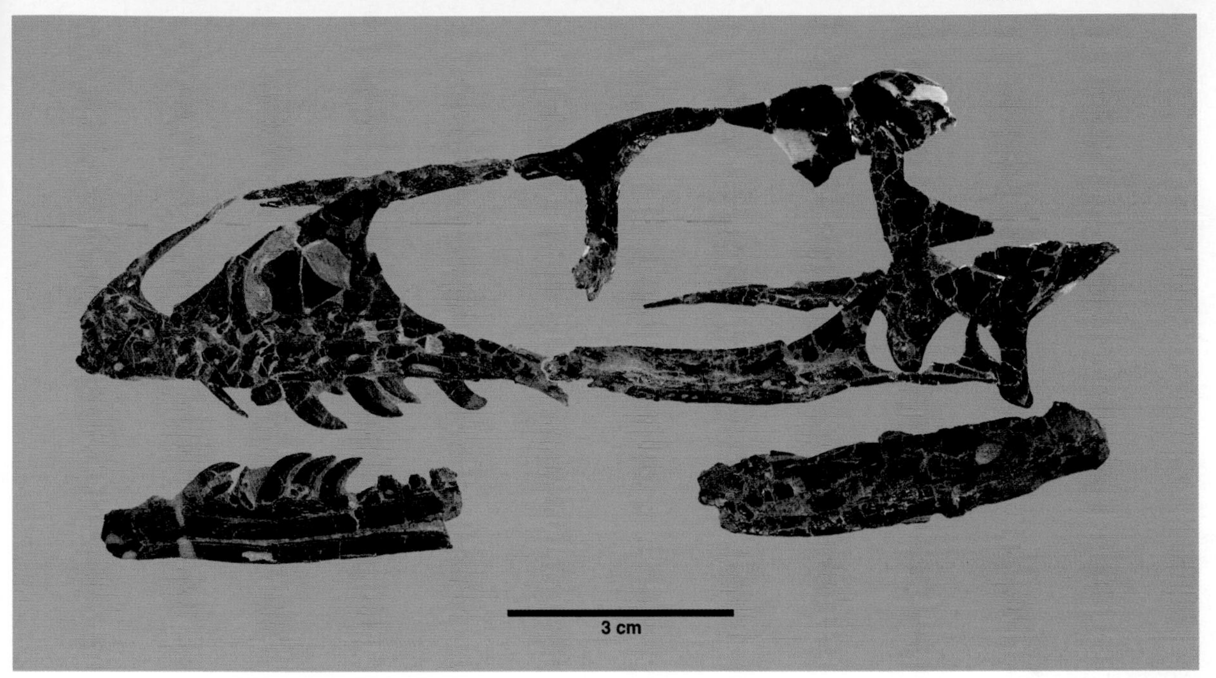

Plate 1: *Bambiraptor feinbergi* (holotype; actual size) AMNH FR: 30556 (Rick Edwards. ©2003 American Museum of Natural History)

Plate 2

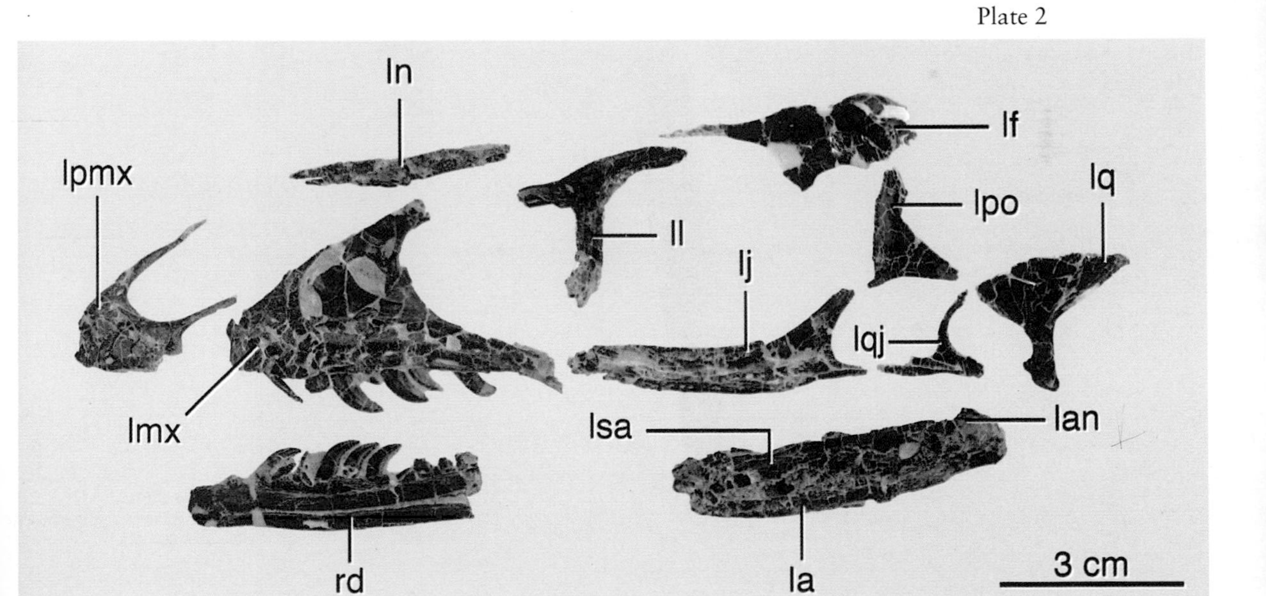

Plate 2: Exploded and labeled view of *Bambiraptor* skull. lpmx=premaxilary; ln=nasal; ll=lacrimal; lf=lacrimal foramen; lpo=postorbital; lq=quadrate; lmx=left maxilla; rd=right dentary; lj=left jugal; lqj=left quadrajugal; lsa=left surangular; la=left angular; lan=left articular. (Rick Edwards. ©2003 American Museum of Natural History)

Plate 3: *Velociraptor mongoliensis* (holotype; actual size) AMNH FR: 6515 (Rick Edwards. ©2003 American Museum of Natural History)

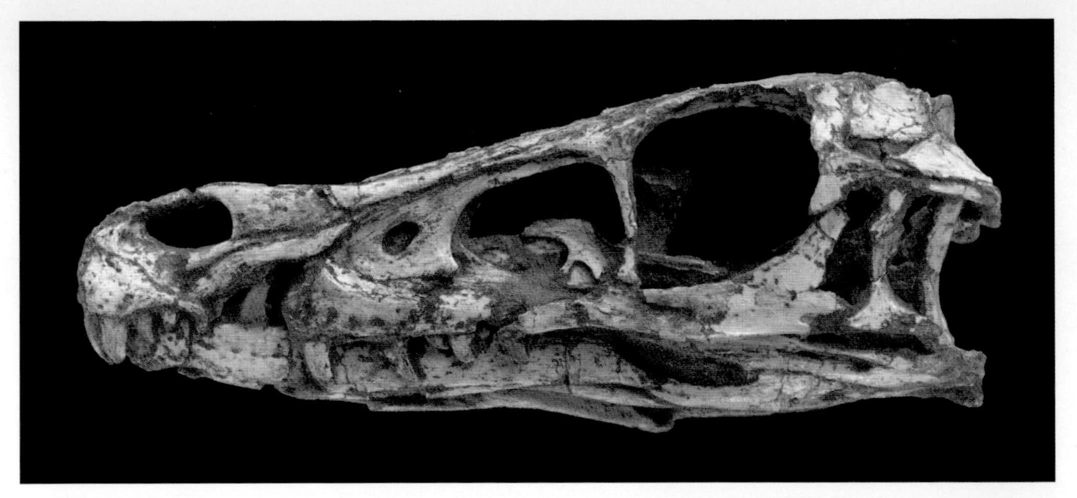

Plate 4

Plate 4: *Velociraptor mongoliensis* GIN 100/25 (Mick Ellison. ©2003 American Museum of Natural History)

Plate 5: *Bambiraptor* tooth (Rick Edwards. ©2003 American Museum of Natural History)

Plate 6: *Bambiraptor* braincase in dorsal view. (Rick Edwards. ©2003 American Museum of Natural History)

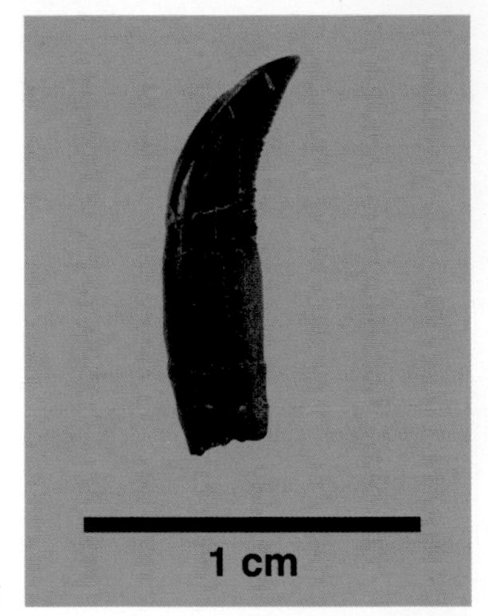

Plate 5

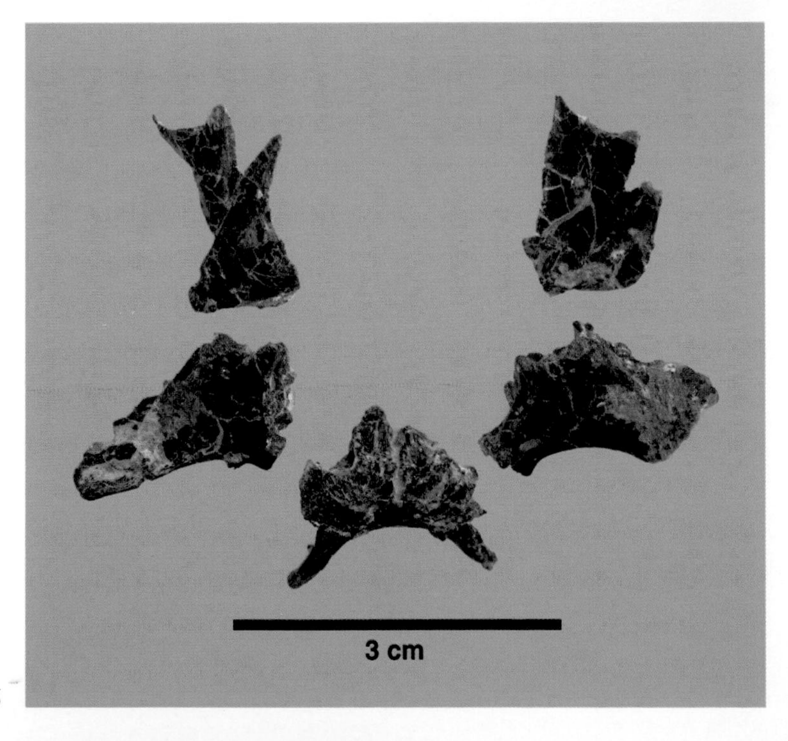

Plate 6

Plate 7

Plate 7: Life reconstruction of *Bambiraptor* by Brian Cooley. (©2003 Brian Cooley)

The plumage of *Bambiraptor*
Restorations by Pat Redman.
(Plates 8–13 all ©2003 Pat Redman)

Plate 8: Restoration with plumage modeled after a juvenile Hoatzin.

Plate 9: With plumage modeled after an adult Hoatzin.

Plate 8

Plate 9

Plate 10

Plate 11

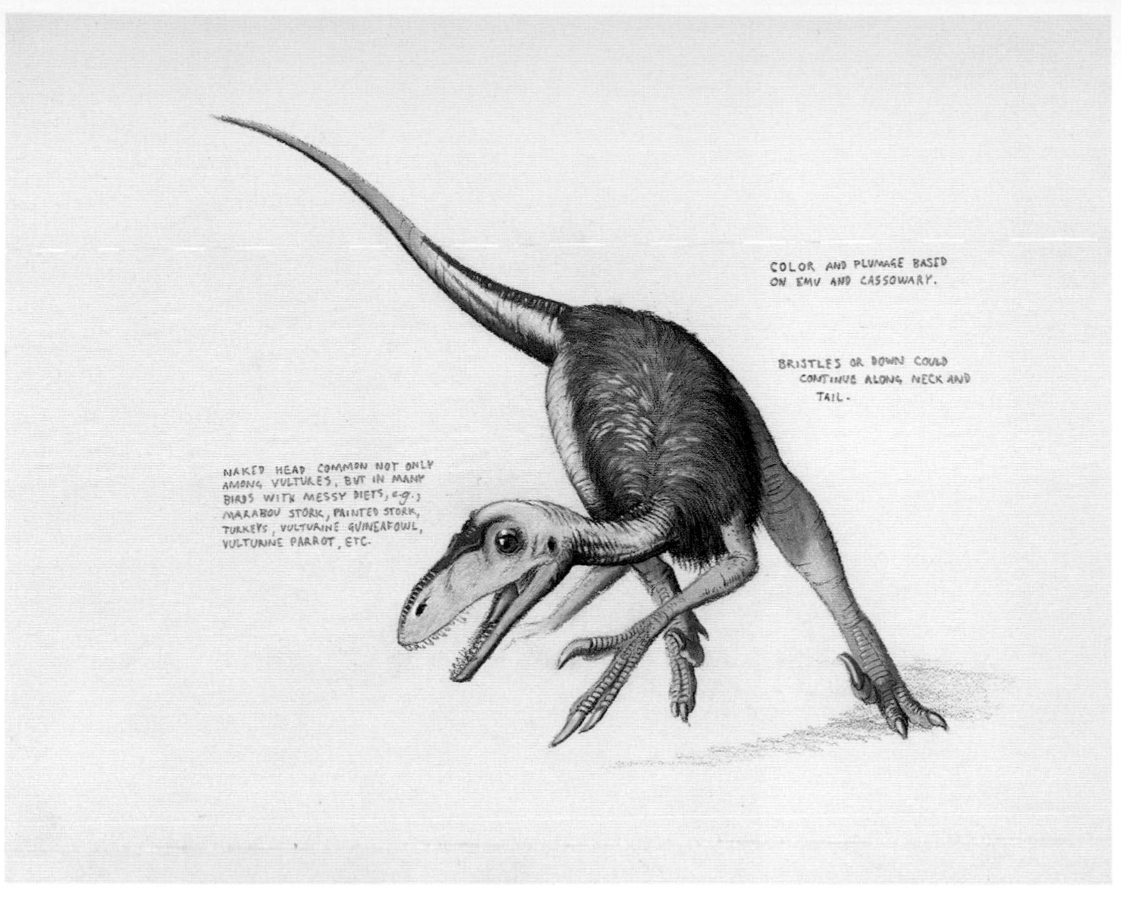

Plate 12

Plate 10: With plumage modeled after an adult Sun bittern.

Plate 11: With a variety of modern plumage.

Plate 12: With a naked head and plumage based on emu and cassowary.

Plate 13: Finished life restoration of *Bambiraptor* by Pat Redman.

Plate 14: Next page. Restoration of *Bambiraptor feinbergi*. (Painting ©2003 Michael W. Skrepnick)

Plate 14

SKREPNICK 2002

Plate 15: Reconstruction of a brooding adult *Troodon*. Recently discovered theropod maniraptoran specimens suggest that feathers might be a more likely skin covering than scales. (Original artwork ©1998 Bill Parsons)

Plate 16

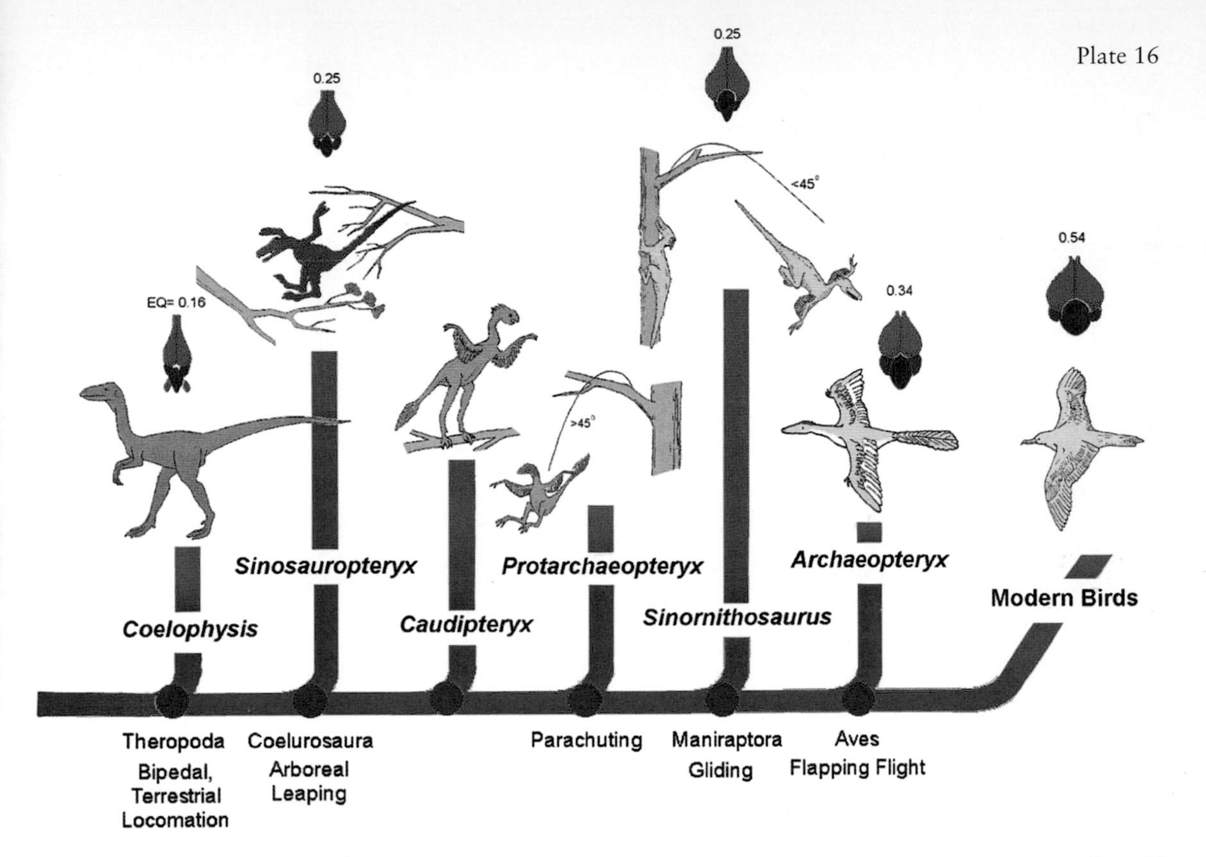

Plate 17

Plate 16: Cladogram showing the paleoecological evolution of birds and their flight (see fig. 12.3).

Plate 17: Small coelurosaurs such as *Sinosauropteryx* climbed trees to elude predators (see fig. 12.5). (Artwork ©2000 Kyle McQuilkin)

Plate 18

Plate 18: Shed crown of hatchling allosaur tooth, 5.5 mm tall. (Courtesy Robert T. Bakker)

Plate 19: Reconstruction based on evidence from the scene at Nail Quarry. Two adult allosaurs drag a carcass toward the lair. (Courtesy Robert T. Bakker)

Plate 19

Plate 20

Plate 20: Nail Quarry.
An adult allosaur delivers prey to hatchlings. (Courtesy Robert T. Bakker)

Plate 21: Prey available at Claw Quarry. Most of the biomass is composed of the giant lung fish, *Ceratodus robustus*. Shed teeth and limb bones represent a *Velociraptor*-size dromaeosaur. Turtles, mammals, and swimming diapsid reptiles are diverse but contribute little to total carcass biomass. (Courtesy Robert T. Bakker)

Plate 21

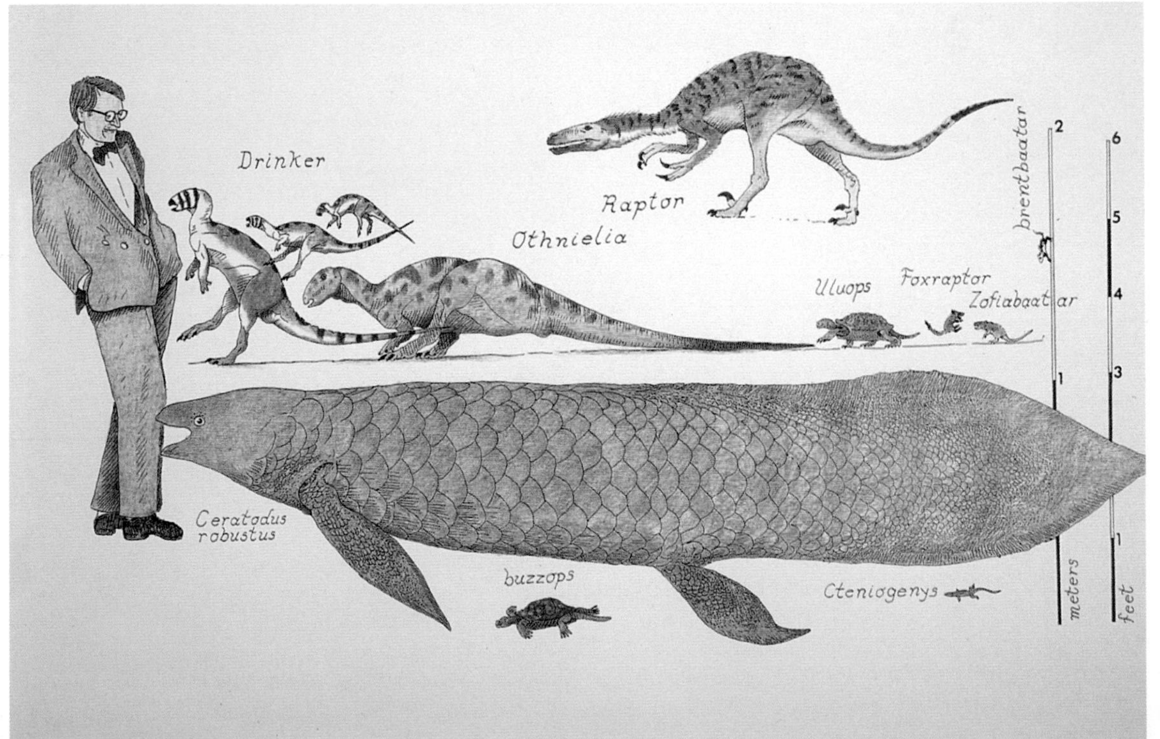

Plate 22: Reconstruction of the head of *Atrociraptor marshalli*
(Painting ©2003 Michael W. Skrepnick)

Acknowledgments. Most of the research leading to this paper was carried out while in receipt of a University of Bristol Postgraduate Scholarship. Subsequent overseas fieldwork was funded by the Geological Society of London, Daniel Pidgeon Fund, and the Palaeontological Association, Sylvester-Bradley Award. The Department of Geology, University of Colorado at Denver covered travel to the symposium and the Graves Museum covered accommodation. I also thank Dr. R. T. Bakker for inviting me to speak at the Symposium. The manuscript was improved by the suggestions of the referees Dr. R. T. Bakker and Dr. J. O. Farlow.

References

Baird, D. 1957. Triassic reptile footprint faunules from Milford, New Jersey. *Bulletin of the Museum of Comparative Zoology, Harvard University* 117: 449–520.

Barsbold, R., and H. Osmólska. 1990. Ornithomimosauria. In D. B. Weishampel, P. Dodson, and H. Osmólska (eds.), *The Dinosauria,* pp. 225–244. Berkeley: University of California Press.

Beckles, S. H. 1851. On supposed casts of footprints in the Wealden. *Quarterly Journal of the Geological Society of London* 7: 117.

———. 1854. On the Ornithoidichnites of the Wealden. *Quarterly Journal of the Geological Society* 10: 456–457.

———. 1862. On some natural casts of reptilian footprints in the Wealden Beds of the Isle of Wight and of Swanage. *Proceedings of the Geological Society* 18: 443–447.

Buffetaut, E., J. Le Loeuff, P. Mechin, and A. Mechin-Salessy. 1995. A large French Cretaceous bird. *Nature* 377: 110.

Calkin, J. B. 1933. *Iguanodon* footprints in Dorset. *Discovery* 14: 157.

Carrano, M. T. 1997. Mammals versus birds as models for dinosaur limb kinematics. *Journal of Vertebrate Paleontology* 17: 36A.

Christiansen, P. 1997. Hindlimbs and feet. In P. J. Currie and K. Padian (eds.), *Encyclopedia of dinosaurs,* pp. 320–328. San Diego: Academic Press.

Colbert, E. H. 1989. The Triassic dinosaur *Coelophysis. Museum of Northern Arizona Bulletin* 57: 1–160.

Currie, P. J., G. Nadon, and M. G. Lockley. 1991. Dinosaur footprints with skin impressions from the Cretaceous of Alberta and Colorado. *Canadian Journal of Earth Sciences* 28: 102–115.

Deane, J. 1861. *Ichnographs from the sandstone of Connecticut River.* Boston: Little, Brown and Company.

Delair, J. B. 1960. The Mesozoic reptiles of Dorset: part 3. *Proceedings of the Dorset Natural History and Archaeological Society* 81: 57–85.

Ensom, P. C. 1987. A remarkable new vertebrate site in the Purbeck Limestone Formation on the Isle of Purbeck. *Proceedings of the Dorset Natural History and Archaeological Society* 108: 205–206.

———. 1995. Dinosaur footprints in the Purbeck Limestone Group (?Upper Jurassic—Lower Cretaceous) of southern England. *Proceedings of the Dorset Natural History and Archaeological Society* 116: 77–104.

Farlow, J. O. 2001. *Acrocanthosaurus* and the maker of Comanchean large-theropod footprints. In D. H. Tanke and K. Carpenter (eds.), *Mesozoic vertebrate life,* pp. 408–427. Bloomington: Indiana University Press.

Farlow, J. O., and Chapman, R. E. 1997. The scientific study of dinosaur

footprints. In J. O. Farlow and M. K. Brett-Surman (eds.), *The complete dinosaur,* pp. 519–553. Bloomington: Indiana University Press.
Farlow, S., M. Gatesy, T. R. Holtz Jr, J. R. Hutchinson, and J. M. Robinson. 2000. *American Zoologist* 40: 640–663.
Feduccia, A. 1996. *The origin and evolution of birds.* New Haven: Yale University Press.
Hitchcock, E. 1836. Ornithichnology. description of the footmarks of birds (ornithichnites) on New Red Sandstone in Massachusetts. *American Journal of Science* 29: 307–340.
———. 1841. *Final report on the geology of Massachusetts.* Northampton, Massachusetts: J. H. Butler.
———. 1858. *Ichnology of New England. A report of the sandstone of the Connecticut Valley, especially its fossil footmarks.* Boston: William White.
Holtz, T. R., Jr. 1996. Phylogenetic taxonomy of the Coelurosauria (Dinosauria: Theropoda). *Journal of Paleontology* 70: 536–538.
Hutchinson, J. R., and K. Padian. 1997. Arctometatarsalia. In P. J. Currie and K. Padian (eds.), *Encyclopedia of dinosaurs,* pp. 24–27. San Diego: Academic Press.
Lockley, M. G., and A. P. Hunt. 1995. *Dinosaur tracks and other fossil footprints from the western United States.* New York: Columbia University Press.
Lockley, M. G., S. Y. Yang, M. Matsukawa, F. Fleming, and S. K. Lim. 1992. The track record of Mesozoic birds: evidence and implications. *Philosophical Transactions of the Royal Society, London* B336: 113–134.
Lockley, M. G., A. P. Hunt, and S. G. Lucas. 1996. Vertebrate track assemblages from the Jurassic Summerville Formation and correlative deposits. *Museum of Northern Arizona Bulletin* 60: 249–254.
Lockley, M. G., C. A. Meyer, and J. J. Moratalla. 1998. *Therangospodus:* trackway evidence for the widespread distribution of a Late Jurassic theropod with well-padded feet. *Gaia* 15: 339–353.
Lull, R. S. 1904. Fossil footprints of the Jura-Trias of North America. *Memoirs of the Boston Society of Natural History* 5: 461–557.
Madsen, J. H., Jr. 1993. *Allosaurus fragilis:* a revised osteology. *Utah Geological Survey, Bulletin* 109: 1–163.
Mansell-Pleydell, J. C. 1888. Fossil reptiles of Dorset. *Proceedings of the Dorset Natural History and Antiquarian Field Club* 9: 1–40.
———. 1896. On the footprints of a dinosaur (?*Iguanodon*) from the Purbeck beds of Swanage. *Proceedings of the Dorset Natural History and Antiquarian Field Club* 17: 115–122.
Molnar, R. E., S. M. Kurzanov, and Z. M. Dong. 1990. Carnosauria. In D. B. Weishampel, P. Dodson, and H. Osmólska (eds.), *The Dinosauria,* pp. 169–209. Berkeley: University of California Press.
Moratalla, J. J. 1993. Restos indirectos de dinosaurios del registro español: Paleoicnología de la Cuenca de Cameros (Jurásico superior—Cretácico inferior) y paleozoología del Cretácico superior. PhD thesis, Universidad Autónoma de Madrid.
Norman, D. B. 1980. On the ornithischian dinosaur *Iguanodon bernissartensis* of Bernissart (Belgium). *Mémoirs de l'Institut Royal des Sciences Naturelle de Belgique* 178: 1–105.
———. 1985. *The illustrated encyclopaedia of dinosaurs.* London: Salamander Books.
Osmólska, H., and R. Barsbold. 1990. Troodontidae. In D. B. Weishampel,

P. Dodson, and H. Osmólska (eds.), *The Dinosauria*, pp. 259–268. Berkeley: University of California Press.
Ostrom, J. H. 1969. Osteology of *Deinonychus antirropus*, an unusual theropod from the Lower Cretaceous of Montana. *Peabody Museum of Natural History, Bulletin* 30: 1–165.
Padian, K., and P. E. Olsen. 1989. Ratite footprints and the stance and gait of Mesozoic theropods. In D. D. Gillette and M. G. Lockley (eds.), *Dinosaur tracks and traces*, pp. 231–241. New York: Cambridge University Press.
Raath, M. A. 1969. A new coelurosaurian dinosaur from the Forest Sandstone of Rhodesia. *Arnoldia* 4: 1–25.
Radley, J. D. 1994a. Field Meeting, 24–5 April, 1993: the Lower Cretaceous of the Isle of Wight. *Proceedings of the Geologists' Association* 105: 145–152.
———. 1994b. Stratigraphy, palaeontology and palaeoenvironment of the Wessex Formation (Wealden Group, Lower Cretaceous) at Yaverland, Isle of Wight. *Proceedings of the Geologists' Association* 105: 199–208.
Rowe, T. 1989. A new species of the theropod dinosaur *Syntarsus* from the Early Jurassic Kayenta Formation of Arizona. *Journal of Vertebrate Paleontology* 9: 125–136.
Rowe, T., and J. Gauthier. 1990. Ceratosauria. In D. B. Weishampel, P. Dodson, and H. Osmólska (eds.), *The Dinosauria*, pp. 151–168. Berkeley: University of California Press.
Sarjeant, W. A. S. 1974. A history and bibliography of the study of fossil vertebrate footprints in the British Isles. *Palaeogeography, Palaeoclimatology, Palaeoecology* 16: 265–378.
Thulborn, R. A. 1990. *Dinosaur tracks*. London: Chapman and Hall.
Thulborn, R. A., and M. Wade. 1979. Dinosaur stampede in the Cretaceous of Queensland. *Lethaia* 12: 275–279.
Wilson, M. C., and P. J. Currie. 1995. *Stenonychosaurus inequalis* (Saurischia: Theropoda) from the Judith River Formation of Alberta: new findings on metatarsal structure. *Canadian Journal of Earth Sciences* 22: 1813–1817.
Wright, J. L. 1996. Fossil terrestrial trackways: preservation, taphonomy and palaeoenvironments. PhD thesis, University of Bristol.
———. 1999. Ichnological evidence for the use of the forelimb in iguanodontid locomotion. *Special Papers in Palaeontology* 60: 209–219.
Wright, J. L., D. M. Unwin, M. G. Lockley, and E. C. Rainforth. 1997. Pterosaur tracks from the Purbeck Limestone Formation of Dorset, England. *Proceedings of the Geologists' Association* 108: 39–48.
Wright, J. L., P. M. Barrett, M. G. Lockley, and E. Cook. 1998. A review of the Early Cretaceous terrestrial vertebrate track-bearing strata of England and Spain. *New Mexico Museum of Natural History Bulletin* 14: 143–153.
Wright, J. L., and M. G. Lockley. 2002. New vertebrate tracks from the Laramie/Arapahoe Formations (Late Cretaceous), near Denver, Colorado. *Cretaceous Research* 22: 365–376.

Section III. Eggs, Nests, Feathers, and Flight

9. Dinosaur Eggs and Nesting: Implications for Understanding the Origin of Birds

Gerald Grellet-Tinner and
Luis M. Chiappe

Abstract

Recent discoveries of dinosaur eggs containing embryos have furnished data to confidently anchor oological characters and in some instances reproductive behavioral strategies to specific dinosaurian taxa. Here we examine the egg morphology and nesting behaviors of hard-shelled turtles, crocodilians, and a variety of dinosaurs (hadrosaurid, titanosaur, troodontid, and oviraptorid), and present comparisons to those found in extant birds. Our study documents the greatest similarity between the eggs and nest attendance of non-avian theropod dinosaurs (i.e., troodontids and oviraptorids) and those of birds, thus adding support to the hypothesis that birds originated among non-avian theropods.

Introduction

The origin of birds has long captured the attention of natural philosophers and evolutionary biologists (Chiappe 2001). With their unique plumage, winged bodies, and bipedal gait, birds greatly differ from other groups of extant organisms, a fact that has kept the search for their origins under debate. Ever since the advent of cladistic analysis, numerous phylogenetic hypotheses have identified different taxa as

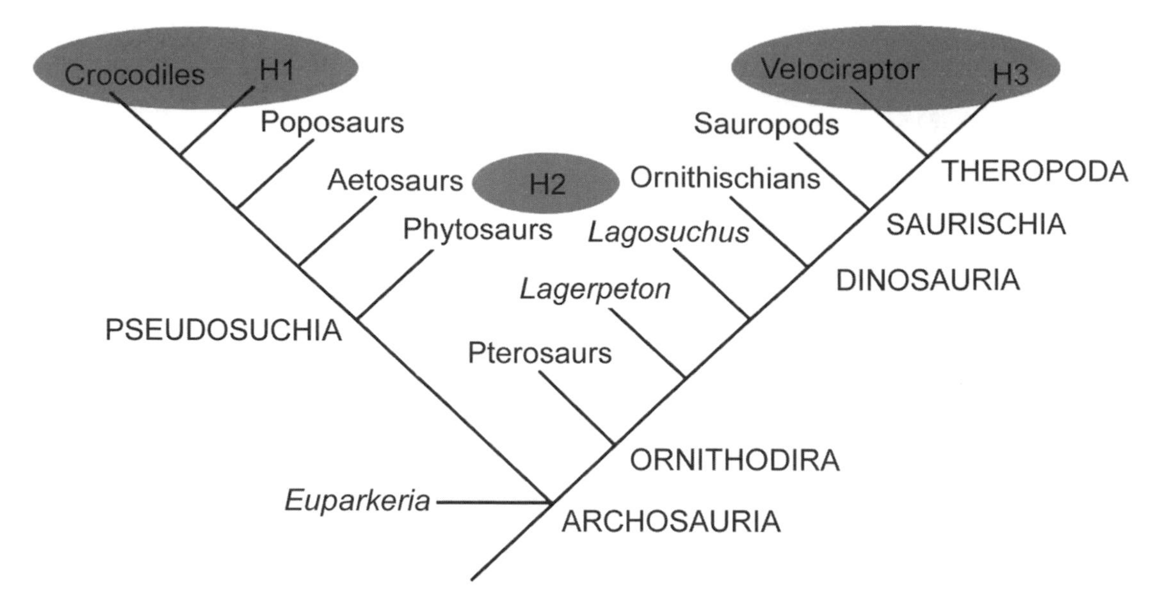

Figure 9.1. Three general hypotheses of bird origins. (From Padian and Chiappe 1998)

the closest relatives of birds (Witmer 1991). Among many alternative hypotheses (most of which have been abandoned), three are still held (fig. 9.1): (1) those identifying crocodiles as the closest relatives of birds (Martin et al. 1980; Martin 1991; Martin and Stewart 1999); (2) those supporting a relationship between birds and basal archosaurs (Welman 1995) or archosauromorphs (Feduccia and Wild 1993; Feduccia 1996); and (3) the most agreed-upon, those regarding birds as descendants of non-avian theropod dinosaurs (Ostrom 1976; Gauthier 1986; Witmer 1991; Padian and Chiappe 1998; Sereno 1999; Chiappe 2001). The latter alternative, in fact, is composed of many specific hypotheses proposing several different non-avian theropod lineages as closest relatives of birds (Witmer 1991).

For decades, since the beginning of comparative zoology, evidence for the ancestry of birds has been based entirely on osteological characters. Discoveries of critical fossils, such as *Compsognathus, Archaeopteryx,* and *Deinonychus,* played paramount roles in the debate on bird ancestry (Desmond 1982; Witmer 1991). More recently, a burst of extraordinary fossil finds has brought additional non-osteological data to the debate (Chiappe 2001). Among these new lines of evidence is oology, the study of egg morphology and eggshell microstructure.

Eggs can be identified taxonomically with certainty only by either observing an egg-laying female or by the embryos they contain. In the fossil record, the former would require the unlikely preservation of a female with eggs in its reproductive system, an event that has not yet been conclusively documented (see Griffiths [2000] for a purported occurrence, however). Fossil occurrences of diagnostic embryonic remains in ovo (inside an egg), albeit rare, are more abundant. Although no oological data is yet available for basal, extinct archosauromorphs, recent findings of a variety of fossil eggs containing embryonic material (e.g., Currie and Horner 1988; Horner and Weishampel 1988; Horner

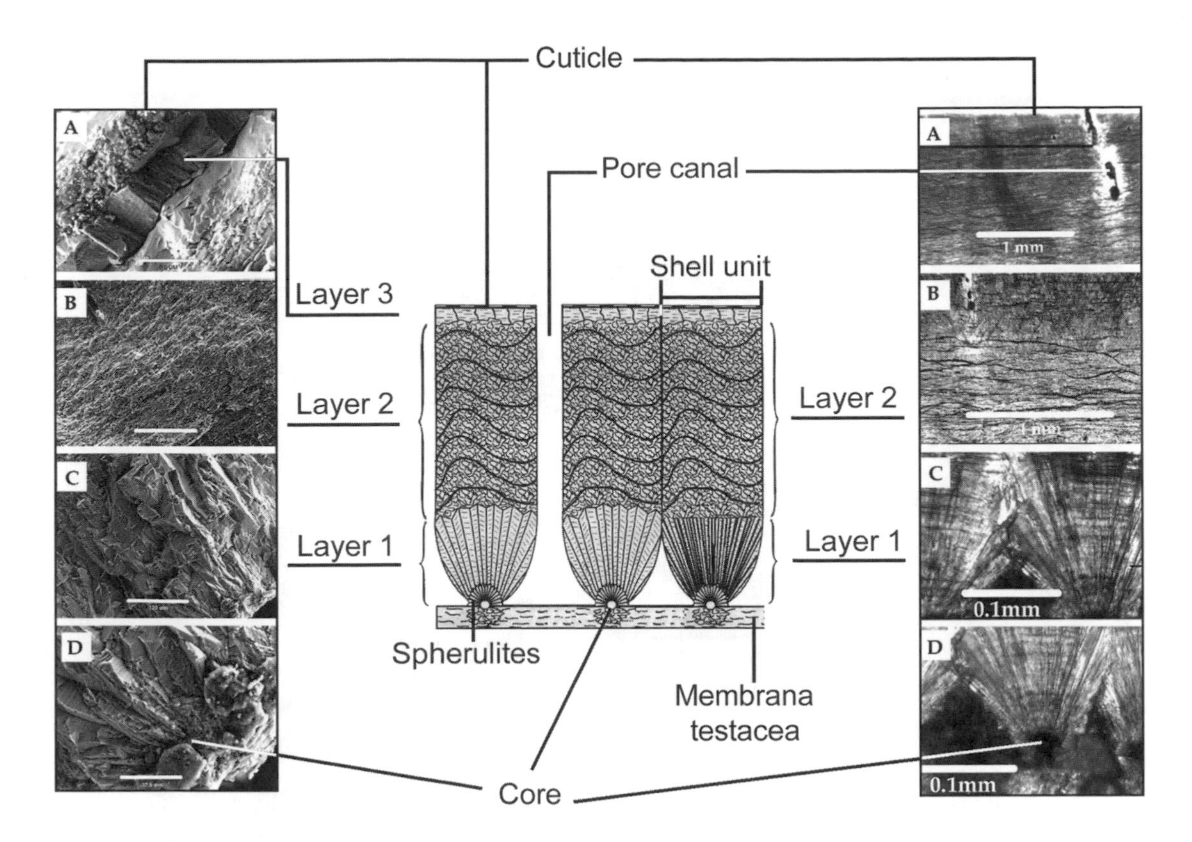

Figure 9.2. Eggshell microstructural characters. SEM images (left), schematic drawings (center), and thin sections (right) illustrating characters from various eggshells.
A (left) shows layer 3 and the organic cuticle that covers it;
A (right) shows layers 3, 2 and a pore canal;
B (right) shows the structure of layer 2;
C, D (left) display bladed crystals in layer 1;
C, D (right) display acicular crystals in layer 1.

and Currie 1994; Norell et al. 1994, 2001; Horner 1997; Manning et al. 1997; Mateus et al. 1997; Chiappe et al. 1998) have provided reliable documentation of the egg morphology of several extinct dinosaur lineages. These discoveries have also made possible the association of eggs of some extinct dinosaurs with specific nest structures (e.g., Varricchio et al. 1997; Chiappe and Dingus 2001) and in some instances, with specific nesting behaviors (e.g., Norell et al. 1995; Dong and Currie 1996; Clark et al. 1999; Varricchio et al. 1999).

It has been convincingly argued that nest structures and behaviors (Wenzel 1992; Clark et al. 1999, Varricchio et al. 1999) as well as egg morphology (Grellet-Tinner 2000a,b; Makovicky and Grellet-Tinner 2000) can provide information pertinent for phylogenetic inference. In this paper, we provide a synopsis of the egg morphology and nesting behavior of those groups of archosaurs for which these data are confidently known (appendix 9.1) and discuss, in light of a cladistic analysis, the significance of this information for understanding the origin of birds.

Institutional Abbreviations: AMNH, American Museum of Natural History, New York, U.S.; GIN, Institute of Geology, Ulaan Baatar, Mongolia; LACM, Natural History Museum of Los Angeles County, Los Angeles, California, U.S.; MCF, Museo Carmen Funes, Plaza Huincul, Neuquén, Argentina; MCZ, Museum of Comparative Zoology, Cambridge, Massachusetts, U.S.; MOR, Museum of the Rockies, Boze-

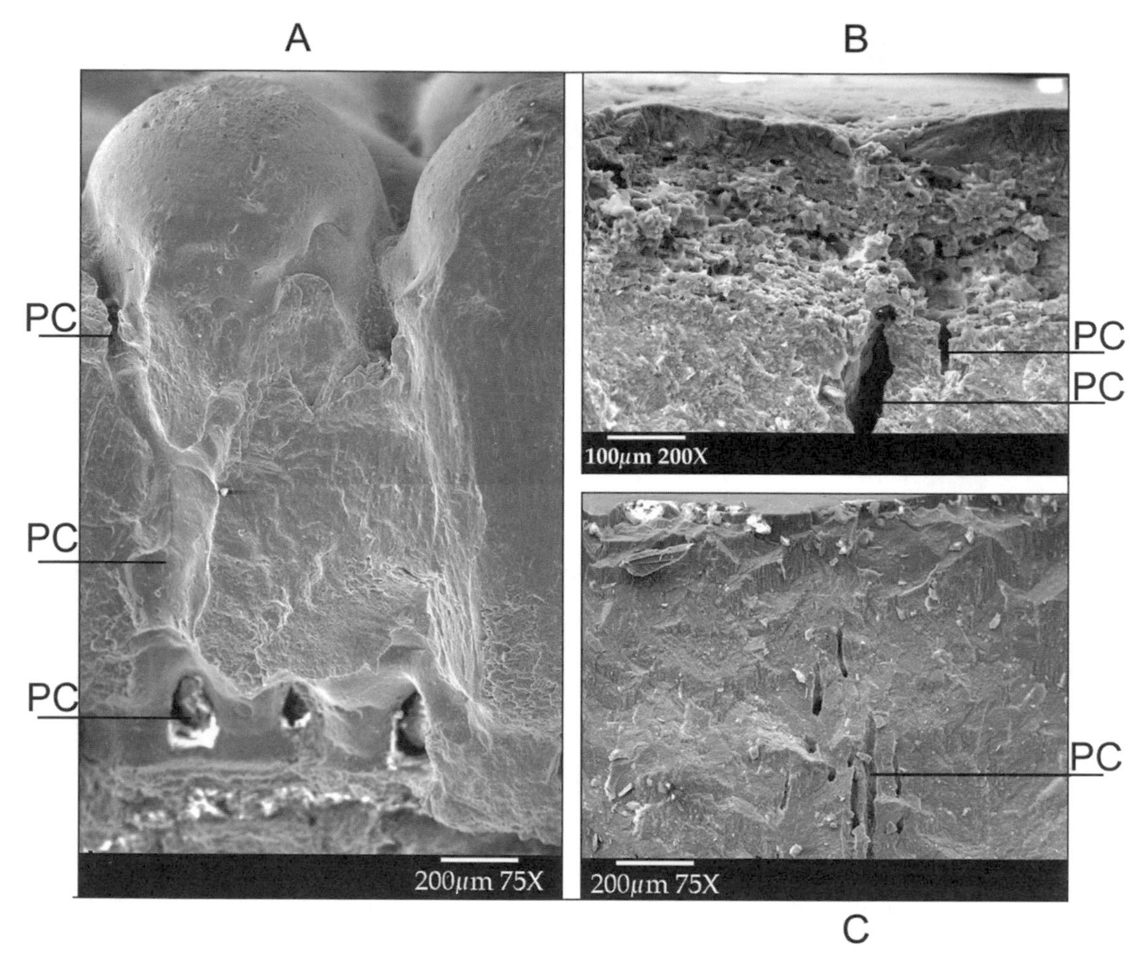

Figure 9.3. Pore canal geometry (SEM). A, titanosaur eggshell. B, Dromaius novaehollandiae. *C,* Struthio camelus.

man, Montana, U.S.; SAMP, South Australian Museum, Adelaide, South Australia, Australia; RTMP, Royal Tyrrell Museum of Palaeontology, Drumheller, Alberta, Canada; UTDGS, University of Texas Department of Geological Sciences, Austin, Texas, U.S.; YPM, Yale Peabody Museum of Natural History, New Haven, Connecticut, U.S.

Terminology and Methods

The descriptive terminology used here relies as much as possible on the published literature on eggs and eggshells (e.g., Mikhailov 1987, 1991, 1997; Hirsch et al. 1997). A few new terms that describe and better define characters are also implemented (Grellet-Tinner 2000a, Makovicky and Grellet-Tinner 2000). By egg morphology we consider all aspects (e.g., egg shape, surficial attributes, eggshell microstructure) of eggs. Eggshell radial sections were examined not only to count the number of layers constituting the total thickness (fig. 9.2), but also to observe the crystallographic arrangement of the shell units (fig. 9.2), the shape of pore canals (fig. 9.3), and presence and geometry of

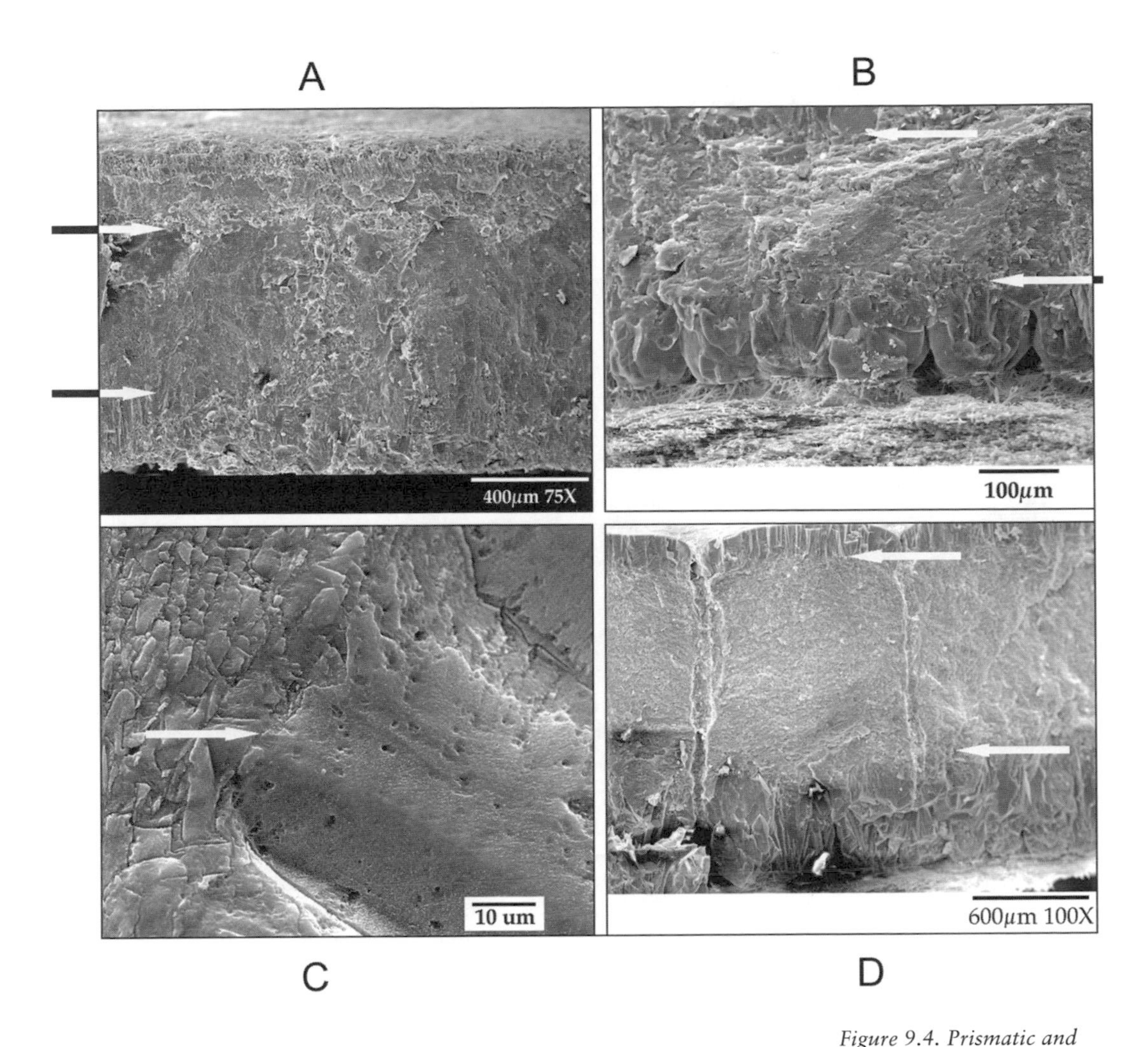

Figure 9.4. Prismatic and aprismatic conditions (SEM). A, prismatic condition of Troodon formosus. *B, prismatic condition of* Anser anser. *C, aprismatic condition between layers 1 and 2 of* Dromaius novaehollandiae. *D, aprismatic condition of* Rhea americana. *White arrows indicate layer boundaries.*

spherulites. Layers are sequentially numbered from the innermost (layer 1) to the outermost. Each layer is regarded as an independent character; their boundaries are either prismatic or aprismatic (Grellet-Tinner 2000a; Makovicky and Grellet-Tinner 2000). A prismatic condition means that the shell unit originating at the membrana testacea (MT) is visible throughout the entire eggshell thickness (fig. 9.4). An aprismatic condition means that the shell unit is visible in layer 1, but is concealed in successive layers because of differences in crystallographic orientation of these layers (fig. 9.4). Layers are more discernible in eggshells displaying an aprismatic condition than those exhibiting a prismatic one. Surficial observations record the varying ornamentation (fig. 9.5) and pore aperture (fig. 9.6).

Eggshells (see appendix 9.1) were washed in distilled water and dried on a warm ceramic plate. Once dried, the eggshell was broken in two pieces to obtain a fresh radial fracture. One of the pieces was used for scanning electron microcopy (SEM) imaging and the other for thin section. The eggshell used for SEM was glued slightly tilted on an aluminum stub with its freshly broken radial side upward to permit

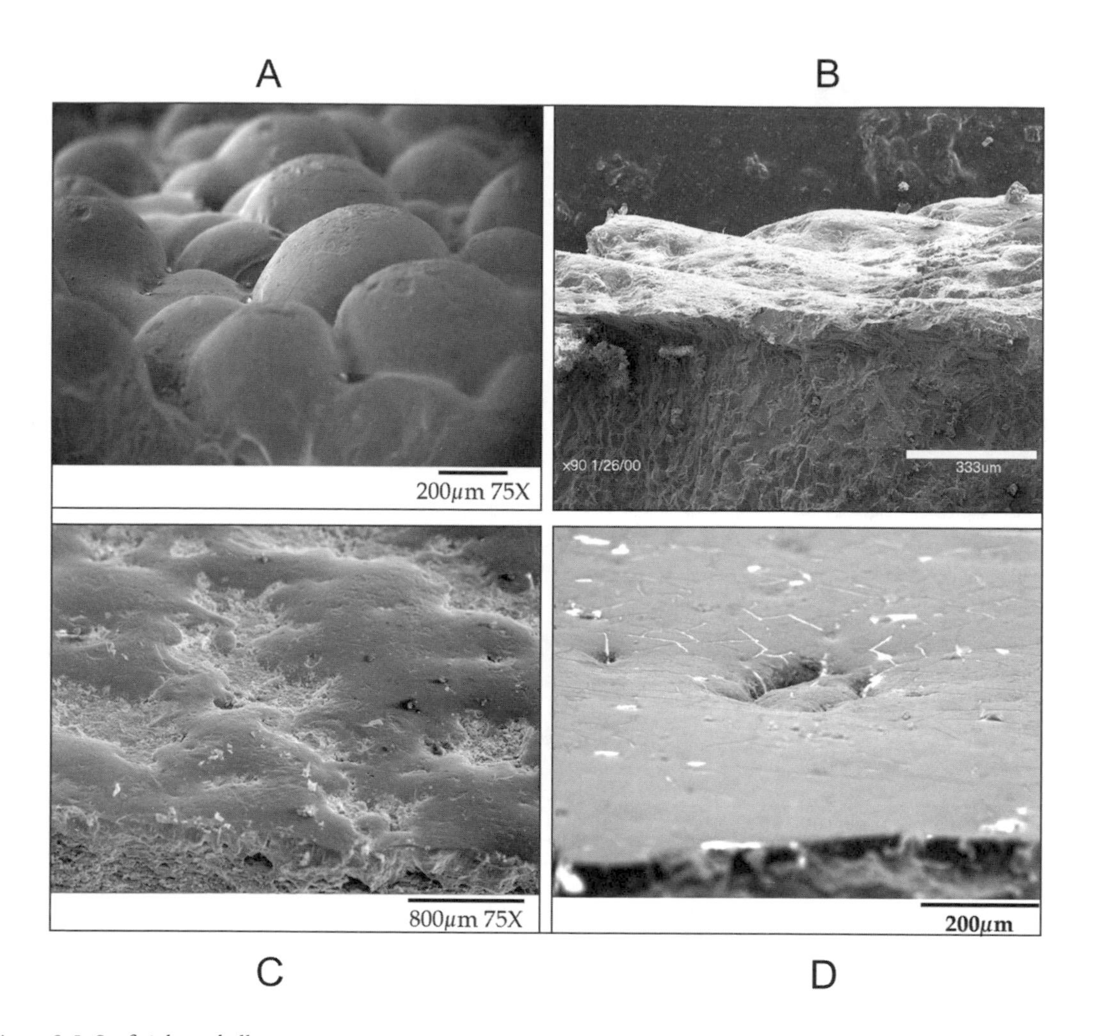

Figure 9.5. Surficial eggshell ornamentations (SEM).
A, nodular ornamentation of titanosaur.
B, ridges parallel to the long axis of an oviraptorid egg.
C, granulated surface of Dromaius novaehollandiae.
D, smooth surface of Struthio camelus *(notice the round pore aperture).*

tangential observation. SEMs were conducted at 30 kilovolts and a scale bar was embedded in each image file during digital acquisition. Adobe Photoshop software (version 5.5) was used to enhance image contrast and brightness, and to measure designated features. Measurements of eggshell thickness were obtained by counting the number of pixels in the embedded scale bar and the observed structure. These measurements were conducted from the base of layer 1 to the outside of the external layer. Mean values only were recorded.

On fossil eggshells, cathodoluminescence (CL) was performed when necessary in order to discriminate diagenetic from original features (Korhing and Hirsch 1996; Barbin 2000). Diagenetic calcite or other minerals often invade pores and blanket the eggshell outer surface, concealing the surficial ornamentation (Hirsch and Quinn 1990) and/or pervading the eggshell inner side (fig. 9.7). Other factors that may show variation between the eggs of different taxa, and hence may be used as characters, include pore distribution, egg size, and latitudinal (position in the egg) eggshell variation. While pore distribution and number have been shown to vary within fossil eggs (Hirsch and Quinn

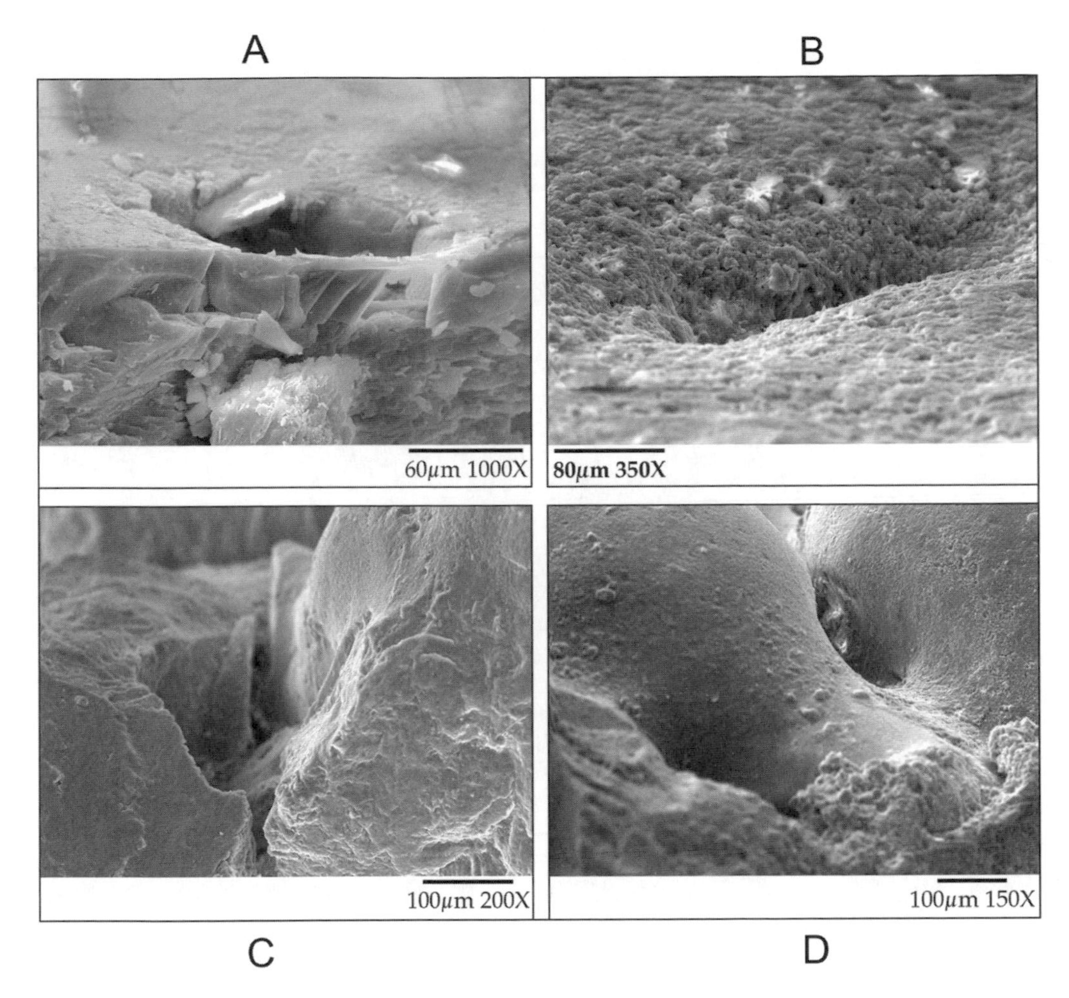

Figure 9.6. Pore apertures (SEM). A, round pore aperture of Rhea americana *and* Struthio camelus. *B, slit-like pore shape of* Genyornis, *an Australian Cenozoic bird. C, radial view of a pore of a titanosaur. D, pore apertures surround individual nodes of titanosaur.*

1990; Varricchio et al. 1997), pore canal geometry and aperture shape appear to remain constant (Grellet-Tinner 2000a), thus making these characters more reliable for phylogenetic analysis. Assessing variation in egg size and shell latitudinal position is difficult and we have been unable to evaluate them in the extinct taxa. Much of this variation, however, was assessed for the extant taxa studied here. In their first nesting season, birds lay smaller eggs that may display aberrant eggshell structures (personal observation). The size of the extant eggs used here agrees with those of second and later nesting seasons. Latitudinal eggshell variation relates to the fact that eggshell thickness, texture, and structure may vary according to the position of the shell within an egg. Our study of *Dromaius novaehollandiae* (emu) eggshell sampled entire eggs (fig. 9.8). We were unable, however, to perform this test in either the remaining extant species or in any of the extinct ones.

A cladistic analysis using 11 characters of egg morphology and nesting behavior (only one multistate character, here treated as an ordered transformation) was conducted using the computer program Hennig 86 (Farris 1988) (appendixes 9.1 and 9.3). Chelonia was used

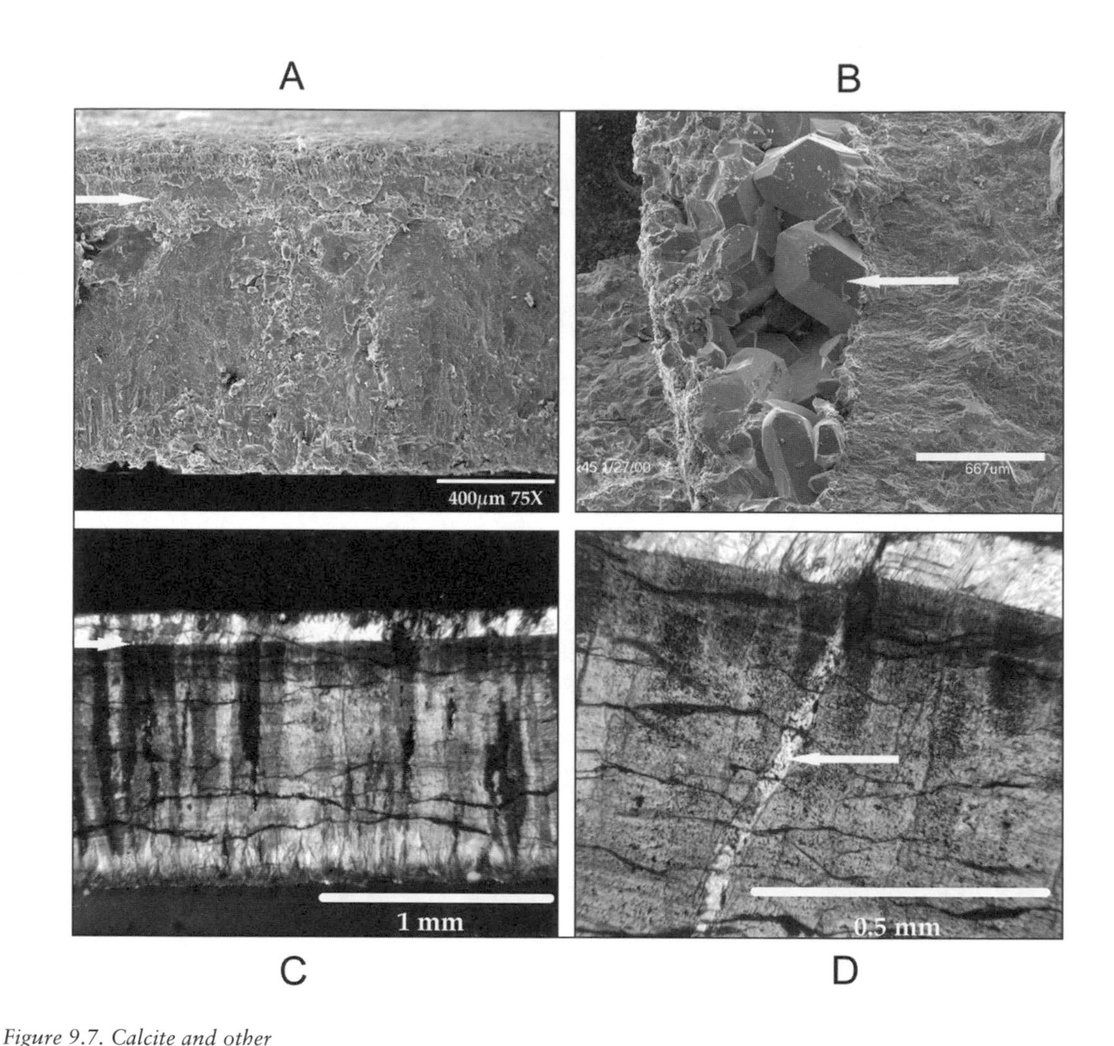

Figure 9.7. Calcite and other minerals invading pores. A, SEM of Troodon formosus *eggshell. B, SEM of crystal growth in layer 1 of an oviraptorid eggshell. C, like A but in thin section. D, calcite invades pores in a* Troodon formosus *eggshell. White arrows point to the designated features.*

as the outgroup and the 10 archosaur taxa here described were included in the ingroup. In order to exclude uninformative characters from the data matrix (e.g., unique autapomorphies), only characters having derived states shared by two or more ingroup taxa were included in the analysis. Applicable character-states were scored as "?" following Strong and Lipscomb (1999). Only the absence of more than one eggshell layer (character 4, appendix 9.1) in Chelonia, Crocodylia, Ornithischia, and Sauropoda made scoring of character 5 (aprismatic versus prismatic delimitation between eggshell layers) inapplicable for the latter taxa. Due to the presence of an inapplicable character-state in the outgroup (i.e., Chelonia), it was uncertain which of the two conditions (i.e., aprismatic or prismatic) is plesiomorphic. We conducted two separate analyses of the data matrix in which the aprismatic and prismatic conditions were alternately scored as primitive ("0") for character 5. The effect of this character-state substitution was found to be nil; both analyses resulted in identical tree topologies and tree lengths.

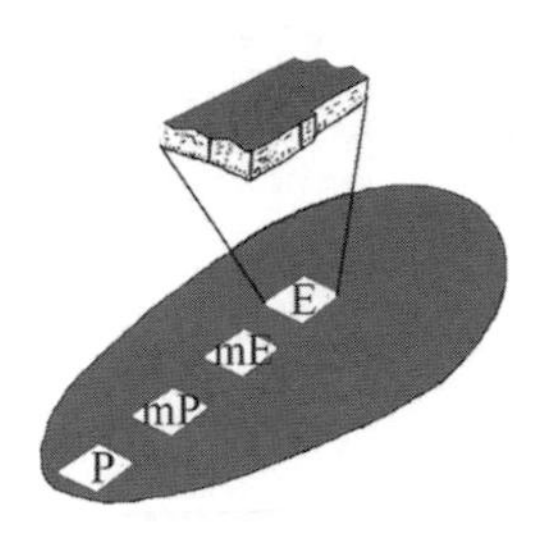

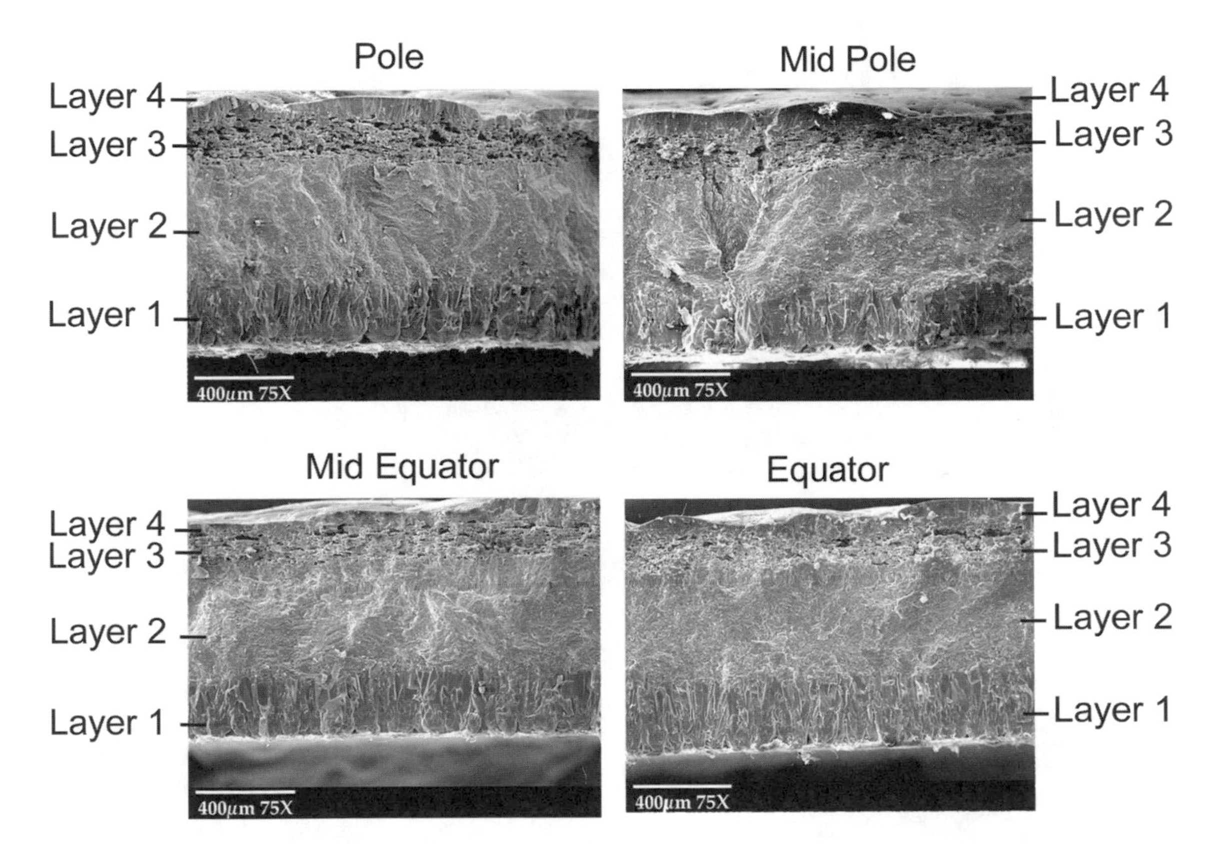

Figure 9.8. Structural homogeneity within a Dromaius novaehollandiae *egg. SEM images of eggshell taken at four intervals on a single* Dromaius novaehollandiae *egg did not reveal any structural differences.*

Egg Morphology and Nest Attendance

Chelonia

Many turtles produce hard-shelled eggs. Nonetheless, it is well known that turtle eggshell may also be soft and leathery (Pritchard 1979), a variation that is interspecific. Although leathery eggs are more advantageous for the female because they require a lesser amount of calcium carbonate than hard-shelled eggs, they offer little diffusive barrier between the egg and the environment of the nest. Hard-shelled turtle eggs vary in shape from elongate to spherical. Their equatorial diameter may range from slightly less than 2 to 7 cm, depending on the species and the maturity of the egg-laying female (Pritchard 1979). Turtle eggshell is mostly smooth to the naked eye and consists of a

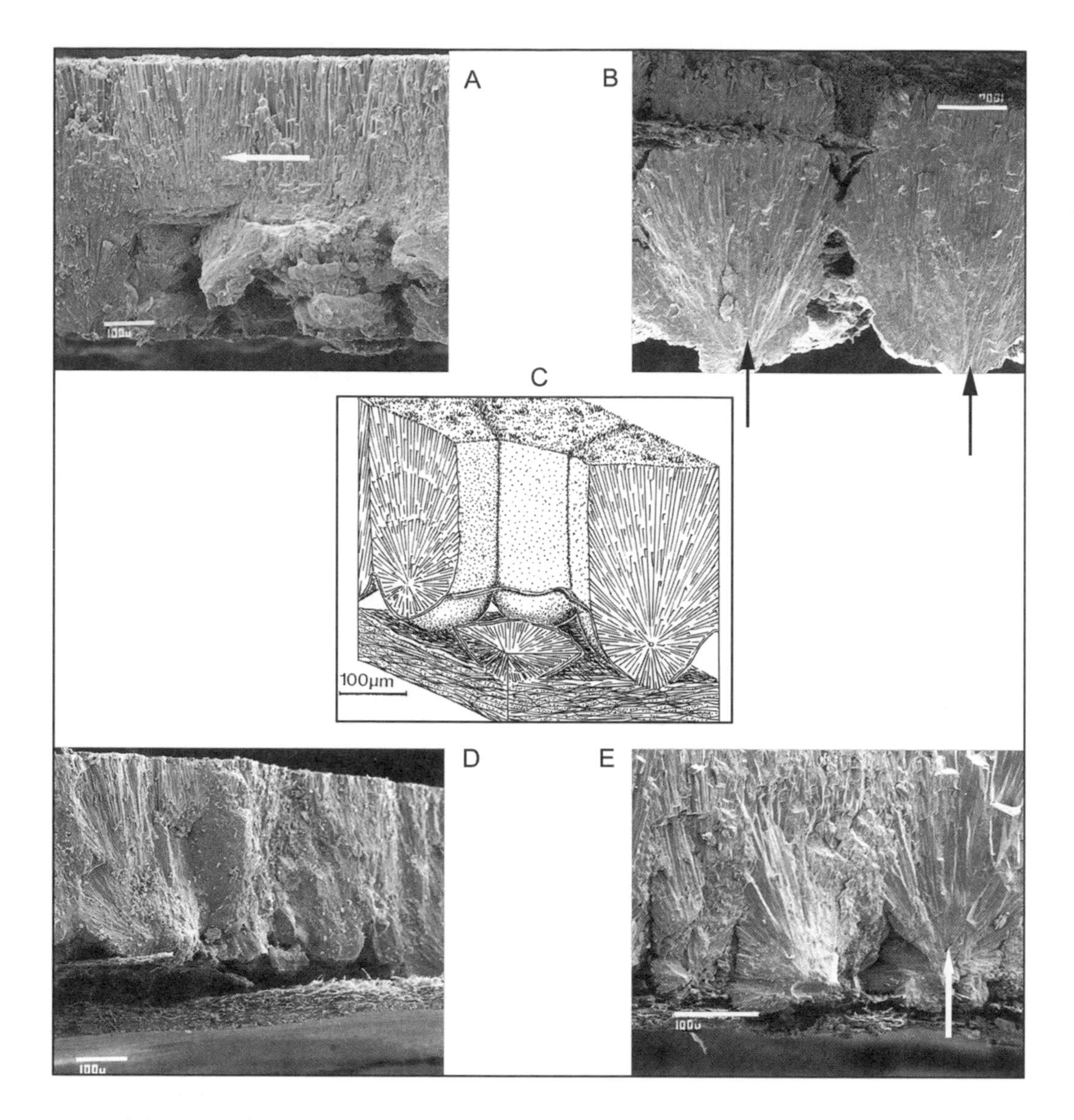

Figure 9.9. Chelonian eggshells. SEM images from Schleich and Kästle (1988).
A, aragonitic acicular crystallographic arrangement of Chelonoidis denticulata.
B, eggshell units fanning out from the core of Chelonoidis carbonaria.
C, acicular crystals and eggshell units of Chelus fimbriatus.
D, total eggshell thickness of Chelus fimbriatus.
E, radiating crystals around the core of Testudo hermanni.
Arrows point to the designated features.

single layer of juxtaposed eggshell units, which grow out of the MT and originate from organic cores (Schleich and Kästle 1988) (fig. 9.9). Turtle eggshell is unique among reptiles in that it is composed of metastable aragonitic acicular crystals (fig. 9.9C). Pores, when present, develop as straight canals between the eggshell units.

Most turtle nests consist of dug-out holes in sand or other substrates where the female lays from a couple to more than 200 eggs (according to the species) in a single ovideposition (Pritchard 1979). In marine species large numbers of females congregate on the same nesting shore within a short period of time to lay their eggs. No parental care has ever been observed among turtles (Pritchard 1979); the young have to fend for themselves when they synchronously hatch.

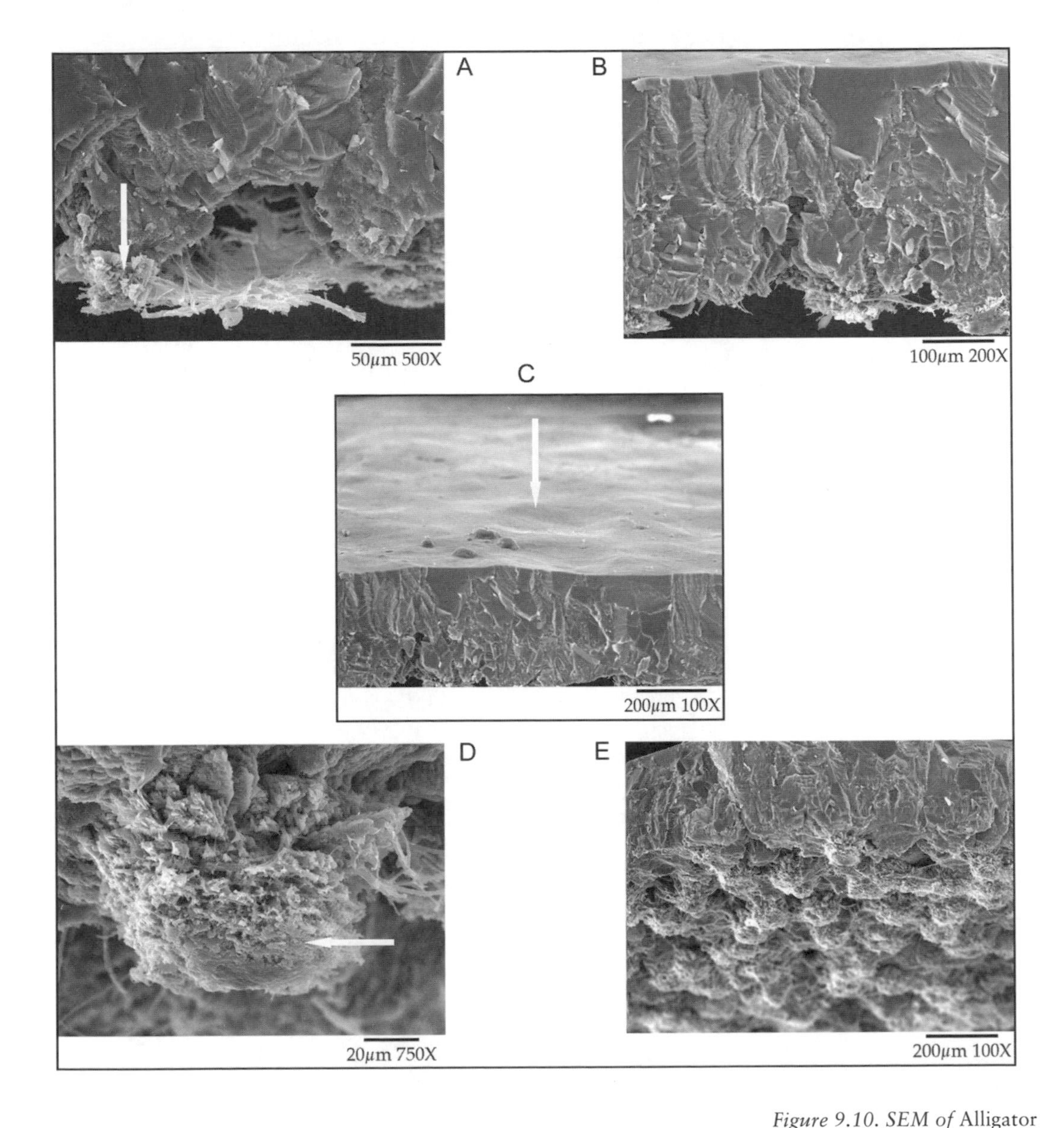

Figure 9.10. SEM of Alligator mississippiensis *eggshell.*
A, calcitic units anchored in the membrana testacea.
B, eggshell units are differentiated one from another.
C, faint surficial undulations.
D, the approximately 0.09-mm rosette is twice as wide as high. White arrows point to the designated features.
E, flat, petal-like rosettes surrounded by protein fibers of the MT form the base of the eggshell units.

Crocodylia

The American alligator (*Alligator mississippiensis*) lays white oblong eggs averaging 73 by 43 mm with an aspect ratio of 1.7. Their surface appears smooth although faint undulations are visible when viewed in SEM cross sections (fig. 9.10C). The one-layered shell of these eggs consists of calcitic units anchored in the MT (fig. 9.10A). These eggshell units display distinct polygons with a horizontal crystal orientation (fig. 9.10B), a geometry also typical of other members of Crocodylia (Schleich and Kästle 1988, Kohring and Hirsch 1996). Eggshell units are well differentiated from one another (fig. 9.10B), and developing pores at the innermost side (sometimes referred to lacunae) pervade the contacts between them. The base of the eggshell unit consists of a flat, petal-like rosette surrounded by protein fibers of the MT (figs. 9.10D, E). This approximately 0.09-mm-wide rosette is twice as wide as high. The eggshell thickness measured from the rosette to the eggshell surface equals 0.71 mm.

In addition to the distinct polygonal unit geometry and lacunae development, the horizontal microscopic calcitic arrangement of crocodilian eggshell further contrasts with the vertical crystallographic arrangement of single-layered eggshell seen in other reptiles.

Alligator mississippiensis lays clutches of 45 eggs on average at a rate of one egg every thirty seconds in nests made of vegetation mounds (Ross 1989). Crocodile parental care involves guarding the nest by the female, which also responds to hatchling cries and provides assistance by transporting them from the nest to the nearest water pool.

Ornithischia

Hadrosaurid eggshells with in ovo embryonic remains were reported by Horner and Currie (1994) from the Late Cretaceous locality of Devil's Coulee in southern Alberta. Similar eggs were found from other localities in southern Alberta and northern Montana. Although initially identified as *Hypacrosaurus stebingeri,* the fragmentary nature of these undoubtedly hadrosaurid embryos makes such specific identification tentative (Horner 2000). Thus we regard the *Hypacrosaurus stebingeri* eggs as hadrosaur eggs without more specific determination (appendix 9.1). Hadrosaurid eggshell microstructure is characterized by one single layer (fig. 9.11C). Pore canals of the studied specimens are filled with secondary calcite. These canals do not display a constant geometry (fig. 9.11A); they all seem to widen in the mid-section of the eggshell and constrict when reaching the outer and inner surfaces. The degree of diagenesis of the studied specimens hinders observation of most pore surficial apertures, but the few visible ones are round. Although not discernible in any of our samples from specimen TMP 87.79.85 (fig. 9.11B,E), Hirsch and Quinn (1990) mentioned the presence of nucleation points, here interpreted as organic cores, in hadrosaurid eggshell of Devil's Coulee. Hirsch and Quinn (1990) performed HCl-etching of *Hypacrosaurus stebingeri* eggshells to enhance the visibility of shell unit boundaries, a procedure we did not follow in the context of this study. Nonetheless, the eggshell units of TMP 87.79.85 appear to fan out from the MT and seem to be nearly twice as high as wide. Surficial ornamentation of this eggshell is vermiculated. The total thickness of the observed samples is 0.71 mm.

Among the several clutches from Devil's Coulee, three contain eggs with in ovo embryos assigned at that time to *Hypacrosaurus stebingeri.* These three clutches were used for comparison with others from this and other localities. In spite of the presence of numerous clutches, the exact number of eggs in them remains unclear given the extent of erosion and other taphonomic factors. Horner and Currie (1994) reconstructed the size and shape of the eggs as subspherical, measuring 18.5 by 20 cm and with a volume of 3,900 ml.

Parental care has been inferred for the hadrosaurid *Maiasaura peeblesorum* (Horner and Makela 1979; Horner 1982, 1984) on the

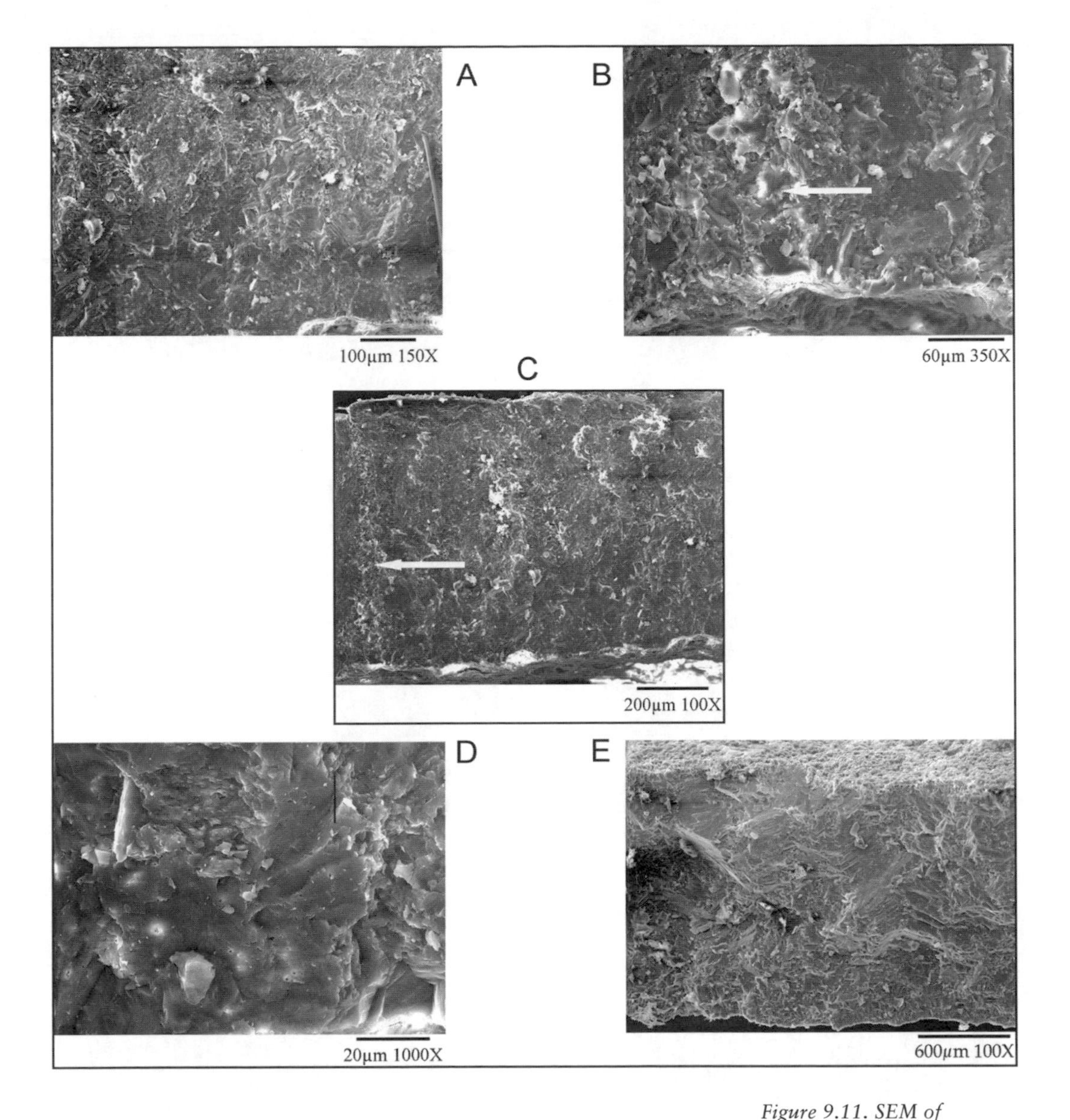

Figure 9.11. SEM of Hypacrosaurus stebingeri *eggshell.*
A, highly diagenetic single-layered eggshell.
B, pore canal.
C, this eggshell is characterized by one single layer.
D, pore canals filled with secondary calcite.
E, boundary between the eggshell units is not clear. White arrows point to the designated features.

basis of discovering a group of juveniles contained within an alleged nest structure. The grouping of these neonates led these authors to argue in favor of nidicolous behavior and thus parental care, given the required dependence on adults to bring food to the nest. Horner and Makela's (1979) interesting behavioral inference hinges upon the recognition of a nest, because otherwise, there is no reason to argue for nidicolous behavior and therefore, nest attendance. Their inference was supported by the presence of a small lens of green mudstone, representing the putative nest structure embedded within beds of red mudstone. Nonetheless, difference in color of the substrate does not seem to be an adequate criterion for recognizing fossil nest structures because such a difference could well be the result of secondary sedimentary processes. Dinosaur nests are best recognized on the basis of textural differences

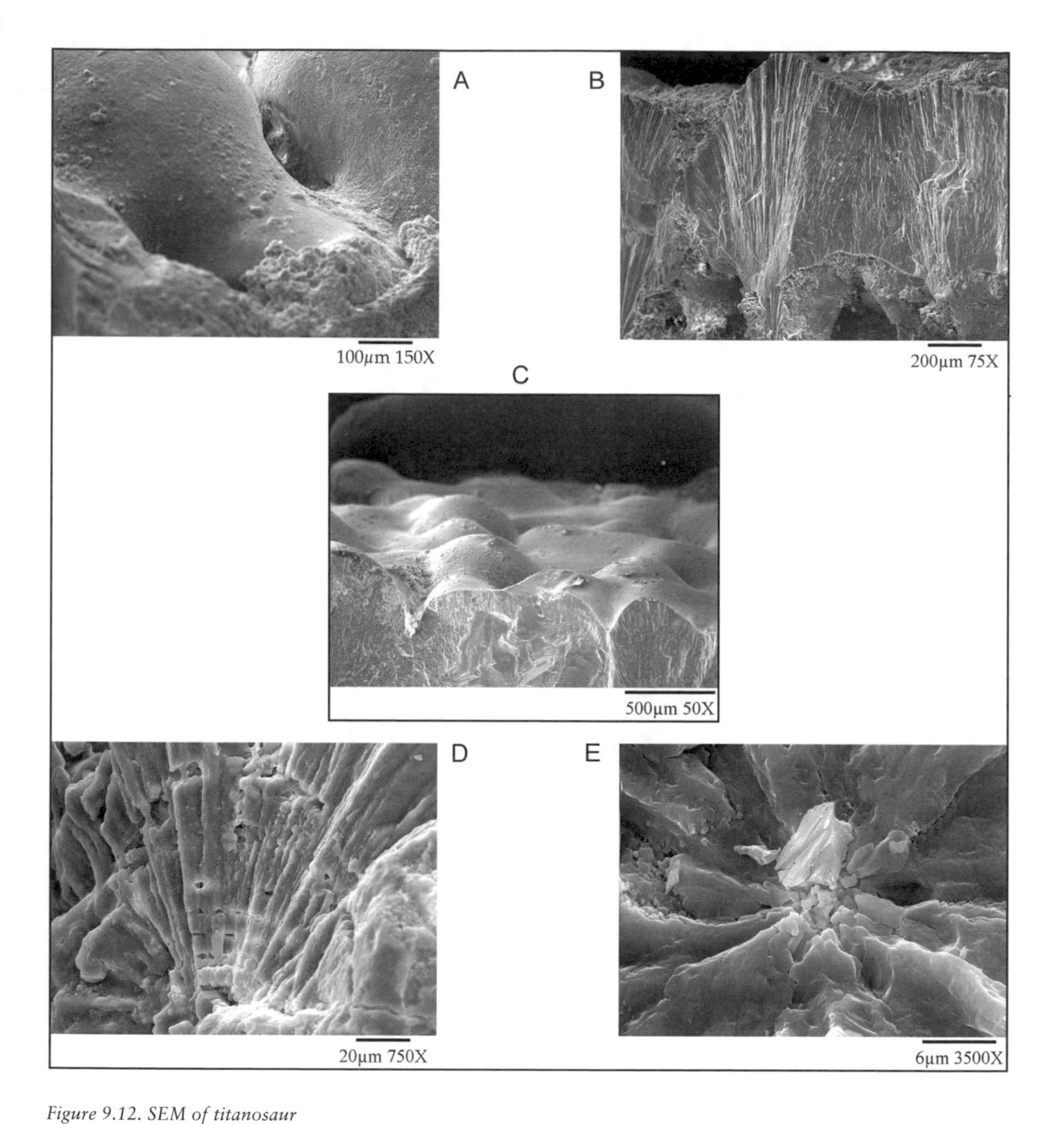

Figure 9.12. SEM of titanosaur eggshell.
A, nodes surrounded by two to four sub-circular pores.
B, single-layered eggshell.
C, these eggs display a nodular surficial ornamentation.
D, E, organic core surrounded by minute semicircular spherulites.

between the host substrate and the overlaying deposits (Sander et al. 1998). To the best of our knowledge, no reliable evidence documenting parental care in hadrosaurid or in any other ornithischians has so far been discovered.

Sauropoda

Among sauropod dinosaurs, only titanosaur eggs from the Patagonian locality of Auca Mahuevo have been identified by embryonic remains in ovo (Chiappe et al. 1998, 1999, 2000; Chiappe and Dingus

2001). Titanosaur eggs are spherical to sub-spherical, with an average diameter of 150 mm. The average shell thickness is approximately 1.8 mm (Chiappe et al. 1998). Titanosaur eggshell is composed of only one layer (fig. 9.12B). This layer consists of laterally juxtaposed shell units originating from an organic core and terminating at a node on the outer surface (fig. 9.12B). The organic core is surrounded by minute semicircular spherulites (figs. 9.12D, E). These eggs display a nodular surficial ornamentation, the nodes being surrounded by two to four sub-circular pores (figs. 9.12A,C) (Grellet-Tinner 2000b). Pore canals are predominantly vertical and straight, but they can bifurcate at mid-section forming a Y pattern (Grellet-Tinner 2000b) (fig. 9.3A). Titanosaur egg-shell units consist of radiating calcitic rhombohedric crystals clearly distinct at high magnification (fig. 9.12D). No "ghost" crystal is detectable at the tip of each of these rhombohedrons, thus eliminating the possibility of aragonite to calcite diagenesis. The lack of cathodoluminescence (Barbin 2000) corroborates the primary nature of the calcite of these eggshells.

A variety of spatial arrangements have been reported for eggs alleged to be of sauropods (Kérourio 1981; Breton et al. 1986; Faccio 1994; Moratalla and Powell 1994; Sahni et al. 1994). None of these eggs, however, contain embryonic remains and therefore their taxonomic identification remains uncertain. The titanosaur eggs from the Late Cretaceous site of Auca Mahuevo were laid in dug-out nests on sandy or muddy substrates (Chiappe and Dingus 2001). Textural differences between the substrate sediment and the overlaying deposit indicate the nests were not buried in the substrate (Chiappe and Dingus 2001). No evidence of parental care has yet been recovered from Auca Mahuevo.

Theropoda

Troodon formosus: The eggs of the Late Cretaceous *Troodon formosus* are oval with a pointed pole (Horner 1982, 1984; Horner and Weishampel 1988; Varricchio et al. 1997). SEM reveals that *Troodon formosus* eggshell exhibits two distinct layers in radial section (figs. 9.13A, B), which in the studied specimens are overlaid by a diagenetic calcite blanket nearly 0.12 mm thick (figs. 9.7C, D). A thin section (fig. 9.7D) reveals a pore filled with the same diagenetic calcite. This pore is slightly oblique in relation to the eggshell inner and outer surfaces. The 0.038-mm diameter of this pore appears constant throughout its length. Varricchio et al. (1997) reported that the pore canals of *Troodon formosus* are concentrated in the upper pole (exposed in the nest) of these eggs. This could explain the paucity of pores in the studied sample, whose topology in the egg is unknown. Owing to the presence of the calcitic surficial blanket, surficial ornamentation is not visible. Nonetheless, one SEM radial view suggests a linearituberculate ornamentation.

Eggshell thickness (diagenetic blanket excluded) is 1.02 mm. The boundary between layers 1 and 2 is prismatic (fig. 9.13B). Layer 1 displays a compact vertical arrangement of blade-shaped calcite crys-

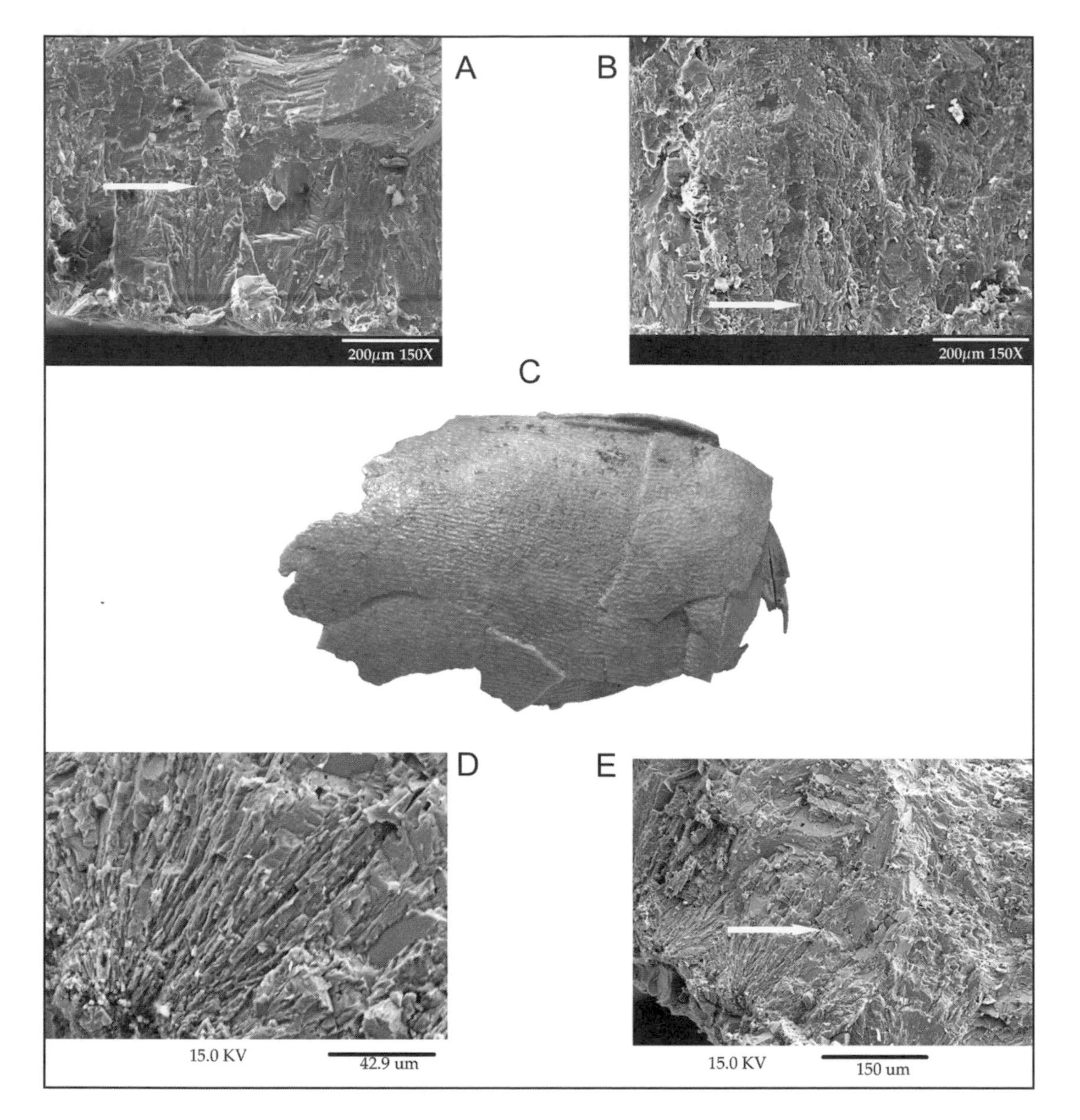

Figure 9.13. SEM of Troodon formosus *eggshell.*
A, two distinct layers.
B, the prismatic boundary between layers 1 and 2.
C, surficial ornamentation of oviraptorid eggshells shows ridges arranged in lines parallel to the long axis of the egg. SEM of oviraptorid eggshell.
D, E, two aprismatic layers of oviraptorid eggshells. White arrows point to the designated features.

tals (fig. 9.13A). Layers 1 and 2 approximate 0.26 and 0.76 mm, respectively (fig. 9.15). Although the prismatic condition of the eggshell of *Troodon formosus* could contribute with a micrometric error to our measurements of their thickness (i.e., prismatic condition often hinders the precise boundary between layers), a 0.40 layer 2:layer 1 ratio appears diagnostic of this species.

Varricchio et al. (1999) described a rimmed, 1-m^2 bowl-shaped structure with 24 eggs (MOR 963) morphologically similar to others containing in ovo remains of *Troodon formosus*. Another clutch of morphologically similar eggs (MOR 748) was found underlying bones of an adult *Troodon formosus*, thus suggesting the existence of brood-

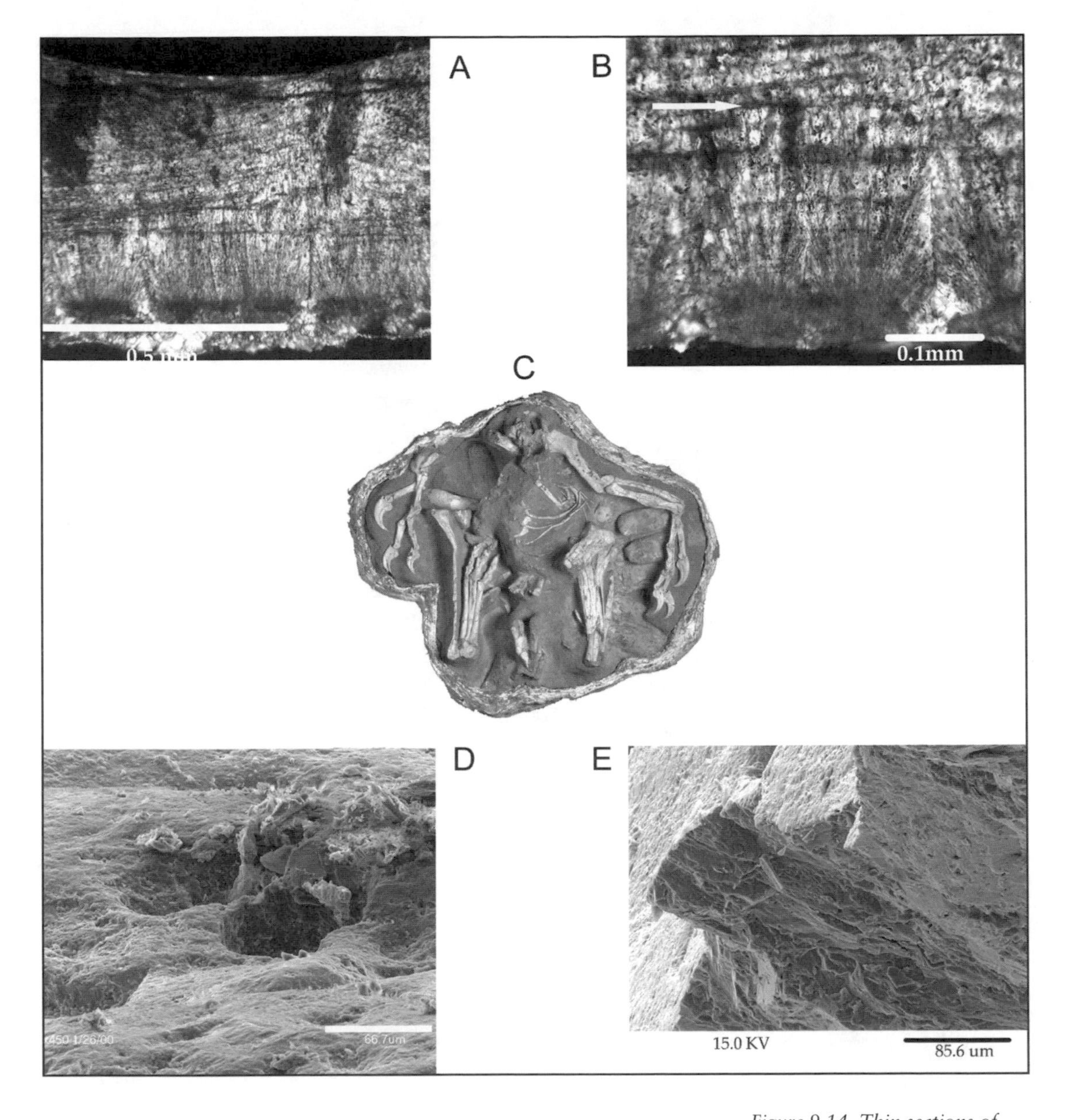

Figure 9.14. Thin sections of oviraptorid eggshell: A, wavy ornamentation of oviraptorid eggshell; B, two aprismatic layers; C, picture of egg clutch in contact with adult oviraptorid. SEM of oviraptor eggshell: D, round pore apertures; E, pore canals are straight or slightly oblique. White arrows point to the designated features.

ing behavior (Varricchio et al. 1997, 1999). According to these authors, the eggs were partially buried and vertically oriented at a slight angle and show a statistically significant paired arrangement in bottom view. This arrangement, however, is not evident when the clutch is viewed from the top. The fact that the upper portion of these eggs remained unburied implies that the eggs were in partial contact with the brooding adult (Varricchio et al. 1999). In spite of this evidence, whether *Troodon formosus* incubated its eggs or simply protected them remains uncertain.

Oviraptoridae: The nearly symmetrical, oval eggs of oviraptorids are approximately 6–7 cm by 18–19 cm. Surficial ornamentation is

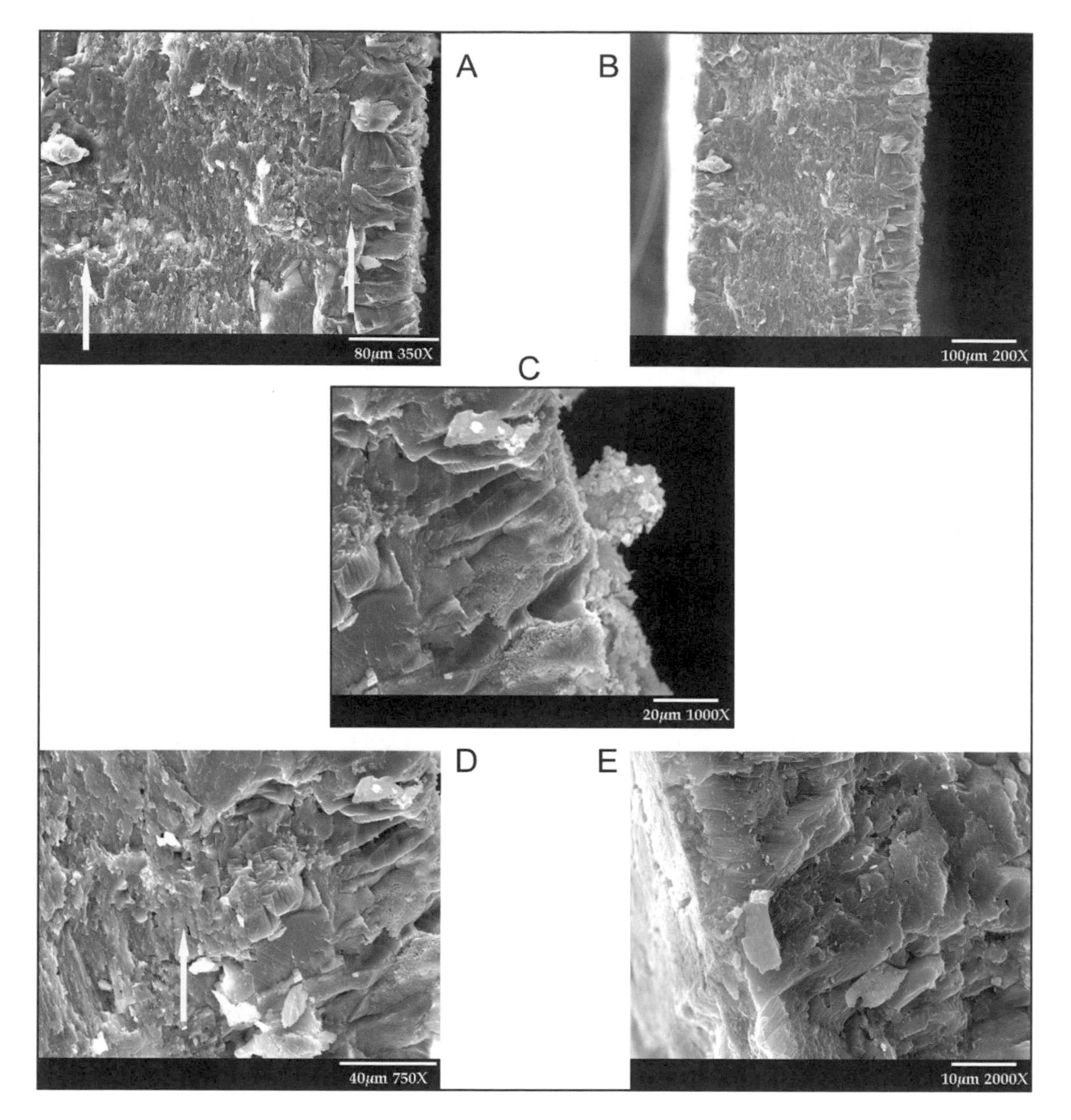

Figure 9.15. SEM images of Meleagris gallopavo *eggshell. A, three prismatic layers. B, smooth eggshell surface. C, bladed calcite crystals of layer 1. D, blurry demarcation between layers 1 and 2 due to the prismatic condition. E, thin layer 3, the hardest one to determine. White arrows point to the designated features, and SEM images are orientated sideways.*

characterized by ridges arranged in lines parallel to the long axis of the egg (fig. 9.13C). This ornamentation gives a wavy appearance to the upper surface when radially viewed in SEM or thin sections (fig. 9.5B). Eggshell thickness and the height of surficial ridges and troughs vary according to the longitudinal position of the sample within the egg. Eggshell thickness ranges between 0.50 and 1.01 mm. Pore canals are straight or slightly oblique (figs. 9.14A, E), with round surficial apertures (fig. 9.14D). SEM observation of radial eggshell sections reveals two aprismatic layers (figs. 9.13E; 9.14A, B). Layer 1 is composed of fan-shaped, radiating, acicular rhombohedric crystals of calcite (figs.

9.13D; 9.14A, B). This layer retains a 0.15 mm thickness independently of the total eggshell variation. The thickness of layer 2, however, fluctuates from 0.35 to 0.86 mm in accordance with the presence of ridges or troughs.

Clark et al. (1999) described the articulated skeleton of a Late Cretaceous oviraptorid (GIN 100/979) in an avian-like brooding posture on top of a clutch of at least 15 and possibly twice as many eggs (fig. 9.14C). These authors remarked upon the identical microstructure and general morphology of GIN 100/979 to GIN 100/971, a partial egg containing embryonic remains (Norell et al. 1994, 2001) and the specimen used for this study (appendix 9.1). In GIN 100/979, at least eight eggs are arranged in pairs forming a circle in direct contact with the adult bones. The association of eggs and an adult skeleton of GIN 100/979 and the existence of comparable examples led several workers (Norell et al. 1995; Dong and Currie 1996; Clark et al. 1999) to infer the presence of an avian type of nest attendance, namely the behavior of sitting over egg clutches during prolonged periods of time, in oviraptorids.

Aves, Neognathae

Meleagris gallopavo (wild turkey): This bird lays oval eggs with one distinct pointed pole. They average 49.60 mm in diameter (at the equator) by 64.46 mm in length, with a diameter/length ratio of 0.77. The eggshell surface is smooth. In the examined sample, eggshell thickness equals 0.34 mm and is composed of three prismatic layers (figs. 9.15A, B, D). SEM radial sections reveal that layer 1 consists of bladed calcite crystals that extend through the entire eggshell (figs. 9.15A, C). Layers 1, 2, and 3 average 0.06, 0.24, and 0.032 mm, respectively. Layer 2 is four times thicker than layer 1. *Meleagris gallopavo* eggs display straight pores with round apertures.

As in many other ground-dwelling birds, *Meleagris gallopavo* builds bowl-shaped nests in secluded areas (Del Hoyo et al. 1992). This simple nest might consist only of flattened vegetation rimmed with available vegetal debris. *Meleagris gallopavo* females lay one egg a day. Ovideposition stops as soon as a predetermined number of eggs is reached in the nest. The female then incubates them by sitting on the eggs until they hatch. After the eggs have hatched, the female rears the chicks, sometimes up to the following laying season.

Anser anser (greylag goose): Eggs are oval with one pointed pole and an average diameter (at the equator) of 48.34 mm and length of 69.93 mm. The diameter/length ratio is 0.70. Eggshell surface is smooth. In the measured sample, eggshell thickness equals 0.41 mm and SEM radial sections reveal three distinct prismatic layers (fig. 9.16) (Grellet-Tinner 2000a). The thickness of layer 1 averages 0.17 mm. This layer consists of blocky calcite crystals with radiating, short, bladelike spherulites at their base (figs. 9.16A, B). Layer 2 is 28 percent thicker than layer 1, with a thickness of 0.22 mm. Layer 3 represents 22 percent and 29 percent of the thickness of layers 2 and 1, respectively. *Anser anser* eggs display straight pores with a round aperture.

The nesting behavior and parental care of *Anser anser* resemble

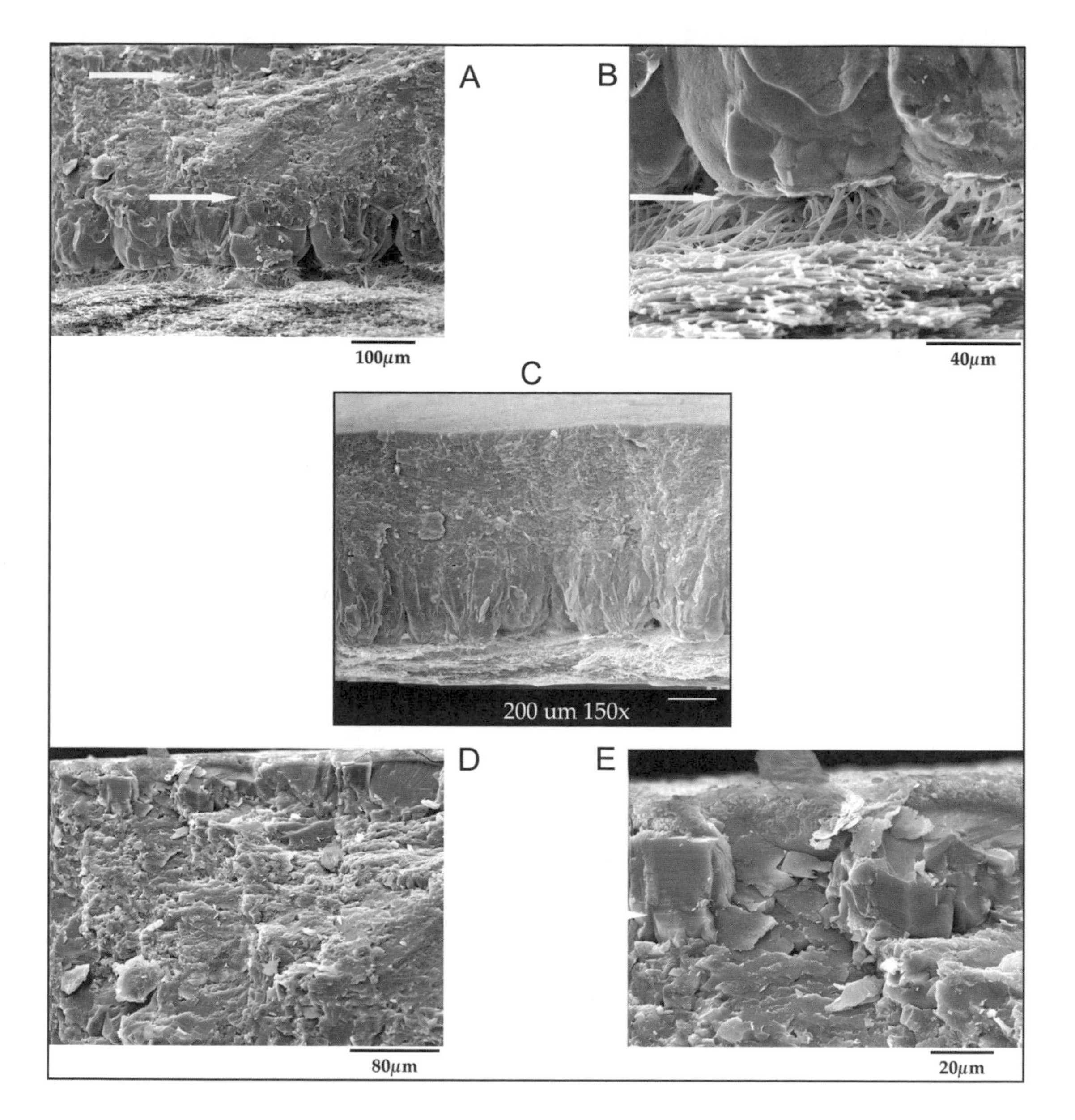

Figure 9.16. SEM images of Anser anser *eggshell.*
A, three distinct prismatic layers.
B, protein filaments of the membrana testacea penetrate the calcitic eggshell.
C, smooth eggshell surface.
D, the columns of the eggshell units continue well into layer 3.
E, thin layer 3 with a slightly different crystallization. White arrows point to the designated features.

those of *Meleagris gallopavo*, although they are more elaborate. Nest structures differ according to the riparian conditions in which *Anser anser* lives. These birds often build their nests with reed stems and grasses, and they line them with down feathers (Del Hoyo et al. 1992).

Aves, Paleognathae

Struthio camelus (ostrich): Eggs are white to creamy-yellow in color and nearly spherical. They average 157.50 and 133.75 mm in length and diameter, respectively. Their diameter/length ratio is 0.85. The egg surface lacks any type of ornamentation visible with the naked

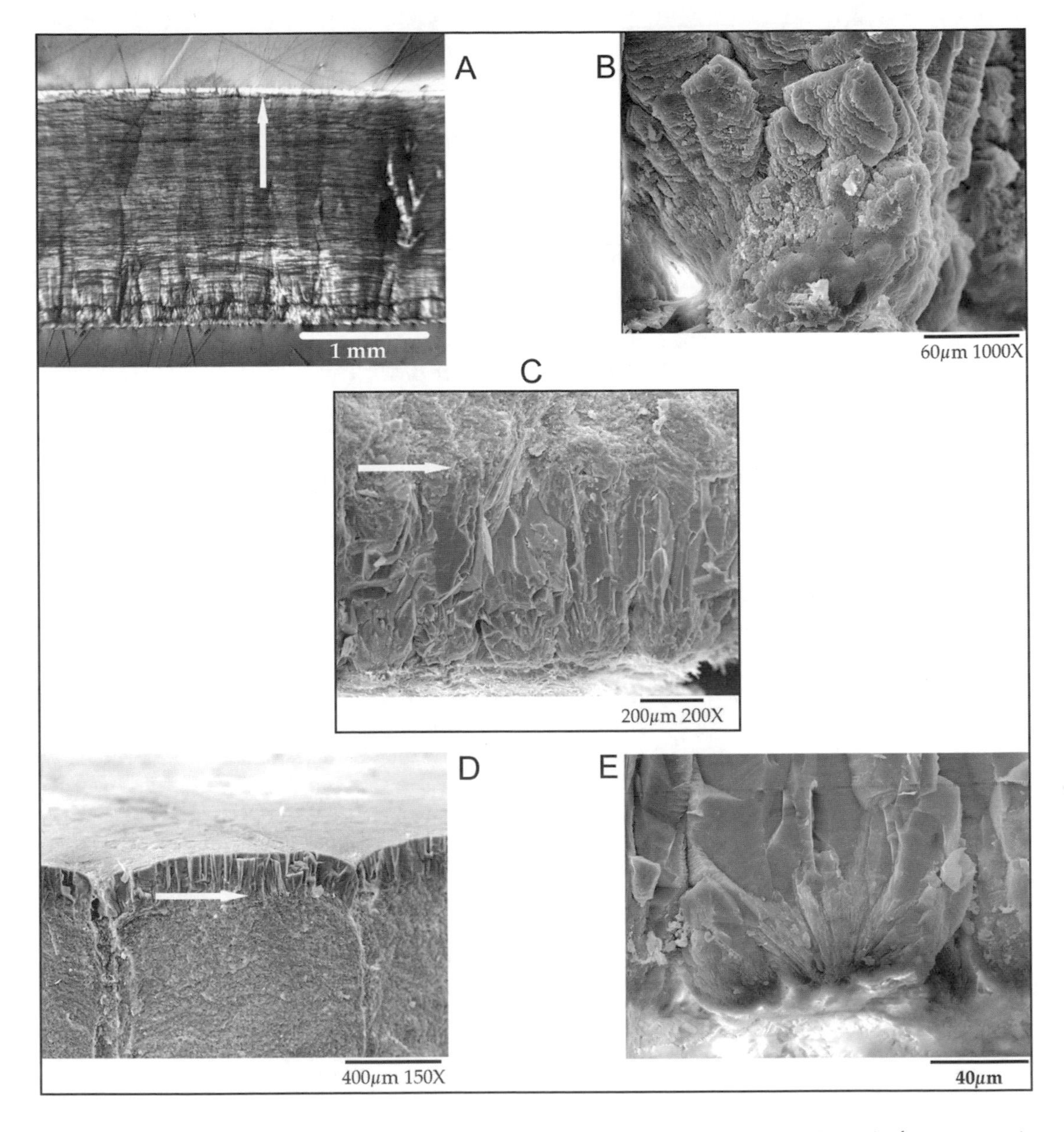

Figure 9.17. A, three aprismatic layers of Struthio camelus *eggshell (thin section).*
B, blocky calcite crystals with radiating, elongated bladelike spherulites at their base of layer 1 (SEM). SEM images of Rhea americana *eggshell.*
C, three aprismatic layers, two of which are visible in this image.
D, thin but noticeable layer 3.
E, blocky calcite crystals with radiating bladelike spherulites at their base of layer 1. White arrows point to the designated features.

eye (fig. 9.5D). In the measured sample, eggshell thickness averages 1.86 mm. SEM radial sections reveal three aprismatic layers (fig. 9.17A) (Grellet-Tinner 2000a). Layer 1 consists of blocky calcite crystals with radiating, elongated, bladelike spherulites at their base (fig. 9.17B). Layer 2 is three times thicker than layer 1, and layer 3 equals 5 percent of layer 2. Thicknesses of layers 1, 2, 3 measure 0.46, 1.26, and 0.06 mm, respectively. *Struthio camelus* eggshell presents numerous branching pores with a round aperture (fig. 9.3C). Pores start as a single canal from the MT region and branch out within layer 2. This character is unique to *Struthio camelus* among other extant paleognaths.

Nest structures of *Struthio camelus* are almost non-existent, being often limited to a shallow scrape in the ground. The male does most of the brooding at night while the more cryptically feathered female often incubates during the day. Male ostriches may either pair during their reproductive season (monogamy) or mate with many females (polyandry). The number of eggs in a given nest varies according to the contributing females. This number could vary from 18 (single female) to more than 50 eggs (harem). Captive-bred ostriches lay an egg every two days. The rearing of the young appears to be equally shared between the male and the single or dominant female (Del Hoyo et al. 1992), often prolonged until the following laying season.

Rhea americana (greater rhea): This bird lays oval eggs averaging 117 by 91 mm, having a diameter/length ratio of 0.63. These eggs typically have smooth shells (fig. 9.17D). SEM radial sections of the eggshell show three aprismatic layers (fig. 9.4D). In our sample, eggshell thickness averages 0.92 mm. Layer 1 consists of blocky calcite crystals with radiating bladelike spherulites at their base (figs. 9.17C, D) (Grellet-Tinner 2000a). Layer 1 is 0.32 mm thick and layers 2 and 3 are 0.57 and 0.03 mm, respectively. Layer 2 is 66 percent thicker than layer 1, and layer 3 equals 10 percent of layer 1. *Rhea americana* eggshell displays oblique and straight pore canals, with a round surficial aperture (fig. 9.17D).

Male rheas are solely responsible for brooding and incubating the eggs as well as rearing the chicks (Del Hoyo et al. 1992). Like *Struthio camelus,* the mating behavior of *Rhea americana* is polygynous. Numerous females each lay an egg every two days near the nest protected by the brooding male. The male then drags the egg with its beak into the nest. The nesting structure consists of a shallow, dug-out depression, which at best contains a small amount of vegetal and other debris. *Rhea americana* nests contain up to 40 eggs (Del Hoyo et al. 1992).

Dromaius novaehollandiae (emu): Eggs are oval, averaging 134 by 86 mm in length and diameter, with a diameter/length ratio of 0.65. The eggshell surface is typically granulated (fig. 9.5C), a condition that approaches *Dromaius novaehollandiae*'s close relative *Casuarius casuarius* (Grellet-Tinner 2000a). SEM radial sections reveal four distinct aprismatic layers (figs. 9.4C; 9.5C), the same ones recognized by Penner (1984) and Grellet-Tinner (2000a). In our sample, these layers add up to a total thickness of 0.96 mm. Layer 1 consists of blocky calcite crystals with radiating bladelike spherulites at their base. Layer 3 has a porous appearance (fig. 9.9), a condition not yet reported in any dinosaur other than in the paleognath *Casuarius casuarius.* The thickness of layers 1, 2, 3, and 4 averages 0.26, 0.45, 0.18, and 0.07 mm, respectively. *Dromaius novaehollandiae* eggs are also characterized in that the pore canals connecting the inner surface to the porous layer 3 are not aligned with those connecting layer 3 to the outer surface (fig. 9.3B).

Dromaius novaehollandiae mating and nesting behaviors differ somewhat from those of *Struthio camelus* and *Rhea americana.* The female may seek two consecutive partners during a mating season and lay approximately twelve eggs in each nest (Del Hoyo et al. 1992).

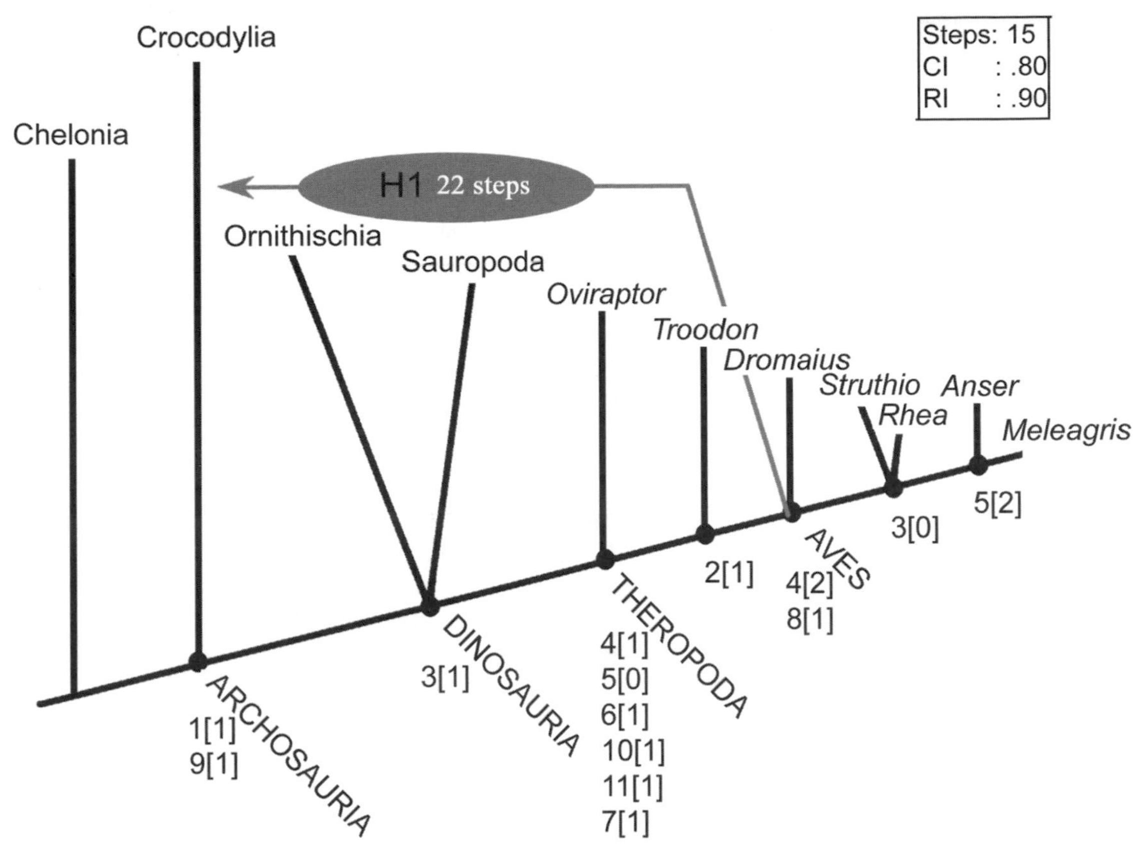

Figure 9.18. Strict consensus cladogram (length: 15; CI: 0.80; RI: 0.90) summarizing three most parsimonious trees. An additional seven steps would be required to support a sister taxon relationship between crocodiles and birds. Synapomorphies diagnosing nodes: 1[1], calcitic eggshell; 9[1], parental nest attendance; 3[1], presence of eggshell surficial ornamentation; 4[1], two eggshell layers; 5[0], aprismatic delimitation between layers; 6[1], asymmetrical transversal plane symmetry of egg; 10[1], presence of parental brooding behavior; 11[1], monoautochronic ovulation (egg are laid at daily interval); 7[1], open nest morphology; 2[1], blade-shaped shell unit crystallization; 4[2], three or more eggshell layers; 8[1], one functioning ovary; –3[1], absence (reversal) of eggshell surficial ornamentation; 5[2], prismatic delimitation between layers.

Ovideposition happens every three days. Like rheas, *Dromaius novaehollandiae* males will brood and incubate the eggs, and rear the young. *Dromaius novaehollandiae* nest structure resembles more that of *Rhea americana* than that of *Struthio camelus.* Prior to the clutch reaching minimal incubation number, the male tends to camouflage the eggs with vegetal debris.

Phylogenetic Analysis of Eggs and Nesting

The cladistic analysis of the data matrix shown in appendix 9.1 resulted in three most parsimonious cladograms (length: 15; CI: 0.80; RI: 0.90) obtained using Hennig 86's command "implicit enumeration" or "ie" (exhaustive search; Farris 1988). A strict consensus cladogram (length: 15; CI: 0.80; RI: 0.90) summarizing these most parsimonious trees is shown in Figure 9.18. Because no oological data is known for any extinct basal archosauromorph, the egg morphologies and nest attendances described above can only be used to contrast two of the three current general hypotheses of bird origins, namely that birds are most closely related to crocodiles or that they are the descendants of non-avian theropod dinosaurs (fig. 9.1).

In the resultant consensus cladogram, birds are nested within Archosauria on the basis of two unambiguous synapomorphies: the calcitic eggshell mineralogy (character state 1[1]) and the presence of parental nest attendance (character state 9[1]). In contrast to other known hard-shelled egg-laying taxa, turtles lay eggs which have shell that consists of aragonitic radiating acicular crystals. Dinosaurs are unambiguously diagnosed by the presence of eggshell surficial ornamentation (character state 3[1]). The optimization of this character in the resultant consensus tree interprets the smooth shell of most bird eggs as a reversal condition evolved from ornamented eggs. The existence of several avian taxa (e.g., *Dromaius novaehollandiae* and some other paleognaths; Grellet-Tinner 2000a) with eggshell surficial ornamentation suggests that additional homoplasy for this character is present within birds. Our data on eggshell morphology and nest attendance do not resolve the relationships between ornithischian, sauropod, and theropod dinosaurs, which are grouped in a trichotomy (fig. 9.18).

Birds are clustered within Theropoda on the basis of five unambiguous synapomorphies. The first one is the presence of two eggshell layers (character state 4[1]). Other synapomorphies include the presence of an aprismatic condition among eggshell layers (character state 5[1]), asymmetrical eggs (character state 6[1]), brooding behavior (character state 10[1]), and monoautochronic ovulation (character state 11[1]). Recent discoveries associating adult skeletons of oviraptorids and *Troodon formosus* with their eggs indicate that brooding behavior had already evolved within maniraptoriform theropods (Norell et al. 1995; Dong and Currie 1996; Clark et al. 1999; Varrichio et al. 1997, 1999). The egg pairing observed in clutches of oviraptorids and *Troodon formosus* suggests that these dinosaurs may have had two functional ovaries (Varricchio et al. 1997, 1999), a primitive condition when compared to the single functional ovary of extant birds. Yet the spatial arrangement of the eggs, coupled with the statistically significant spacing between the paired eggs of MOR 363, led Varricchio et al. (1997) to suggest that at least *Troodon formosus* had monoautochronic ovulation. This inference indicates that it might have taken several days for *Troodon formosus* to lay a clutch of eggs, a derived condition shared with living birds. The presence of open nests (character state 7[1]) also diagnoses Theropoda. However, the lack of reliable data on this aspect of nest morphology for either ornithischians or sauropods results in an ambiguous optimization for this character state. Additional information on the nest morphology of these taxa is needed in order to evaluate whether the presence of open nests is an unambiguous synapomorphy of Theropoda or a synapomorphy of a more inclusive taxon.

Within Theropoda, the present cladistic analysis supports a sister group relationship between troodontids and birds. This node is diagnosed by the presence of shell units with blade-shaped crystals (character state 2[1]). Aves is unambiguously diagnosed by the presence of three or more eggshell layers (character state 4[2]) and one functioning ovary (character state 8[1]).

On the basis of the descriptions presented, it is clear that the morphology of eggs and the nesting strategies of the parents possess distinct attributes that are suitable for phylogenetic inference. While behavioral characters have been commonly used for inferring phylogenetic relationships (e.g., Wenzel 1992; Clark et al. 1999), studies of egg morphology have been greatly limited to alpha-taxonomy with new descriptions of oospecies (Erben 1970; Mikhailov 1987; Hirsch and Quinn 1990; Vianey-Liaud et al. 1994; Zelenitsky et al. 1996). A few exceptions are Mikhailov's (1992) mapping of dinosaurian eggshell attributes on a cladogram and Grellet-Tinner's (2000a) cladistic analysis of paleognath eggshell. Although Mikhailov (1992) recognized few derived similarities grouping different dinosaurian eggshells, his study was based on samples derived from eggs lacking embryos *in ovo*, thus of uncertain taxonomic identity.

Aside from the calcitic composition of their eggshell, no further morphological similarity can be identified uniting crocodilian and avian eggs. Likewise, the nest attendance and strategies seen between crocodiles and birds do not go beyond the presence of a basic type of parental care. These similarities, however, are shared by other (and likely all) archosaurs and are most parsimoniously interpreted as archosaurian synapomorphies (i.e., character states 1[1] and 9[1]). The results of the present cladistic analysis indicate that seven additional steps would be required to support a sister-group relationship between Crocodylia and Aves (fig. 9.18). This implies that at least seven unambiguous synapomorphies of crocodiles and birds would need to be discovered if such a hypothesis is to be entertained. On the contrary, this study documents how egg and nesting behavior attributes of living birds can be seen as transformations of those found in non-avian theropod dinosaurs, hence adding support to the hypothesis that birds are evolutionarily nested within Theropoda.

Acknowledgments. Gerald Grellet-Tinner is grateful to K. Milliken and A. Klaus respectively from the UT Department of Geological Sciences and AMNH, for trusting him for countless hours of SEM use. This research would not have been possible without the help of many who have supplied him with needed specimens, particularly R. Coria, P. Currie, J. Horner, M. Norell, N. Pledge, P. Rich, and D. Zelenitsky. Gerald Grellet-Tinner would like to express his gratitude to the AMNH Study Grant Program, the Infoquest Foundation, the Jurassic Foundation, and USC Department of Earth Sciences for their financial support of this research and its publication. The Natural History Museum of Los Angeles County provided additional support for this research.

APPENDIX 9.1
Samples used for eggshell microstructural studies

Egg and eggshells	Specimen	Preservation
Alligator mississippiensis	Langston's collection	excellent
Hadrosauridae	TMP 877985	strong diagenesis
Titanosauria	PVPH 263	excellent
Titanosauria	PVPH 264	excellent
Titanosauria	PVPH 272	excellent
Troodon formosus	MOR DU 2258	good with diagenesis
Oviraptoridae	GIN 100/971	excellent
Meleagris gallopavo	TMM(M) 758	excellent
Anser anser	TMM(M) 7592	excellent
Anser anser	TMM(M) 7588	excellent
Struthio camelus	TMM(M) 7585	excellent
Rhea americana	TMM(M) 7587	excellent
Rhea americana	MSC 704	excellent
Dromaius novaehollandiae	TMM(M)7594	excellent

APPENDIX 9.2
List of characters and character states used in the present cladistic analysis. Character states present only in one ingroup terminal taxon were excluded. The only multistate character, the number of eggshell layers (character 4), was treated as ordered. The transformation from one character state to another was specified according to the increasing number of layers.

1. Eggshell mineralogy: aragonite (0), calcite (1).
2. Shell unit crystallization: acicular (0), blade-shaped (1).
3. Eggshell surficial ornamentation: absent (0), present (1).
4. Number of eggshell layers: one (0), two (1), three or more (2).
5. Delimitation between layers: aprismatic (0), prismatic (1).
6. Transversal plane symmetry of egg: symmetrical (0), asymmetrical (1).
7. Nest morphology: closed (egg completely covered by substrate or vegetation) (0), open (1).
8. Number of functioning ovaries (extant birds have only one functioning ovary): two (0), one (1).
9. Nest attendance: absent (0), present (1).
10. Brooding behavior: absent (0), present (1).
11. Monoautochronic ovulation (eggs are laid at daily intervals): absent (0), present (1).

APPENDIX 9.3
Character matrix used in the present analysis
(see Appendix 9.2 for character list)

Chelonia	0000000000
Alligator mississippiensis	1?00000100
Hadrosauridae	1?1000?0?00
Titanosauria	101000?0?00
Oviraptoridae	1011111011?
Troodon formosus	11112110111
Anser anser	11022111111
Meleagris gallopavo	11022111111
Struthio camelus	11021011111
Rhea americana	11021111111
Dromaius novaehollandiae	11121111111

References

Barbin, V. 2000. Cathodoluminescence of carbonate shells: biochemicals vs. diagenesis processes. In M. Pagel, V. Barbin, and P. Blanc (eds.), *Cathodoluminescence in geosciences,* pp. 303–329. Berlin: Springer Press.

Breton, G., R. Fourniet, and J-P. Watté. 1986. Le lieu de ponte de dinosaures de Rennes-le-Chateau (Aude): Premiers résultats de la campagne de fouilles 1984. *Annales du Museum du Havre* 32: 1–13.

Chiappe, L. M. 2001. The rise of birds. In D. Briggs and P. Crowther (eds.), *Paleobiology II*, pp. 102–106. Oxford: Blackwell.

Chiappe, L. M., R. Coria, L. Dingus, F. Jackson, A. Chinsamy, and M. Fox. 1998. Sauropod dinosaur embryos from the Late Cretaceous of Patagonia. *Nature* 396: 258–261.

Chiappe, L. M., L. Dingus, F. Jackson, G. Grellet-Tinner, and R. Coria. 1999. Auca Mahuevo, an extraordinary dinosaur nesting ground from the Late Cretaceous of Patagonia. *Journal of Vertebrate Paleontology* 19 (Supplement to Number 3): 37A.

Chiappe, L. M., L. Dingus, F. Jackson, G. Grellet-Tinner, R. Aspinal, J. Clark, R. Coria, A. Garrido, and D. Loope. 2000. Sauropod eggs and embryos from the Late Cretaceous of Patagonia. In A. M. Bravo and T. Reyes (eds.), *First International Symposium on Dinosaur Eggs and Babies,* pp. 23–29. Isona I Conca Dellà Catalonia. Spain.

Chiappe, L. M., and L. Dingus. 2001. *Walking on eggs: the astonishing discovery of thousands of dinosaur eggs in the badlands of Patagonia.* New York: Scribner.

Clark, J. M., M. A. Norell, and L. M. Chiappe. 1999. An oviraptorid skeleton from the Late Cretaceous of Ukhaa Tolgod, Mongolia, preserved in an avianlike brooding position over an oviraptorid nest. *American Museum Novitates* 3265: 1–36.

Currie, P. J., and J. R. Horner. 1988. Lambeosaurine hadrosaur embryos (Reptilia: Ornithischia). *Journal of Vertebrate Paleontology* 8 (Supplement to Number 3): 13A.

Del Hoyo, J., A. Elliot, and J. Sargatal. 1992. *Handbook of the birds of the world,* Volumes 1, 2, 3, and 4. Barcelona: Lynx Edicions.

Desmond, A. 1982. *Archetypes and ancestors.* Chicago: University of Chicago Press.

Dong, Z., and P. J. Currie. 1996. On the discovery of an oviraptorid skeleton on a nest of eggs at Bayan Mandahu, Inner Mongolia, People's Republic of China. *Canadian Journal of Earth Sciences* 33: 632–636.

Erben, H. 1970. Ultrastrukturen und Mineralisation rezenter und fossiler Eischalen bei Vögeln und Reptilien. *Biomineralisation Forschungsberichte* 1: 1–66.

Faccio, G. 1994. Dinosaurian eggs from the Upper Cretaceous of Uruguay. In K. Carpenter, K. F. Hirsch, and J. R. Horner (eds.), *Dinosaur eggs and babies,* pp. 47–55. New York: Cambridge University Press.

Farris, J. 1988. Hennig 86 references. Documentation for version 1.5. Privately published.

Feduccia, A. 1996. *The origin and evolution of birds.* New Haven: Yale University.

Feduccia, A., and R. Wild. 1993. Birdlike characters in the Triassic archosaur *Megalancosaurus. Naturwissenschaften* 80: 564–566.

Gauthier, J. A. 1986. Saurischian monophyly and the origin of birds. *Memoirs of the California Academy of Sciences* 8: 1–55.

Grellet-Tinner, G. 2000a. Phylogenetic interpretation of eggs and eggshells. In A. M. Bravo and T. Reyes (eds.), *First international symposium on dinosaur eggs and babies,* pp. 61–75. Isona I Conca Dellà Catalonia. Spain.

———. 2000b. A study of titanosaurid eggshell from Auca Mahuevo. *Journal of Vertebrate Paleontology* 20 (Supplement to Number 3): 46A.

Griffiths, G. 2000. *Compsognathus* eggs revisited. In A. M. Bravo and T. Reyes (eds.), *First international symposium on dinosaur eggs and babies,* pp. 77–83. Isona I Conca Dellà Catalonia. Spain.

Hirsch, K. F., and B. Quinn. 1990. Eggs and eggshell fragments from the Upper Cretaceous Two Medicine Formation of Montana. *Journal of Vertebrate Paleontology* 10: 491–511.

Hirsch, K. F., A. Kihm, and D. Zelenistky. 1997. New eggshell of ratite morphotype with predation marks from the Eocene of Colorado. *Journal of Vertebrate Paleontology* 17: 360–369.

Horner, J. R. 1982. Evidence of colonial nesting and "site fidelity" among ornithischian dinosaurs. *Nature* 297: 675–676.

———. 1984. The nesting behavior of dinosaurs. *Scientific American* 250: 130–137.

———. 1997. Rare preservation of an incompletely ossified fossil embryo. *Journal of Vertebrate Paleontology* 17: 431–434.

———. 2000. Dinosaur reproduction and parenting. *Annual Review of Earth and Planetary Science* 28: 19–45.

Horner, J. R., and R. Makela. 1979. Nest of juveniles provides evidence of family structure among dinosaurs. *Nature* 282: 296–298.

Horner, J. R., and D. B. Weishampel. 1988. A comparative embryological study of two ornithischian dinosaurs. *Nature* 332: 256–257.

Horner, J. R., and P. J. Currie. 1994. Embryonic and neonatal morphology and ontogeny of a new species of *Hypacrosaurus* (Ornithischia, Lambeosauridae) from Montana and Alberta. In K. Carpenter, K. F. Hirsch, and J. R. Horner (eds.), *Dinosaur eggs and babies,* pp. 312–336. New York: Cambridge University Press.

Kérourio, P. 1981. Nouvelles observations sur le mode de nidification et de

ponte chez les dinosauriens du Crétacé terminal du Midi de la France. *Compte Rendu Sommaire des Séances de la Société Géologique de France* 1: 25–28.
Kohring, R., and K. F. Hirsch. 1996. Crocodilian and avian eggshells from the Middle Eocene of the Geiseltal, Eastern Germany. *Journal of Vertebrate Paleontology* 16: 67–80.
Makovicky, P., and G. Grellet-Tinner. 2000. Association between a specimen of *Deinonychus antirrhopus* and theropod eggshell. In A. M. Bravo and T. Reyes (eds.), *First international symposium on dinosaur eggs and babies,* pp. 123–128. Isona I Conca Dellà Catalonia. Spain.
Manning, T. W., K. A. Joysey, and A. R. I. Cruickshank. 1997. Observations of microstructures within dinosaur eggs from Henan Province, Peoples' Republic of China. In D. L. Wolberg, E. Stump, and G. D. Rosenberg (eds.), *Dinofest International,* pp. 287–290. Philadelphia: The Academy of Natural Sciences.
Martin, L. D. 1991. Mesozoic birds and the origin of birds. In H. P. Schultze and L. Trueb (eds.), *Origins of the higher groups of tetrapods,* pp. 485–540. Ithaca: Comstock Publishing Associates.
Martin, L. D., J. D. Stewart, and K. N. Whetstone. 1980. The origin of birds: structure of the tarsus and teeth. *Auk* 97: 86–93.
Martin, L. D., and J. D. Stewart. 1999. Implantation and replacement of bird teeth. *Smithsonian Contributions to Paleobiology* 89: 295–300.
Mateus, I., H. Mateus, M. T. Antunes, O. Mateus, P. Taquet, V. Ribeiro, and G. Manuppella. 1997. Couvée, oeufs et embryons d'un dinosaure théropode du Jurassique supérieur de Lourinhã (Portugal). *Comptes Rendus de l'Académie des Sciences de Paris,* Série 2, 325(A): 71–78.
Mikhailov, K. E. 1987. Some aspects of the structure of the shell of the egg. *Palaeontological Journal* 3: 54–61.
———. 1991. Classification of fossil eggshells of amniotic vertebrates. *Acta Palaeontologica Polonica* 36: 193–238.
———. 1992. The microstructure of avian and dinosaurian eggshell: phylogenetic implications. *Natural History Museum of Los Angeles County,* Science Series 36: 361–373.
———. 1997. Fossil and Recent eggshell in amniotic vertebrates: fine structure, comparative morphology and classification. *Special Papers in Palaeontology* 56: 1–80.
Moratalla, J. J., and J. E. Powell. 1994. Dinosaur nesting patterns. In K. Carpenter, K. F. Hirsch, and J. R. Horner (eds.), *Dinosaur eggs and babies,* pp. 37–46. New York: Cambridge University Press.
Norell, M. A., J. M. Clark, D. Dashzeveg, R. Barsbold, L. M. Chiappe, A. R. Davidson, M. C. McKenna, A. Perle, and M. J. Novacek. 1994. A theropod dinosaur embryo and the affinities of the Flaming Cliffs dinosaur eggs. *Science* 266: 779–782.
Norell, M. A., J. M. Clark, L. M. Chiappe, and D. Dashzeveg. 1995. A nesting dinosaur. *Nature* 378: 774–776.
Norell, M. A., J. M. Clark, and L. M. Chiappe. 2001. An embryonic oviraptorid (Dinosauria: Theropoda) from the Upper Cretaceous of Mongolia. *American Museum Novitates* 3325: 1–17.
Ostrom, J. H. 1976. *Archaeopteryx* and the origin of birds. *Biological Journal of the Linnean Society, London* 8: 91–182.
Padian, K., and L. M Chiappe. 1998. The origin and early evolution of birds. *Biological Reviews* 73: 1–42.
Penner, M. M. 1984. Singularités de la coquille des oeufs des Casuarii-

formes par rapport à celle des autres oiseaux. *Comptes Rendus de l'Académie des Sciences de Paris,* Série 3, 299: 293–296.
Pritchard, P. C. H. 1979. *Encyclopedia of turtles.* Neptune City, N.J.: T. F. H. Publications, Ltd.
Ross, C. A. 1989. *Crocodiles and alligators.* New York: Facts On File.
Sahni, A., S. K. Tandon, A. Jolly, S. Bajpai, A. Sood, and S. Srinivasan. 1994. Upper Cretaceous dinosaur eggs and nesting sites from the Deccan Volcano-Sedimentary of Peninsular India. In K. Carpenter, K. F. Hirsch, and J. R. Horner (eds.), *Dinosaur eggs and babies,* pp. 204–226. New York: Cambridge University Press.
Sander, P. M., C. Peitz, J. Gallemi, and R. Cousin. 1998. Dinosaur nesting on a red beach. *Comptes Rendus de l'Académie des Sciences de Paris* 327: 67–74.
Schleich, H., and W. Kästle. 1988. *Reptile egg-shells SEM atlas.* Stuttgart: Gustav Fisher Verlag.
Sereno, P. C. 1999. The evolution of dinosaurs. *Science* 284: 2137–2147.
Strong, E. E., and D. Lipscomb. 1999. Character coding and inapplicable data. *Cladistics* 15: 363–371.
Varricchio, D. J., F. Jackson, J. J. Borkowski, and J. R. Horner. 1997. Nest and egg clutches of the dinosaur *Troodon formosus* and the evolution of avian reproductive traits. *Nature* 385: 247–250.
Varricchio, D. J., F. Jackson, and C. N. Trueman. 1999. A nesting trace with eggs for the Cretaceous theropod dinosaur *Troodon formosus. Journal of Vertebrate Paleontology* 19(1): 91–100.
Vianey-Liaud, M., P. Mallan, O. Buscail, and C. Montgelard. 1994. Review of French dinosaur eggshells: morphology, structure, mineral, and organic composition. In K. Carpenter, K. F. Hirsch, and J. R. Horner (eds.), *Dinosaur eggs and babies,* pp. 151–183. New York: Cambridge University Press.
Welman, J. 1995. *Euparkeria* and the origin of birds. *South African Journal of Science* 91: 533–537.
Wenzel, J. 1992. Behavioral homology and phylogeny. *Annual Review of Ecology and Systematics* 23: 361–381.
Witmer, L. M. 1991. Perspectives on avian origins. In H.-P. Schultze and L. Trueb (eds.), *Origins of the higher groups of tetrapods: controversy and consensus,* pp. 427–466. Ithaca: Cornell University Press.
Zelenitsky, D., L. V. Hills, and P. J. Currie. 1996. Parataxonomic classification of ornithoid eggshell fragments from the Oldman Formation (Judith River Group; Upper Cretaceous), southern Alberta. *Canadian Journal of Earth Sciences* 33: 1655–1667.

10. Two Eggs Sunny-Side Up: Reproductive Physiology in the Dinosaur *Troodon formosus*

David J. Varricchio and
Frankie D. Jackson

Abstract

The Campanian Two Medicine Formation of Montana has produced an excellent sample of *Troodon* nests, egg clutches, and embryos. Complete clutches contain up to 24 eggs. In bottom view eggs show a statistically significant paired arrangement. In contrast, eggs in top view appear tightly packed. One clutch contains embryos, all apparently at an equivalent stage of development. These specimens provide the basis for interpreting *Troodon* reproduction: *Troodon* had monoautochronic ovulation and lacked egg retention, producing two eggs at daily or greater intervals. Eggs were placed subvertically to vertically in the ground with their upper portions exposed and remained unincubated until completion of the clutch. At that time the eggs were gathered into a tighter configuration with an exposed area of ~0.5 m^2. This obscured the original paired pattern in top view. The adult likely incubated the eggs directly with body heat. The precocial young hatched synchronously. This pattern of iterative egg-laying, delayed incubation, and brooding differs little from the behavior of most extant birds. Physiological interpretation of *Troodon* is based upon the direct products of reproductive physiology (egg size) or behavior (egg/clutch arrangements, nesting traces). Extant avian reproductive behaviors evident in *Troodon* also imply a strong physiological similarity to modern

A B

Figure 10.1. Eggs.
A, two Troodon *eggs, MOR 702, from the Two Medicine Formation of Montana. Both had weathered out of a sandstone block. Differential lithostatic crushing accounts for the apparent shape variation.*
B, one of the thousands of titanosaur eggs from the Auca Mahuevo locality, Neuquen Province, Argentina. Upper Cretaceous, Rio Colorado Formation. Scale bar equals 10 cm.

Figure 10.2. (opposite page) Egg clutches.
A, B, Troodon *clutches from the Egg Mountain locality, Montana. Upper Cretaceous, Two Medicine Formation.*
A, MOR 963 in top view showing the tight placement of the 24 eggs.
B, MOR 393 in bottom view showing the paired arrangement of the 22 eggs.
C, MOR 770, a clutch of 20 lambeosaurine eggs from the Upper Cretaceous, Judith River Formation of Montana.
(continued on page 219)

birds: 1) Egg pairs, with each egg roughly half the predicted egg size for an adult bird of similar body weight, suggest a reproductive output comparable to that of modern birds. 2) Iterative laying with synchronous hatching is possible only with an embryonic developmental temperature greater than that of the ambient environment. 3) Brooding in combination with this laying-hatching pattern also implies higher than ambient adult body temperatures. This suggests *Troodon* maintained a metabolism and body temperature similar to those of modern birds. Preservable features significant to the recognition of this reproductive mode in *Troodon* include large relative egg size in comparison to those of modern reptiles, asymmetrical egg shape, egg pairing, and partially exposed clutches. Except for an increase in egg size in the lineage leading to dinosaurs and in the Lambeosaurinae, these physical characteristics are restricted to Coelurosauria or possibly only to Maniraptora. Thus, the pattern of iterative egg-laying, delayed incubation by brooding, and synchronous hatching likely originated within one of these theropod clades.

Introduction

Troodon formosus is a small (less than 50 kg) theropod from the Upper Cretaceous of North America. Some notable features of this dinosaur include long, slender hindlimbs with a gracile foot, an enlarged and retractable pedal claw, a high EQ (encephalization quotient), and an unusual dentition (Russell 1969; Russell and Seguin 1982; Wilson and Currie 1985; Currie 1987). Recent speculation about the diet of *Troodon* has suggested the possibility of both herbivory (Holtz et al. 1994) and insectivory (Varricchio 1997). Nevertheless, this

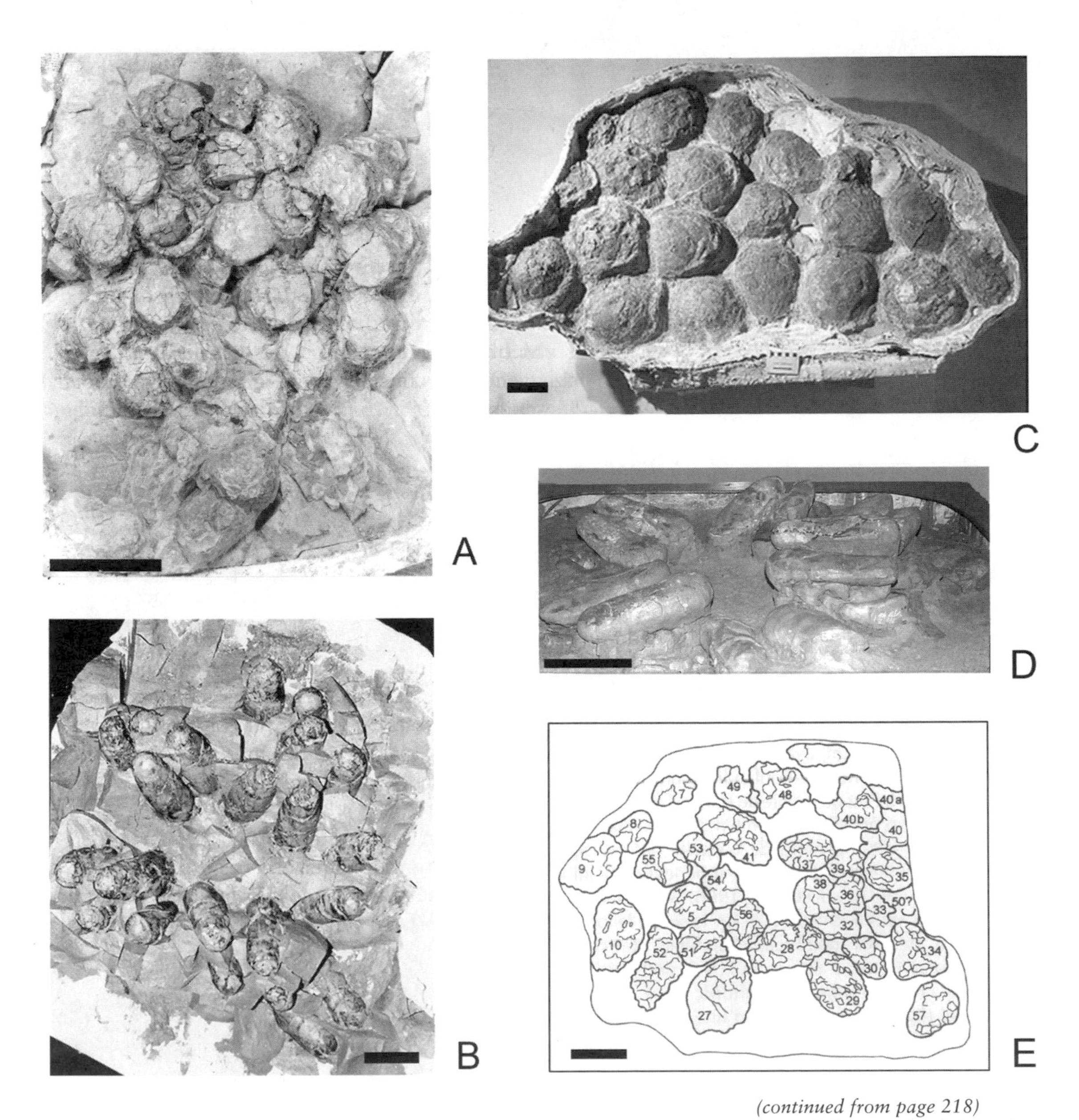

(continued from page 218)
D, an Oviraptor *clutch from the Upper Cretaceous of Mongolia with horizontally oriented elongate eggs showing a paired arrangement.*
E, clutch of 34 eggs for the allosauroid Lourinhanosaurus, *from the Late Jurassic of Portugal. Scale bar in all equals 10 cm. (Figure taken from Mateus et al. 1997)*

possibly omnivorous dinosaur apparently preyed upon small vertebrates such as hypsilophodonts (Sheetz 1999; Varricchio et al. 2002) and young hadrosaurs (Ryan et al. 2000). *Troodon* belongs to the Troodontidae, a family of theropods known from Asia and North America (Osmólska and Barsbold 1990). Although some doubt remains (Holtz 2000), several recent phylogenetic analyses favor a close relationship between troodontids, dromaeosaurids, and birds (Forster et al. 1998; Makovicky and Sues 1998; Sereno 1999).

The Campanian Two Medicine Formation of Montana and the Dinosaur Park Formation of Alberta have produced a diverse assemblage of *Troodon* specimens. Reproduction-related fossils include an excellent array of eggshell, eggs (fig. 10.1A), clutches (fig. 10.2A, B), nest structures, and embryos. In previous papers (Varricchio et al. 1997, 1999, 2002), some of these specimens have been described and a hypothesis of *Troodon* reproductive behavior has been outlined. Placed within an extant phylogenetic bracket, *Troodon* shows a suite of reproductive characters that are primitive (shared with crocodilians), derived (shared with birds), or unique to the species or perhaps to troodontids.

Some caution is warranted before delving into speculations on *Troodon* reproductive behavior. Two levels of inference are involved: behavior interpreted from body and trace fossils, and physiology inferred from behavior. This tiering of interpretation provides a greater opportunity for error (fig. 10.2.1 in Witmer 1995). Nevertheless, the subject of *Troodon* reproductive physiology merits consideration for the following reasons:

1) Direct assessment of metabolism, body temperature, rate of O_2 consumption, etc., currently remains impossible in extinct vertebrates. Therefore, any physiologic interpretation must rely on inference.

2) The *Troodon* specimens represent the direct products of either reproductive behavior or physiology. Features like egg position, clutch arrangement, and nest structure are reproductive trace fossils and thus the products of adult *Troodon* behavior. Other specimens, namely eggshell and eggs, are the direct physiological output of breeding females.

3) In living vertebrates, reproductive behavior is inherently linked to physiology. The number of offspring produced, the mode of egg incubation, and the amount of parental care all reflect varying energy investments by the parent or parents (Farmer 2001). A behavioral reproductive hypothesis for *Troodon* will naturally carry with it physiological implications.

4) Finally, we feel the proposed hypotheses for *Troodon* are robust, for they incorporate a wide range of specimens and concur with established phylogenies.

Troodon Reproductive Behavior

Specimen Descriptions

A number of papers contain descriptions of *Troodon* eggshell, eggs, clutches, nests, nesting grounds, and embryos: Horner (1982, 1984, 1987), Horner and Weishampel (1988), Hirsch and Quinn (1990), Geist and Jones (1996), Zelenitsky and Hills (1996), Varricchio et al. (1997, 1999, 2002), Carpenter (1999), and Horner et al. (2001). Note that pre-1997 references, those published prior to the correction of Horner and Weishampel (1996), refer to these specimens as belonging to the hypsilophodont *Orodromeus makelai.* The following description represents a summary.

Troodon eggs belong to the oospecies *Prismatoolithus levis* (Zelenitsky and Hills 1996). The elongate and asymmetrical eggs (fig. 10.1A) range in length from approximately 12 to 16 cm and have a volume of approximately 440 cm^3. Exterior surfaces of the eggs vary from smooth to faintly striated. Round to oval pores are unevenly distributed over the shell surface with a denser concentration occurring toward the broader end of the egg (Hirsch and Quinn 1990).

In radial view, *Troodon* eggshell exhibits distinct structural layering characteristic of all theropods including extant birds: an interior mammillary layer exhibiting barrel-shaped cones, a prismatic layer consisting of vertically oriented calcite columns (Hirsch and Quinn 1990; Zelenitsky and Hills 1996), and a possible third structural layer similar to the external layer found in neognathid birds (Varricchio et al. 2002). Eggshell from hatched clutches exhibits a cratered interior surface (Hirsch and Quinn 1990; Varricchio et al. 2002). The prismatic to mammillary layer thickness ratio is approximately 6:1 (Zelenitsky and Hills 1996; Varricchio et al. 2002).

The clutches are oval and contain up to 24 eggs. The eggs stand vertically to subvertically, narrow-end down within the sediment, and are inclined toward the clutch center (Horner 1984, 1987; Zelenitsky and Hills 1996). In top view the eggs appear closely packed with the total clutch area not exceeding 0.5 m^2 (fig. 10.2A). In bottom view clutches display a distinct paired-egg pattern (fig. 10.2B). Egg pairs are distinguishable by their closer spacing and near-parallel long axes, and pairing is supported by statistical analysis (Varricchio et al. 1997, 1999). Often two sedimentary lithologies are associated with unhatched clutches. Typically, limestone entombs the bottom two-thirds of the eggs whereas mudstone covers the top (Horner 1984, 1987; Varricchio et al. 1997, 1999). Hatched clutches consist only of the upright bottoms (Horner 1984, 1987). Nests consist of a shallow bowl-shaped depression surrounded by a raised rim. In the most nearly complete specimen (MOR 963), the exterior dimensions measure 1.6 by 1.7 m. The clutch occupies less than half of the enclosed area (Varricchio et al. 1997, 1999). Additional fossils associated with nesting horizons include isolated to articulated *Orodromeus* remains, rare small vertebrates such as lizards and mammals (Horner 1984, 1987; Varricchio et al. 2002), pupae cases that are likely assignable to Hymenoptera (Martin and Varricchio 2001), small invertebrate burrows, and root traces (Varricchio et al. 1999). Embryos exhibit well-ossified bone (Horner and Weishampel 1988; Geist and Jones 1996; Horner et al. 2001) and show a similar level of development within a single clutch (Varricchio et al. 2002).

Behavioral Interpretation

The interpretation of *Troodon* reproduction is briefly summarized based on previous work (Varricchio et al. 1997, 1999, 2002). *Troodon* constructed earthen nests usually within well-drained alkaline soils. Iterative egg-laying at daily or greater intervals produced pairs of eggs within the clutches. This resulted from *Troodon* possessing two avian-like reproductive tracts and having monoautochronic ovulation (a

Figure 10.3. Reconstruction of a brooding adult Troodon. *Original artwork by Bill Parsons. Recent maniraptoran specimens suggest that feathers might be a more likely skin covering than scales (Xu et al. 1999).*

synchronous release of a single ovum from each ovary). As the eggs were laid, the adult partially embedded them within the sediments. Incubation did not begin until completion of the clutch, at which time the adult may have gathered the upper, exposed portions of the eggs into a slightly tighter configuration. Then one or both adults likely incubated the eggs by brooding them with direct body contact (fig. 10.3). Based on the requirements of modern taxa (Rahn and Ar 1974; Stewart 1989), egg-laying and incubation may have required around two months.

For *Troodon,* primitive reproductive features—those shared with crocodilians and other living reptiles—include two functional ovaries and oviducts, some burial of the clutch within sediments, a lack of egg rotation, and presumably the proteinaceous egg structures (chalazae) that allow such rotation in birds. Avian-like oviducts producing a maximum of one egg per day, some exposure of the clutch, and brooding represent more derived features shared with modern birds. The unusual, near-vertical orientation of the eggs within the sediments is unknown among living vertebrates. Among dinosaurs this appears unique to either *Troodon* or troodontids.

Troodon Reproductive Physiology

Iterative Egg-Laying and Synchronous Hatching

The most primitive groups of living birds, Palaeognathes, Galliformes, and Anseriformes, share several reproductive features (Sibley and Ahlquist 1990; Sillen-Tullberg and Temrin 1994). As with most extant birds, these groups exhibit iterative egg-laying and delayed incubation. Most lay at a maximum rate of one egg per day, but some with particularly large eggs (e.g., ostriches and emus) require several days to produce a single egg (Stewart 1989; Terres 1995). Consequently, some eggs may be many days older than other eggs within the same clutch. For example, in a grouse clutch of 15 eggs, two weeks may separate the first and last laid eggs.

In these birds, brooding begins only after completion of the clutch, and the delayed incubation results in synchronous hatching. During egg-laying, the embryo experiences a drop in temperature from the high body temperature of the adult to a lower ambient condition. The cooler external temperatures halt development and the embryo remains in stasis. The inception of brooding raises the clutch temperature to approximately that of the adult and restarts development (Gill 1989). Except for the megapodes (Jones et al. 1995), this pattern is critical for the protection of the hatchlings. All of these birds possess precocial, self-feeding, and nidifugous (i.e., nest-fleeing) young (Starck and Ricklefs 1998). The delayed incubation ensures that all the chicks, despite some being days or even weeks older than others, hatch and leave the nest together. Because brooding ceases simultaneously for all eggs, adults are free to accompany the young.

Troodon appears to exhibit a similar pattern of behavior. Clutches exhibit a paired arrangement in bottom view but a close-packed configuration in top view. This suggests *Troodon* laid eggs iteratively, as modern birds do, but primitively retained two functional ovaries and oviducts. The tightly packed upper portions (fig. 10.2A) indicate that incubation either masked or was not dependent upon the original paired pattern (fig. 10.2B). Embryonic bone clearly demonstrates that *Troodon* had precocial hatchlings (Geist and Jones 1996; Horner et al. 2001). Finally, the embryos of MOR 246, the only clutch known with young, exhibit a similar level of development (Varricchio et al. 2002).

If these reproductive patterns observable in extant primitive birds occur in *Troodon,* they likely impart one or perhaps two physiological constraints. First, synchronous hatching for iteratively laid eggs can only occur if embryonic developmental temperatures significantly differ from those of the environment. Otherwise, embryonic development would never experience stasis, and eggs laid days apart would hatch days apart (White and Kinney 1974; O'Connor 1984; Gill 1989). Thus, no matter how *Troodon* eggs were incubated, their developmental temperature would need to exceed ambient temperatures. Further, the hypothesized behavior implies that if adult *Troodon* brooded their clutches, their body temperatures also must have at least temporarily surpassed those of the environment.

TABLE 10.1
Egg and clutch data for two extant archosaurs and those dinosaurs with identifiable eggs

Taxon	Adult Weight	Egg Pairs	Clutch	Egg Shape	Egg Mass	Er	Eb
Alligator	70 kg[1]	No[2]	Buried[2]	Ellipsoid[2]	84 g[3]	44 g	1,400 g
Maiasaura	4,000 kg[4]	No[5]	?	Sphere[5]	980 g[5]	243 g	31,000 g
Lambeosaurine	4,000 kg[6]	No[5]	?	Sphere[5]	4,600 g[5]	243 g	31,000 g
Sauropod	7,000 kg[7]	No[8,9]	?	Sphere[8]	870 g[8]	307 g	48,000 g
Lourinhanosaurus	176 kg[6]	No[10]	?	Ellipsoid[10]	645 g[11]	65 g	2,800 g
Oviraptor	40 kg[7,12]	Yes[13-15]	Exposed[15]	Ellipsoid[15]	423 g[16]	35 g	910 g
Troodon	50 kg[12,17]	Yes[18]	Exposed[19]	Asymmetrical[20,21]	485 g[21]	39 g	1100 g
Meleagris	4.1 kg[22]	No[22,23]	Exposed[23]	Asymmetrical[23]	88 g[24]	13 g	160 g

The unidentified lambeosaurine comes from the Judith River Formation of Montana (Horner 1999), the sauropod is a possible titanosaur from the Rio Colorado Formation of Argentina (Chiappe et al. 2000), and the allosauroid, Lourinhanosaurus, comes from the Late Jurassic of Portugal (Mateus et al. 1997, 2001; Mateus 1998). Er and Eb are the predicted egg masses for an adult reptile and bird of equivalent adult body size. See text for calculations of egg masses.

References: 1, Rootes et al. 1991; 2, Magnusson et al. 1989; 3, Ferguson 1985; 4, Dunham et al. 1989; 5, Horner 1999; 6, Anderson et al. 1985; 7, Peczkis 1995; 8, Chiappe et al. 1998; 9, Chiappe et al. 2000; 10, Mateus et al. 1997; 11, Antunes et al. 1998; 12, Paul 1988; 13, Norell et al. 1995; 14, Dong and Currie 1996; 15, Clark et al. 1999; 16, Carpenter 1999; 17, Varricchio 1993; 18, Varricchio et al. 1997; 19, Varricchio et al. 1999; 20, Hirsch and Quinn 1990; 21, Varricchio et al. 2002; 22, Terres 1995; 23, Bent 1932; 24, Ar et al. 1974.

Egg Size

Blueweiss et al. (1978) provide allometric equations relating individual egg weight to maternal body weight for both birds and reptiles:

(Equation 1) $Eb = 0.26\ W^{0.77}$

(Equation 2) $Er = 0.41\ W^{0.42}$

Eb and Er are the bird and reptile egg weights respectively, and W is the maternal body weight. All weights are given in grams. Only at very small body weights (adult weight ~10 g) do the two groups produce similarly sized eggs. For most body sizes, birds produce significantly larger eggs. For example, egg masses for emu and ostrich are 578 g and 1,480 g, respectively (Ar et al. 1974). Eggs of some of the largest living reptiles, for example *Alligator mississippiensis* and *Crocodylus porosus,* average only 84 g and 113 g (Ferguson 1985). This drastic difference in egg size may reflect the varying modes of egg production: iterative, one-at-a-time formation in birds versus en masse clutch formation in most reptiles.

Because most *Troodon* eggs have undergone lithostatic crushing or erosion, volume can not be measured directly. Instead volume measurements were made on a three-dimensional model based upon several specimens. Estimated volume is 440 cm^3 (Table 10.1). Using an egg density of 1.09g/cm^3 (Williams et al. 1984), *Troodon* eggs would have had a mass of 480 g. Adult *Troodon* weighed approximately 50 kg (Paul 1988; Varricchio 1993).

Calculated with the equations of Blueweiss et al. (1978), the expected egg sizes for a 50-kg bird and reptile are 1,100 g and 39 g, respectively. *Troodon* eggs are over ten times larger than the expected reptilian egg. Further, the mass of a *Troodon* egg pair (960 g) approximates that of a single avian egg. Thus, if *Troodon* laid two eggs at daily or greater intervals as hypothesized, its reproductive output would approximate that of a modern bird. The asymmetrical shape and avian-like microstructure of *Troodon* eggs, features not typical of modern reptiles, further support the hypothesis that its reproductive tracts functioned in a similar manner to those of modern birds (Varricchio et al. 1997, 2002).

Blueweiss et al. (1978) also presented equations relating overall clutch weight to adult body weight for birds and reptiles. Unlike individual egg size, clutch weight does not differ radically (i.e., by an order of magnitude) between birds and reptiles. Predicted clutch weights for a 50-kg bird and reptile are 3.7 kg and 4.7 kg, respectively. *Troodon* clutches with 24 eggs would, in contrast, weigh about 12 kg. These drastically larger clutches suggest the possibility that more than one female may have contributed to a single clutch.

Brooding

Although the term "brooding" is often restricted to egg-warming activities (Shine 1988), it may refer to any form of egg attendance by an adult after oviposition (Somma 1988). In the broad sense, brooding behavior is fairly widespread among amniotes. Among some lizards and snakes brooding serves to deter potential egg predators, prevent disturbance by other ovipositing females, and regulate the hydric conditions of the clutch (Shine 1988). Taxa that actively warm their eggs include pythons which generate heat through muscular shivering and the five-lined skink, *Eumeces fasciatus,* which returns to the clutch after basking (Shine 1988). For birds, brooding generally implies incubation of eggs with body heat, a behavior considered primitive for at least Neornithes (Clark et al. 1999). In birds, adult egg attendance may also incorporate other functions such as predator protection and cooling of eggs by shading (Terres 1995).

Reproductive specimens suggest brooding in the strict sense (sensu Shine 1988) as the most likely form of incubation in *Troodon.* Supporting evidence in part includes: the exposed and compact upper portion of eggs within the clutch; the broad, open nest structure; concentration of eggshell pores in the upper portions of the eggs; and the occurrence of an adult fossil on top of an egg clutch (Varricchio et al. 1997, 1999). In the latter, the adult was not preserved in a brooding position and therefore the evidence for *Troodon* brooding remains equivocal.

If brooding in the strictest sense can be demonstrated for *Troodon,* it may carry the implication of elevated body temperature and metabolism. Although among living amniotes brooding behavior in the broad sense (sensu Somma 1988) serves a wide range of functions and is not universally linked with elevated metabolisms, brooding sensu stricto by body contact appears primitive for modern birds. This implies that incubating eggs with body heat evolved prior to or with Neornithes.

The close sister relationship of *Troodon* and Aves (e.g., Forster et al. 1998) suggests that brooding behavior in *Troodon* would be not just homoplastic but also homologous to that of modern birds. Furthermore, brooding incubation in the context of iterative laying and synchronous hatching necessitates an elevated adult body temperature. Adult temperatures must be sufficiently high both to trigger the onset of and maintain embryonic development.

Reproduction within Dinosauria

Reviews of dinosaur reproduction (Coombs 1989; Paul 1994; Weishampel and Horner 1994; Carpenter 1999; Horner 2000; Varricchio 2000) have been limited by a lack of good specimens. Relatively few dinosaur discoveries have identifiable nests, eggs, or even eggshell. Consequently, speculations about dinosaur reproductive behavior and its evolution have relied largely on the use of possible modern analogs. Witmer (1995) outlined the Extant Phylogenetic Bracket (EPB), a methodology for assessing the soft tissue attributes of extant taxa using osteological correlates of the soft tissues within their living relatives. This method may be applied to aspects of dinosaur reproduction. Modern birds and reptiles, in particular crocodilians, represent the extant bracket, while the physical attributes of eggs, clutches, and reproductive behaviors are substituted for the osteological correlates and their soft tissue attributes respectively. The physical attributes examined include: egg shape, egg size relative to adult body size, paired versus massed egg arrangements, and exposed versus buried clutches (table 10.1; fig. 10.4).

Two modern archosaurs, the American alligator *Alligator mississippiensis* and the wild turkey *Meleagris gallopavo,* are included in Table 10.1 simply as comparative examples. Only six dinosaur taxa have identifiable eggs (table 10.1). Three additional dinosaur taxa may also be represented by identifiable eggs; these include a possible unidentified therizinosaur from the Henan Province of China, *Sinosauropteryx,* and *Compsognathus* (Ostrom 1978; Griffiths 1993; Manning et al. 1997, 2000; Chen et al. 1998). Although the Henan eggs contain well-preserved embryos, their taxonomic identification has not been worked out in detail (Manning et al. 1997, 2000). Chen et al. (1998) describe a specimen of *Sinosauropteryx prima* with two ellipsoidal objects within its gastral basket. Given their location, they interpret these 37-mm by 26-mm bodies as eggs. They do not provide any indication of whether they have identified eggshell, and suggest the possibility of additional hidden eggs. A number of spherical objects occur in association with the type specimen of the very closely related *Compsognathus longipes.* These relatively small bodies (diameters equal 10 mm) have been interpreted as the dinosaur's eggs (Griffiths 1993), but could instead represent invertebrate remains or inorganic structures (Ostrom 1978). Further analysis of any of these three specimens should be relevant to the interpretation of dinosaur reproduction.

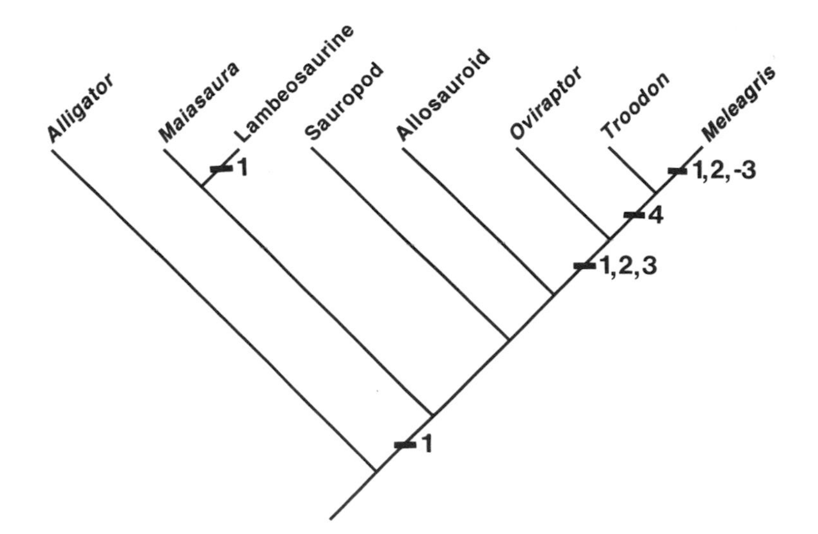

Figure 10.4. Phylogeny of the two extant archosaurs and those dinosaurs with identifiable eggs from Table 1, showing the acquisition and loss of four reproductive features:
1, larger relative egg size;
2, increase in clutch exposure;
3, egg pairs; and
4, asymmetric egg shape. Note the multiple increases in egg size and the concentration of acquisitions within the Maniraptora.

Relative Egg Size

The relative sizes of dinosaur eggs can be assessed in part by using equations 1 and 2 listed previously (Blueweiss et al. 1978). The egg mass for extinct taxa was estimated using an average egg density of 1.09 g/cm^3 (Williams et al. 1984), and egg volumes were either taken directly from the literature or calculated from published egg dimensions. These equations are based on adult body sizes that reach maximums of roughly 1,000 kg in reptiles and 100 kg in birds. Consequently, application of these allometric equations is most meaningful when applied to dinosaurs with smaller body sizes than these maxima. The heaviest modern reptile eggs include those of the Indopacific crocodile, *Crocodylus porosus,* 113 g (Ferguson 1985); the Komodo dragon, *Varanus komodoensis,* 125 g; and *Python molurus,* 226–309 g (Iverson and Ewert 1987). All of the identified dinosaur eggs (table 10.1) exceed these in size as well as the predicted egg masses (Er) for reptiles of equivalent body size. Still, none of the dinosaur eggs approximates the predicted egg mass (Eb) for birds. Dinosaurs with the largest relative egg sizes include the Lambeosaurinae, *Oviraptor,* and *Troodon.* The sizable lambeosaurine eggs with an estimated mass of 4,600 g (Horner and Currie 1994) exceed the largest eggs among modern birds, but do not approach the largest known eggs, the roughly 8-kg eggs of *Aepyornis,* the extinct elephant bird. *Oviraptor* and *Troodon* eggs compare most closely with the predicted avian eggs, with estimated egg masses being 0.46 and 0.44 of the Rb values, respectively.

Egg Shape

Reptile eggs with few exceptions are spherical or ellipsoidal. In contrast the typical avian egg is asymmetrical and tapered at one end (Iverson and Ewert 1987). Although the shape of avian eggs may serve several functions, the asymmetry apparently results from varying constrictive forces within the oviduct wall. This activity only works when a single egg occupies the reproductive tract (Iverson and Ewert 1987; Smart 1987). Thus, for birds asymmetrical egg shape appears associated with iterative egg formation. Among the identifiable dinosaur eggs, only those of *Troodon* are asymmetrical (fig. 10.1). The remainder are either spherical or ellipsoidal (table 10.1). Within the Extant Phylogenetic Bracketing (EPB) framework, the assumption in *Troodon* of both an avian-like reproductive tract and iterative laying represents a level II inference. Several Cretaceous egg types have been identified as avian on the basis of embryos, and at least three varieties show an asymmetry typical of modern birds. These eggs, from Argentina and Mongolia, belong to primitive ornithothoracines and enantiornithines (Mikhailov 1997; Schweitzer et al. 2002).

Egg Arrangement

While some dinosaurs (fig. 10.2), including hadrosaurs, an Argentine titanosaur, and the Portuguese allosauroid *Lourinhanosaurus*, have massive clutches apparently lacking any internal arrangement (Mateus et al. 1997; Horner 1999; Chiappe et al. 2000), those of the maniraptorans clearly exhibit a paired-egg pattern (Norell et al. 1995; Dong and Currie 1996; Varricchio et al. 1997; Clark et al. 1999). Several other dinosaur ootaxa also display clutches with egg pairs. These include the giant, 40-cm-long to 50-cm-long *Macroelongatoolithus xixia* (Carpenter 1999); the small, bumpy-surfaced *Continuoolithus canadensis* (Horner 1987); and the Mongolian *Protoceratopsidovum* (Mikhailov et al. 1994). All of these unidentified Cretaceous egg taxa possess eggshell microstructures that have only been associated with non-avian and avian theropods (Carpenter 1999). Thus, egg-pairing appears to be restricted to a clade within Theropoda, perhaps Coelurosauria (fig. 10.4).

Among the closest extant relatives of dinosaurs, crocodilians, and birds, large clutches with egg pairs are unknown. Dinosaur clutches exhibiting pairedness have been interpreted in two ways (Clark et al. 1999). The first hypothesis proposes that the entire clutch was laid at one sitting with two eggs emerging from the cloaca simultaneously as the female rotated within the nest (Dong and Currie 1996; Larson 2000). The apparent spiral pattern within the clutch is cited as necessitating the near continuous egg-laying (Clark et al. 1999; Larson 2000). Egg pairs result from the active participation of both oviducts (Sabath 1991; Mikhailov et al. 1994; Larson 2000) and either an adherence of the eggs with mucus (Sabath 1991; Mikhailov et al. 1994) or a simultaneous expulsion due to the pressure of a subsequent egg pair (Larson 2000). Most extant vertebrates display polyautochronic ovulation, where ovulation of multiple ova occurs essentially simultaneously from both ovaries (Smith et al. 1973). These vertebrates do not exhibit egg pairing. Thus, simply involving two reproductive tracts without specifying the mode of ovulation cannot fully explain dinosaur egg pairing. Additionally, the methods hypothesized for forming pairs, adhering with mucus or forced by a subsequent pair, are unknown among amniotes. This hypothesis falls outside of the EPB framework in that a completely novel adaptation is invoked to explain the observed feature and would represent a level III inference. The second hypothesis (Varricchio et al. 1997) invokes monoautochronic ovulation, where a single egg is produced synchronously from each ovary and oviduct (Smith et al. 1973). Egg pairs would not necessitate simultaneous emersion from the cloaca. Pairing instead would reflect both the synchronous formation of eggs and the significant hiatus between subsequent pairs. As suggested (Varricchio et al. 1997), two eggs would be laid one at a time during a single sitting, and then a day or more would pass before the laying of the next pair. An entire clutch would be laid via multiple sittings of the adult female. A somewhat similar pattern occurs in most species of gekkonid lizards: During the breeding season females lay multiple clutches, each consisting of a pair of monoautochronically produced eggs (Smith et al. 1973; Dunham et al. 1988; Zug 1993; Punzo and Turingan 2001). Egg formation is also similar in birds, where eggs are ovulated and formed one at a time (Gill 1989). However, avian egg pairing does not occur due to the lack of a second functional reproductive tract. The interpretation of monoautochronic ovulation and its resulting iterative laying in coelurosaurian theropods would be a level II inference. The *Sinosauropteryx prima* specimen described as containing two eggs (Chen et al. 1998) would seem to strongly support this argument. But more work to verify this and to understand the rather different eggs associated with the closely related *Compsognathus* is needed. Nevertheless, the lack of modern analogs for the first hypothesis (entire clutch laid at one sitting), and the great similarity between the maniraptoran eggs and those of modern birds in size, shape, and microstructure would favor the second hypothesis of monoautochronic ovulation. It is possible that a more inclusive clade within dinosaurs produced eggs in pairs but subsequent nesting or incubation behavior (e.g., massing eggs for burial) obscured this original pattern. Demonstrating such an ovulation mode would require exceptional preservation as in *Sinosauropteryx,* because statistically significant egg pairs cannot be recognized in massed egg arrangements (Fig. 10.2C,E).

Clutch Arrangement

A variety of nesting modes have been proposed for dinosaur eggs: buried within sediments, covered by vegetation mounds, or exposed within bowl-like structures (Coombs 1989; Cousin et al. 1994; reviewed by Carpenter 1999). Ideal preservation would include a nesting trace associated with eggs, i.e., a modification of the surrounding sediments by the nest maker. This trace would record the size and nature of the nest at the time of incubation. Likely, post-egg-laying events, such as predation, hatching of young, past and recent erosion, sedimentation, and diagenesis, all contribute to the loss of nesting traces from the fossil record. Some dinosaur nests may simply not have been preserved because they were elevated or not associated with sediments. Despite the numerous discoveries of dinosaur clutches, relatively little is known about specific dinosaur nests. Such remains the case for *Maiasaura,* the lambeosaurine, and *Lourinhanosaurus* (table 10.1). The egg-filled depressions for the Argentine sauropod suggest excavated nests where eggs were not buried by sediments (Garrido et al. 2001). Whether the eggs remained fully unburied or covered by vegetation remains to be determined. Strong evidence exists for both *Oviraptor* and *Troodon* to suggest that their clutches were in part free from sediment (fig. 10.2A, D). In the former, skeletons of adults occur in contact or nearly in contact with significant portions of the eggs (Dong and Currie 1996; Clark et al. 1999). In most *Troodon* clutches with whole eggs, two distinct lithologies encase the upper and lower portions of the upright eggs. The upper lithology suggests rapid burial of the exposed eggs by a flooding event (Horner 1984, 1987; Varricchio et al. 1999). Among both modern reptiles and birds, exposed eggs are nearly universally associated with brooding behavior in the broad sense (Shine 1988; Gill 1989). Brooding behavior in these two maniraptorans would represent an EPB level I inference. Further, the assumption of typical avian brooding where adults warm their eggs with body heat would be a level II inference. The occurrence of adults in contact with clutches in both dinosaurs (Norell et al. 1995; Dong and Currie 1996; Varricchio et al. 1997; Clark et al. 1999) minimally confirms brooding in the broad sense.

Discussion

The paucity of taxonomically identifiable dinosaur eggs allows recognition of only a few trends (fig. 10.4): 1) In general dinosaurs have relatively larger egg sizes than those of their modern reptilian outgroups, but smaller than those of modern birds. At least two additional increases in egg size occur within non-avian dinosaurs, in the lambeosaurine lineage and within Maniraptora. 2) Most dinosaur eggs retain a typical reptilian shape, being either spherical or ellipsoidal. The only identifiable, asymmetrical eggs belong to *Troodon.* 3) Egg pairing within clutches occurs in Maniraptora or perhaps Coelurosauria, if the internal eggs of *Sinosauropteryx* have been correctly identified. 4) Partially exposed clutches occur in maniraptorans. Whether they have a broader distribution among dinosaurs remains to be demonstrated. A better understanding of dinosaur nests will require more detailed descriptions of the sediments surrounding egg clutches. 5) Several attributes of modern birds, such as large eggs relative to body size, asymmetrical egg shapes, and exposed clutches, may correlate with iterative egg-laying and brooding. Because *Troodon* shares these physical attributes, the assumption of these two behaviors represents a level II inference within the EPB framework. Even greater similarities exist between *Troodon* and some Cretaceous birds. Eggs from the Upper Cretaceous of Mongolia have been identified as avian on the basis of embryos (Elzanowski 1981). These asymmetrical eggs occur subvertically to vertically within sediments, blunt-end up (Sabath 1991). Although this position may reflect depositional hydraulics (Mikhailov et al. 1994), it could also reflect an incubation style (Sabath 1991) very similar to that observed in *Troodon.*

Conclusions

Troodon appears to share a number of reproductive features with modern birds, particularly with groups like Paleognathes, Galliformes, and Anseriformes. Shared features include iterative egg-laying, delayed incubation by brooding, and synchronous hatching of precocial young. For modern birds, these activities are inherently linked with and dependent upon the avian endothermic physiology. Consequently, if the hypothesized reproductive behaviors are true for *Troodon,* that implies that *Troodon* maintained an elevated metabolism and body temperature, as in these living birds. A variety of morphologic data are consistent with this interpretation: upright stance (Bakker 1972), high EQ (Hopson 1980), bone histology (Varricchio 1993), and the high likelihood of feathers within troodontids (Xu et al. 1999). The most significant physical attributes for the interpretation of *Troodon* reproduction include relatively large eggs, asymmetrical egg shape, egg pairs, and partially exposed clutches. Except for some increase in egg size common to all dinosaurs and again to the lambeosaurine lineage, these physical characteristics are restricted to Coelurosauria or possibly only Maniraptora. Thus, many reproductive features considered typically avian among extant vertebrates likely originated within one of these theropod clades.

Acknowledgments. We thank Jack Horner, the Paleontology Department at the Museum of the Rockies, Jim Schmitt, and the Department of Earth Sciences at Montana State University. We also thank the Graves Museum for inviting our participation in the 2000 Bambiraptor conference, Bruce Selyem for *Troodon* clutch photos, and Greg "Bubba" Erickson. A Jurassic Foundation grant provided partial funding for this study.

References

Ar, A., C. V. Paganelli, R. B. Reeves, D. G. Greene, and H. Rahn. 1974. The avian egg: water vapor conductance, shell thickness, and functional pore area. *Condor* 76: 153–158.

Bakker, R. T. 1972. Anatomical and ecological evidence of endothermy in dinosaurs. *Nature* 238: 81–85.

Blueweiss, L., H. Fox, V. Kudzma, D. Nakashima, R. Peters, and S. Sams. 1978. Relationships between body size and some life history parameters. *Oecologia* 37: 257–272.

Carpenter, K. 1999. *Eggs, nests, and baby dinosaurs: a look at dinosaur reproduction.* Bloomington: Indiana University Press.

Chen, P.-J., Z.-M. Dong, and S.-N. Zhen. 1998. An exceptionally well-preserved theropod dinosaur from the Yixian Formation of China. *Nature* 391: 147–152.

Chiappe, L. M., L. Dingus, F. Jackson, G. Grellet-Tinner, R. Aspinall, J. Clarke, R. Coria, A. Garrido, and D. Loope. 2000. Sauropod eggs and embryos from the Late Cretaceous of Patagonia. In A. M. Bravo and

T. Reyes (eds.), *First international symposium on dinosaur eggs and babies*, pp. 23–29.
Clark, J. M., M. A. Norell, and L. M. Chiappe. 1999. An oviraptorid skeleton from the Late Cretaceous of Ukhaa Tolgod, Mongolia, preserved in an avian-like brooding position over an oviraptorid nest. *American Museum Novitates* 3265: 1–36.
Coombs, W. P. 1989. Modern analogs for dinosaur nesting and parenting behavior. In J. O. Farlow (ed.), *Paleobiology of the dinosaurs. Geological Society of America* Special Paper 238: 21–54.
Cousin, R., G. Breton, R. Fournier, and J.-P. Watte. 1994. Dinosaur egg-laying and nesting in France. In K. Carpenter, K. Hirsch, and J. R. Horner (eds.), *Dinosaur eggs and babies,* pp. 56–74. Cambridge University Press.
Currie, P. J. 1987. Bird-like characteristics of the jaws and teeth of troodontid theropods (Dinosauria, Saurischia). *Journal of Vertebrate Paleontology* 7: 72–81.
Dong Z.-M., and P. J. Currie. 1996. On the discovery of an oviraptorid skeleton on a nest of eggs at Bayan Mandahu, Inner Mongolia, People's Republic of China. *Canadian Journal of Earth Sciences* 33: 631–636.
Dunham, A. E., D. B. Miles, and D. N. Reznick. 1988. Life history patterns in squamate reptiles. In C. Gans and R. B. Huey (eds.), *Biology of the Reptilia,* Volume 16: 441–522. New York: Wiley.
Elzanowski, A. 1981. Embryonic bird skeletons from the Late Cretaceous of Mongolia. *Palaeontologia Polonica* 42: 147–179.
Farmer, C. G. 2001. A new perspective on the origin of endothermy. In J. Gauthier and L. F. Gall (eds.), *New perspectives on the origin and early evolution of birds,* pp. 389–409. New Haven, Connecticut: Peabody Museum of Natural History.
Ferguson, M. W. J. 1985. Reproductive biology and embryology of the crocodilians. In C. Gans, F. Billett, and P. F. A. Maderson (eds.), *Biology of the Reptilia,* Volume 14 (Development A): 329–491. New York: Wiley.
Forster, C. A., S. D. Sampson, L. M. Chiappe, and D. W. Krause. 1998. The theropod ancestry of birds: new evidence from the Late Cretaceous of Madagascar. *Science* 279: 1915–1919.
Garrido, A. C., L. M. Chiappe, F. Jackson, J. Schmitt, and L. Dingus. 2001. First sauropod nest structures. *Journal of Vertebrate Paleontology* 21: 52–53.
Geist, N. R., and T. D. Jones. 1996. Juvenile skeletal structure and the reproductive habits of dinosaurs. *Science* 272: 712–714.
Gill, F. B. 1989. *Ornithology.* New York: Freeman.
Griffiths, P. 1993. The question of *Compsognathus* eggs. *Revue de Paléobiologie* 7: 85–94.
Hirsch, K. F., and B. Quinn. 1990. Eggs and eggshell fragments from the Upper Cretaceous Two Medicine Formation of Montana. *Journal of Vertebrate Paleontology* 10: 491–511.
Holtz, T. R. 2000. A new phylogeny of the carnivorous dinosaurs. *Gaia* 15: 5–61.
Holtz, T. R., D. L. Brinkman, and C. L. Chandler. 1994. Denticle morphometrics and a possibly different life habit for the theropod dinosaur *Troodon. Journal of Vertebrate Paleontology* 14: 30A.
Hopson, J. A. 1980. Relative brain size in dinosaurs: implications for endothermy. In D. K. Thomas and E. C. Olson (eds.), *A cold look at the warm-blooded dinosaurs,* pp. 287–300. Washington, D.C.: American Association for the Advancement of Science.

Horner, J. R. 1982. Evidence for colonial nesting and "site fidelity" among ornithischian dinosaurs. *Nature* 297: 675–676.
———. 1984. The nesting behavior of dinosaurs. *Scientific American* 250: 130–137.
———. 1987. Ecological and behavioral implications derived from a dinosaur nesting site. In S. J. Czerkas and E. C. Olson (eds.), *Dinosaurs past and present,* Volume II: 51–63. Los Angeles: Natural History Museum of Los Angeles County.
———. 1999. Egg clutches and embryos of two hadrosaurian dinosaurs. *Journal of Vertebrate Paleontology* 19: 607–611.
———. 2000. Dinosaur reproduction and parenting. *Annual Review of Earth and Planetary Science* 28: 19–45.
Horner, J. R., and P. J. Currie. 1994. Embryonic and neonatal morphology and ontogeny of a new species of *Hypacrosaurus* (Ornithischia, Lambeosauridae) from Montana and Alberta. In K. Carpenter, K. F. Hirsch, and J. R. Horner (eds.), *Dinosaur eggs and babies,* pp. 312–336. New York: Cambridge University Press.
Horner, J. R., and D. B. Weishampel. 1988. A comparative embryological study of two ornithischian dinosaurs. *Nature* 332: 256–257.
———. 1996. A comparative embryological study of two ornithischian dinosaurs—a correction. *Nature* 383: 103.
Horner, J. R., K. Padian, and A. de Ricqlés. 2001. Comparative osteohistology of some embryonic and perinatal archosaurs: developmental and behavioral implications for dinosaurs. *Paleobiology* 27: 39–58.
Iverson, J. B., and M. A. Ewert. 1987. Physical characteristics of reptilian eggs and a comparison with avian eggs. In D. C. Deeming and M. W. J. Ferguson (eds.), *Egg incubation: its effect on embryonic development in birds and reptiles,* pp. 87–100. New York: Cambridge University Press.
Jones, D. N., R. W. R. J. Dekker, and C. S. Roselaar. 1995. *The Megapodes.* New York: Oxford University Press.
Larson, P. L. 2000. The theropod reproductive system. *Gaia* 15: 389–398.
Makovicky, P. J., and H.-D. Sues. 1998. Anatomy and phylogenetic relationships of the theropod dinosaur *Microvenator celer* from the Lower Cretaceous of Montana. *American Museum Novitates* 3240: 1–27.
Manning, T. W., K. A. Joysey, and A. R. I. Cruickshank. 1997. Observations of microstructure within dinosaur eggs from Henan Province, People's Republic of China. In D. L. Wolberg, E. Stump, and G. D. Rosenberg (eds.), *Dinofest International,* pp. 287–290. Philadelphia: The Academy of Natural Sciences.
———. 2000. *In ovo* tooth replacement in a therizinosaurid dinosaur. In A. M. Bravo and T. Reyes (eds.), *First international symposium on dinosaur eggs and babies,* pp. 129–134. Isona, Catalunya, Portugal.
Martin, A. J., and D. J. Varricchio. 2001. Evidence for insect nesting and its paleoenvironmental significance in the Two Medicine Formation (Late Cretaceous), Choteau, Montana. *North American Paleontological Convention, Program and Abstracts, PaleoBios* 21: 88.
Mateus, I., H. Mateus, M. T. Antunes, O. Mateus, P. Tacquet, V. Ribeiro, and G. Manuppella. 1997. Couvée, oeufs et embryons d'un dinosaure théropode du Jurassique supérieur de Lourinhã. *Comptes Rendus de l'Académie des Sciences de Paris, Sciences de la terre et des planets* 325: 71–89.
Mikhailov, K. E. 1997. Fossil and recent eggshell in amniotic vertebrates: fine structure, comparative morphology and classification. *Special Papers in Paleontology* 56: 1–80.

Mikhailov, K. E., K. Sabath, and S. Kurzanov. 1994. Eggs and nests from the Cretaceous of Mongolia. In K. Carpenter, K. Hirsch, and J. R. Horner (eds.), *Dinosaur eggs and babies,* pp. 88–115. New York: Cambridge University Press.

Norell, M. A., J. M. Clark, L. M. Chiappe, and D. Dashzeveg. 1995. A nesting dinosaur. *Nature* 378: 774–776.

O'Connor, R. J. 1984. *The growth and development of birds.* New York: Wiley.

Osmólska, H., and R. Barsbold. 1990. Troodontidae. In D. B. Weishampel, P. Dodson, and H. Osmolska (eds.), *The Dinosauria,* pp. 259–268. Berkeley: University of California Press.

Ostrom, J. H. 1978. The osteology of *Compsognathus longipes. Zitteliana Abhandlungen der Bayerischen Staatssammlung für Paläontologie und historische Geologie* 4: 73–118.

Paul, G. S. 1988. *Predatory dinosaurs of the world.* New York: Simon and Schuster.

———. 1994. Dinosaur reproduction in the fast lane: implications for size, success, and extinction. In K. Carpenter, K. Hirsch, and J. R. Horner (eds.), *Dinosaur eggs and babies,* pp. 244–255. New York: Cambridge University Press.

Punzo, F., and R. G. Turingan. 2001. The Mediterranean gecko, *Hemidactylus turcicus:* life in an urban landscape. *Florida Scientist* 64: 56–66.

Rahn, H., and A. Ar. 1974. The avian egg: incubation time and water loss. *Condor* 76: 147–152.

Russell, D. A. 1969. A new specimen of *Stenonychosaurus* from the Oldman Formation (Cretaceous) of Alberta. *Canadian Journal of Earth Science* 6: 595–612.

Russell, D. A., and R. Séguin. 1982. Reconstruction of the small Cretaceous theropod *Stenonychosaurus inequalis* and a hypothetical dinosauroid. *Syllogeus* 37: 1-43.

Ryan, M. J., P. J. Currie, J. D. Gardner, M. K. Vickaryous, and J. M. Lavigne. 2000. Baby hadrosaurid material associated with an unusually high abundance of *Troodon* teeth from the Horseshoe Canyon Formation, Upper Cretaceous, Alberta, Canada. *Gaia* 15: 123–133.

Sabath, K. 1991. Upper Cretaceous amniotic eggs from the Gobi Desert. *Palaeontologia Polonica* 36: 151–192.

Schweitzer, M. H., F. D. Jackson, L. M. Chiappe, J. O. Calvo, and D. E. Rubilar. 2002. Cretaceous avian eggs and embryos from Argentina. *Journal of Vertebrate Paleontology* 22: 191–195.

Sereno, P. C. 1999. The evolution of dinosaurs. *Science* 284: 2137–2147.

Shine, R. 1988. Parental care in reptiles. In C. Gans and R. B. Huey (eds.), *Biology of the Reptilia*, Volume 16: 275–329. New York: Wiley.

Sibley, C. G., and J. E. Ahlquist. 1990. *Phylogeny and classification of birds.* New Haven: Yale University Press.

Sillen-Tullberg, B., and H. Temrin. 1994. On the discrete characters in phylogenetic trees with special reference to the evolution of avian mating systems. In P. Eggleton and D. Vane-Wright (eds.), *Phylogenetics and Ecology,* pp. 311–322. London: Academic Press.

Smart, I. H. M. 1987. Egg-shape in birds. In D. C. Deeming and M. W. J. Ferguson, *Egg incubation: its effect on embryonic development in birds and reptiles,* pp. 101–116. New York: Cambridge University Press.

Smith, H. M., G. Sinelnik, J. D. Fawcett, and R. E. Jones. 1973. A survey of the chronology of ovulation in anoline lizard genera. *Transactions of the Kansas Academy of Science* 75: 107–120.

Somma, L. A. 1988. Comments on the use of the term brooding to describe parental behaviour in squamate reptiles. *Amphibia-Reptilia* 9: 89–91.

Starck, J. M., and R. E. Ricklefs. 1998. Patterns of development: the altricial-precocial spectrum. In J. M. Starck and R. E. Ricklefs (eds.), *Avian growth and development,* pp. 3–30. New York: Oxford University Press.

Stewart, J. S. 1989. Husbandry and medical management of ostriches. *Proceedings Association of Avian Veterinarians* 208–212.

Terres, J. K. 1995. *The Audubon Society encyclopedia of North American birds.* New York: Wings Books.

Varricchio, D. J. 1993. Bone microstructure of the Upper Cretaceous theropod dinosaur *Troodon formosus. Journal of Vertebrate Paleontology* 13: 99–104.

———. 1997. Troodontidae. In P. J. Currie and K. Padian (eds.), *Encyclopedia of dinosaurs,* pp. 749–754. San Diego: Academic Press.

———. 2000. Reproduction and parenting. In G. Paul (ed.), *The Scientific American book of dinosaurs,* pp. 279–293. New York: St. Martins Press.

Varricchio, D. J., F. Jackson, J. Borkowski, and J. R. Horner. 1997. Nest and egg clutches of the dinosaur *Troodon formosus* and the evolution of avian reproductive traits. *Nature* 385: 247–250.

Varricchio, D. J., F. Jackson, and C. N. Trueman. 1999. A nesting trace with eggs for the Cretaceous theropod dinosaur *Troodon formosus. Journal of Vertebrate Paleontology* 19: 91–100.

Varricchio, D. J., J. R. Horner, and F. D. Jackson. 2002. Embryos and eggs for the Cretaceous theropod *Troodon formosus. Journal of Vertebrate Paleontology* 22: 564–576.

Weishampel, D. B., and J. R. Horner. 1994. Life history syndromes, heterochrony, and the evolution of Dinosauria. In K. Carpenter, K. Hirsch, and J. R. Horner (eds.), *Dinosaur eggs and babies,* pp. 229–243. New York: Cambridge University Press.

White, F. N., and J. L. Kinney. 1974. Avian incubation. *Science* 186: 107–115.

Williams, D. L. G., R. S. Seymour, and P. Kerourio. 1984. Structure of fossil dinosaur eggshell from Aix Basin, France. *Palaeogeography, Palaeoclimatology, Palaeoecology* 45: 23–37.

Wilson, M. C., and P. J. Currie. 1985. *Stenonychosaurus inequalis* (Saurischia: Theropoda) from the Judith River (Oldman) Formation of Alberta: new findings on metatarsal structure. *Canadian Journal of Earth Sciences* 22: 1813–1817.

Witmer, L. M. 1995. The extant phylogenetic bracket and the importance of reconstructing soft tissues in fossils. In J. Thomason (ed.), *Functional morphology in vertebrate paleontology,* pp. 19–33. New York: Cambridge University Press.

Xu X., Tang Z.-L., and Wang X.-L. 1999. A therizinosauroid dinosaur with integumentary structures from China. *Nature* 399: 350–354.

Zelenitsky, D. K., and L. V. Hills. 1996. An egg clutch of *Prismatoolithus levis* oosp. nov. from the Oldman Formation (Upper Cretaceous), Devils Coulee, southern Alberta. *Canadian Journal of Earth Sciences* 33: 1127–1131.

Zug, G. R. 1993. *Herpetology: An introductory biology of amphibians and reptiles.* San Diego: Academic Press.

11. Dinosaur Brooding Behavior and the Origin of Flight Feathers

Thomas P. Hopp and Mark J. Orsen

Abstract

The origin of birds from within the group of theropod dinosaurs has been controversial because it is difficult to understand how wing feathers evolved through short intermediate stages before becoming long enough to generate adequate power and lift for flight. The "ground up" concept of flight evolution among cursorial dinosaurs can be criticized because there is no apparent selective pressure to drive the forelimb feather lengthening process through its earliest stages. Feather functions such as insect-trapping, hunting, and display have been proposed, but none of these require the feather length and shape that evolved by the time *Archaeopteryx* appeared.

We propose a mechanism to account for the forelimb and tail feather lengthening process, based on a behavior that exists in living birds—namely, brooding. Interestingly, despite the many known examples of modern birds that use their wing feathers in nesting and chick-rearing, there has been no previous proposal of brooding as a selective pressure in the evolution of flight feathers. We present fossil evidence that nesting and care of hatchlings could have been responsible for the development of long feathers on the forelimbs and tails of pre-avian theropod dinosaurs. It has been noted that oviraptorids incubated their nests in a posture strikingly similar to that of many modern birds, with breast and feet in contact with the eggs. However, gaps in the animal's coverage of its eggs were sufficiently large to allow solar-heating, wind-cooling, or rain-wetting effects on the exposed eggs. Comparisons to modern birds demonstrate that these gaps could have been covered by wing and tail feathers. Thus, the evolution of long feathers on the forelimbs and tail base of theropod bird ancestors could have been driven, not by flight requirements, but by the advantage of decreased environmental stress on eggs and hatchlings. Significantly, this evolutionary process would have provided brooding advantages at every increment of feather lengthening. Even the first relatively short feathers would have offered increased protection for the young.

To assess whether brooding feathers could have originated among early non-avian dinosaurs, we undertook a comparative study of dinosaur and bird skeletal anatomy with emphasis on modern birds' nesting and brooding postures. We determined the extent to which theropod dinosaurs could adopt birdlike postures while incubating eggs and tending hatchlings, and concluded that the use of long forelimb and tail feathers for brooding could readily have existed even among early theropods. Furthermore, because the skeletons of these older theropods were conducive to brooding but not flying, forelimb and tail quill feathers may have evolved in these animals to the sizes and shapes seen in *Archaeopteryx* in the absence of flight, whereupon they were subsequently co-opted by *Archaeopteryx* or a similar creature for the additional purpose of flying.

Introduction

The theory of bird origins among the ground-dwelling, cursorial theropod dinosaurs (Ostrom 1973, 1974, 1976) has a long history of controversy. A central problem is that of providing an explanation for evolution of the form and function of flight feathers, because it has been hard to see a selective advantage in the early stages of forelimb feather development in a ground-dwelling, cursorial theropod. Since the discovery of the first *Archaeopteryx* fossil in 1865, numerous hypotheses have been advanced to explain the origin of flight feathers. A number of these hypotheses have concentrated on nonflight selective pressures as the initial driving force. For example, Mayr (1960) suggested that display and sexual selection were the driving forces behind wing and tail feather development. Regal (1975) proposed that feathers were developed primarily as heat shields, and Ostrom (1979) suggested insect trapping. Dyck (1985) pointed out that the water repellency of feathers might have been of primary importance. Thulborn and Hamley (1985) proposed shading while hunting, where the wings were spread as a canopy in the fashion practiced by some modern egrets. It is plausible that all of these disparate functionalities may have, at one point or another, played a role in feather evolution. However, these proposals still leave open the possibility that other mechanisms were involved in the evolution of bird flight (Feduccia 1994).

Heilmann (1927) and Savile (1962) proposed the distinct alternative theory that feathers developed primarily for aerodynamic uses in an arboreal, gliding animal. Feduccia (1980) and others have since elaborated and advocated this concept. Such feather features as the asymmetrical vanes have been cited to indicate that aerodynamic function was critical early in feather development (Feduccia and Tordoff 1979). The proposal that pennaceous feathers evolved before downy feathers (Parkes 1966; Feduccia 1995) has been cited as a further example of flight, rather than insulation, as the original function of feathers. These as well as other arguments have been raised as objections to the cursorial origin of flight (reviewed by Ostrom 1979, 1997). Proponents of the cursorial theory have been hard-pressed to explain the driving force behind feather lengthening that must have preceded the use of wings for flight (Feduccia 1995). Advocates of the arboreal theory point out that their concept features a single consistent selective pressure, feathers as airfoils, while the cursorial theory requires a switch from one use to another at some indeterminate point in feather evolution (for example, from sun-shielding to flying).

Remarkably, both sides in this debate have neglected one of the most important aspects of avian life—brooding. Here we address the issue of flight evolution among cursorial theropods once more, this time with emphasis on a possibly critical role played by brooding. Because modern birds still, almost universally, use their wing feathers for brooding, this is the only cursorial theory that does not require a change in selective pressure. As we see it, pressure to optimize brooding has been continuous from the ancient past until today. In what follows, fossil evidence and modern bird behavior and anatomy will be cited as guides for understanding a possible mechanism for the brooding-driven evolution of wing and tail feathers.

Brooding Postures and Feather Evolution

Brooding is a vital and widespread activity of modern birds that involves egg incubation and covering of chicks by adults to provide warmth, shielding from solar heating, shelter from rain and wind, and protection from predators (Skutch 1976). With few exceptions, brooding adults assume characteristic postures in which the feet are placed medially and the wing feathers are used to cover surrounding eggs or chicks (Wallace and Mahan 1975). These postures most often involve extension of the wings from their normal "folded" or fully flexed posture (figs. 11.1–11.3). Furthermore, the extended wings are often "drooped" or lowered toward the ground (Howell et al. 1974; Johnsgard 1993) and sometimes drawn backward at the shoulder (figs. 11.2, 11.3). These movements provide optimal cover for the chicks, which through their size or numbers may present a considerable challenge for parents to cover (e.g., the large chicks in fig. 11.3).

The discovery and description of nesting oviraptorids implied that dinosaurs may have incubated their eggs much like modern birds (Norell et al. 1995; Dong and Currie 1996; Clark et al. 2001). Unfortunately, lack of any evidence for feathers on the oviraptorid specimens made it impossible to verify whether feathers could have played a part in such incubations. Nevertheless, consideration of the postural details of nesting oviraptorids underscores the similarity of dinosaur and avian incubation. In the most complete specimen (Norell et al. 1995) the animal lies with its breast, belly, and feet in contact with its eggs (fig.

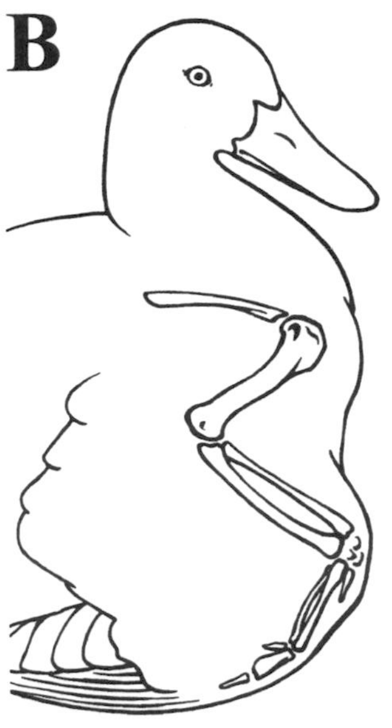

Figure 11.1. (above) Mother duck and ducklings. A. This female mallard has lowered her wings until her wrists are near the ground, and fanned her secondary feathers to cover her brood. The light-and-dark-patterned, downy ducklings are just visible beneath the rear quarter, partially protected by the adult's backward-directed primary feathers and the tail feathers. B. Diagrammatic representation of the bone structure underlying the wing in panel A. Note the partial extension of the forelimb. (Photograph from Hosking and Kear 1985, p. 82. The original caption reads, "Mother's wings provide shelter for a brood of mallard ducklings Anas platyrhynchos.*" Used with permission)*

Figure 11.2. (left) Bonelli's Eagle shading its brood. The outstretched wing and fanned-out primary and secondary feathers are characteristic of brooding postures used by many modern birds to prevent solar overheating of eggs and chicks. (Photograph from Nicolai 1973, p. 75. The original caption reads, "Wings outstretched, an African Bonelli's eagle Hieraeetus fasciatus spilogaster *provides shade for its young." Used with permission)*

11.4), in a posture very similar to modern birds such as the ostrich (Sauer and Sauer 1966). However, forelimbs of *Citipati* are not in the folded posture that many modern birds use while resting or covering a nest. Rather, they are in an extended, rearward-directed orientation. This also appears to be the orientation in a second, less complete oviraptorid specimen (Dong and Currie 1996). As mentioned above, this posture is similar to one employed by birds brooding large chicks

Figure 11.3. Brooding peregrine falcon Falco peregrinus *shelters her two large, downy nestlings from a rain shower. The wings are held in an extended posture similar to the forelimbs of fossil nesting oviraptorids. If deprived of parental cover, chicks often die of exposure-related hypothermia. (Photograph from Olsen 1995, p. 157. The original caption reads, in part, "Day 15: chicks no longer require constant brooding except at night and in cool weather." Used with permission)*

or by birds that produce large clutches of eggs. A covering of enlarged wing feathers and tail feathers would greatly improve the ability of an oviraptorid to shield its eggs from sun, wind, and rain (figs. 11.4, 11.5). Such feathers would also provide protection for chicks after hatching.

A Selective Pressure for Feather Lengthening

More significantly, the forelimb position of the nesting oviraptorid fossils suggests a selective mechanism for the elongation of wing feathers. Whereas it has been hard to determine a selective pressure on a cursorial hunting animal that would drive the initial phases of evolving long forelimb feathers, there is no difficulty in seeing the selective advantage that would result from even small increases in the length of forelimb feathers on a brooding animal. Small increments of feather lengthening would have offered a cumulative selective advantage by gradually increasing the area of cover around the adult (fig. 11.6). Starting with a short-feathered oviraptorid ancestor (at left in fig. 11.6A) there would be a survival advantage in any mutation that improved the animal's coverage of larger clutches of eggs, larger broods, or both. Thus, even small increments of feather lengthening on the forelimbs would enhance breeding success.

Over time, a series of small elongation increments would have

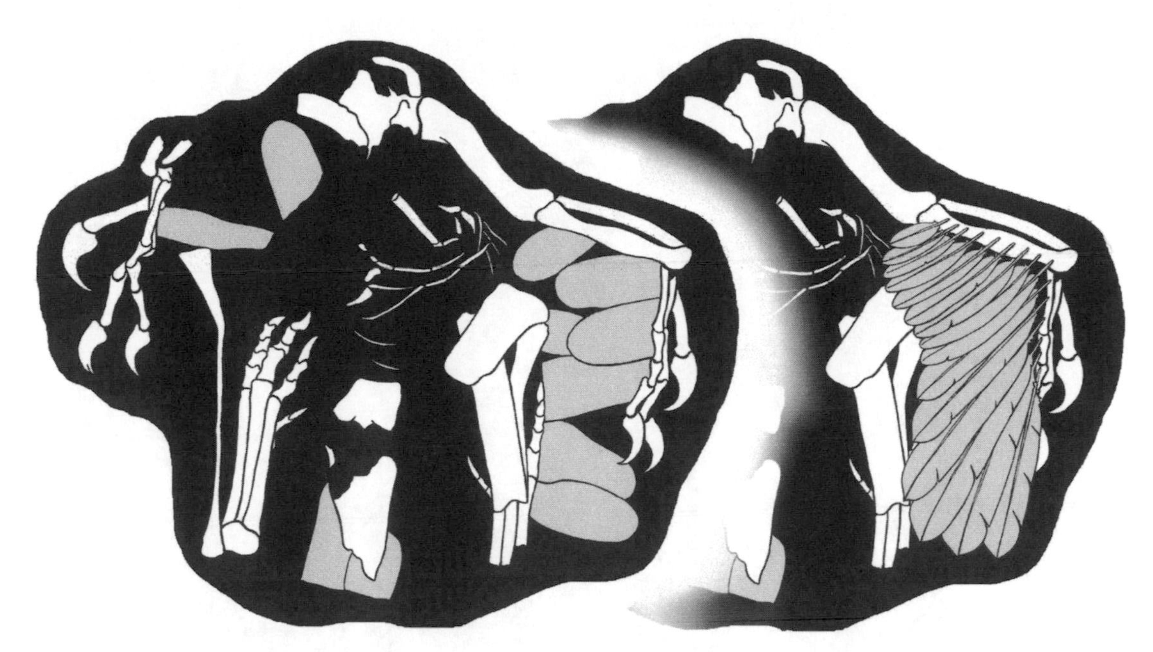

Figure 11.4. Oviraptorid nest fossil and proposed feathers. The diagram on the left represents the nesting Citipati *fossil (Norell et al. 1995; Clark et al. 2001), specimen IGM 100/979, with eggs shaded gray. The animal's breast, forelimbs, and feet were in contact with the eggs. A closely similar posture has been described for another nesting oviraptorid, IVPP V9608 (Dong and Currie 1996). Note the gap in egg coverage between the body and the forelimbs, and the supinated (thumb-outward) position of the manus. Although no traces of feathers were found in the fossil, it is noteworthy that this manus orientation would make it possible for a birdlike set of primary (originating on the manus) and secondary (originating on the forearm) feathers to cover the exposed eggs. A hypothetical set of primary and secondary feathers are illustrated on the right. Also note the* Archaeopteryx-*like juxtaposition of the metacarpals and phalanges of digits 2 and 3, suggesting that, like* Archaeopteryx, *this manus did indeed bear long feathers.*

resulted in forelimb feathers of considerable length. We propose that these brooding feathers ultimately became long enough to be exapted for the additional use of flying, presumably in an animal smaller than *Oviraptor* (for example, *Archaeopteryx*-sized). A need to cover chicks or eggs near the rear quarter of the parent might additionally have resulted in lengthening of the forelimbs themselves and perhaps some degree of tail feather lengthening (fig. 11.6). Both of these traits would also have helped to pre-adapt *Archaeopteryx*'s ancestors for a transition from brooding to flight. While tail feather evidence is rare, the trait of forelimb lengthening is seen clearly among those theropods related to *Archaeopteryx*'s ancestors. Thus, forelimbs lengthened more or less continuously in going from *Coelophysis* to *Ornitholestes*, to dromaeosaurs, and to *Archaeopteryx* (Dingus and Rowe 1998). In our view, brooding may have been the primary driving force behind both forelimb and feather lengthening among pre-avian theropods.

Among the oviraptorid ancestors, the need for egg and chick coverage behind the forearm and manus could have led to the exact pattern of forelimb feathers commonly found in modern birds. Coverage of the exposed oviraptorid eggs would have required the longest feathers on the animal's manus, whereas shorter feathers would have sufficed along the ulna. The greater length of primary versus secondary feathers has been cited as proof that wing feathers are optimized for flight (Feduccia 1996). However, it would appear that the wing feather lengths of modern birds are equally fine-tuned for brooding. Note again how a mallard (fig. 11.1) uses its longest primary feathers to cover chicks at its rear quarter.

Figure 11.5. Nesting oviraptorid restored with or without feathers. A set of birdlike "wing" feathers enables this oviraptorid to cover its entire nest, with help from elongated feathers at the base of the tail and a flap of skin (propatagium) between the shoulder and wrist (painting by Mark Orsen). Inset, a featherless oviraptorid is shown in the orientation dictated by the original fossil. (M. Novacek 1996, drawing used with permission from AMNH. Notable are the areas where eggs are exposed in front of and behind the forelimb, and at the rear quarter)

Evolution of Feather Form

In proposing brooding as the driving force behind flight feather evolution, it is necessary to explain why these feathers developed into flat, aerodynamically useful structures rather than downy or plumaceous forms seen in modern flightless birds. Some authors believe that the flattened pennaceous type of feather was the first to evolve, based on morphological grounds (Dyck 1985); and Parkes (1966) and Feduccia (1996) consider the flight feather itself as the primordial type, with body contour feathers and downy or plumaceous feathers being derived from it. This contention is not universally accepted, and the recent discovery of filamentous featherlike structures on the theropod *Sinosauropteryx* casts further uncertainty on the timing of the origins of various feather types (Chen et al. 1998). We consider it likely that the requirements of brooding would have led to the present form of the flight feather, whether starting from scales or from a pre-existing downy, filamentous, or contour feather type.

We envision that the feathered brooding forelimb evolved as an integrated unit, shaped by the need to shelter chicks and to carry the sheltering structure compactly when not in use. The "flight" feather is a lightweight, easily manipulated component of this overall structure,

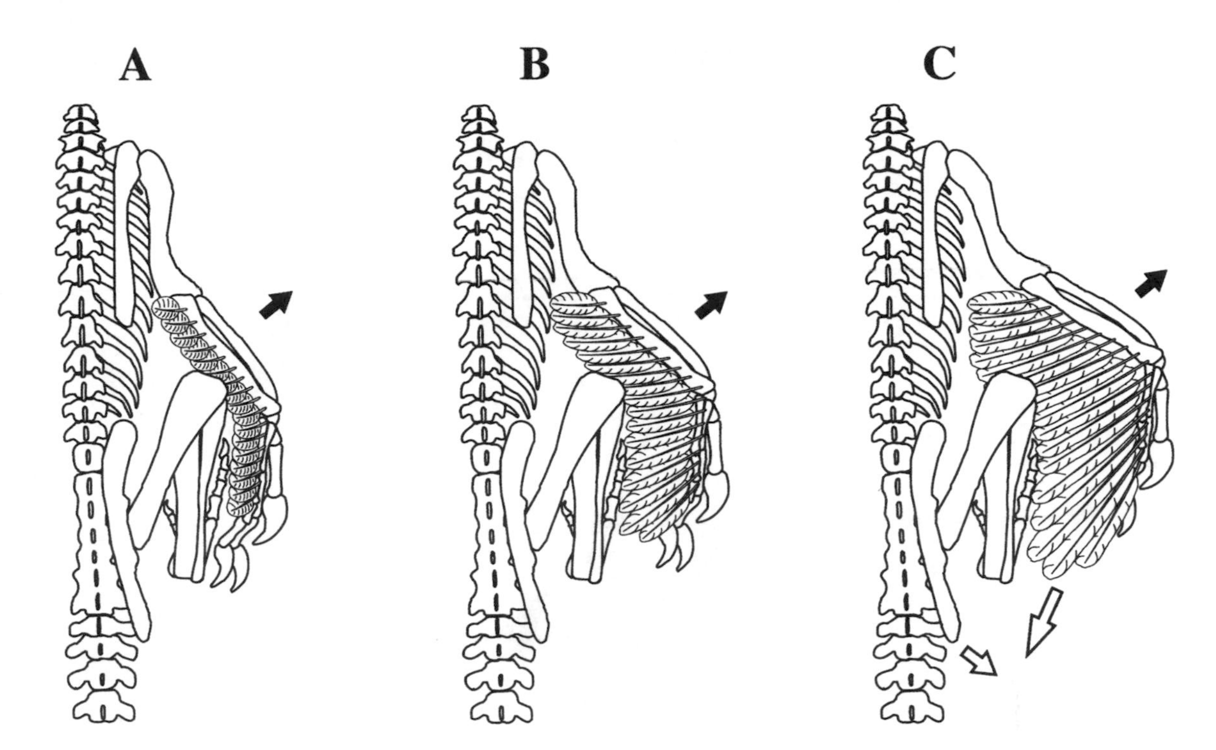

Figure 11.6. Incremental feather lengthening. Oviraptor *(C) is compared to a hypothetical short-feathered ancestor (A) and an intermediate evolutionary stage (B). As the feathers lengthen through a series of mutations, the forelimb is able to move outward (solid arrows) to accommodate larger or more numerous eggs and chicks. The open arrows indicate the area at the rear quarter where longer primary feathers, as well as tail feathers, would improve the coverage of the brood (compare to the mallard in fig. 11.1).*

and is highly optimized for its brooding-related purposes. For example, the stout quill (rachis) is required for directing each feather to its unique position, and modern birds exhibit a remarkable range of feather orientations while brooding (cf. figs. 11.1–11.3). Feathers may be either fanned out or compactly folded back against the body, depending on the particular need. The flattened, bladelike form of the feather vane facilitates smooth sliding back and forth during fanning or folding adjustments of the wing shelter, and the barbule-and-hooklet vane microstructure adds rigidity that helps the feather resist the rigors of nestling care, where feathers are trodden upon and come in contact with a variety of nest-building materials. Thus, resistance to damage and reparability, functions of the barbule-and-hooklet system, are required not only for flight, but also for allowing the adult bird to maintain an adequate umbrella-like cover for its chicks.

Feduccia and Tordoff (1979) have made much of the asymmetrical vanes in the feathers of *Archaeopteryx,* arguing that the narrow leading edge is a flight-related condition that favors an arboreal, gliding history among the ancestors of *Archaeopteryx*. However, we note that these authors cited exceptions to their own rule; for example, the retention of asymmetry in the wing feathers of the flightless grebe, *Centropelma micropterum*. Furthermore, there may be nonflight origins of such an asymmetry. The narrow-vaned, leading feather edges are nearest the ground during brooding (note fig. 11.1), suggesting that feather asymmetry could have arisen in a brooding ground-dwelling theropod in order to keep the edges from fraying through contact with soil or vegetation.

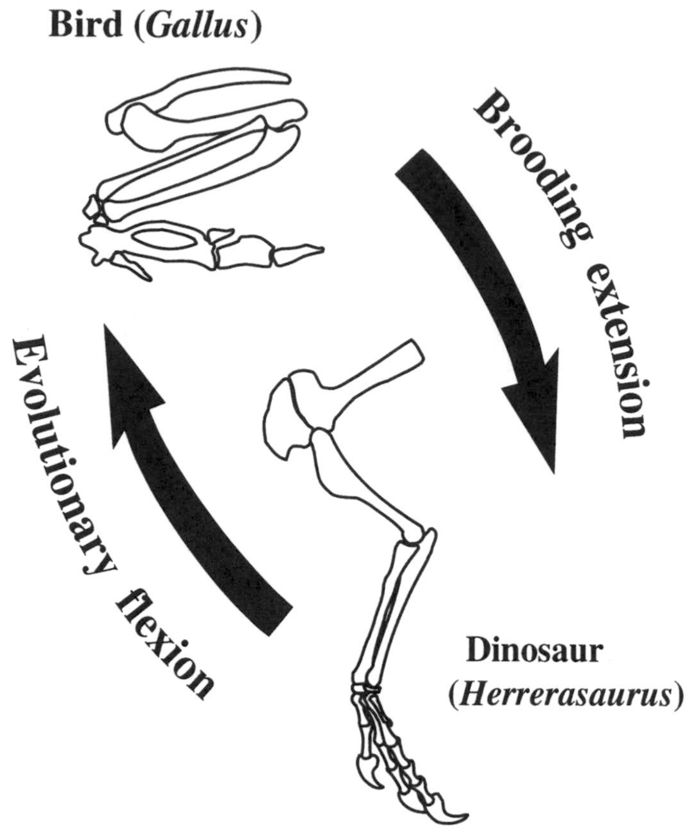

Figure 11.7. Folded vs. extended forelimb (wing). The folded wing of modern birds (top) is an advanced derived condition that is needed to keep the long flight feathers off the ground. However, even modern birds lower their wings to a more primitive posture similar to that of Citipati *while covering chicks (bottom). The relatively inflexible forelimbs of* Herrerasaurus *could attain the postures used by brooding birds (compare to figs. 11.1B, 11.2, and 11.3).*

Viewed as an integrated unit, the feathered wing offers a highly adaptable shelter that is useful either for warming or shading the young, yet which can be tucked away tightly when not needed for these purposes. The value of such an adaptable, foldable shelter may have influenced the evolution of the bones of the wing and the flight stroke as well.

Evolution of the Folded Wing

In the evolution of bird flight, an important anatomical development is the folded forelimb posture found among modern birds (fig. 11.7, top). While this is the normal resting posture for most birds, it also approximates the up (or recovery) stroke used in flight (Ostrom 1997). We propose that a need to manipulate brooding feathers may have led to the folding anatomical adaptations of the forelimb before flight arose, and that a brooding-adapted forelimb may have dictated aspects of the anatomy and geometry of the forelimb's use in flight. That is, brooding may have pre-adapted the forelimb for the motions seen in the modern flight stroke, rather than the reverse, where brooding is considered secondary to flight.

Among the theropod ancestors of birds, the evolution of longer brooding feathers on the forelimbs eventually would have led to problems—the feathers might be stepped on by the hind feet, seized by a predator, or damaged by contact with the ground or vegetation. The solution to this problem is to hold them close to the body, which in turn requires the joints of the forelimb to be flexed sharply. We suggest that a tightly folded forelimb structure is the consequence of the need to manage the evolving set of brooding forelimb feathers, and that this fold was not, at its outset, dictated in any way by flight requirements. The widespread occurrence of a tightly folding wing, found among many flightless birds, confirms that the way a wing folds, of itself, represents a critical function.

Therefore, most of the anatomical adaptations that occurred during the progression from the forelimbs of theropods to those of early birds may have taken place without any contribution from flight requirements, being driven solely by the need to lift the forelimb feathers. These adaptations include the elevation of the glenoid fossa, development of the uniquely shaped articular surfaces of the elbow that allow its tight folding, and the semilunate carpal bone and wrist architecture that allows acute sideways flexion of the wrist. Most importantly, our concept reverses the order of things: long brooding feathers came first, necessitating the evolution of the elevated, folded forelimb and semilunate carpal, which in turn were co-opted (exapted) for flight.

Although the wrist and elbow joints of the Triassic dinosaur *Herrerasaurus* were far less flexible than those of the maniraptors and *Archaeopteryx,* they nevertheless suggest the beginnings of the folding characteristics of birds' wings (Paul 1988). Sereno (1993) noted moderate flexion capacity in the forelimb of *Herrerasaurus,* including some sideways wrist flexion (eversion). In particular, we note that the ulnare of *Herrerasaurus* and its wide articulation surface on the ulna (Sereno 1993) suggest a primitive form of birdlike wrist eversion. To us, this implies that this early dinosaur might already have possessed lengthened primary feathers, because this eversion is required for folding such feathers. Novas and Puerta (1997) noted that the extremely birdlike Cretaceous theropod *Unenlagia* could fold its forelimb tightly even though it appears to have been a nonflying species. Thus, forelimb feather lengthening and forelimb folding may have been interrelated processes that started in the Triassic Period and continued among theropods into the Cretaceous.

We note that, while modern birds are capable of assuming tightly folded forelimb postures, many of the orientations used in brooding are reminiscent of those available to their ancestors (fig. 11.7, bottom). Thus, when an adult bird covers a large nest of eggs or a brood of partially grown chicks, it unflexes the wing and extends it outward, downward, and sometimes backward. While the degree to which the wing is extended varies considerably, it is nevertheless true that all brooding uses of the wing move the bones of the forelimb into postures that are more like those that were attainable by primitive dinosaurs such as *Herrerasaurus* (Sereno 1993) and *Coelophysis* (Paul 1988). Some brooding postures represent nearly complete extension of the forelimb into an orientation that should have been available even to

Figure 11.8. Brooding gray gull Larus modestus *with one wing extended above its chick. Brooding encompasses many postural variations, and forelimb extension and fanning of wing feathers over the young is important in the hostile environment where these birds nest. Gray gull chicks may die if left unshaded through the heat of a single afternoon in their dry, high desert habitat. (Photograph from Howell et al. 1974, plate 13. The original caption reads, "Gray gull shading chick." Used with permission)*

theropods with relatively inflexible forelimbs (fig. 11.8). This behavior of modern birds leaves open the possibility of a truly ancient origin for wing feather brooding, because it defines a function for which no flexion capacity in bone structure is required, in contrast to flight, which requires both feathers and forelimb flexion.

Brooding and the Origin of Flight

In over 100 years of theorizing about bird evolution, the concept of *Archaeopteryx* as a brooding parent has not to our knowledge been discussed. However, although it is still debated whether *Archaeopteryx* was a good or a poor flier, there can be little doubt that it was a good parent, or it would not have survived long enough to leave fossils at Solnhofen. Whether or not it engaged in brooding behavior as portrayed (fig. 11.9) remains speculative, but is worth bearing in mind when considering the transition from brooding to flight that probably occurred shortly before the appearance of *Archaeopteryx*.

Figure 11.9. Archaeopteryx *parent brooding its chicks. In this restoration, the adult has lowered its forelimbs and spread its wing feathers to provide shelter for its young. Only two chicks are shown for clarity, but we make no assumption regarding the actual brood size for* Archaeopteryx. *(Painting by Mark Orsen)*

Assuming that *Archaeopteryx* arose from a small, nonflying, cursorial theropod ancestor, it is instructive to use the roadrunner *Geococcyx californianus* (fig. 11.10) as a model for the transition to flight. Although roadrunners differ from theropods in that they can fly, they are good models because they spend most of their time on the ground and are cursorial hunters of lizards and snakes (Meinzer 1993). While moving through brush or attacking prey with its beak, a roadrunner keeps its wings tightly folded, only opening them momentarily to confuse its victims or to augment its leaps away from prey counterstrikes. In the latter maneuver, flapping increases the duration of the leap, keeping the roadrunner off the ground for a longer period.

In preceding sections of this paper, we have explained how a creature very much like *Archaeopteryx* could have evolved long feathers on its forelimbs for brooding purposes, and developed the bone structure necessary to manage the orientation of these feathers. This leads to an animal closely similar to *Archaeopteryx* in bone and feather structure, from which a roadrunner-like lifestyle could, in turn, lead to the transition to flight. As is true for the roadrunner, the cursorial nonflying theropod could have benefited from brief flapping of the forelimbs to lengthen the distance or to increase the duration of leaps while hunting or escaping predators. Note that such brief flight maneuvers start from the folded wing position and return to it after landing, providing an explanation for the unique geometry of the flight stroke, which includes a close approach to the folded position during recovery before the next stroke begins (Ostrom 1997).

Based on the above rationale, we propose a revised sequence of events in the evolution of bird flight:

Figure 11.10. Roadrunner Geococcyx californianus *at full stride. This cursorial hunter is a good model for a feathered, brooding theropod. In the theropod ancestors of birds, the shapes of brooding forelimb feathers and the tight folding of the forelimbs may have been dictated by the requirements of this type of foraging, where the feathers would have been subject to trampling or entanglement in vegetation. (Photograph from Meinzer 1993; used with permission)*

1. Long quill feathers with modern flight-feather architecture developed on the forelimbs of nonflying theropod dinosaurs for brooding purposes.
2. The ability to fold the forelimb arose next, or concurrently, to streamline the animal and protect the feathers.
3. Flapping maneuvers developed next, starting from the folded posture, and evolved into the flight stroke.

In our view, the folding geometry of the theropod forelimb, which developed for brooding-feather management, subsequently influenced the mechanism and pattern of the flight stroke. Features of the stroke, such as the simultaneous flexion-extension movements of the wrist and elbow (Ostrom 1997) and the ability to spread the primary feathers into an airtight foil, arose first for brooding and were subsequently exapted for flight. Furthermore, the seemingly prescient development among early dinosaurs of anatomical features used in flight (clavicles, folded forelimbs, elevated glenoids, robust sterna) may instead represent a progression of refinements for handling the ever-lengthening sets of forelimb feathers whose primary purpose was, and is, brooding.

Evolutionary Timing Considerations

In previous models of flight evolution, the development of wing feathers and flight are assumed to be tightly coupled because they are causally related. However, in our model, long forelimb feathers exist independently of flight (or gliding). This allows the possibility that such feathers may have arisen much earlier in time, much more gradually, or

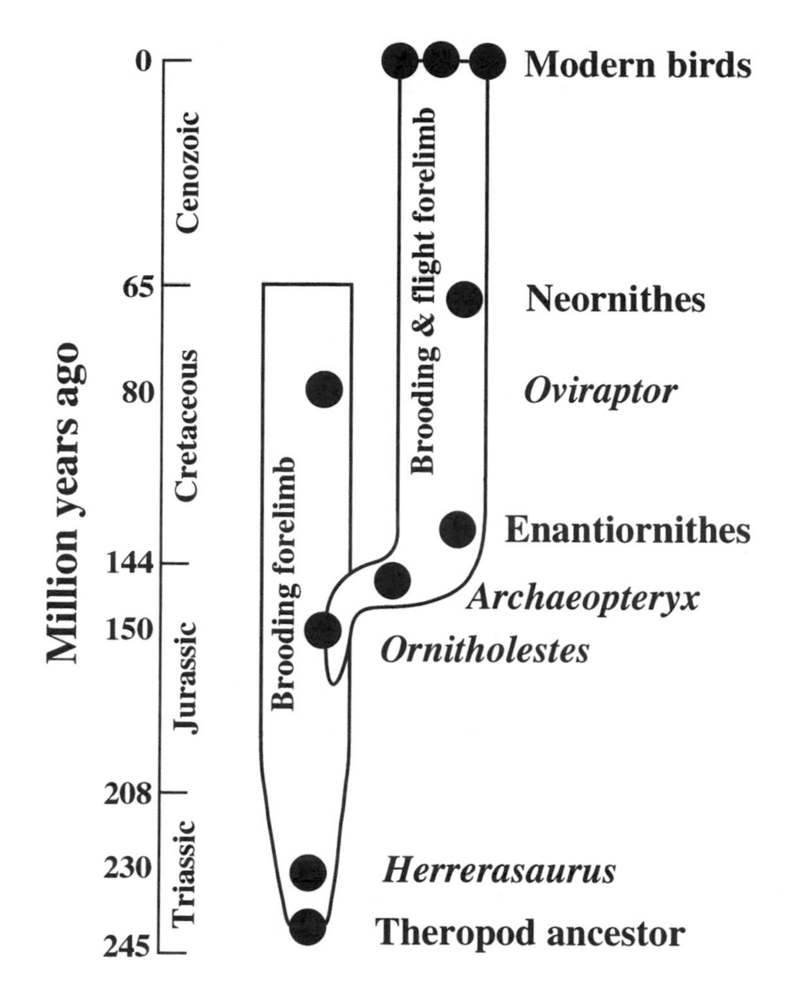

Figure 11.11. Time diagram for feather evolution. The two most critical uses of wing feathers are brooding and flight. If brooding preceded flight as we propose, then the use of a feathered forelimb for this purpose may extend much farther back in time than flight does. Even the basal theropod Herrerasaurus *shows a moderate degree of forelimb flexibility, implying that feathered brooding forelimbs may have existed throughout the history of the theropods from their origins in the Triassic to their demise at the end of the Cretaceous Period (left-hand bar). Bird flight arose secondarily as an exaptation of the already large brooding forelimb feathers (right-hand bar). Brooding remains to this day a major use for wing feathers.*

both. Given that wings are used for brooding among all the major phylogenetic groups of modern birds, it is most parsimonious to consider wing-brooding a basal rather than a repeatedly derived trait. Therefore it must have existed in the common ancestor of paleognaths and neognaths, whose divergence occurred some time in the Late Cretaceous (Chiappe 1995) or earliest Tertiary (Feduccia 1996). This pushes back the origin of wing feather brooding to a time nearly contemporaneous with *Oviraptor.* Furthermore, if oviraptorids did indeed possess brooding feathers, then the divergence point would revert to an even earlier common ancestor with *Archaeopteryx,* one that must have predated the Jurassic appearance of the latter.

It is not possible to extrapolate further back based on modern birds or oviraptorids, but it is interesting to speculate on how early the forelimb feather brooding trait might have arisen. No definite feather fossils have been found before *Archaeopteryx,* but just as the posture of the oviraptorid forelimbs (fig. 11.4) suggests the presence of brooding feathers, other bone evidence may be relevant as well. As mentioned, the wrist and elbow joints of *Herrerasaurus* suggest the beginnings of the folding that is characteristic of birds' wings, which in our view may relate to an already existing need in *Herrerasaurus* to lift brooding feathers away from the ground. This suggests that even primitive theropods assumed postures conducive to the kind of feather lengthening portrayed in fig. 11.6.

Based on the points just mentioned, fig. 11.11 portrays graphically the timing of events involved in our "brooding came first" scenario for wing and flight evolution. In this scenario, forelimb feathers arose among the early theropods, and then underwent a substantial period of selection for brooding throughout the Late Triassic and into the Early and Middle Jurassic. Once large forelimb feathers were present, they were eventually exapted for flight as a secondary use at around the time *Archaeopteryx* appeared in the Late Jurassic.

Conclusions

Recent descriptions of the feathered theropods *Caudipteryx* and *Protarchaeopteryx* (Ji et al. 1998), *Microraptor* (Xu et al. 2000), *Sinornithosaurus* (Xu et al. 1999; Xu et al. 2001), and the unnamed dromaeosaur NGMC 91 (Ji et al. 2001), which appeared in the literature after our presentation of this concept, are consistent with what we proposed. The forelimb feathers of all these creatures are insufficient for flight, but are large enough and distributed appropriately along the forelimb to have functioned in brooding. Prum (1999) recently proposed a model of feather evolution based on the ontogeny of feather forms. His model is entirely consistent with our concept, and in fact, brooding may have been the primary driving force for the evolution of increased feather complexity that Prum described. Ostrom recently published a discussion of limb-bone flexibility in the evolution of the flight stroke (Ostrom et al. 1999) that is also consistent with our proposal. Recent suggestions that oviraptorosaurs may be secondarily flightless birds (Lu 2000; Maryanska et al. 2002) do not negate our arguments regarding nesting in theropods, because similar nest arrangements are found widely among dinosaurs.

We propose that the need to provide better cover for eggs and chicks was a powerful selective pressure that drove much of the early evolution of flight feathers. In addition, the development of folded wings may be explained by the need to manage the forelimb feathers during different activities ranging from brooding to cursorial hunting. Fossils of nesting oviraptorids provide a snapshot of a primitive, extended-forelimb posture in which forelimb feathers would have been useful for covering either eggs or a brood of chicks. Given that the forelimbs of basal theropods were capable of attaining similar brooding positions, it is possible that forelimb feather lengthening might have begun as much as 80 million years before *Archaeopteryx*. Modern birds retain brooding postures related to those of theropods, suggesting that flight may represent a later secondary adaptation of long feathers that have been used over a much longer period of time for brooding.

Acknowledgments. We thank John Ostrom for encouragement and helpful suggestions during the development of this concept and the preparation of this chapter. This material was presented as a lecture and accompanying poster at the Dinofest '98 Symposium in Philadelphia (Hecht 1998).

References

Chen P.-J., Dong Z.-M., and Zhen S. 1998. An exceptionally well-preserved theropod dinosaur from the Yixian Formation of China. *Nature* 391: 147–152.

Chiappe, L. M. 1995. The first 85 million years of avian evolution. *Nature* 378: 349–355.

Clark, J. M., M. A. Norell, and R. Barsbold. 2001. Two new oviraptorids: (Theropoda: Oviraptorosauria), Upper Cretaceous Djadokhta Formation, Ukhaa Tolgod, Mongolia. *Journal of Vertebrate Paleontology* 21: 209–213.

Dingus, L., and T. Rowe. 1998. *The mistaken extinction.* New York: Freeman.

Dong Z. M., and P. J. Currie. 1996. On the discovery of an oviraptorid skeleton on a nest of eggs. *Canadian Journal of Earth Science* 33: 631–636.

Dyck, J. 1985. The evolution of feathers. *Zoologica Scripta* 14: 137–154.

Feduccia, A. 1980. *The age of birds.* Cambridge: Harvard University Press.

———. 1994. The great dinosaur debate. *Living Bird* 13: 29–33.

———. 1995. The aerodynamic model for the evolution of feathers and feather misinterpretation. *Courier Forschungsinstitut Senckenberg* 181: 65–77.

———. 1996. *The origin and evolution of birds.* New Haven: Yale University Press.

Feduccia, A., and H. B. Tordoff. 1979. Feathers of *Archaeopteryx:* asymmetric vanes indicate aerodynamic function. *Science* 203: 1021–1022.

Hecht, J. 1998. Let me take you under my wing. *New Scientist* 158: 22.

Heilmann, G. 1927. *The origin of birds.* Appleton, New York.

Hosking, E., and J. Kear. 1985. *Wildfowl.* New York: Facts On File.

Howell T. R., B. Araya, and W. R. Millie. 1974. Breeding biology of the Gray Gull. *University of California Publications in Zoology* 104: 1–57.

Ji Q., P. J. Currie, M. A. Norell, and Ji S.-A. 1998. Two feathered dinosaurs from northeastern China. *Nature* 393: 753–761.

Ji Q., M. A. Norell, Gao K.-Q., Ji S.-A., and Ren D. 2001. The distribution of integumentary structures in a feathered dinosaur. *Nature* 410: 1084–1088.

Johnsgard, P. A. 1993. *Cormorants, darters, and pelicans of the world.* Washington: Smithsonian Institute Press.

Lu, J. 2000. Oviraptorosaurs compared to birds. *Vertebrata PalAsiatica* 38: 18.

Maryanska, T., H. Osmólska, and M. Wolsan. 2002. Avialan status for Oviraptorosauria. *Acta Palaeontologica Polonica* 47: 97–106.

Mayr, E. 1960. The emergence of evolutionary novelties. In S. Tax (ed.), *The evolution of life,* pp. 349–380. Chicago: University of Chicago Press.

Meinzer, W. 1993. *The roadrunner.* Lubbock: Texas Tech University Press.

Nicolai, J. 1973. *Bird life.* New York: Putnam.

Norell, M. A., J. M. Clark, L. M. Chiappe, and D. Dashzeveg. 1995. A nesting dinosaur. *Nature* 378: 774–776.

Novacek, M. 1996. *Dinosaurs of the flaming cliffs*. New York: Doubleday.
Novas, F. E., and P. F. Puerta. 1997. New evidence concerning avian origins from the Late Cretaceous of Patagonia. *Nature* 387: 390–392.
Olsen, P. 1995. *Australian birds of prey.* Baltimore: Johns Hopkins University Press.
Ostrom, J. H. 1973. The ancestry of birds. *Nature* 242: 136.
———. 1974. *Archaeopteryx* and the origin of flight. *Quarterly Review of Biology* 49: 27–47.
———. 1976. *Archaeopteryx* and the origin of birds. *Biological Journal of the Linnaean Society* 8: 91–182.
———. 1979. Bird flight: how did it begin? *American Scientist* 67: 46–56.
———. 1997. How bird flight might have come about. In D. L. Wolberg, E. Stump, and G. Rosenberg (eds.), *Dinofest International,* pp. 301–310. Philadelphia: The Academy of Natural Sciences.
Ostrom, J. H., S. O. Poore, and G. E. Goslow, Jr. 1999. Humeral rotation and wrist supination: Important functional complex for the evolution of powered flight in birds? *Smithsonian Contributions to Paleobiology* 89: 301–309.
Parkes, K. C. 1966. Speculations on the origins of feathers. *Living Bird* 5: 77–86.
Paul, G. S. 1988. *Predatory dinosaurs of the world.* New York: Simon and Schuster.
Prum, R. O. 1999. Development and evolutionary origin of feathers. *Journal of Experimental Zoology* 285: 291–306.
Regal, P. 1975. The evolutionary origin of feathers. *Quarterly Review of Biology* 50: 35–66.
Sauer, E. G. F., and E. M. Sauer. 1966. The behavior and ecology of the South African ostrich. *Living Bird* 5: 45–75.
Savile, D. B. O. 1962. Gliding and flight in vertebrates. *American Zoologist* 2: 161–166.
Sereno, P. C. 1993. The pectoral girdle and forelimb of the basal theropod *Herrerasaurus ischigualastensis*. *Journal of Vertebrate Paleontology* 13: 425–450.
Skutch, A. F. 1976. Parent birds and their young. Austin: University of Texas Press.
Thulborn, R. A., and T. L. Hamley. 1985. A new palaeoecological role for *Archaeopteryx*. In M. K. Hecht, J. H. Ostrom, G. Viohl, and P. Wellnhoffer (eds.), *The beginnings of birds,* pp. 81–89. Eichstätt: Freunde des Jura-Museums.
Wallace, G. J., and H. D. Mahan. 1975. *An introduction to ornithology.* New York: Macmillan.
Xu X., Wang X.-L., and Wu X.-C. 1999. A dromaeosaurid dinosaur with a filamentous integument from the Yixian Formation of China. *Nature* 401: 262–266.
Xu X., Zhou Z.-H., and Wang X.-L. 2000. The smallest known non-avian theropod dinosaur. *Nature* 408: 705–708.
Xu X., Zhou Z.-H., and R. O. Prum. 2001. Branched integumentary structures in *Sinornithosaurus* and the origin of feathers. *Nature* 410: 200–203.

12. Feathered Coelurosaurs from China: New Light on the Arboreal Origin of Avian Flight

SANKAR CHATTERJEE AND R. J. TEMPLIN

Abstract

The origin of avian flight is one of the most contentious issues in evolutionary biology. Between two competing theories, cursorial and arboreal, the latter is more parsimonious on the basis of ecological and aerodynamic constraints. The recent discovery of a series of exquisite feathered coelurosaurs from the Early Cretaceous of China offers a rare glimpse of the transitional stages of avian flight and reinforces the arboreal theory. It suggests that the direct ancestors of birds went through arboreal, climbing, parachuting, and gliding stages before they acquired powered flight. By climbing trees to elude predators, small coelurosaurs such as *Sinosauropteryx* invaded arboreal habitats, exploited new resources, and modified locomotory and neurosensory control. This climbing adaptation put new demands on forelimbs for locomotor function. The forelimbs became progressively longer to facilitate climbing and thus exapted for flight. Arboreal life in three-dimensional space promoted enlargement of the brain and increased visual acuity. Downy feathers evolved at this stage, probably for insulation in the cooler environments of trees, and for streamlining the body for leaping. Leaping from branch to branch with outstretched forelimbs conferred balance, coordination, and bodily control. Symmetrical contour feathers appeared in the remiges and rectrices in *Caudipteryx* and *Protarchaeopteryx*, providing a large lifting surface for parachuting. Climbing ability was enhanced in *Sinornithosaurus* with the modification of shoulder and pelvic girdles, elongation of forelimbs, and the development of the swivel wrist joint and a rigid tail. The sequence of a strong, synchronized forelimb cycle during climbing was a precursor to flight stroke: up-and-forward and down-and-backward. A wing folding mechanism, necessary for gliding, was achieved at this stage. Once gliding was perfected, asymmetrical and large contour feathers appeared in *Archaeopteryx*. Increased wing area, reduced wing loading, and asymmetrical flight feathers led to flapping flight. In this stage, the tail became shorter and formed an integral part of flight to control pitch and increase lift. For *Archaeopteryx*, takeoff from a perch would have been more efficient and cost-effective than takeoff from the ground. *Archaeopteryx* may have made short flights among trees, utilizing a novel method of phugoid gliding. Heterochrony appears to have been the dominant factor in the evolution of feathers. Both ontogenetic and phylogenetic evidence indicates that downlike feathers may have formed initially as an insulating cover of arboreal coelurosaurs. Subsequently this fluffy blanket was modified to plumaceous, contour, and flight feathers with increased morphological complexity during the evolution of flapping flight.

Figure 12.1. (opposite page) A, the cursorial (ground up) theory is based on the idea that the protobirds were essentially small running theropods that leaped up into the air and became active fliers without any gliding stage. B, the arboreal (trees down) theory begins with a climbing protobird that started to parachute from treetops and eventually began to glide as the angle of descent decreased from 45°. Gliding would increase maneuverability and slow landing. Once gliding was perfected, flapping would begin to prolong flight. In the arboreal theory, gravity is used to convert potential energy into kinetic energy.

Introduction

The evolution of flight was key to the success of early birds in the dinosaur worlds, enabling them to escape from predators, to exploit new resources, to search for mates, and to migrate cheaply and rapidly to more favorable habitats. The primary adaptation to this purpose lies in the modification of the forelimbs into wings to be used as organs of flight. The wings provide both upward lift against gravity and forward propulsion through air. Lift and thrust are provided by the proximal and distal wings respectively. In this respect, birds are rather like helicopters, which depend on their rotors for both lift and propulsion (Alexander 1992). Bird flight appears very graceful and artistic. It is a complex performance that combines three independent movements of the wings to exchange energy from the air: dorso-ventral flapping about the gleno-humeral joint, a rotation of the wing about its long axis, and a complex wrist movement that shortens the wing during the upstroke but locks the manus during the downstroke (Rayner 1991; Vasquez 1992). The evolution of feathers is linked closely to the insulation and aerodynamics of protobirds. A complex adaptation like flight had to evolve in a sequence of small stages, over many generations, with each stage in the development of wings and feathers becoming functional and adaptive; each level was contingent on the preceding stage (Bock 1986, 2000).

The origin of powered flight in birds has been debated for over a century. It is often equated with the origin of *Archaeopteryx*, the primitive Late Jurassic bird from Germany. Ostrom (1976, 1979, 1986) suggested that *Archaeopteryx* was a ground dweller similar to quail and roadrunners, without any flying power. Others pointed out that *Archaeopteryx* possessed features such as asymmetrical primary feath-

ers (Feduccia and Tordoff 1979), a robust furcula for the origin of pectoralis muscle (Olson and Feduccia 1979), and a claw geometry for perching and climbing typical of perching and tree-climbing birds (Feduccia 1993), which suggest it was capable of short-distance flapping flight (Norberg 1990; Rayner 1991; Feduccia 1996; Chatterjee 1997). Assuming *Archaeopteryx* was already a flier, as most workers now believe, it was far removed from the lineage of the early evolution of flight. We have to search for the closest extinct relatives, the sister groups of *Archaeopteryx*, to understand the morphology and aerodynamic constraints of the early attempts of incipient flights.

Traditionally, two opposing theories have been championed concerning the origin of powered flight in birds. Each theory relies heavily upon speculation about the paleoecology and functional adaptation of protobirds and *Archaeopteryx*. The cursorial (ground up) theory maintains that flight evolved in running bipeds as they leaped into the air after prey (Williston 1879; Nopsca 1923; Ostrom 1979, 1986; Gauthier and Padian 1985; Balda et al. 1985; Padian and Chiappe 1998; Sumida and Brochu 2000). This theory centers on the interpretation that the immediate theropod ancestors of birds were strictly terrestrial. These bipedal cursors employed their wings as bilateral stabilizers during a jump in the air in pursuit of prey. As jumps became extended, the wings were used for balance and propulsion, and the animals began to fly actively without passing through an intermediate gliding stage (fig. 12.1A). The cursorial theory is biomechanically untenable. The shortcomings of the cursorial theory have been discussed elsewhere (Bock 1986; Rayner 1991; Norberg 1990; Chatterjee 1997; Tarsitano et al. 2000). The cursorial theory works against the downward pull of gravity. Its mechanical approach is weak and the purported outcome is energetically expensive. It is marred by functional and adaptive gaps, from preflight to the active, flapping flight stage. It does not explain the origin of feathers, the origin of brain enlargement, or the evolution of flight stroke in the ancestors of birds (Bock 1985, 1986; Chatterjee 1997).

The arboreal (trees down) theory (Marsh 1880; Heilmann 1926; Bock 1986; Rayner 1991; Norberg 1990; Feduccia 1996; Chatterjee 1997; Tarsitano et al. 2000) states that flight originated in tree-living bipeds that leaped from branch to branch or tree to ground, steadying themselves with outstretched forelimbs (fig. 12.1B). The argument for an arboreal ancestor, as opposed to a strictly cursorial ancestor, seems obvious from modern analogs where virtually all gliding and flying animals live or move in trees. It is likely that arboreal ancestors of birds began to parachute, glide, and then fly downward from the heights as feathers became larger over time. Arboreal life is key to the origin of feathers, elongated forelimbs, enlargement of the brain, three-dimensional perceptual control, and the evolution of flight stroke (Chatterjee 1997; Bock 1986, 2000). The arboreal theory is supported by the fact that the acceleration of gravity accommodates rather than hinders the process of flight (Norberg 1990; Rayner 1991). It uses gravity to create the airflow over the body surfaces, making it more energy efficient (Tarsitano et al. 2000). The arboreal theory also provides a complete series of plausible transitional stages of adaptations, from preflight to flapping flight, as is evident from the Chinese feathered coelurosaurs.

A modified arboreal theory has been proposed by Garner et al. (1999), where the pouncing protobirds leaped from elevated sites, such as a rock ledge, using a drag-based control mechanism. However this ambush model from elevation is not energetically efficient because it explains the origin of flight as a feeding adaptation at a high cost (Elzanowski 2000). For a cursorial predator, capturing prey would be

A

Figure 12.2. A, (left) Liaoning fossil site showing two farmers exposing a feathered coelurosaur from the ash bed.
B, (below) plate tectonic reconstruction for the Early Cretaceous (Hauterivian, 126 M.Y.B.P.) showing the paleo-positions of North and South China blocks and the location of the Liaoning fossil locality. (Chris Scotese pers. comm.)

B

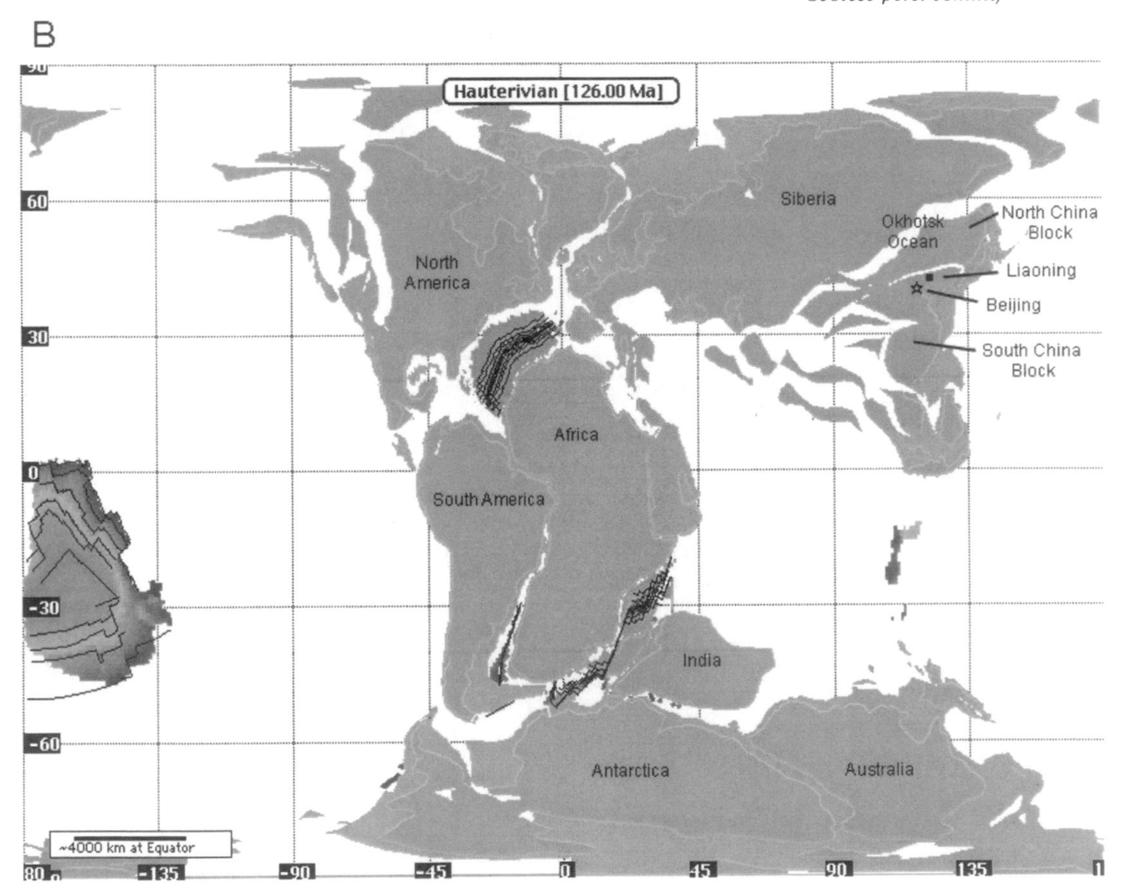

more efficient and cost-effective by pursuing the prey on the ground rather than climbing, waiting, and jumping. The pouncing protobird, according to this model, lived in an almost two-dimensional world. In this environment there was no selection pressure for the enlargement of the brain as a processing center for visual information, integrated with auditory acuity. The low encephalization quotient (EQ) of pterosaurs has been attributed to their two-dimensional, cliff-dwelling habitats (Jerison 1973). This pouncing theory fails to account for the origin of feathers, the development of small body size, and the acquisition of climbing adaptation in early protobirds.

Ecological Phylogeny

It is now widely accepted that birds are a subset of theropod dinosaurs and that they evolved from a small, unknown animal closely related to dromaeosaurs (Ostrom 1976; Gauthier 1986; Padian and Chiappe 1998). The theropod origin of birds has been criticized by other workers who championed various groups of non-dinosaurian archosaurs as putative relatives to birds (Feduccia 1996; Geist and Feduccia 2000; Ruben and Jones 2000; Tarsitano et al. 2000). However, their arguments are plagued by lack of a convincing alternative for bird ancestry and are not based on phylogenetic relationships. This alternative view has been countered forcefully by Sumida and Brochu (2000) in a detailed phylogenetic framework.

It is ironic that most proponents of the arboreal theory tend to dismiss the theropod-*Archaeopteryx* link. To resolve the long-standing impasse between these two competing theories, Chatterjee (1997) proposed a new model that is both biomechanically and phylogenetically congruent. He accepts the cladistic relationship of theropods and birds, but argues that many small coelurosaurs were arboreal as evidenced from their small body size, large brain size, visual acuity, grasping claws, and gradual lengthening of the forelimbs. Flying birds arose from these small, arboreal coelurosaurs through successive stages. This view of the origin of avian flight is strongly supported by the interpretations of the phylogenetic relationships and inferred ecology of a series of missing-link, feathered coelurosaurs from China.

Recent discoveries of several chicken-sized, feathered coelurosaurs from the Early Cretaceous Yixian and Jiufotang Formations of Liaoning Province of northeastern China shows that theropods went through different stages of plumage and wing development that led to powered flight. The site has been dubbed the Cretaceous Pompeii where, on an ancient lake bed, hundreds of birds, feathered dinosaurs, pterosaurs, mammals, fish, insects, and plants were preserved exquisitely, entombed by occasional volcanic eruptions. Perhaps these organisms were engulfed in a cloud of ash or poisonous gas from recurrent volcanic activity around 125 million years ago from the west, from what is now Inner Mongolia (Ackerman 1998). Plants and animals died instantly, dropped to the lake bottom, and were buried by fine ash (fig. 12.2A). Their rapid burial in the ash bed kept the coelurosaur skeletons intact with skin and feather impressions, as there was little bacterial activity.

During Late Jurassic and Early Cretaceous times, China was an isolated, island continent (fig. 12.2B). When Pangea broke up during the Early Jurassic, the North and South China blocks, together with Indochina and Southeast Asia, drifted northward. A narrow seaway, the Mongol-Okhotsk Ocean, separated northern China from Siberia

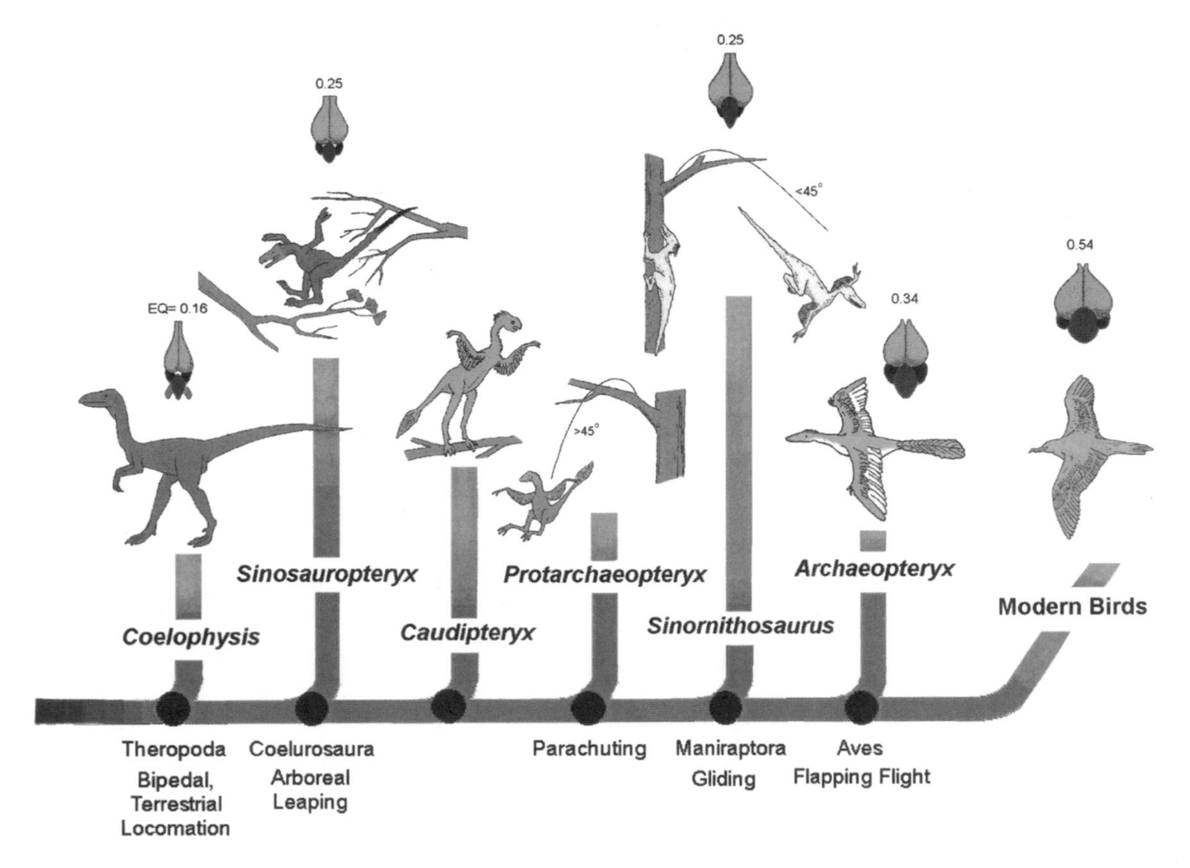

Figure 12.3. Cladogram showing the paleoecological evolution of birds and their flight. The sequence of events, fully adaptive at each stage, led from arboreal coelurosaurs such as Sinosauropteryx, *to parachuters such as* Caudipteryx *and* Protarchaeopteryx, *to gliders such as* Sinornithosaurus, *to true fliers such as* Archaeopteryx. *Arboreal adaptations led to the evolution of downy feathers for insulation, lengthening of forelimbs for climbing, enlargement of the brain for three-dimensional orientation, and evolution of flight feathers.*

(Scotese 1991). Several vertebrates from the Liaoning site appear to be closely related to the Late Jurassic fauna of Europe and North America, indicating that Liaoning fauna evolved in splendid isolation as long-lived relicts (Luo 1999). These feathered coelurosaurs were "living fossils" in the Early Cretaceous forests of China. They represent vestiges of older ancestral lineages that gave rise to the birds. It is likely that similar animals had lived on other continents during the Jurassic Period, prior to *Archaeopteryx*, but they are not yet documented by the fossil record. The extraordinary preservation of Liaoning coelurosaurs provides a window on early bird evolution, from the origin of feathers to the evolution of wings and flight.

So far, six genera of feathered coelurosaurs have been reported from northeastern China. Phylogenetic analyses indicate that these feathered coelurosaurs, such as *Sinosauropteryx* (Chen et al. 1998), *Beipiaosaurus* (Xu et al. 1999a), *Caudipteryx* and *Protarchaeopteryx* (Ji et al. 1998), *Microraptor* (Xu et al. 2000), and *Sinornithosaurus* (Xu et al. 1999b) are successively closer to *Archaeopteryx*. Of these six taxa, *Beipiaosaurus* is the largest known coelurosaur (2.2 m long) from the Liaoning locality. The larger body size indicates that it became secondarily terrestrial, like modern flightless birds, and is not analyzed here for aerodynamic analysis. The remaining five genera—*Sinosauropteryx, Caudipteryx, Protarchaeopteryx, Microraptor,* and *Sinornitho-*

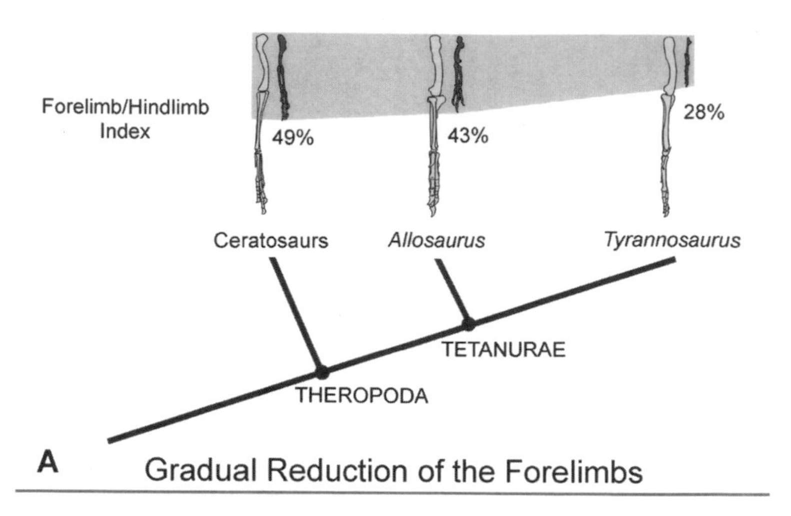

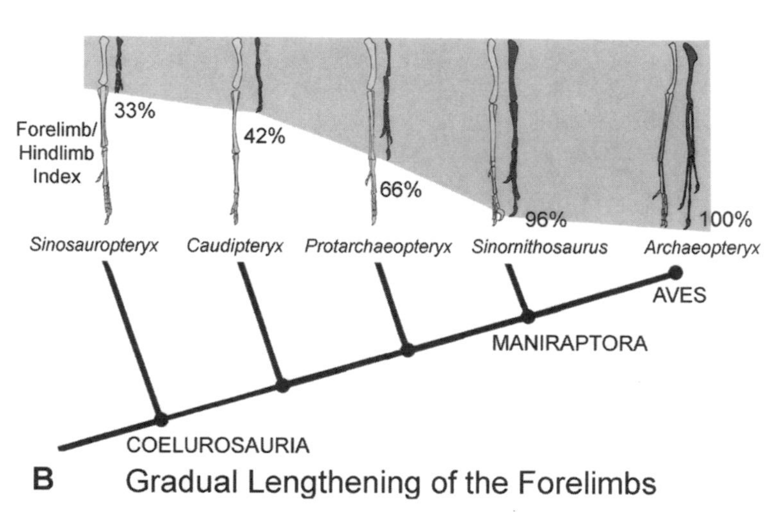

Figure 12.4. Comparison of forelimb/hindlimb index between terrestrial and arboreal modes. A, in terrestrial theropods, the forelimb becomes progressively shorter in relation to the hindlimb, as represented by ceratosaurs, allosaurs, and tyrannosaurs.
B, in arboreal theropods, on the other hand, the forelimb becomes progressively longer to facilitate climbing. Starting from arboreal coelurosaurs such as Sinosauropteryx, *there is a steady increase of the forelimb length in the cladogram. The forelimb becomes as long as the hindlimb in maniraptorans such as* Sinornithosaurus *and the early bird* Archaeopteryx.

saurus—are chicken-sized, and adopted arboreal adaptations, as discussed later. The phylogeny of the Chinese coelurosaurs, based on osteological characters, offers a framework to trace the sequence of modifications of the brain, wing development, and feathers in an arboreal setting during the early evolution of avian flight (fig. 12.3). The ecological phylogeny accords well with the character-based phylogeny.

The Evolution of Flight

Birds acquired two independent and specialized methods of locomotion: walking with hindlimbs and flying with forelimbs. The former is primitive, inherited from theropod ancestors; the latter is derived (Gatesy and Dial 1996). Basal theropods such as ceratosaurs and allosaurs were obligatory bipeds, where lightweight skeletons were supported by the hindlimbs. The forelimbs, being decoupled from terrestrial locomotion, were used for prey capture and manipulation.

Supporting additional weight, the ilium became elongated and was strengthened by five coalesced sacral vertebrae; the neck and head were counterbalanced by a long and slender tail. The tail also served as a primary mechanism for femoral retraction during walking and running (Gatesy 1990). They had a single locomotor module composed of the hindlimb and tail (Gatesy and Dial 1996). The pes became tridactyl and digitigrade for cursorial adaptation. The basal theropods were essentially terrestrial cursors, as evidenced from their relatively small brains, where the encephalization quotient (EQ) was around 0.15. They lived in a neurologically less demanding, two-dimensional world (Jerison 1973; Chatterjee 1997). In terrestrial theropods, there is a gradual reduction of forelimbs relative to hindlimb length, as documented by ceratosaurs, allosaurids, and tyrannosaurs respectively (fig. 12.4A). The reduction of forelimbs in terrestrial theropods is compensated by the development of large heads equipped with enlarged teeth that become the primary capturing and killing machine.

From this ancestral terrestrial stage, a sequence of four phylogenetic and adaptive changes of coelurosaurs in an arboreal environment leading to powered flight is presented here. These coelurosaurs probably invaded a set of niches like those of living tree-dwelling primates. Small body size and claw geometry support the arboreal lifestyle of Chinese coelurosaurs.

Stage 1. Arboreal leaping. Small coelurosaurs, such as *Sinosauropteryx* (0.28–0.63 kg body mass) were agile, lightly built animals that probably invaded arboreal habitats as an escape strategy (fig. 12.5). They could not outrun much larger theropods and their best strategy was to climb to the safety of the heights (Elzanowski 2000). The forelimbs regained locomotory function during climbing. The gradual lengthening of forelimbs in protobirds can be explained in an arboreal setting (fig. 12.4B). Small coelurosaurs were more versatile in scansorial adaptation, as they had a larger surface-to-volume ratio. What was the selection pressure that led them to occupy a new ecospace? We can speculate from modern analogs. An arboreal biped, such as the tree kangaroo, is adept at climbing trees with its recurved claws. These arboreal acrobats have developed various techniques for moving from branch to branch and tree to tree. It is likely that during the Mesozoic, small coelurosaurs occupied a similar arboreal niche, which allowed them access to previously unavailable habitats. They were sprinting and jumping predators that were probably well adapted to foraging on the ground and preying on insects, small amphibians, reptiles, and mammals. Stomach contents of the holotype of *Sinosauropteryx* (fig. 12.6B) reveal that its last supper consisted of lizards and mammals (Chen et al. 1998).

There are other advantages to an arboreal lifestyle. To avoid predators, an arboreal form can climb to attain a safe perch where it cannot be followed or attacked (fig. 12.5). It is likely that these small coelurosaurs could climb trees to escape from predators, to hide, to sleep, to seek food, and to nest (Chatterjee 1997; Bock 2000). Climbing may be initiated by juvenile coelurosaurs as a defensive mechanism to hide from predators (Elzanowski 2000). The predator pressure on the juve-

Figure 12.5. Escape strategy of small feathered coelurosaurs. Large predators could easily overrun small coelurosaurs to capture them for food. Small coelurosaurs such as Sinosauropteryx *climbed trees to elude predators. Downy feathers probably evolved in arboreal forms for insulation in the cooler environments of trees.*

niles favored a fast escape route and efficient use of both limbs. Ecological separation from terrestrial relatives resulted in a wide range of sympatric speciation among coelurosaurs.

There are several anatomical evidences that support the arboreal lifestyle of small coelurosaurs. They had a large brain (EQ = 0.25), and large eyes with stereoscopic vision to coordinate locomotion and foraging among complex, three-dimensional branches. These protobirds relied mainly on their vision to locate prey. Front-facing eyes enhanced visual acuity in a confusingly mottled background of leaves, branches, and other foliages. Life in trees provided a strong selection pressure for enlargement of the brain for proprioception, improvement of sense organs, and external morphology for insulation (Jerison 1973; Bock 1986). The movements of the hindlimbs during leaping were similar to those during hopping. When nearing a landing place, the forelimbs were outstretched to grab a branch or to balance. The outstretched forelimbs enhanced stability and bodily control during leaping from branch to branch (Tarsitano et al. 2000). The flexible branches

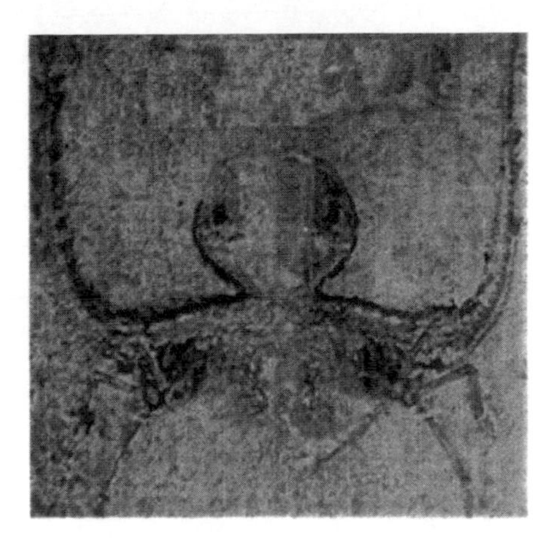

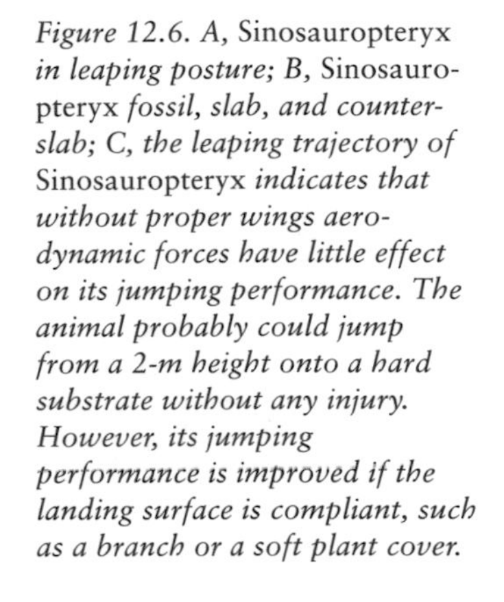

Figure 12.6. A, Sinosauropteryx *in leaping posture; B,* Sinosauropteryx *fossil, slab, and counterslab; C, the leaping trajectory of* Sinosauropteryx *indicates that without proper wings aerodynamic forces have little effect on its jumping performance. The animal probably could jump from a 2-m height onto a hard substrate without any injury. However, its jumping performance is improved if the landing surface is compliant, such as a branch or a soft plant cover.*

Sinosauropteryx

(Small Individual)
mass = 0.28 kg
wingspan = 0.11 m

(Large Individual)
mass = 0.63 kg
wingspan = 0.19 m

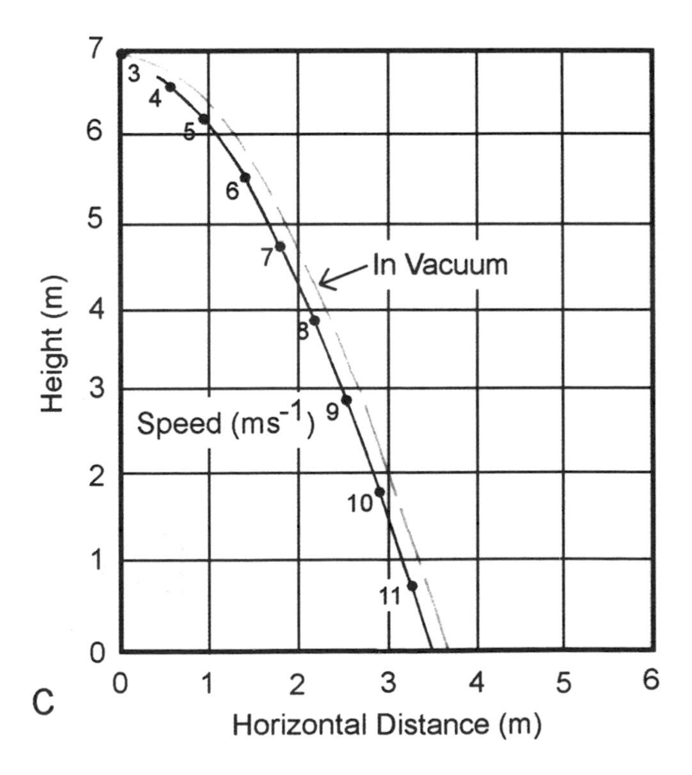

or soft ground cover may have acted as a cushion during landing. The downy covering in *Sinosauropteryx* may have offered additional protection from an accidental fall from a height.

Templin (1977) has pointed to the necessity of slowing down to about 6 m s^{-1} before impact in order to avoid injury. *Sinosauropteryx* could jump from a 2-m height onto a hard substrate without sustaining injury. Figure 12.7B shows a calculated leaping trajectory of *Sinosauropteryx,* at two values of the estimated mass (0.28 and 0.63 kg), leaping horizontally from a perch at 3 m s^{-1} at an initial height of 7 m. Only a single curve is shown, because mass variation within this range has little effect. The jumping speeds, from a standing start, are about the same for animals of all sizes (Templin 2000). Also shown for comparison is the theoretical trajectory in a vacuum, indicating that, for an animal of this size (or larger) with no wings, aerodynamic forces have little effect on its jumping performance. Speed rapidly builds up to 6 m s^{-1} after falling from about 2 m, and to 10 m s^{-1} after about a 5-m drop. The latter speed is probably too high for a safe landing unless it ends on a compliant surface such as a flexible tree branch or a soft ground cover. We have calculated the effects of landing on compliant surfaces such as branches or soft, pliable vegetation. These well-damped, springy surfaces can probably accept speeds of 10 m s^{-1} or more, provided that their effective inertial masses are small in comparison with the animal's mass, and that their rigidities also are small compared with the animal's leg stiffness.

Life in trees would favor smaller animals (Bock 1985). *Sinosauropteryx* was a small predator that had several advantages over larger animals in an arboreal setting: It could be supported by delicate branches; it had greater agility in negotiating difficult leaps among branches; and it needed less energy to climb vertical trunks. If *Sinosauropteryx* fell accidentally from a tree, it suffered less impact from the fall.

The body of *Sinosauropteryx* was covered by simple, hairlike filaments which may be the precursor to feathers (Chen et al. 1998). These protofeathers had some aerodynamic initial effects (fig. 12.6A). The coat of protofeathers conferred smooth and streamlined body contours in *Sinosauropteryx* to reduce friction drag during leaping among tree branches with folded forelimbs (fig. 12.6B). The streamlined body must have played a critical role in the evolution of avian flight in an arboreal setting, where protofeathers acted as pressure and turbulence sensors (Homberger and de Silva 2000).

These protofeathers probably evolved in *Sinosauropteryx* to provide insulation in the cooler microclimate of a tree. The insulation allowed it to maintain a constant body temperature independent of the external environment (Regal 1975; Bock 1986). Because trees are considerably cooler than the ground, an insulating body cover was advantageous for the arboreal *Sinosauropteryx*. These protofeathers were thin, long, cylindrical, numerous, and pliable like mammalian hair and provided significant insulation. Large patches of long, filament-like integuments with hollow cores and a branching pattern also are known

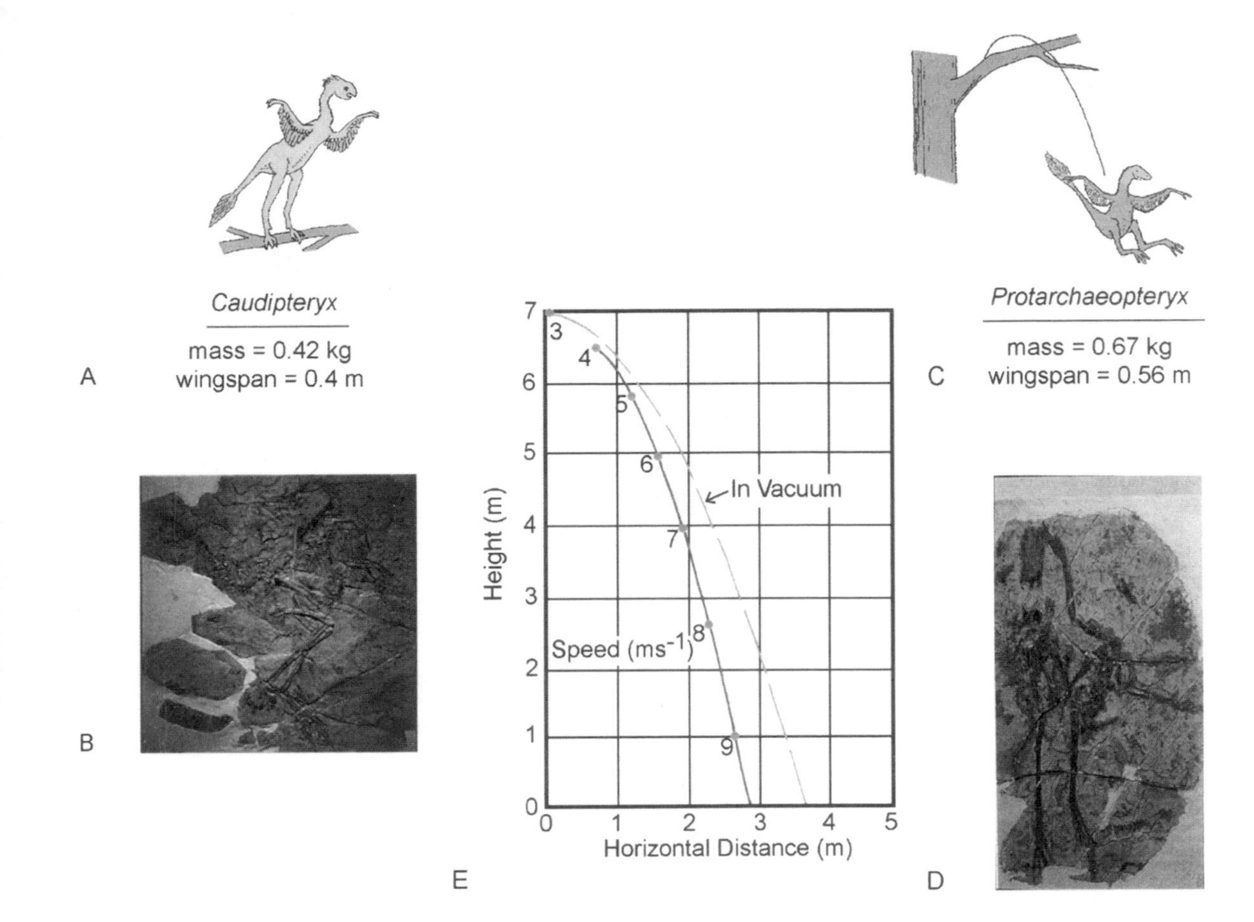

Figure 12.7. A, Caudipteryx *on a tree branch;*
B, Caudipteryx *fossil;*
C, Protarchaeopteryx, *parachuting from a perch;*
D, Protarchaeopteryx *fossil;*
E, the parachuting trajectory of feathered Caudipteryx *and* Protarchaeopteryx. *Only a single curve is shown because their trajectories are essentially the same. With the development of narrow wings, these animals could parachute and retard their fall from a height.*

in *Beipiaosaurus,* a therizinosauroid coelurosaur from China (Xu et al. 1999a).

Stage 2. Parachuting. Jumping proavians used their outstretched forelimbs and tails to maximize drag forces. Eventually, the stiff and collapsible flight surface evolved in the forelimbs and tails of protobirds. In *Caudipteryx* (figs. 12.7A, B) and *Protarchaeopteryx* (figs. 12.7C, D), there is evidence of contour feathers on forelimbs and at the end of the tail, and downlike feathers on the body (Ji et al. 1998). Although these contour feathers are long, vaned, and barbed, they are symmetrical, indicating that these animals could not fly or glide effectively (Feduccia and Tordoff 1979); their arms and feathers were not large enough for flight. *Caudipteryx* and *Protarchaeopteryx* had a high wing loading and low aspect ratio, as appeared from our study. However, their small protowings evolved as a safety device for arboreal maneuvering, allowing parachuting that increased drag and retarded a fall. Drag is the main force responsible for parachuting. *Caudipteryx* was a small animal. We estimated its body mass to be around 0.43 kg and its wing span, 0.4 m. *Protoarchaeopteryx* was somewhat larger, with a body mass of 0.67 kg and a wing span of 0.56 m. We have

Figure 12.8. The function of the forelimbs in dromaeosaurs. Gauthier and Padian (1985) suggested that the predatory motion of the forelimbs (1–5) led to the evolution of flight stroke. However, the forelimbs developed an automatic linkage system between the elbow and the wrist joints as in modern birds. As a result, the hands could not be used for catching and eating prey because of restricted wrist movement. Moreover, the predatory movement is essentially forward and backward whereas the flight stroke is essentially dorso-ventral, with a rotary component. It is unlikely that the propalinal motion of the arms would give rise to the flight stroke in birds.

computed the parachuting trajectory of *Caudipteryx* and *Protarchaeopteryx* with narrow wings using the "animal flight simulation program" ANFLTSIM (Templin 2000). Only a single curve is drawn in Figure 12.7, because separate trajectories were essentially the same. The graph has an auxiliary plot of speed versus height loss, with the theoretical curve shown for a trajectory in vacuum (at the same initial condition). The difference between it and the animal's actual curve is an effect of aerodynamic forces. Comparison between the two curves shows that even small wings may have a sufficient effect on the fall of animals of this size. The terminal velocity in this case is about 11 m s^{-1}, and is 2 m s^{-1} less after a fall of 7 m than after leaping (see fig. 12.7E).

The ability of arboreal protobirds to leap from a great height without injury was a critical adaptation. These animals would launch from a perch with outstretched wings to maximize the drag forces, fall in a descent path greater than 45°, and land on their feet. Parachuting allowed for vertical foraging—from the treetop, to low branches, and down to the ground without taxing the animal's muscular effort.

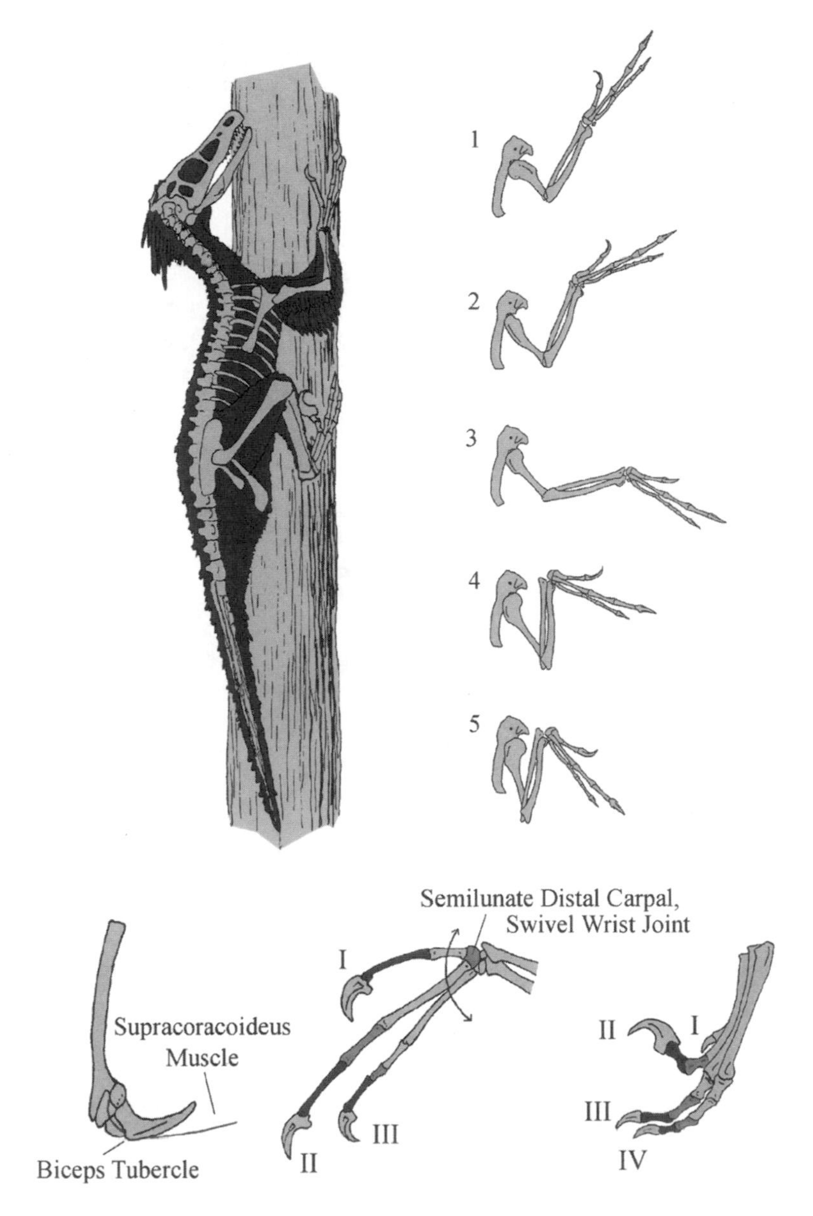

Figure 12.9. Climbing adaptations of dromaeosaurs and evolution of the flight stroke. Small dromaeosaurs such as Sinornithosaurus *show several climbing traits such as a laterally facing glenoid for dorso-ventral humeral excursion, a biceps tubercle for the inception of the supracoracoideus pulley, elongated forelimbs for grasping trunks, ossification of the sternum, a furcula for the attachment of the pectoralis and supracoracoideus muscles, swivel wrist joint, longer penultimate phalanx and recurved claws for hooking and clinging to trunks, a caudally directed pubis for flattening the body, and a stiffened tail as a supporting prop. The sequence of a strong, synchronized forelimb cycle (1–5) during climbing was a precursor to a flight stroke: up-and-forward and down-and-backward.*

Stage 3. Gliding. One of the arguments cited by the proponents of the cursorial hypothesis is that the immediate ancestors of birds (such as dromaeosaurs) were fully terrestrial; their forelimbs were used for subduing and killing prey (fig. 12.8). These predatory motions later gave rise to the flight stroke (Ostrom 1979; Gauthier and Padian 1985; Padian and Chiappe 1998). There are several inconsistencies with these terrestrial scenarios. With the development of the semilunate carpal, a linkage system between the elbow and wrist joints was developed in dromaeosaurs in avian fashion. As the elbow folded, the hand would have moved farther away from the mouth. Like those of modern birds, the hands of dromaeosaurs could not be used for predation and feeding. Moreover, the predatory movements of the forelimbs would be forwardly directed to grasp prey, while the flight movement is essentially dorso-ventral. It is inconceivable that this propalinal motion of the hand would give rise to the dorso-ventral flight stroke. The hands of dromaeosaurs were doing something other than catching prey. They could be folded on the side of the body, as in extant birds, during terrestrial locomotion. If so, what was the main function of these hands, equipped with the swivel wrist joint and sharp claws? The elongated forelimbs and swivel wrist joint in dromaeosaurs can be interpreted as features evolved for climbing.

Recent functional analysis suggests that small dromaeosaurs acquired a suite of climbing traits, such as a powerful shoulder girdle, biceps tubercle, elongated coracoid, laterally facing coracoid, ossification of the sternum and furcula, elongated forelimbs, evolution of the swivel wrist joint, and opisthopubic pelvis (Chatterjee 1997). The climbing adaptation was refined further in *Sinornithosaurus* (Xu et al. 1999b), a small dromaeosaur weighing about 1.5 kg, with a wing span of 0.92 m. The forelimbs were as long as the hindlimbs (fig. 12.4B). The elongated forelimbs facilitated climbing. The hands, especially the thumbs, were effective grasping structures for holding branches, and later evolved into the alula. An arboreal setting provided opportunities for considerable locomotor diversity and opened up three locomotor modules for maniraptorans: the forelimb, hindlimb, and tail. The wing and tail formed the aerial locomotor system. The manual phalangeal proportions also attest to the climbing capability of *Sinornithosaurus*. In the second finger, the penultimate phalanx is longer than the proximal phalanx, a condition also present in the climbing hoatzin (fig. 12.9). In digit I, the penultimate phalanx is longer than metatarsal I. The lengthening of the penultimate phalanges in the inner two digits increased the diameter of the available grasp between the opposing fingers, I and II. This facilitated grasping broad tree trunks during climbing. As in modern birds, such as the woodpecker, these dromaeosaurs probably used their stiffened tail as a supporting prop as they climbed trees (Chatterjee 1997). Presumably, dromaeosaurs used their limbs alternately in unison, first the paired forelimbs and then the paired hindlimbs. Modern flying squirrels climb rapidly in this manner. The sequence of a strong, synchronized forelimb cycle during climbing was a precursor to flight stroke: up-and-forward and down-and-backward. The beauty of this theory is that the same nervous circuits used for climbing in unison with the forelimbs in protobirds were exapted to control the wings in birds. With the development of an opisthopubic pelvis, the abdomen could be flattened during gliding into an aerodynamic shape (contra Tarsitano et al. 2000).

Figure 12.10. (opposite page) Contrasting locomotory styles in terrestrial and arboreal theropods. A, the terrestrial bipedal theropods had a single locomotor module in which the hindlimb and tail functioned as a unit. In birds, three locomotor modules can be seen: hindlimb, wing, and tail. The tail is decoupled from the hindlimb during terrestrial locomotion and has become an integral part of the wing during flight (simplified from Gatesy and Middleton 1997). However, this shift from one to three locomotor modules in a terrestrial setting is abrupt without any transitional stages. B, climbing adaptations put new demands on forelimbs for locomotory function. In an arboreal setting, the gradual evolution of limb modules can be seen from one, to two, to three, where both forelimbs and hindlimbs, and later a stiff tail, participate in climbing.

Birds possess three locomotor modules: the ancestral hindlimb module for terrestrial locomotion and wing-tail modules for flight (Gatesy and Dial 1996; Gatesy and Middleton 1997). However, it is not clear how the evolution of these three avian locomotor modules from the one locomotor module of theropods developed in a phylogenetic context without transitional stages (fig. 12.10A). Dromaeosaurs bridge

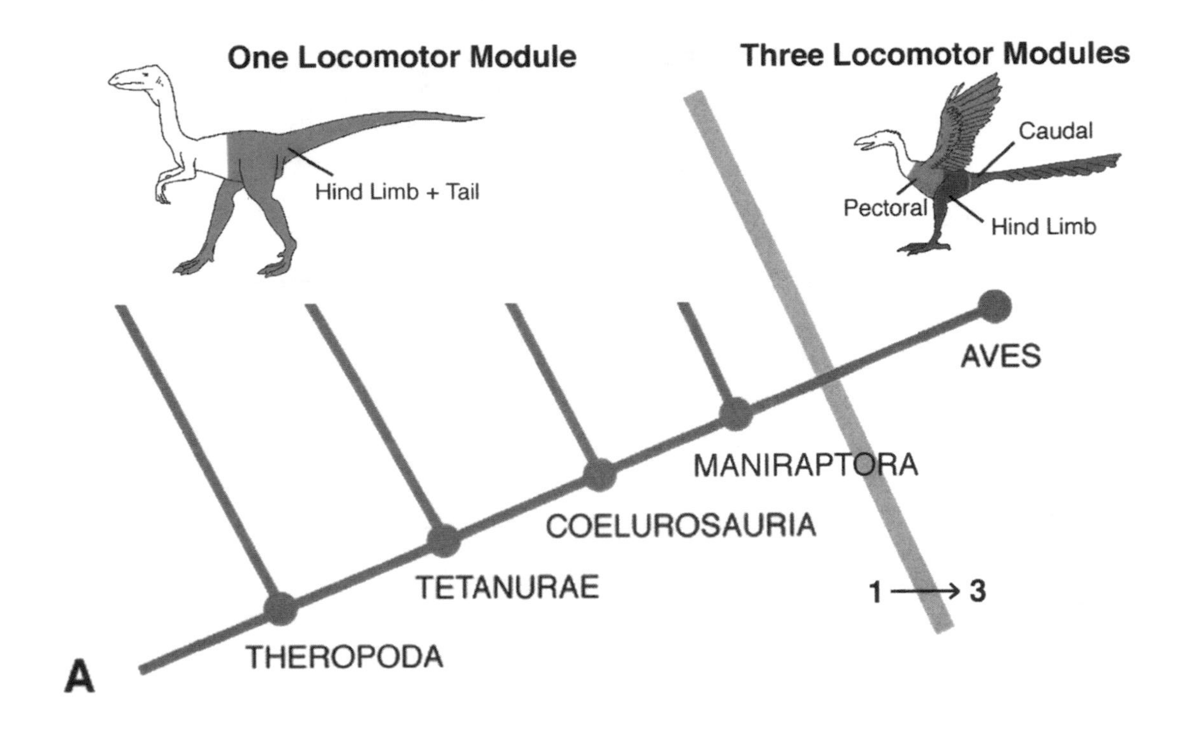
One Locomotor Module
Three Locomotor Modules
Hind Limb + Tail
Caudal
Pectoral
Hind Limb
AVES
MANIRAPTORA
COELUROSAURIA
TETANURAE
THEROPODA
1 → 3
A

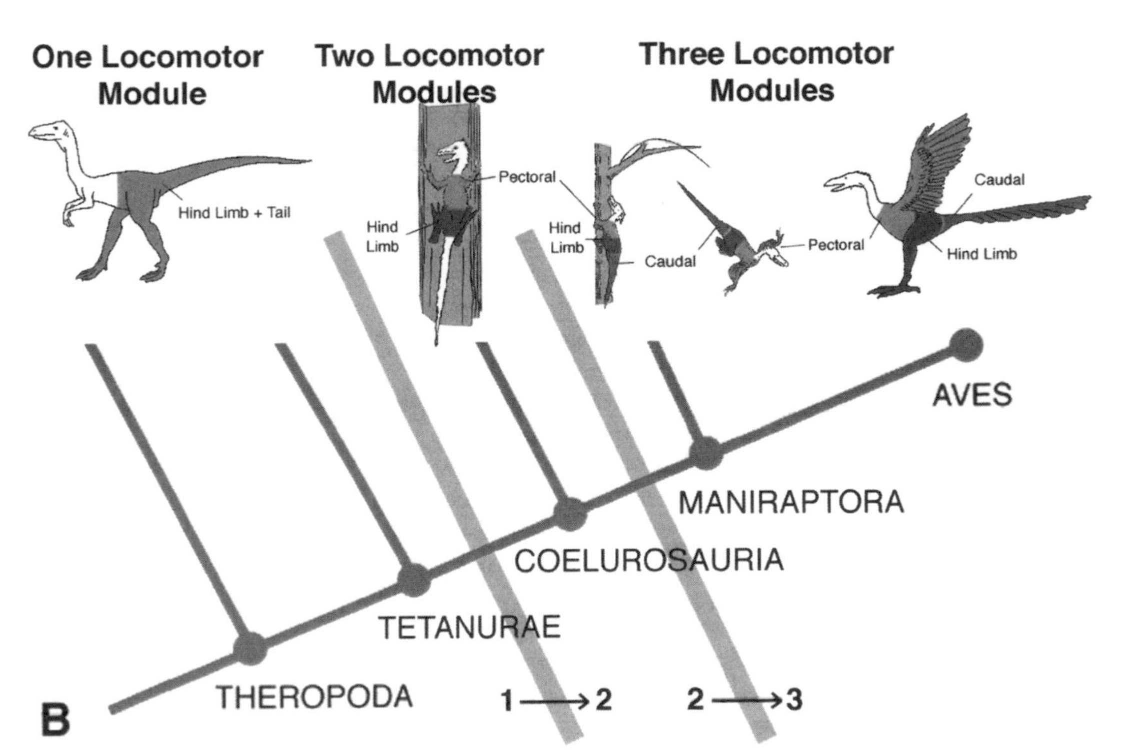
One Locomotor Module
Two Locomotor Modules
Three Locomotor Modules
Hind Limb + Tail
Pectoral
Hind Limb
Hind Limb
Caudal
Pectoral
Caudal
Hind Limb
AVES
MANIRAPTORA
COELUROSAURIA
TETANURAE
THEROPODA
1 → 2
2 → 3
B

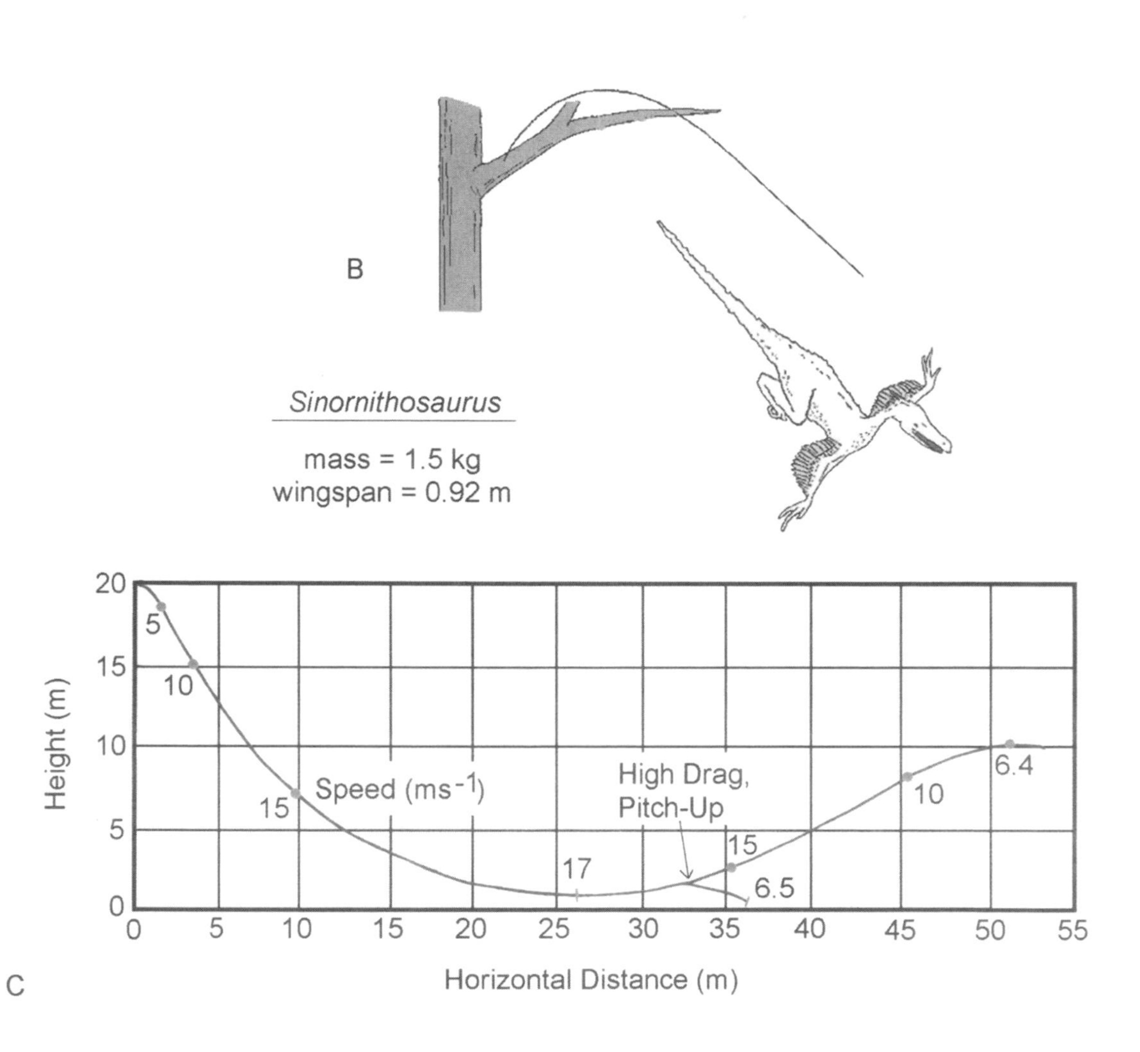

A
B
Sinornithosaurus
mass = 1.5 kg
wingspan = 0.92 m
C
Height (m)
Horizontal Distance (m)
Speed (ms-1)
High Drag,
Pitch-Up
5
10
15
17
15
6.5
10
6.4

this structural gap of evolution in an avian locomotor module. In the dromaeosaur, a stiffened tail was freed from femoral retraction and was used as a supporting prop during climbing. The inception of the coordinated wing and tail movement was exapted during climbing, and further refined during flying in early birds (fig. 12.10B).

The body of *Sinornithosaurus* was covered by downy feathers, but contour feathers in the hand and tail are not preserved in the fossil record (fig. 12.11A). Because *Sinornithosaurus* is more derived than *Caudipteryx* and *Protarchaeopteryx* (Xu et al. 1999b), it is likely that it had acquired remiges and rectrices to form a gliding surface (fig. 12.11B). Recent discovery of additional material suggests that *Sinornithosaurus* developed two types of branching structures: filaments joined in a basal tuft, and a feather with a central rachis and serially fused barbs (Xu et al. 2001). Ji et al. (2001) reported a nearly complete, *Sinornithosaurus*-like dromaeosaur, most of whose body is densely covered with featherlike structures. The contour feathers in the arms and tails show a herringbone pattern around a central rachis and serially attached barbs. Another recent dramatic find from China is *Microraptor*, a feathered dromaeosaur smaller than *Archaeopteryx* with climbing ability (Xu et al. 2000). Unfortunately, osteological data is not available to calculate its mass and gliding ability. *Microraptor* was small enough (~0.2 kg) and had fairly elongated forelimbs, indicating that wing loading was within the realm of gliding.

Some osteological evidence (fig. 12.9) suggests that *Sinornithosaurus* probably expanded its locomotor repertoire from parachuting to gliding. The shoulder girdle is similar to that of *Archaeopteryx*, where the glenoid faces laterally, allowing dorso-ventral excursion of the wing (Xu et al. 1999b). Moreover, the development of the swivel wrist joint indicates that *Sinornithosaurus* achieved an automatic wing folding mechanism for changing the shape of the wing to control gliding. It could supinate the distal part of the wing during an upstroke to minimize the drag.

We have calculated the flight path of a winged *Sinornithosaurus* (fig. 12.11C). It shows the typical undulatory motion of a glider that holds a fixed aerodynamic attitude (fixed lift and drag coefficients) from an initial horizontal launch at 3 m s^{-1} from a height of 20 m. *Sinornithosaurus* has an estimated mass of about 1.5 kg, and for this simulation its wingspan was taken to be 0.92 m. Its wing area was assumed arbitrarily from the wingspan to be about 0.15 m^2 (wing aspect ratio = approximately 5.6). The flight path has a large-amplitude undulation, the so-called phugoid oscillation. This occurs when a powered or unpowered flying device is launched at a speed or altitude far away from its eventual equilibrium condition and when no adjustment is made to its "controls" after launch (Templin 2000). The undulating, phugoid flight of *Sinornithosaurus* reaches a maximum speed of 17 m s^{-1} at the bottom of the glide, and then recovers about half the original height at a speed of 6.4 m s^{-1} at the top before descending again. A landing could be made safely at this speed, but not at the highest speed. However, the animal has at least one other choice: near the bottom of the glide it can make a high-drag pitchup, using its wing, body, and tail

Figure 12.11. (opposite page) A, fossil of Sinornithosaurus; *B,* Sinornithosaurus, *gliding from a perch; C, gliding of a winged* Sinornithosaurus. *The flight path shows the typical undulating, phugoid motion from an initial horizontal launch at 3 m s^{-1} from a height of 20 m. A landing could be made safely at this speed (6.4 m s^{-1}).*

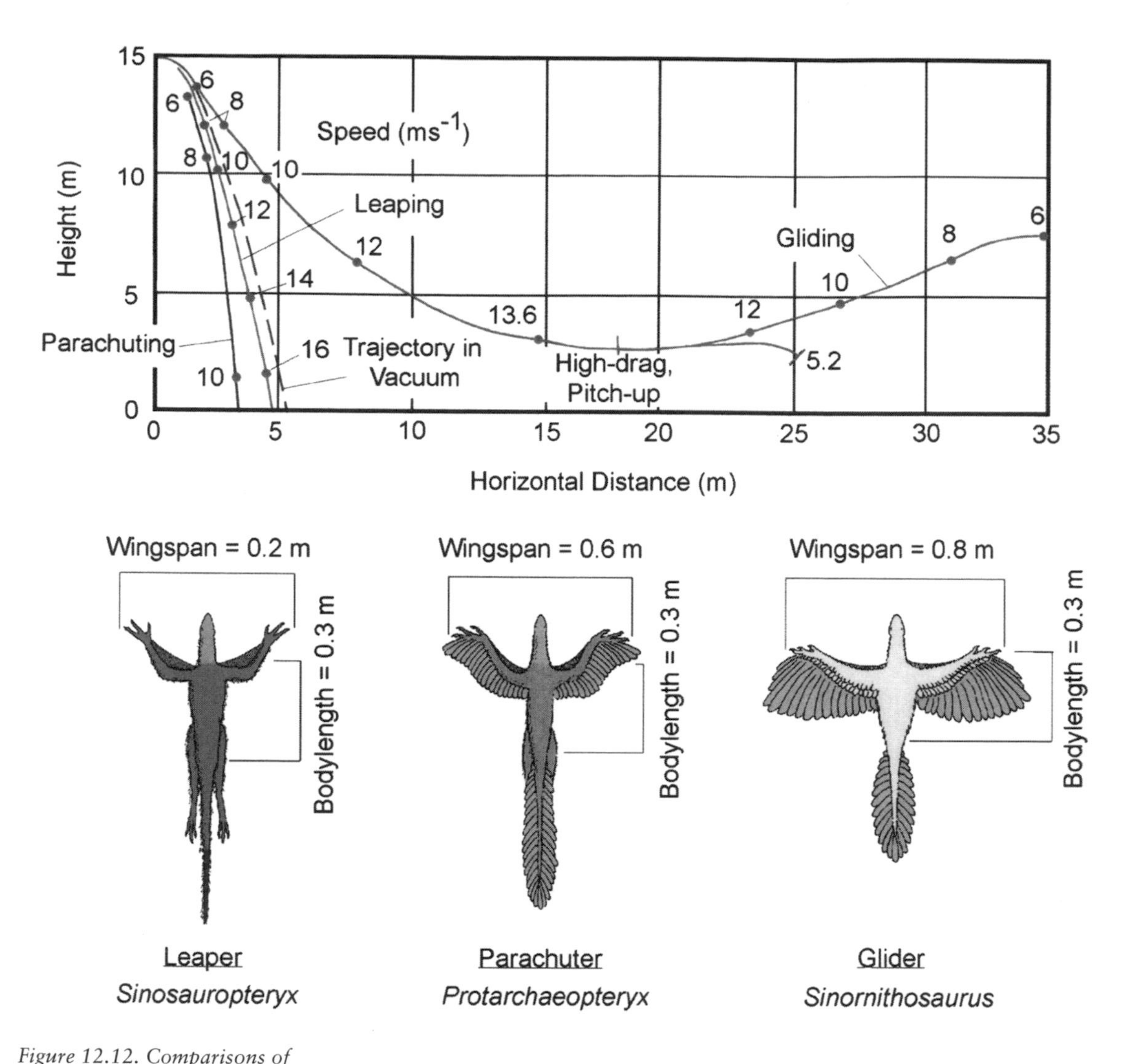

Figure 12.12. Comparisons of flight paths of three protobirds, Sinosauropteryx, Caudipteryx *(or* Protarchaeopteryx*), and* Sinornithosaurus. *To remove mass as an independent variable three groups of protobirds—leaper, parachuter, and glider—were scaled to 1 kg mass and their flight paths were calculated.*

as a kind of horizontal parachute to sharply decelerate to a lower speed (6.5 m s^{-1} in this case) for possible landing on solid ground. Note that this type of braking maneuver, similar to that used by flying squirrels or modern hang gliders, does not involve significant loss in altitude before the minimum speed is reached. *Sinornithosaurus* might have used gliding as a cheap way of moving between foraging sites.

The simulated trajectories (figs. 12.6, 12.7, and 12.11) were computed for the Chinese coelurosaurs that ranged in mass from 0.28 to 1.5 kg. In order to remove mass as an independent variable, three trajectories for hypothetical leaping, parachuting, and gliding theropods (all having 1 kg mass) are plotted in Figure 12.12. In this model, all three coelurosaurs are shown leaping from a height of 15 m; their wing shapes were based on available estimates. One can see the progressive improvement of flight performance to retard the fall from a perch during the progression from leaping, to parachuting, to gliding. The figure confirms that the theoretical flight modes available to evolving fliers would depend on wing development rather than size, within the range represented by Chinese fossil specimens.

Stage 4. Flapping flight. This stage is represented by *Archaeopteryx,* with the development of large asymmetrical contour feathers on the wing and tail to provide large lift and flight surface (fig. 12.13A). The wingspan has been increased considerably to spread the vortices of the wing tips apart and reduce the lift-dependent drag. The climbing and arboreal adaptations of *Archaeopteryx* have been widely accepted from anatomical evidence (Yalden 1985; Feduccia 1993, 1996; Chatterjee and Templin in press). All three locomotor modules—hindlimb, pectoral, and tail—were fully functional in *Archaeopteryx* (Gatesy and Middleton 1997; fig. 12.10B).

The flight capabilities of *Archaeopteryx* have been debated for more than a century. The consensus is that it could fly actively; however, it was incapable of taking off from the ground (Rayner 1991; Norberg 1990; Feduccia 1996). Its lack of a supracoracoideus (SC) pulley, the primary elevator of the wing, would prevent *Archaeopteryx* from executing humeral rotation during the upstroke, a condition necessary for a cursorial takeoff (Poore et al. 1997). Hindlimb proportions support the view that *Archaeopteryx* was not adapted for cursorial locomotion and had great difficulty in reaching a ground speed (7 m sec^{-1}) necessary to become airborne (Hopson in press; Chatterjee and Templin in press; but see Burgers and Chiappe 1999 for a contrary view).

To calculate the flight capability of *Archaeopteryx,* we selected three size groups of individuals (M = 0.2, 0.4, and 0.6 kg) representing the Eichstätt, London, and Solnhofen specimens, respectively (Chatterjee and Templin in press). We used a flight simulator to calculate the flight capability of *Archaeopteryx*. The simulation is adapted from the helicopter momentum streamtube theory (Templin 2000), which indicates that most flight performance items depend almost entirely on two parameters: mass (M) and wing span (b). Flight is controlled by variations in power and the momentum wake deflection angle (Ø), which controls aerodynamic lift. Long-period oscillations are suppressed by a subroutine called PitchDamper, which generates increments in Ø proportional to longitudinal acceleration until equilibrium flight is established.

For mass = 0.2 kg, the wing span was set at 0.55 m and the maximum continuous power at 3.4 W, which were varied as $M^{1/3}$ and as $M^{2/3}$ respectively, as mass increased. Figure 12.13B shows glide path simulations of the three sizes of *Archaeopteryx,* all taking off at maximum continuous power from a perch at V_0 = 2 m s^{-1} horizontally, and then pulling up at maximum continuous power. As expected, the heaviest birds lose the most height, but at M = 0.2 kg this loss is about 5.5 m. The pitch damper was used to suppress violent wave motion. The height losses were used to define the minimum safe heights for "trees down" launching. The energy required for cursorial versus arboreal takeoffs for varying masses of *Archaeopteryx* is listed in Table 12.1. For example, the energy required for a smaller individual (M = 0.2 kg) in an arboreal takeoff is only 7.1 J after 1.9 s. As the animal becomes larger, the cost/benefit ratio becomes apparent in the arboreal mode.

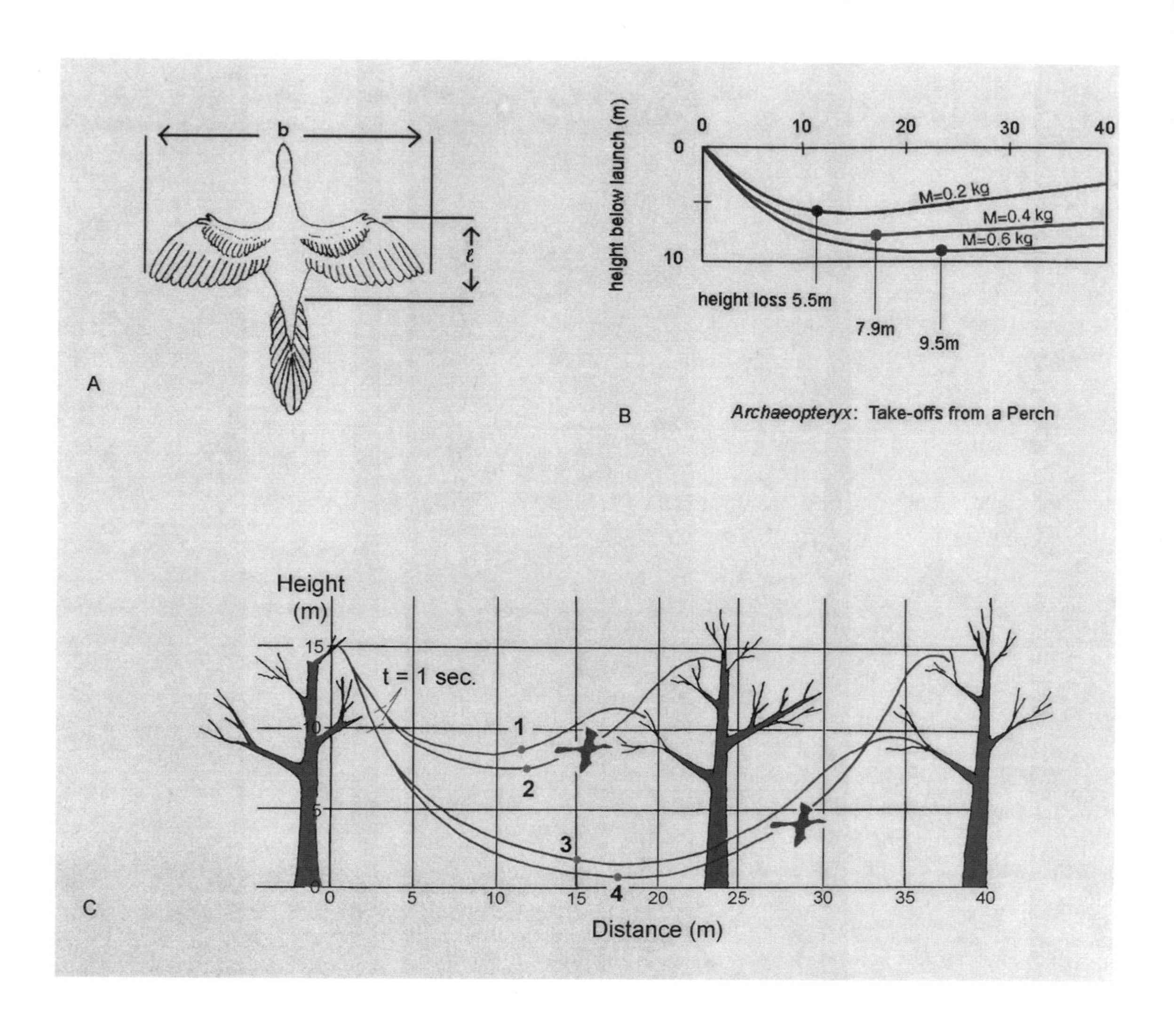

Table 12.1

Comparison of running takeoffs of *Archaeopteryx* with its takeoffs from a perch

Mass M (kg)	Distance to Liftoff (m)	Liftoff speed (m s^{-1})	Liftoff Time (sec)	Climb Angle (deg)	Energy to Lift Off (J)	Energy to Level flight (J)	Minimum Perch Height (m)
0.2	13.4	7.6	2.75	6.2	8.2	7.1	5.5
0.4	28.1	8.6	4.7	2.9	24.01	3.3	7.9
0.6	47.3	9.1	7.0	1.4	48.1	9.5	9.5

Our flight simulator study suggests that *Archaeopteryx* was capable of powered flights when launching from a height (< 10 m), even without a headwind. When compared with the ground takeoff model, the energy expended in takeoff by an arboreal *Archaeopteryx* is more cost-effective.

Three flight paths of *Archaeopteryx* (M = 0.2, 0.4, and 0.6 kg), all taking off from a perch at 2 m s^{-1} and then pulling up at maximum continuous power, are shown in Figure 12.13B. Because the initial launch speed is far below the equilibrium speed, the flight path at first falls and would become oscillatory, but is controlled by the pitch damper, imitating a built-in automatic control of lift, such as by wings or the long, feathered tail. It appears from our analysis that arboreal takeoff saves considerable energy from the cursorial mode, especially when the animal is large. An adult *Archaeopteryx* probably needed a height (~10 m) from which to launch itself safely in still air. A headwind also would have helped. There were tall trees (>10 m) such as cycads, ginkgos, seed ferns, and conifers in the Central European islands where *Archaeopteryx* lived (Barthel et al. 1990). These trees might have provided places to perch and launch, as well as trunks to climb.

Flying squirrels travel through a forest by climbing the trunks of trees and gliding between trunks (Norberg 1990). When crows take off from a tree, they do not seem to use excess power; they lose height at first and then swoop up to swing between two perches. This occurs whenever any winged object (aircraft, model glider, or flying animal) finds itself in a non-equilibrium situation, such as when launched without sufficient wing lift to balance weight (Templin 2000). The result is an initial loss of height at an increasing speed. Lift increases as speed squares (if controls are not moved), and the subsequent motion is an undulation with potential and kinetic energy being periodically exchanged. In gliding flight, the motion is eventually damped to a steady glide, and in fact, the rate of dampening is inversely proportional to the lift-drag (L/D) ratio. Objects with high L/D configuration, such as modern aircraft, have low phugoid damping, but because the period of motion is proportional to speed, control is not difficult. *Archaeopteryx* probably used a similar strategy to move from tree to tree using phugoid gliding without expending much energy. Similar intermittent bounding flight has been advocated by Homberger and de Silva (2000) in the evolution of flight without a gliding stage.

Four glide paths of *Archaeopteryx* were computed from ANFLTSIM with launch speed 2 m sec^{-1} (horizontal) from an arbitrary height of 15 m (fig. 12.13C). The mass was fixed at 0.2 kg (representing the Eichstätt specimen) and the wing span was 0.545 m. No pitch damping was used, in order to encourage the height recovery as much as possible. Curves 1 and 3 show unpowered phugoid oscillations of *Archaeopteryx,* while curves 2 and 4 are with 3.42 W of flapping power (the estimated maximum available continuous power), beginning one second after jumpoff when air speed is high enough to lift wings during the upstroke without muscular effort. These curves show how *Archaeopteryx* could travel from treetop to treetop without expending much muscular energy.

Figure 12.13. (opposite page) *A, dorsal view of* Archaeopteryx, *showing two principal parameters for flight aerodynamics, wing span (b), and body length (l).* *B, flight paths of* Archaeopteryx *(for Mass = 0.2, 0.4, and 0.6 kg) taking off horizontally from perch at 2 m s^{-1}, then pulling up at a maximum continuous power (3.42, 5.43, and 7.11 Watts respectively). (After Chatterjee and Templin in press).* *C, phugoid gliding paths of* Archaeopteryx *at M = 0.2 kg, launching from a height of 15 m with launch speed 2 m s^{-1} (horizontal). Curve #1 shows an unpowered glide at relatively high lift coefficient (C_L ~1.13). One wave length later, about 18 m from launch, height passes through a maximum of 11.5 m (a loss of 3.5 m from launch). No pitch damping was used, in order to encourage the height recovery as much as possible. Curve #2 is the same initial glide up to 1 second after launch, when speed has increased to 8.4 m s^{-1}. Flapping flight is at maximum continuous power (3.42 W), and it continues for a further 2.5 seconds. A final short ascending glide puts the trajectory over the treetop at nearly original height, with speed reduced to 3.3 m s^{-1}. The horizontal distance covered is 22.3 m for an average horizontal component of speed of 5.95 m s^{-1}. The energy expended during 2.5 s of powered flight is 3.42 × 2.5 = 8.55 Joule. Curve #3 is another unpowered phugoid, at half the lift coefficient assumed in Curve #1. Consequently the height loss is considerably greater, but the horizontal distance covered is increased. The height recovery at the phugoid peak is less than for Curve #1. Curve #4 corresponds to #2, but with 3.42 W of flapping wing power on for a duration of 3.5 s. The energy expended is 12 Joule, and full height is recovered with a final minimum speed of 2.7 m s^{-1}.*

Heterochrony: The Key to Feather Evolution

Among all living vertebrates, feathers are a unique evolutionary novelty of birds. They provide the aerodynamic surface area necessary

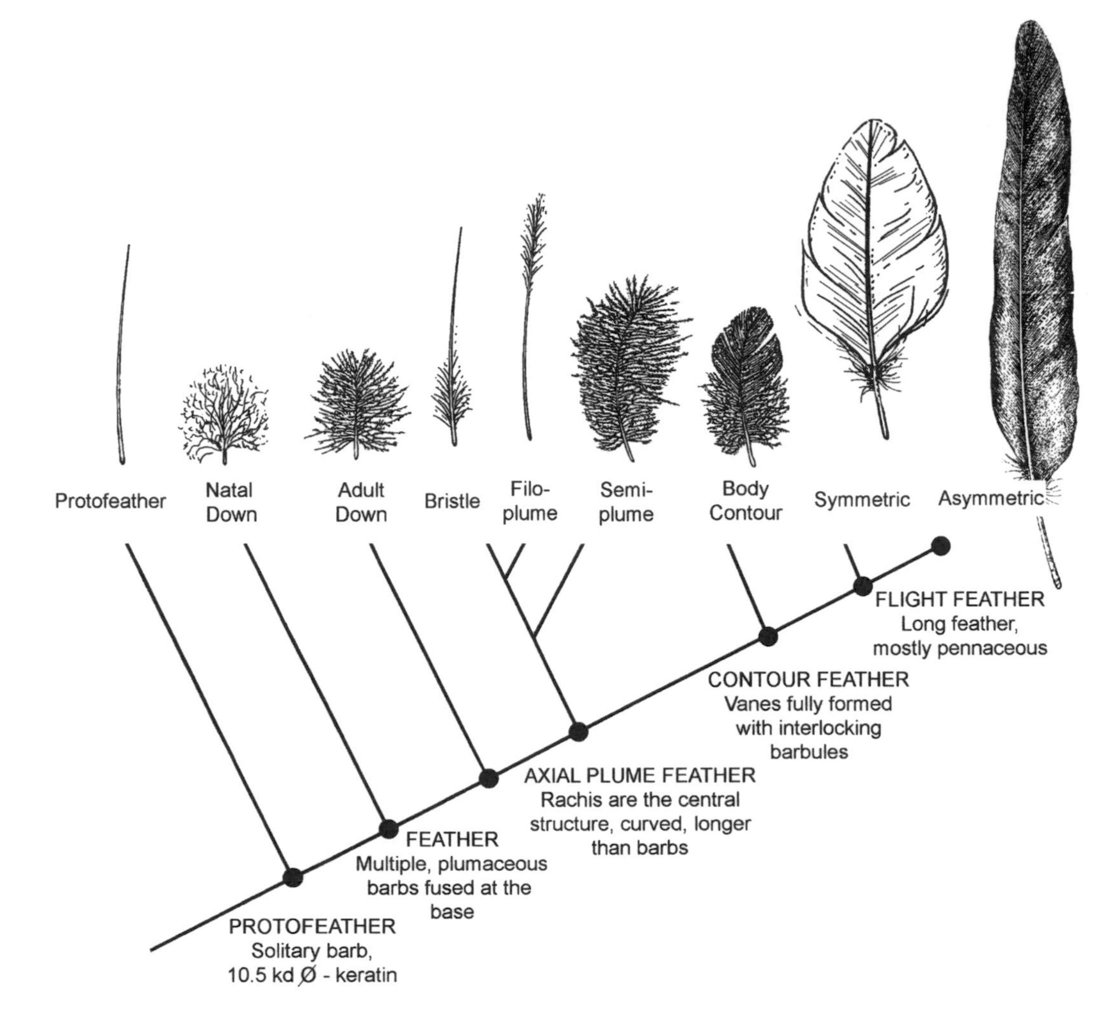

Figure 12.14. Heterochrony of feather evolution. Ontogenic evolution of feathers with increased morphological complexity from natal down to asymmetric flight feather. The most primitive structure may have been a protofeather similar to a solitary barb with 10.5 kd Ø-keratin. (Modified from Brush 2000)

for flight, insulation for maintaining body temperature, and coloration for courtship and camouflage. They also serve for streamlining of the body, cleaning of the plumage, sound production, water repellency, and chemical defenses (Bock 2000). There are six major types of feathers in volant forms: contour feathers, down feathers, semiplumes, filoplumes, powder feathers, and bristles (fig. 12.14A). Each group of feathers has a different biomechanical property and different design, and each feather is built to play a distinct role in the group. A typical contour feather combines a long central shaft with a broad flexible vane on either side of the shaft. It has a complicated branching pattern like fractals, where the central rachis gives rise to hundreds of parallel rows of barbs on either side, which are sloped toward the tip of the feather. The barb, in turn, branches out to a tiny row of very fine fibers (or barbules) on each side. Hooks on the barbules link neighboring barbs

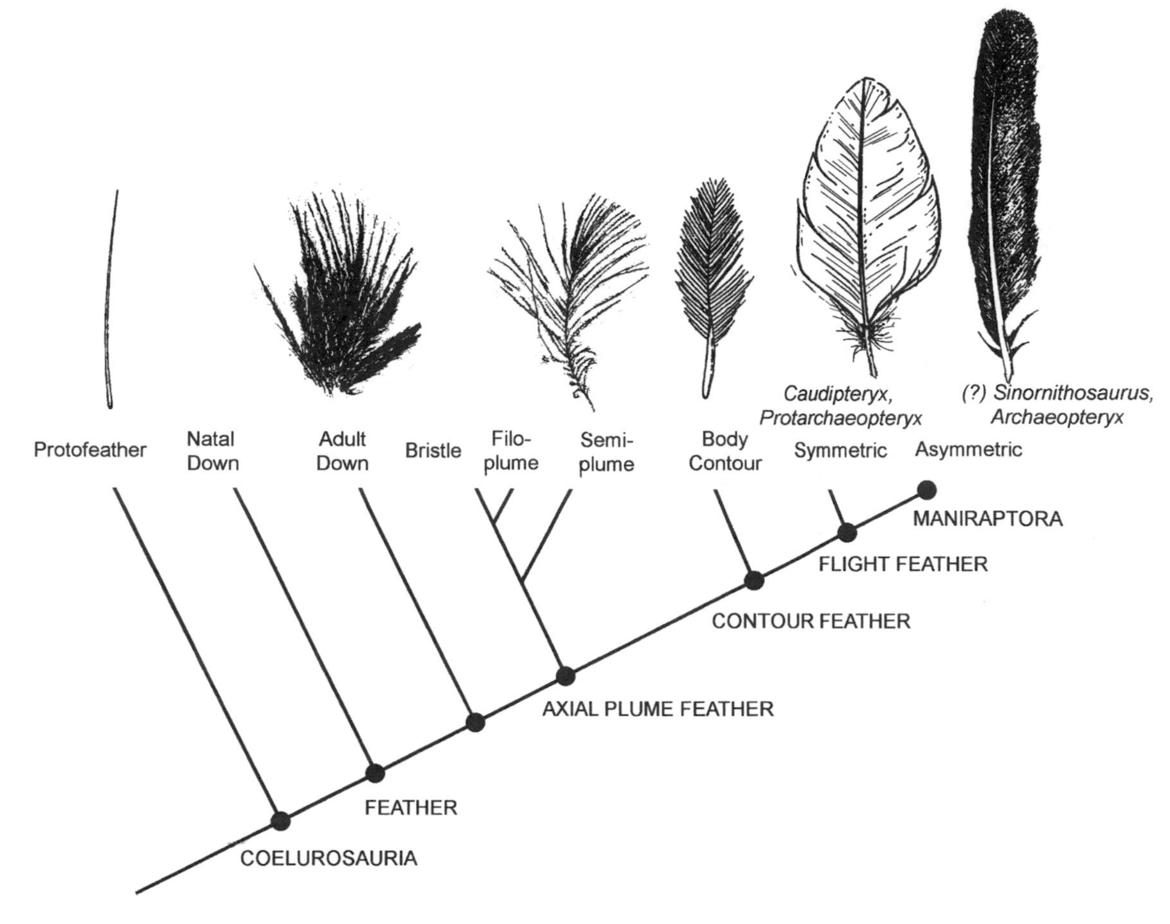

Figure 12.14. Heterochrony of feather evolution (cont.). Phylogenic evolution of feathers as reconstructed from fossil record. Future discovery of fossil feathers from non-avian dinosaurs may fill in important morphologic and phylogenetic gaps. (Data from Chatterjee 1997; Xu et al. 2001)

to form a web similar to a Velcro fastener. This gives the feather both strength and flexibility. This interlocking system characterizes the pennaceous feather structure in the distal part of the contour feather, making it a firm, bladelike, and airtight flight surface. Proximally, the contour feather loses this pennaceous structure and becomes plumaceous (i.e., soft, downy). Contour feathers are the basic vaned feathers on the body, wing (remiges), and tail (rectrices), with a central rachis; they grow only in the feather tracts (pterylae). Body contour feathers are smaller than the flight feathers with symmetrical vanes. Vanes are asymmetrical in primary flight feathers (remiges and rectrices), where the leading edge is narrower and stronger than the trailing edge. Moreover, the feather is convex on the top surface, but concave below. This asymmetrical design and differential curvature gives the flight feather an airfoil cross section for aerodynamic function. The semiplumes are intermediate in form between the more pennaceous contour feathers and the plumaceous downy feathers. They have a distinct rachis that is longer than the barb; the barb, in turn, lacks hooklets. The semiplumes usually lie under the contour feathers. The filoplumes are long, hairlike feathers at the base of the remiges and rectrices and act as sensors for their movements. It has a long shaft, but barbs are located more distally. The bristles are contour feathers without vanes, and are generally found around the eye. The barbs are located more proximally, opposite to the condition seen in filoplumes. The down feathers of an adult have short shafts and a loose, fluffy texture so that they may bend in all directions. They are usually placed under the contour feathers to provide insulation. Their soft, spreading plumes and many tiny enclosed airspaces make them fine insulators. The natal down, which covers the hatchling, is simpler in structure, with numerous barbs converging at a common point. The powder feathers are special vaned feathers with barbs that disintegrate into a fine powder which helps in grooming and waterproofing. In flightless birds, the vanes are symmetrical and simplified for insulation. These feathers have become degenerate, lost their pennaceous structure and aerodynamic design, and become fluffy and hairlike in structure, similar to down feathers.

Fossil feathers are rare in the paleontological record. Because feathers decay and disintegrate easily, and are lighter than water, the preservation of this soft, delicate structure requires an unusual setting such as the calm water of a lacustrine (Liaoning, China) or a lagoonal (Solnhofen, Germany) environment, and subsequent rapid burial without turbulence. The fossil record indicates that feathers covered the entire body in protobirds, with regional differentiation. The oldest possible feather impression is known from the Early Jurassic Portland Formation of Massachusetts in the form of semiplume, a feather type intermediate between contour feather and down (Gierlinski 1997). Unfortunately, these feather impressions are associated with squatting footprints of a non-avian theropod, whose identity is unknown. Chatterjee reviewed (1997) other types of isolated feathers from the Jurassic and Cretaceous sediments, such as adult down feathers from Brazil; semiplume feathers from Brazil, Lebanon and New Jersey; and body contour feathers from Australia. Asymmetrical flight feathers are known in *Archaeopteryx, Confuciusornis,* and *Ambiortus*. The absence of bristles in the fossil record may be a preservational artifact.

Feathers are the most complex form of integuments known in vertebrates characterized by structural diversity at hierarchical development. As a keratinous product, they grow from pits (or follicles) in the epidermal layer of the skin. Keratin is a mechanically a tough, lightweight material that makes an ideal flight surface. However, feathers are dead, horny structures that deteriorate with time; once a year birds must renew their contour feathers, usually after breeding season, to make them airworthy.

The biological and phylogenetic origin of feathers remains enigmatic. It is generally believed that feathers originated from reptilian scales, because they share similar keratinous tissue (Rawles 1963; Maderson 1972; Maderson and Alibardi 2000). However, recent molecular data suggest that reptilian scales and avian feathers are not homologous. The latter contains a unique protein molecule, such as a larger (14.4 kd) o-keratin and a follicular mechanism for production and assembly (Brush 1996, 2000). The partial or complete replacement of avian foot scales by down feathers (ptilopody) during growth may indicate that down feathers are primitive and that contour feathers are derived (Rawles 1963). Recent experiments indicate that blockage of bone morphogenetic protein (BMP) causes the foot scales of the chick to develop into feathers (Zou and Niswander 1996).

Traditionally, feathers have been used as a defining characteristic of birds. In fact, the avian identity of *Archaeopteryx* has never been questioned since its discovery because both the London and Berlin specimens contain beautiful feather impressions of remiges, rectrices, downs, and coverts. They are similar in microstructure to those of modern birds in every detail. The rachis supports the asymmetrical vane on either side; each vane forms interlocking barbules that end in hooks. As a result, the *Archaeopteryx* feather does not provide any clues to the precursor of feathers. However, recent discovery of feathers in non-avian coelurosaurs clearly suggest that Aves cannot be diagnosed by the presence of feathers. Several interesting conclusions can be made from the Chinese discoveries: (1) feathers are not unique to birds (Aves), but evolved earlier in their lineage, perhaps near the base of Coelurosauria; (2) the origin of feathers predates the origin of avian flight; (3) feathers were exapted for other functions such as insulation, parachuting, and gliding before flight; (4) different stages of feather evolution must have preceded before the *Archaeopteryx*-stage.

Two major theories for the evolution of feathers have been presented: (1) Feathers evolved originally in connection with temperature control by providing an insulating layer of plumage in the cooler environment of trees, and they were later co-opted for flight (Regal 1975; Ostrom 1979; Bock 1986). (2) Feathers evolved in connection with flight by providing an increased airfoil (Heilmann 1926; Parkes 1966; Feduccia 1996). Because feathers are designed for flight, it is not surprising that much speculation about the origin of feathers is linked to the origin of flight. However, recent discovery of a series of feathered coelurosaurs from China settles the issue that feathers originally evolved for insulation and were later co-opted for flight.

Heterochrony may provide an additional clue to the origin of feathers. It is the mechanism by which a species gives rise to a descendant species through changes in the timing or rate of developmental events relative to the ancestral condition (McKinney and McNamara 1991). Heterochrony provides the raw material upon which natural selection works. Lucas and Stettenheim (1972) show that the development of different kinds of feathers occurs in a hierarchical fashion in successive generations. The natal down appears in the first generation during embryonic life, which is followed by a second-generation adult down, then plumaceous, and finally contour feathers. Recently, Brush (2000) argues that heterochrony may have played a key role in feather evolution. He suggests that feather evolution may have been a peramorphic trend with increased complexity where ontogeny of feathers may recapitulate their phylogeny (fig. 12.14). Evidence of ptilopody, ontogeny, and the fossil record suggests that primitive feathers were relatively simple structures resembling a natal down feather.

Six stages of development of feathers can be reconstructed from ontogeny and phylogeny (fig. 12.14B). The most primitive structure of protofeathers may have been a simple, conical, cylindrical filament, as documented in *Sinosauropteryx* and *Beipiaosaurus*. The latter had a central core and branching pattern at the distal end, similar to a solitary barb. A thick coat of protofeathers could have functioned for insulation and streamlining of the body contours. From this ancestral condition, the second stage of feather evolution may correspond to natal down, the first generation of feathers in ontogeny. This stage is represented by *Sinornithosaurus* in the fossil record (Xu et al. 2001). The third stage of feather evolution may resemble adult down, with the development of a small rachis and barbs. This stage is reported in *Protarchaeopteryx, Caudipteryx,* and *Sinornithosaurus* (Xu et al. 2001). The fourth stage of feather development is the formation of a central shaft such as in bristle, semiplume, and filoplume with plumose barbs, where the rachis is longer than the barbs. This design increases the length and surface area of individual feathers to give a smooth, continuous, and streamlined body cover with aerodynamic design. This stage is represented by *Sinornithosaurus* (Xu et al. 2001). The fifth stage is the development of symmetrical contour feathers where vanes are fully formed with an interlocking mechanism. The fossils of *Protarchaeopteryx* and *Caudipteryx* have symmetrically vaned feathers in remiges and rectrices (Ji et al. 1998). Recent discovery of a *Sinornithosaurus*-like dromaeosaur shows similar contour feathers in rectrices and remiges (Ji et al. 2001). The sixth stage is the development of asymmetrical vane with robust rachis in rectrices and remiges for aerodynamic function. This stage is represented by *Archaeopteryx, Confuciusornis* and various other Chinese fossil birds. Using this ontogenetic relationship as a guide, the phylogeny of feathers is reconstructed in Figure 12.14B. The cladogram indicates that there are some structural and morphological gaps in the fossil record, which may indicate preservational artifacts.

Acknowledgments. The invitation and hospitality from Martin Shugar to participate in the Florida Dinosaur-Bird Symposium are greatly appreciated. We thank Xing Xu for allowing examination of the fabulously preserved feathered coelurosaurs from China, and Zhonghe Zhou for arranging a field trip to the famous Liaoning fossil locality. We thank Cecilia Carter, Kyle McQuilkin, and two anonymous reviewers for critically appraising the manuscript, and for making helpful suggestions. We thank Kyle McQuilkin for his skillful illustrations and computer graphics, John Ruben and Xing Xu for providing some slides of feathered coelurosaurs, Alan H. Brush and James A. Hopson for providing copies of their unpublished manuscripts, and Chris Scotese for providing the plate tectonic map. The research is supported by Texas Tech University.

References

Ackerman, J. 1998. Dinosaurs take wing. *National Geographic* 194: 74–99.

Alexander, R. Mc. 1992. *Exploring biomechanics.* New York: Scientific American Library.

Balda, R. P., G. Caple, and R. R. Wills. 1985. Comparisons of the gliding and flapping sequence with the flapping to gliding sequence. In M. K. Hecht, J. H. Ostrom, G. Viohl, and P. Wellnhofer (eds.), *The beginnings of birds,* pp. 267–277. Eichstätt: Freunde des Jura-Museums, Eichstätt.

Barthel, K. W., N. H. M. Swinburne, and S. Conway Morris. 1990. *Solnhofen.* New York: Cambridge University Press.

Bock, W. J. 1985. The arboreal theory for the origin of birds. In M. K. Hecht, J. H. Ostrom, G. Viohl, and P. Wellnhofer (eds.), *The beginnings of birds,* pp. 199–207. Eichstätt: Freunde des Jura-Museums.

———. 1986. The arboreal origin of avian flight. In K. Padian (ed.), *The origin of birds and the evolution of flight,* pp. 57–72. San Francisco: California Academy of Sciences.

———. 2000. Explanatory history of the origin of feathers. *American Zoologist* 40: 478–485.

Brush, A. H. 1996. On the origin of feathers. *Journal of Evolutionary Biology* 9: 131–142.

———. 2000. Evolving a protofeather and feather diversity. *American Zoologist* 40: 631–639.

Burgers, P., and L. M. Chiappe. 1999. The wing of *Archaeopteryx* as a primary thrust generator. *Nature* 399: 60–62.

Chatterjee, S. 1997. *The rise of birds.* Baltimore: Johns Hopkins University Press.

Chatterjee, S., and R. J. Templin (2003). The flight of *Archaeopteryx. Naturwissenschaften* 90: 27–32.

Chen P. J., Dong Z.-M., and Zhen S. 1998. An exceptionally well-preserved theropod dinosaur from the Yixian Formation of China. *Nature* 391: 147–152.

Elzanowski, A. 2000. The flying dinosaurs. In G. S. Paul (ed.), *The Scientific American book of dinosaurs,* pp. 169–182. New York: Byron Press.

Feduccia, A. 1993. Evidence from claw geometry indicating arboreal habits of *Archaeopteryx. Science* 259: 790–793.

———. 1996. *The origin and evolution of birds.* New Haven: Yale University Press.

Feduccia, A., and H. B. Tordoff. 1979. Feathers of *Archaeopteryx:* asymmetric vanes indicate aerodynamic function. *Science* 203: 1021–1022.

Garner, J. P., G. K. Taylor, and A. L. R. Thomas. 1999. On the origin of birds: the sequence of character acquisition in the evolution of avian flight. *Proceedings of the Royal Society of London* B 266: 1259–1266.

Gatesy, S. M. 1990. Caudofemoral musculature and the evolution of theropod locomotion. *Paleobiology* 16: 170–186.

Gatesy, S. M., and K. P. Dial. 1996. Locomotor modules and the evolution of avian flight. *Evolution* 50: 331–340.

Gatesy, S. M., and K. M. Middleton. 1997. Bipedalism, flight, and the evolution of theropod locomotor diversity. *Journal of Vertebrate Paleontology* 17: 308–329.

Gauthier, J. 1986. Saurischian monophyly and the origin of birds. In K. Padian (ed.), *The origin of birds and the evolution of flight,* pp. 1–55. San Francisco: California Academy of Sciences.

Gauthier, J., and K. Padian. 1985. Phylogenetic, functional and aerodynamic analysis of the origin of birds and their flight. In M. K. Hecht, J. H. Ostrom, G. Viohl, and P. Wellnhofer (eds.), *The beginnings of birds,* pp. 185–197. Eichstätt: Freunde des Jura Museums, Eichstätt.

Geist, N. R., and A. Feduccia. 2000. Gravity-defying behaviors: identifying models for Protoaves. *American Zoologist* 40: 665–685.

Gierlinski, G. 1997. What type of feathers could non-avian dinosaurs have,

according to an early Jurassic ichnological evidence from Massachusetts. *Przeglad Geologiczny* 45: 419–422.
Heilmann, G. 1926. *The origin of birds*. London: Witherby.
Homberger, D. G., and K. N. de Silva. 2000. Functional microanatomy of the feather-bearing integument: implications for the evolution of birds and avian flight. *American Zoologist* 40: 553–574.
Hopson, J. A. (2001). Ecomorphology of avian and non-avian theropod phalangeal proportions: implications for the arboreal versus terrestrial origin of flight. In J. Gauthier and L. F. Gall (eds.), *New perspectives on the origin and early evolution of birds*, pp. 211–235. New Haven: Yale University Press.
Jerison, H. L. 1973. *Evolution of the brain and intelligence*. New York: Academic Press.
Ji Q., P. J. Currie, M. A. Norell, and Ji S.-A. 1998. Two feathered dinosaurs from northeastern China. *Nature* 393: 753–761.
Ji Q., M. A. Norell, Gao K.-Q., Ji S.-A., and Ren D. 2001. The distribution of integumentary structures in feathered dinosaurs. *Nature* 410: 1084–1088.
Lucas, A. M., and P. R. Stettenheim. 1972. *Avian anatomy.* Washington, D.C.: U.S. Government Printing Office.
Luo, Z. 1999. A refugium for relicts. *Nature* 400: 23–25.
Maderson, P. F. A. 1972. On how an archosaurian scale might have given rise to an avian feather. *American Naturalist* 176: 424–428.
Maderson, P. F. A., and L. Alibardi. 2000. The development of the sauropsid integument: a contribution to the problem of the origin and evolution of feathers. *American Zoologist* 40: 513–529.
Marsh, O. C. 1880. Odontornithes: a monograph on the extinct toothed birds of North America. *Report of the geological exploration of the fortieth parallel* 7: 1–201.
McKinney, M. L., and K. J. McNamara. 1991. *Heterochrony.* New York: Plenum Press.
Nopsca, F. 1923. On the origin of flight of birds. *Proceedings Zoological Society of London* 1907: 223–236.
Norberg, U. M. 1990. *Vertebrate flight.* Berlin: Springer-Verlag.
Olson, S. L., and A. Feduccia. 1979. Flight capability and pectoral girdle of *Archaeopteryx. Nature* 278: 247–248.
Ostrom, J. H. 1976. *Archaeopteryx,* and the origin of birds. *Biological Journal of the Linnean Society* 8: 91–182.
———. 1979. Bird flight: how did it begin? *American Scientist* 67: 46–56.
———. 1986. The cursorial origin of avian flight. In K. Padian (ed.), *The origin of birds and the evolution of flight,* pp. 73–81. San Francisco: California Academy of Sciences.
Padian, K., and L. M. Chiappe. 1998. The origin of birds and their flight. *Scientific American* 278(2): 38–47.
Parkes, K. C. 1966. Speculations on the origin of feathers. *Living Bird* 5: 77–86.
Poore, S. O., A. Sanchez-Harman, and G. E. Goslow, Jr. 1997. Wing upstroke and the evolution of flapping flight. *Nature* 387: 799–802.
Rawles, M. E. 1963. Tissue interactions in scale and feather development as studied in dermal-epidermal recombinations. *Journal of Embryology and Experimental Morphology* 11: 765–789.
Rayner, J. M. V. 1991. Avian flight and the problem of *Archaeopteryx*. In J. M. V. Rayner and R. J. Wooton (eds.), *Biomechanics and evolution,* pp. 183–212. New York: Cambridge University Press.

Regal, P. J. 1975. The evolutionary origin of feathers. *Quarterly Review of Biology* 50: 35–66.
Ruben, J. A., and T. D. Jones. 2000. Selective factors associated with the origin of fur and feathers. *American Zoologist* 40: 585–596.
Scotese, C. R. 1991. Jurassic and Cretaceous plate tectonic reconstructions. *Palaeogeography, Palaeoclimatology, Palaeoecology* 87: 493–501.
Sumida, S. S., and C. A. Brochu. 2000. Phylogenetic context for the origin of feathers. *American Zoologist* 40: 486–503.
Tarsitano, S. F., A. P. Russell, F. Horne, C. Plummer, and K. Millerchip. 2000. On the evolution of feathers from an aerodynamic and constructional view point. *American Zoologist* 40: 676–686.
Templin, R. J. 1977. Size limits in flying animals. In M. K. Hecht, P. Goody, and B. M. Hecht (eds.), *Major patterns in vertebrate evolution,* pp. 411–421. New York: Plenum Press.
———. 2000. The spectrum of animal flight: insects to pterosaurs. *Progress in Aerospace Sciences* 36: 393–436.
Vasquez, R. J. 1992. Functional osteology of the avian wrist and evolution of flapping flight. *Journal of Morphology* 211: 259–268.
Williston, S. W. 1879. Are birds derived from dinosaurs? *Kansas City Review of Sciences* 3: 457–460.
Xu X., Tang Z.-L., and Wang X. 1999a. A therizinosaurid dinosaur with integumentary structure from China. *Nature* 399: 350–354.
Xu X., Wang X.-L., and Wu X.-C. 1999b. A dromaeosaurid dinosaur with a filamentous integument from the Yixian Formation of China. *Nature* 401: 262–266.
Xu X., Zhou Z., and Wang X.-L. 2000. The smallest known non-avian theropod dinosaur. *Nature* 408: 705–708.
Xu X., Zhou Z., and R. O. Prum. 2001. Branched integumental structures in *Sinornithosaurus* and the origin of feathers. *Nature* 410: 200–204.
Yalden, D. W. 1985. Forelimb function in *Archaeopteryx*. In M. K. Hecht, J. H. Ostrom, G. Viohl, and P. Wellnhofer (eds.), *The beginnings of birds,* pp. 91–97. Eichstätt. Freunde des Jura-Museums, Eichstätt.
Zou H., and L. Niswander. 1996. Requirement for BMP signaling in interdigital apoptosis and scale formation. *Science* 272: 738–741.

13. The Plumage of *Archaeopteryx*: Feathers of a Dinosaur?

PETER WELLNHOFER

Abstract

The discovery of "feathered" dinosaurs from the Early Cretaceous of China revitalized the long disputed question of whether *Archaeopteryx* from the Late Jurassic of Germany was avian or just another feathered dinosaur. In addition to an isolated feather, feathers can be identified in virtually all seven known specimens of *Archaeopteryx;* however, they are in different states of preservation and completeness. A comparative study suggests that in all individuals of *Archaeopteryx* the flight feathers are uniform in structure, pattern, and arrangement and, except for the feathered tail, the plumage was definitely modern and adapted for active, powered flight. This implies that the feathers of *Archaeopteryx* are already highly evolved, advanced avian structures that must have an evolutionary history prior to the Late Jurassic. In contrast to the predominantly theropodan postcranial skeleton of *Archaeopteryx,* some skull bones and especially the ulnar abduction in the wrist can be diagnosed as specifically avian, which supports the conventional view that *Archaeopteryx* was a true bird.

The filamentous integumentary structures preserved in several theropodan taxa from the Lower Cretaceous Yixian Formation of Liaoning, China, may have evolved independently and convergently. They occur in archosaurian lineages, such as the pterosaurs, which were not closely related to dinosaurs, and need not be interpreted as protofeathers. However, whether animals like *Caudipteryx* from the same deposits in China, having short feathered arms and living at least 25 million years after *Archaeopteryx,* should be classified as maniraptoran theropods or as secondarily flightless birds is basically a problem of definition that possibly may never be resolved. No matter which assumption is correct, it does not invalidate the supposed close relationship between theropods and birds, but indicates an early evolutionary history of Aves that is much more complex than hitherto assumed.

Introduction

Until recently, feathers have been considered the key character of birds, a structure unique to the class Aves. It was John Ostrom who has told us repeatedly that if no feathers had been found with the skeletons of *Archaeopteryx* from the Solnhofen limestone, this animal would have been identified not as a bird, but rather as a small coelurosaurian dinosaur. Now, the dogma that feathers diagnose a bird has been challenged. *Protarchaeopteryx, Caudipteryx,* and others (Early Cretaceous of China) display obvious featherlike structures, or even feathers attached to the forelimbs and the long vertebral tail. In contrast, based on osteological analysis, these taxa have been interpreted as maniraptoran theropods (Ji et al. 1998). The exceptional fossilization of soft parts in vertebrates of the Lower Yixian Formation of Liaoning has facilitated the preservation of integumentary body covering, be it called filaments, fibers, or protofeathers, in other taxa, such as *Sinosauropteryx prima* (Ji and Ji 1996; Chen et al. 1998; Currie and Chen 2001; Padian et al. 2001), *Sinornithosaurus millenii* (Xu et al. 1999b, 2001), *Beipiaosaurus inexpectus* (Xu 1999a), and an as yet undetermined dromaeosaur (Ji et al. 2001).

These new findings reopen the question of whether the Late Jurassic *Archaeopteryx,* generally considered to be the oldest bird from the fossil record, can still be classified as a member of the class Aves. In other words, is *Archaeopteryx* a dinosaurlike bird, or rather a birdlike, i.e., feathered, dinosaur? Was Andreas Wagner right after all, when in 1861 he took the first example, later called the London specimen, to be a saurian that he christened *Griphosaurus,* "riddle lizard" (Wagner 1861)? There is no doubt that the skeleton of *Archaeopteryx* principally has a coelurosaurian skeletal bauplan, despite some avian traits. But which bird characters of *Archaeopteryx* should be diagnostic and significant, if not the feathers? Will the structure of its feathers help to solve this problem? And, generally, should the boundary line between birds and reptiles be drawn before or after *Archaeopteryx*?

Feathers in the Fossil Record

Although feathers decay readily, fossil preservation is not uncommon. Davis and Briggs (1995) reported isolated fossil feathers in Jurassic, Cretaceous, and Tertiary deposits from 39 localities. In 77 percent of the sites yielding bird feathers they found that they are in fact the only evidence of an avian presence, and bones are absent. In most cases the fossil feathers are preserved as carbonized traces, as in the lacustrine deposits of the early Eocene Green River Formation in Wyoming and

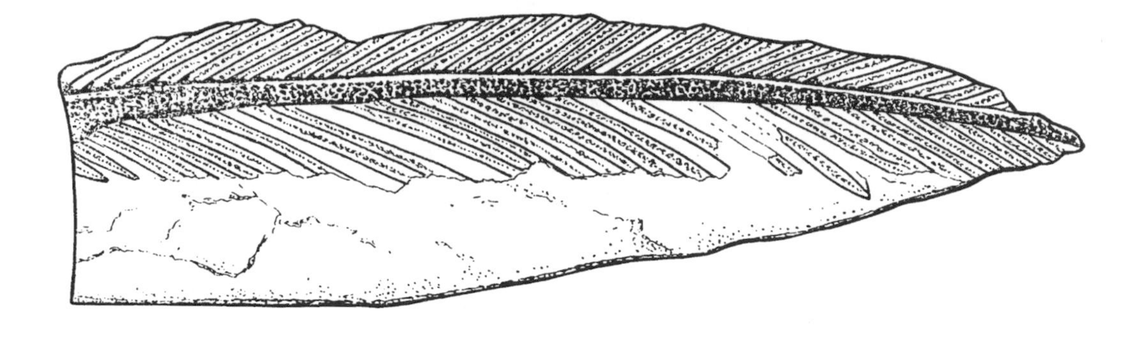

Figure 13.1. Praeornis sharovi *Rautian, Late Jurassic Karabastau Formation, Karatau, Kazakhstan, type specimen, Paleontological Institute Moscow, PIN 2585/32, as figured by Rautian (1978) and interpreted as an avian feather. However, it is considered here as a cycad leaf, an opinion expressed already by Bock (1986) and Doludenko et al. (1990). Scale bar 50 mm.*

Figure 13.2. (opposite page) The seven skeletal specimens of Archaeopteryx *from the Lower Tithonian Solnhofen limestone of Bavaria, Germany: The London, Berlin, "Maxberg," Haarlem, Eichstätt, Solnhofen, and Munich specimens, discovered and/or described in this sequence. All seven skeletal specimens known to date have feathers documented, although in different quality of preservation. Drawn to the same scale.*

the Oligocene of Cereste in France, or in the restricted marine environment of the early Cretaceous Santana formation in northeastern Brazil, where even the original color pattern is preserved (Martill and Frey 1995). A second mode of preservation originates from bacterial autolithification, as is documented in fossil birds from the early Cretaceous Yixian formation in China and the early Eocene lake deposits of Messel in Germany. The best possible preservation of feathers is as inclusions in amber that are known from six amber deposits ranging in age from Early Cretaceous to Early Oligocene.

The geologically oldest feathers are still of Late Jurassic age, documented in *Archaeopteryx*. However, Rautian (1978) described a fossil from the Upper Jurassic lake deposits of the Karatau Range in Kazakhstan as the primitive feather of a bird (fig. 13.1) which he named *Praeornis sharovi,* including it in a new family, order, and even subclass of its own. It is interesting to note that the Karatau fossil sites have become famous for the "hairy" pterosaur *Sordes pilosus* (Sharov 1971). The fragment of the *Praeornis* "feather" is 15 cm long, and shows a central "rachis" and the proximal portion of flattened and broad "barbs." So, the original shape is not known in its outline. From the parts preserved, however, it would appear to be a very large bird feather, much larger than the feathers in *Archaeopteryx* or in any other Mesozoic bird. Considering that the same deposits in the Karatau Range have yielded a rich fossil flora, I believe that the alleged feather of *Praeornis* was in fact a cycad leaf, and I concur with this opinion expressed already by Bock (1986) and Doludenko et al. (1990). Rather than being an avian feather it appears to be morphologically comparable to leaves that have been described as *Paracycas harrisii* by Doludenko and Orlovskaya (1976: pl. 62). This would leave the feathers of *Archaeopteryx* as the only known feathers of Jurassic age, and the oldest from the fossil record.

The Feathers of *Archaeopteryx*

In addition to the isolated feather (Meyer 1861), feathers are documented for all seven skeletal specimens of *Archaeopteryx* (fig.

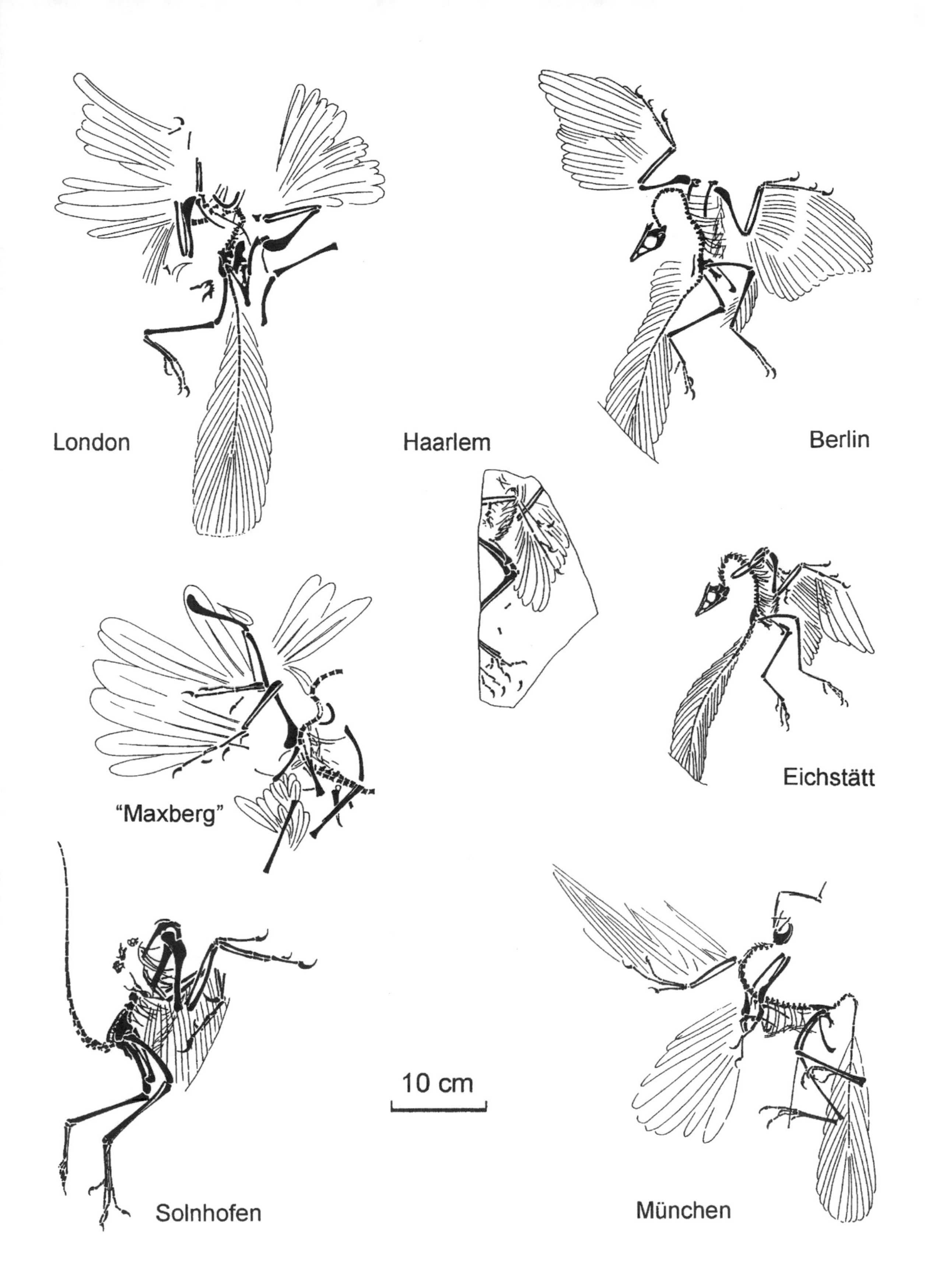
London
Haarlem
Berlin
Eichstätt
"Maxberg"
10 cm
Solnhofen
München

13.2). This should be sufficient evidence against the claim that the feathers of the London and Berlin specimens were forgeries (Hoyle and Wickramasinghe 1986). However, the quality of preservation is different, depending on taphonomic, sedimentological, and diagenetic conditions.

The Isolated Feather

The best-preserved *Archaeopteryx* feather is still the isolated feather (fig. 13.3) found in the Solnhofen quarry in 1860, and described in detail by Hermann von Meyer (1862). The main slab is housed in the Natural History Museum in Berlin, the counterslab in the Bavarian Paleontological State Collections in Munich. Von Meyer introduced the name *Archaeopteryx lithographica* (Meyer 1861). There has been some discussion about whether he intended to name the animal to which this single feather belonged, or the feathered skeleton found shortly after the isolated feather, the London specimen. This taxonomic problem is not relevant here. According to Griffiths (1996) this single feather was a distal secondary flight feather. With its length of about 6 cm it is actually too small to fit in a wing of any of the known skeletal specimens of *Archaeopteryx*. So, it came from a smaller individual, a juvenile, or a different taxon. The latter might be the case, because the feather shows a clearly angled clip distally, not known in the other specimens. The barbs are well developed and have separated at several locations. By close examination barbules also can be identified. At the base of the feather the barbs are arranged in tufts, not hooked together as in the distal vane. Griffiths speculated that this might have served to trap air against the skin, and so to keep the body temperature high, as in modern birds. This supports the idea that *Archaeopteryx* might have been endothermic.

Most unusual, however, is the preservation of this feather. It is not represented as an imprint or as a cast (Rietschel 1985) in the matrix, as is the case for the feathers of the other specimens. But whether it is composed of carbonized matter or of manganese oxide (MnO_2) has not been determined. MnO_2 is rather common in the inorganic dendrites in the Solnhofen lithographic limestone. If this was indeed a flight feather, it can be concluded that it was probably a molted feather and had not separated from the wing following postmortem decay, because “the ligaments that attach the wing contour feathers to the ulna are decay resistant” (Davis and Briggs 1995, p. 783). This is also shown by the preservation of the feathered wings of the other specimens. The difference in feather preservation may be related to the taphonomic and diagenetic history that we cannot yet reconstruct in detail. This single feather is structurally very similar to those of modern birds. Its asymmetrical shape suggests that it was adapted for flight (Feduccia and Tordoff 1979), and the plumaceous base indicates an insulatory function (Griffiths 1996).

The London Specimen

When the first skeleton of *Archaeopteryx* was discovered in 1861,

Figure 13.3. Archaeopteryx lithographica *H. v. Meyer, isolated feather from the Lower Tithonian Solnhofen limestone, found in 1860 near Solnhofen, Bavaria, Germany. Shown is the main slab housed in the Berlin Museum für Naturkunde. Total length 60 mm.*

it was evident that the plumage appeared modern, because the impressions of both feathered wings and a feathered tail were preserved. The structure and arrangement of the plumage of this London specimen were studied in detail by de Beer (1954). He recognized that the wings showed the imprints of the ventral surface. This is documented by the longitudinal groove of the rachis between two sharp ridges which indicate the ventral side of the feathers in modern birds. De Beer interpreted the peculiar "double-struck" impressions of feathers as follows: "the impression of the rachis had been made first and then, by displacement of the feather, the impressions of the barbs had been superimposed on that impression" (de Beer 1954, p. 34). Rietschel (1985) interpreted these "double-struck" shafts as an early diagenetic structure of underlying feathers, and called them "shaft-shadows" of original feathers. This has consequences with respect to the reconstruction of the wing feather count. So, de Beer identified only six primary and ten secondary flight feathers in each wing, in addition to the impressions of seven undercoverts, about half the length of the corresponding primaries. The remiges are about 130 mm long with asymmetrical vanes similar to the single feather of *Archaeopteryx* and to modern flight feathers. In many locations the feather barbs can clearly be distinguished, and their smooth parallel arrangement suggests the presence of barbules with hamuli (or hooklets).

The imprints of the rectrices on the long vertebral tail, which is stiffened by elongate zygapophyses and chevrons, are well preserved. Each pair corresponds to one caudal vertebra attached to the lateral surface, from the 6th to the 20th caudals. The proximal rectrices are about 60 mm long, increase to about 120 mm at the 17th caudal, and decrease thereafter to a length of about 90 mm at the terminal caudal. The feathered tail is about 300 mm long and expands to about 80 mm in width. The rectrices have more or less symmetrical vanes. De Beer also mentioned coverts covering the bases of the rectrices.

The "Maxberg" Specimen

In the "Maxberg" specimen, now unfortunately lost, Heller (1959) reported the presence of the primary and secondary flight feathers in this partly disarticulated and incomplete specimen. The skull and the vertebral tail are missing. However, traces of coverts in the area of the second metacarpal and its digit could be distinguished. The primaries were attached to the second metacarpal and to the first two phalanges of the second digit, leaving only the claw free of feathers. This is confirmed by the other specimens of *Archaeopteryx* in which the second digit is preserved in its natural articulation, i.e., the Berlin, Eichstätt, and Munich specimens. Impressions of smaller contour feathers near the left tibia indicate the presence of crural feathers that covered the tibial portion of the leg, forming so-called "breeches," as is documented also in the London, Berlin, and Eichstätt specimens.

The Haarlem Specimen

The very fragmentary Haarlem specimen of *Archaeopteryx* is known to have initially been described as a pterodactyl by von Meyer (1857) as "*Pterodactylus crassipes.*" Although he was puzzled by the obscure impressions on the surface of the slab, he took them to be those of the wing membrane of a pterosaur. Only Ostrom (1970) recognized them as feather imprints which belonged to the skeletal remains of *Archaeopteryx*. He identified the rachii of the secondary flight feathers of the left wing, radiating out from the left ulna. Finer details, however, are too faint to detect (Ostrom 1972).

The Eichstätt Specimen

In the Eichstätt specimen the impressions of feathers are visible only under low-angle illumination. The wings and the feathered tail appear to be rather blurred in their outlines. Only four primaries, about 80 mm long, can be distinguished. This series is certainly incomplete, but they overlap with the secondaries at a pronounced angle in consequence of the abduction of the hand. Faint impressions of contour feathers in front of the tibia indicate similar "breeches" as observed in the London, Berlin, and Maxberg examples. The vertebral tail seems to have been devoid of rectrices at the proximal 5 or 6 caudals. This matches the section of the caudal vertebral column that had more flexibility than the more distal section of the stiffened vertebral tail. The plumage of the Eichstätt specimen, which is the smallest and, according to Wellnhofer (1974), a juvenile individual of *Archaeopteryx lithographica,* is comparable in structure and arrangement with the other specimens.

The Solnhofen Specimen

In the Solnhofen specimen the evidence of feathers is restricted to mere impressions of feather shafts, which appear as a series of curved ridges originating from the left ulna, indicating the shafts of about 13 secondaries. Under low-angle light imprints subparallel to the second digit seem to mark the outline of the left feathered wing. No feathers can be recognized on the right wing or the tail. This is possibly an artifact of preservation or—more likely—of preparation (Wellnhofer 1988a,b).

The Munich Specimen

The specimen originally introduced as "the seventh specimen of *Archaeopteryx*" or the "Specimen of the Solenhofer Aktien-Verein" in 1993 was purchased by the Bavarian State Collection of Paleontology and Historical Geology (now Bayerische Staatssammlung für Paläontologie und Geologie) in Munich in 1999. Therefore, and according to the unwritten rule that the individual fossils of *Archaeopteryx* should bear the trivial name after the location of their depositories, this example can now be called "The Munich specimen of *Archaeopteryx*." It was discovered in 1992 and subsequently described as a new species, *Archaeopteryx bavarica,* by Wellnhofer (1993), distinct from the other six specimens which, in my opinion, can be assigned to *Archaeopteryx lithographica.* This specimen also shows the impressions of feathers, in places even in fine detail. Primaries and secondaries are present in both wings. Rectrices occur along the vertebral tail. In the right wing the primaries are fanned out; in the left wing they are more closely connected and superimposed on each other, terminating in a narrow, pointed wing tip. The sixth primary feather is about 120 mm in length and seems to have been the longest flight feather. This is almost as long as the flight feathers in the London specimen (maximum length 130 mm), which was a considerably larger individual. It is also relatively longer (overall) than in the Berlin specimen (Rietschel 1985). In the number of primaries the Munich specimen seems to agree with the Berlin specimen in which Helms (1982) and Rietschel (1985) have found 11, and Stephan (1987) possibly 12, primaries. However, in the Munich specimen the number of secondaries cannot be determined unequivocally. In some parts of the right wing the feather shafts are imprinted rather sharply and reveal a longitudinal groove indicating the ventral aspect of the feathers, as well as finer structures of the vanes, showing the barbs.

In the vertebral tail the most proximal pair of feathers visible are attached to the 9th caudal, leaving a relatively long proximal postsacral section of the caudal column (the flexing zone of the tail) free of at least longer and stronger tail feathers. If there were short proximal caudal feathers, as shown in the Berlin specimen, they were not preserved here. The rectrices are about 90 mm in length. They are significantly longer, i.e., 53 percent of the caudal vertebral column, than in the larger London specimen in which they contribute only 48 percent. The complete length of the tail, including the rectrices, is about 260 mm in the Munich specimen, compared to about 300 mm in the London specimen, or 87 percent, whereas its individual size based on the length of the humerus is only 66 percent of the size of the London specimen (Wellnhofer 1993). This suggests that *A. bavarica* was also characterized by feathered wings and tail that were relatively longer than in *A. lithographica.*

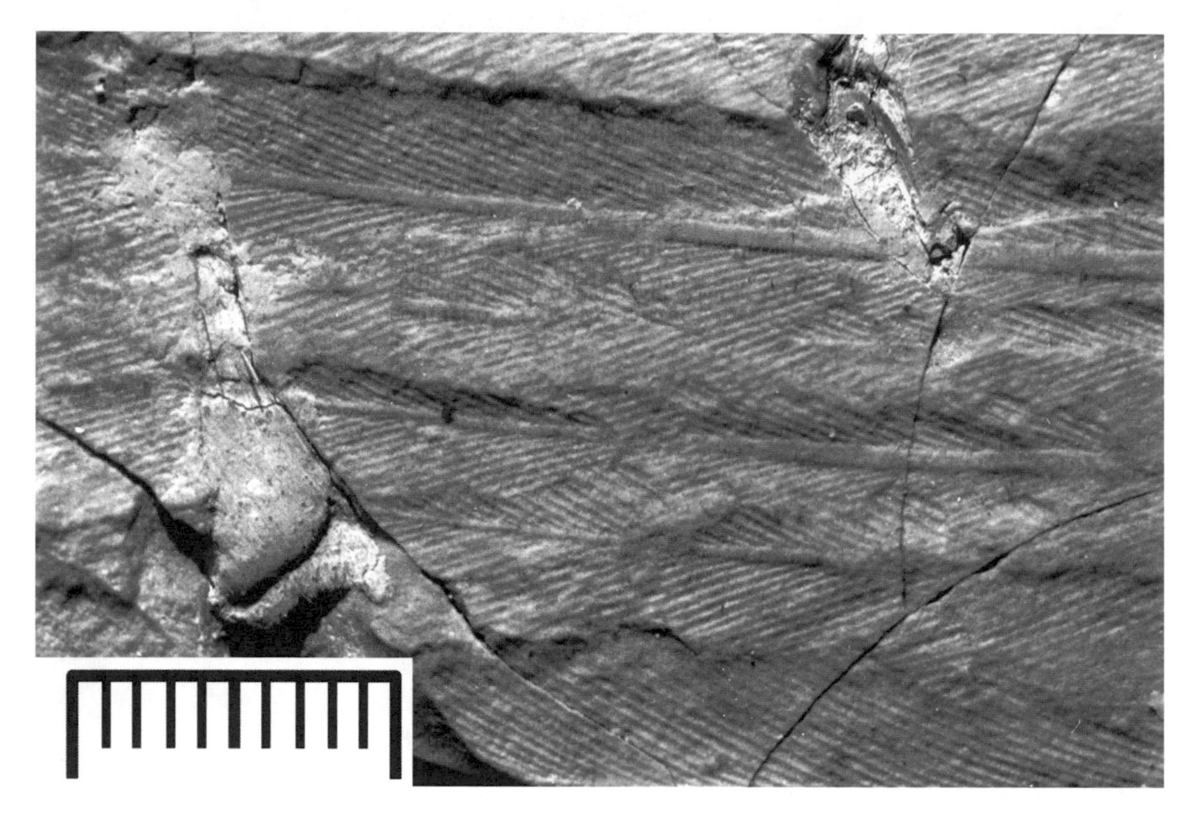

Figure 13.4. Detail of distal primary flight feathers on the right wing of the Berlin specimen of Archaeopteryx lithographica *as preserved on the counterslab (i.e., the underlying slab), showing clearly the rachis, barbs, and traces of the barbules. Scale in millimeters.*

The Berlin Specimen

The Berlin specimen (fig. 13.4) is presented here last because it has the best preserved and most complete plumage of all seven skeletal specimens of *Archaeopteryx*. Rietschel (1976, 1985) studied the taphonomy and preservation of its feathers in detail and concluded that the *Archaeopteryx* fossils are the result of drowned animals, floating for some time on the water surface, until they sank to the bottom of the Solnhofen lagoon. The feathers were subsequently covered by very fine-grained sediment and/or by bacterial overgrowth, resulting in "autolithification" (term from Davis and Briggs 1995). Therefore, the overlying slab (called the "main slab") has on its underside a negative mold of this surface. Because the wings of the Berlin specimen show their ventral side, it follows that the carcass was deposited lying on its back. The main slab contains the solid skeleton and the negative mold of the feathers, and the counterslab, i.e., the underlying slab, shows the negative mold of the skeleton and the feathers as a positive cast from the negative mold of the opposite main slab. Consequently, we are looking at the same surface of the feathers on both slabs and not—as has often been suggested—at the ventral surface on one slab and the dorsal surface on the other. The diagenetic evidence infers that the feathers of *Archaeopteryx* are not preserved as imprints, but as molds and casts of only one surface on both counterslabs. The natural articulation and

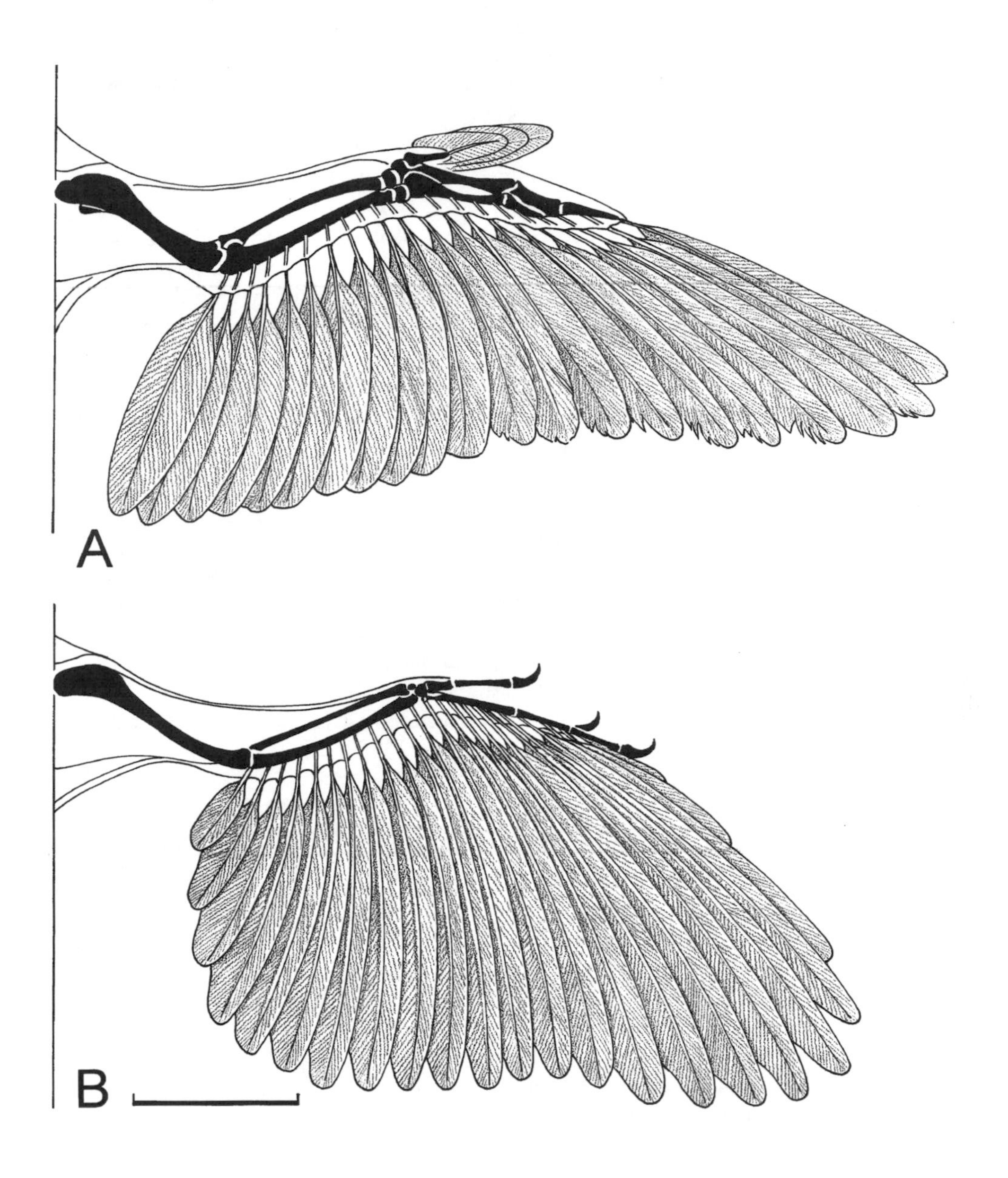

completeness of skeleton and plumage of the Berlin specimen suggest that the carcass could not have floated for a long time, but sank to the bottom relatively rapidly, before major decay processes could take effect.

The interpretation of the flight feathers of the Berlin specimen has been subject of some controversy and discussion, beginning with the first description by Dames (1884), followed by Steiner (1917, 1962), Heinroth (1923), Heilmann (1926), Reichel (1941), Bohlin (1947), Stresemann (1963), Stephan (1970, 1985, 1987), Helms (1982), Yalden (1985), and Rietschel (1985), and need not be repeated here. Disagreement persists, however, about whether the number of primaries was 11 (Rietschel 1985) or 12 (Stephan 1987).

Figure 13.5. Reconstruction of the remiges in the wing of Archaeopteryx *in comparison with a modern bird.*
A, Columba livia *(redrawn after Herzog 1968).*
B, Archaeopteryx lithographica, *based on the Berlin specimen. The wing of the urvogel appears to be of the modern, avian type. Scale bar 50 mm.*

Steiner (1962) identified 12 secondaries in the wing of the Berlin specimen. Stephan (1987) suggested that in *Archaeopteryx* at least 12 and perhaps 15 secondaries might have been developed, a variation known also in recent birds. It was concluded also that the wing of *Archaeopteryx* was eutaxic, not diastataxic, which means that the fifth secondary flight feather was present and therefore left no gap in the secondary series. The eutaxic wing condition seems to have been a basic pattern. Accordingly, the diastataxic condition would be a derived character. Stephan (1987) suggested that there must have been a functional reason for this, as the wing of *Archaeopteryx* could not have been flexed at the wrist to the same extent as in modern birds. In the Berlin specimen the remiges are also overlapped by coverts. Their barbs are partly frayed and preserved in much detail.

The tail feathers show the same preservation as those of the wings. They are exposed from their ventral side, as is indicated by the longitudinal furrows in the shafts. The barbs are clearly visible, especially on the counterslab, which seems not to have been coated by shellac, other chemical substances, and molding material as often as the main slab. Stephan (1987) recognized six pairs of short feathers, about 45 mm in length, attached to the proximal postsacral vertebrae. In the other specimens these caudals are devoid of feather imprints, perhaps due to less favorable fossilization conditions. Distally, there follow 10 or 11 pairs of rectrices with a maximum length of 87 mm, decreasing in size up to the terminal (the 22nd) caudal vertebra.

Contour feathers from other parts of the body can also be identified. These are mainly small feathers in the area of the lower neck, and crural feathers along the tibia forming so-called "breeches" as in recent birds of game.

Discussion and Conclusions

Despite the differences among the individual specimens of *Archaeopteryx,* which are probably biased by different preservational conditions, it can be concluded confidently that the plumage of *Archaeopteryx* was uniform in all specimens. In addition to the asymmetrical flight feathers with 11–12 primaries and 12–15 secondaries, in agreement with the variation in modern birds, there is fossil evidence for coverts on the wings and the tail, and for contour feathers on different parts of the body, particularly the neck and the lower leg. This indicates that *Archaeopteryx* was fully feathered, like modern birds (fig. 13.5). However, there is no alula (bastard wing) as reported in *Eoalulavis hoyasi* from the Late Barremian of Spain (Sanz et al. 1996) and in more advanced birds. The long vertebral tail, stiffened by elongate zygapophyses and chevrons, supported a series of rectrices attached in pairs to the caudals. This pattern of the feathered tail differs significantly from the modern avian condition (fig. 13.6). Together with the unfused vertebral tail composed of 21 to 22 caudals this can be evaluated as a primitive character.

Structurally, the feathers of *Archaeopteryx* are of the modern avian architecture in every detail. Because these highly evolved feathers still

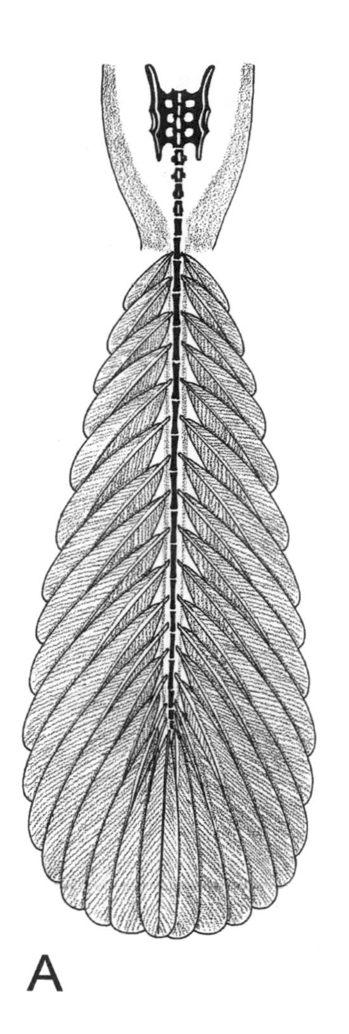

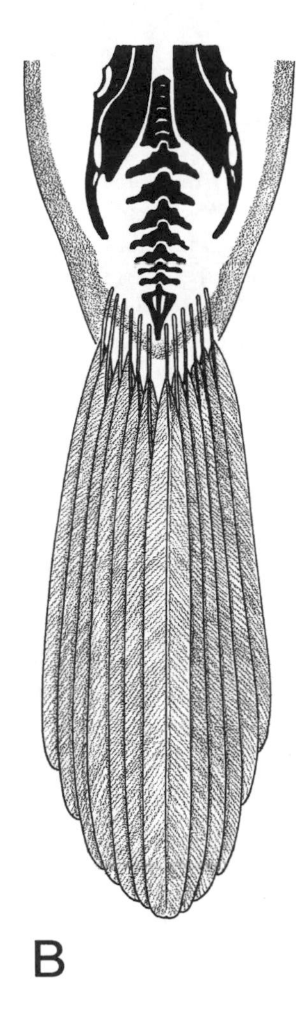

Figure 13.6. Reconstruction of the rectrices in the tail of Archaeopteryx *in comparison with a modern bird.*
A, Archaeopteryx lithographica, *based on the London and Berlin specimens;*
B, Columba livia *(redrawn after Steiner 1938). The long vertebral tail of* Archaeopteryx *with the rectrices attached in pairs to the individual caudals shows the primitive condition, in contrast to the condition in advanced and modern birds. Scale bar 50 mm.*

represent the earliest fossil record, they must have an evolutionary history of some length prior to the Late Jurassic. However, there is no way to determine the time necessary for the evolution from hypothetical protofeathers to the avian-type feathers of *Archaeopteryx*. It is reasonable to conclude that *Archaeopteryx* was an actively flying animal, as a small, ossified sternum (as preserved in *A. bavarica*), a massive furcula, and wide coracoids for the attachment of flight musculature were present. Elzanowski and Wellnhofer (1996) demonstrated that in contrast to the predominantly theropodan postcranial skeleton, the skull of *Archaeopteryx* shows several specifically avian traits adding evidence that *Archaeopteryx* is the most primitive of the unquestionable fossil birds known to date.

If we assume that the filamentous integumentary structures in the compsognathid *Sinosauropteryx prima* (Ji and Ji 1996), the dromaeosaurid *Sinornithosaurus millenii* (Xu et al. 1999b, 2001), or a new, yet undetermined dromaeosaurid (Ji et al. 2001) from the Early Cretaceous of Liaoning, China, might have been some sort of protofeathers, we are faced with a considerable time problem. These dinosaurs lived about 25 million years later than *Archaeopteryx* (according to Swisher et al. 1999, in the Middle Barremian, 124.6±0.1 M.Y.B.P.). They would have preserved this primitive body covering as a plesiomorphic character inherited from some common ancestor. Unless we take these dinosaurs as living fossils of their time, it is equally possible to assume that integumentary filaments may have evolved independently and convergently in a variety of reptilian groups not closely related to each other. This is suggested by the "hairs" or integumentary filaments found in pterosaurs, for example in *Sordes pilosus* from the Late Jurassic of Kazakhstan (Sharov 1971; Unwin and Bakhurina 1994), *Dorygnathus banthensis* from the Liassic of Holzmaden, Germany (Broili 1939), *Rhamphorhynchus* sp. from the Late Jurassic of Solnhofen, Germany (Wellnhofer 1975, 1991), *Jeholopterus ningchengensis* from the Early Cretaceous of China (Wang et al. 2002), and yet undetermined Rhamphorhynchoidea and Pterodactyloidea from the Upper Jurassic or Lower Cretaceous of China (Ji and Yuan 2002). Therefore it could be as well that these filamentous structures of the integument have nothing to do with protofeathers at all.

A quite different problem concerns the clearly identifiable avian-type feathers attached to the metacarpal and the two proximal phalanges of digit II of *Caudipteryx*. Do these indicate the avian status of that taxon, or must it be considered a feathered maniraptoran dinosaur that lived contemporarily with true birds, like *Confuciusornis* and others (Chiappe et al. 1999)? The feathers of *Caudipteryx* have been described as having symmetrical vanes (Ji et al. 1998; Zhou and Wang 2000). However, in my opinion, it does not matter whether these feathers are symmetrical or asymmetrical, because we cannot know if feather symmetry in these animals was a primitive condition or the result of secondary reduction in the course of becoming a flightless bird.

Elzanowski (1999) has raised the interesting question of whether the oviraptorosaurs from the Late Cretaceous of Mongolia might have been the earliest known flightless birds, which branched off after *Archaeopteryx*. This conclusion was based on four cranial characters shared with ornithurine birds. Osteological analyses of both "feathered" dinosaurs from the Early Cretaceous of China, *Protarchaeopteryx robusta* and *Caudipteryx zoui,* have resulted in their assignment to the Maniraptora, indicating that they were more primitive than *Archaeopteryx* (Ji et al. 1998). More recently, Sereno (1999) has interpreted *Caudipteryx* as a basal oviraptorosaur. If Elzanowski's (1999) suggestion that oviraptorosaurs were secondarily flightless birds is correct, *Caudipteryx* could be considered the earliest known flightless bird from the fossil record. This interpretation would be in agreement with the birdlike pygostyle recently discovered in an oviraptorosaur from Mongolia (Barsbold et al. 2000), with other postcranial birdlike structures, such as uncinate processes, and with the avian-like brooding behavior of oviraptorids (Clark et al. 1999). Should we accept that pygostyle-like structures and other avian osteological characters evolved independently in non-avian theropods that had highly differentiated feathers (remiges, rectrices, plumaceous, and downy feathers), and that those animals lived at the same time as existing advanced, flying birds? The logical consequence would be that feathers evolved several times independently. I think this is unacceptable and for good reason.

In this context, one wonders whether comparison of the architecture of the forelimbs in birds and maniraptoran theropods as reflected in the fossil preservation might help in the distinction between birds and dinosaurs. It is a general observation that articulated skeletons of fossil birds normally show the typical avian flexion at the wrist, as is the case in the skeletons of *Archaeopteryx*. The main wing-folding articulation in flying birds is coupled with pronation/supination of the hand (Vasquez 1992; Ostrom 1995; Ostrom et al. 1999). It is different from the position of the forelimb as preserved in fossil theropod skeletons, for example in *Velociraptor* (Ostrom 1995), *Compsognathus* (Ostrom 1978; Michard 1991), or *Scipionyx* (Dal Sasso and Signore 1998), and others. In these specimens, the hand is not flexed at the wrist because there was no feathered wing to be folded back. On the contrary, the three-fingered hands of maniraptoran theropods had a grasping function (Ostrom 1969, 1995), and retro-flexion was not the primary function. As a result, in fossil preservation the hands of theropods are usually rectilinear with the lower arm or even the wrist bent anteromedially. Further, in the type specimens of both *Sinosauropteryx prima* and *Protarchaeopteryx robusta* the lower arms and the hands are preserved rectilinearly, while in both specimens of *Caudipteryx zoui* they form an angle of about 120° (Ji et al. 1998). In *Caudipteryx dongi* this angle is 135° and 150° (Zhou and Wang 2000). In the naturally articulated, undisturbed skeletons of *Archaeopteryx*, the Berlin, Eichstätt, and Solnhofen specimens, this angle ranges between 95° and 115°. In most Cretaceous and Tertiary birds the hand is usually folded against the lower arm at an angle less than 90°. This postmortem position of the arm skeleton can hardly be taken to be accidental. It must be related to the specific biomechanical constraints of the wrist architecture in both birds and theropods (Ostrom 1995). If this observation is correct, *Protarchaeopteryx* could not be classified as a secondary flightless bird. In addition, it had no obvious feathers attached to the forelimbs. However, *Caudipteryx*, and possibly the oviraptorosaurs in general, could be considered to be more closely related to the avian ancestral stock or even as avian. In this view it is interesting that the oviraptorid described by Norell et al. (1995) as the "nesting dinosaur" shows retro-flexed hands in a birdlike position (Clark et al. 1999).

Finally, I want to emphasize that the presence of feathers in tetrapods of basically coelurosaurian osteology is in itself strong positive evidence in favor of a theropod ancestry of birds, or at least of close relationship, no matter whether we consider these vertebrates as feathered dinosaurs or flightless birds. In addition, we must not forget that the Linnaean classification we are using was established almost 250 years ago, and based exclusively on extant animals. While living birds can be recognized easily, even without feathers, we have difficulties with the taxonomic classification of early or evolutionarily intermediate forms such as *Archaeopteryx* or *Caudipteryx*. Obviously we can no longer take for granted certain characters, like feathers, to be significant for the class Aves.

Is there any way out of this dilemma? Can cladistic methodology solve this problem? Of course, we can collect character states and run them through sophisticated computer programs resulting in more or less parsimonious cladograms. However, the phylogenetic systematics obtained reflect merely the degree of relationships, not actual family trees of descent. Therefore, it does not matter where a boundary line is drawn between Aves and Reptilia, or, in cladistic terms, between non-avian and avian theropods. It is realistically not possible either. In strict consequence, classification from a cladistic approach must not use Linnaean units, but only its own terminology and systematics. In this respect, it would generally agree with the theory of Darwinian evolution, postulating gradually evolving taxa. However, it will be very difficult to establish a reliable link to the living world. Furthermore, it will be a different matter whether, for example, a distinction between "non-avian dinosaurs" and "avian dinosaurs" would be generally accepted by the scientific community, not to mention the average human of common sense. As long as cladistics and Linnaean systematics are in mixed use, it is impossible to draw a precise boundary line between reptiles and birds. In this case, paleontologists are in a similar situation as the paleoanthropologists. Based on the available fossil documents the boundary line between nonhuman primates and human primates cannot be determined unambiguously. The process of humanization could have been rather gradual, driven by selective factors, as for example the upright posture and bipedal gait. What remains is to simply accept the existence of taxa in a transitional state of "ornithization," intermediate between the two admittedly anthropogenetic classes of vertebrates, Reptilia and Aves, no matter whether we base our evolutionary hypotheses on a gradualistic or on a punctualistic model. The most effective selective advantage for becoming a bird was certainly the evolution of feathers. There is sufficient evidence now that these integumentary structures must have originated before Late Jurassic times among either theropod dinosaurs or their ancestors, which also gave rise to the avian lineage.

Acknowledgments. I am indebted to John Ostrom (New Haven) for helpful discussions and information, to Evgeny Kurochkin (Moscow) for providing photographs and information on the feather of *Praeornis sharovi,* to David Unwin (Berlin) for unpublished data on the "hairs" of *Sordes pilosus,* and to Winand Brinkmann (Zürich) for making available the publications of Hans Steiner. Finally, I thank Alan Brush (Mystic, Conn., U.S.) who reviewed the manuscript and made many helpful suggestions that have improved the text considerably.

References

Barsbold, R., P. J. Currie, N. P. Myhrvold, H. Osmólska, K. Tsogtbaatar, and M. Watabe. 2000. A pygostyle from a non-avian theropod. *Nature* 403: 155–156.

Bock, W. 1986. The arboreal origin of avian flight. In K. Padian (ed.), *The*

origin of birds and the evolution of flight, pp. 57–72. *Memoirs of the California Academy of Sciences* 8.
Bohlin, B. 1947. The wing of the *Archaeornithes*. *Zoologisk Bidrag, Uppsala* 25: 328–334.
Broili, F. 1939. Ein *Dorygnathus* mit Hautresten. *Sitzungsberichte der Bayerischen Akademie der Wissenschaften, mathematisch-naturwissenschaftliche Klasse:* 129–132.
Chen P.-J., Z.-M. Dong, and S.-N. Zhen. 1998. An exceptionally well-preserved theropod dinosaur from the Yixian Formation of China. *Nature* 391: 147–152.
Chiappe, L. M., S.-A. Ji, Q. Ji, and M. A. Norell. 1999. Anatomy and systematics of the Confuciusornithidae (Theropoda: Aves) from the Late Mesozoic of Northeastern China. *Bulletin of the American Museum of Natural History* 242: 1–89.
Clark, J. M., M. A. Norell, and L. M. Chiappe. 1999. An oviraptorid skeleton from the Late Cretaceous of Ukhaa Tolgod, Mongolia, preserved in an avianlike brooding position over an oviraptorid nest. *American Museum Novitates* 3265: 1–36.
Currie, P. J., and Chen P. J. 2001. Anatomy of *Sinosauropteryx prima* from Liaoning, northeastern China. *Canadian Journal of Earth Sciences* 38: 1705–1727.
Dal Sasso, C., and M. Signore. 1998. Exceptional soft-tissue preservation in a theropod dinosaur from Italy. *Nature* 392: 383–387.
Dames, W. 1884. Über *Archaeopteryx*. *Palaeontologische Abhandlungen* 2: 119–196.
Davis, P. G., and D. E. G. Briggs. 1995. Fossilization of feathers. *Geology* 23(9): 783–786.
De Beer, G. R. 1954. *Archaeopteryx lithographica*. London: British Museum (Natural History).
Doludenko, M. P., and E. R. Orlovskaya. 1976. Jurassic flora of the Karatau. *Transactions of the Academy of Sciences of the USSR* 284: 1–262 (in Russian).
Doludenko, M. P., G. V. Sakulina, and A. G. Ponomarenko. 1990. *Geology of the unique deposits of the fauna and flora from the Late Jurassic of Aulie (Karatau, South Kazakhtsan)*. Academy of Sciences of the USSR, Geological Institute. Moscow (in Russian).
Elzanowski, A. 1999. A comparison of the jaw skeleton in theropods and birds, with a description of the palate in the Oviraptoridae. *Smithsonian Contributions to Paleobiology* 89: 311–323.
Elzanowski, A., and P. Wellnhofer. 1996. Cranial morphology of *Archaeopteryx:* evidence from the seventh specimen. *Journal of Vertebrate Paleontology* 16(1): 81–94.
Feduccia, A., and H. B. Tordoff. 1979. Feathers of *Archaeopteryx:* asymmetrical vanes indicate aerodynamic function. *Science* 203: 1021–1022.
Griffiths, P. J. 1996. The isolated *Archaeopteryx* feather. *Archaeopteryx* 14: 1–26.
Heilmann, G. 1926. *The origin of birds*. Witherby, London.
Heinroth, O. 1923. Der Flügel von *Archaeopteryx*. *Journal für Ornithologie* 71: 277–283.
Heller, F. 1959. Ein dritter *Archaeopteryx*-Fund aus den Solnhofener Plattenkalken von Langenaltheim/Mfr. *Erlanger Geologische Abhandlungen* 31: 1–25.
Herzog, K. 1968. *Anatomie und Flugbiologie der Vögel*. Stuttgart: Gustav Fischer Verlag.

Hoyle, F., and C. Wickramasinghe. 1986. *Archaeopteryx, the primordial bird: a case of fossil forgery.* Swansea: C. Davies.

Ji Q., and Ji S.-A. 1996. On the discovery of the earliest known bird fossil in China and the origin of birds. *Chinese Geology* 233: 30–33 (in Chinese).

Ji Q., P. J. Currie, M. A. Norell, and Ji S.-A. 1998. Two feathered dinosaurs from northeastern China. *Nature* 393: 753–761.

Ji Q., M. A. Norell, Gao K.-Q., Ji S.-A., and Ren D. 2001. The distribution of integumentary structures in a feathered dinosaur. *Nature* 410: 1084–1088.

Ji Q., and Yuan C. 2002. Discovery of two kinds of protofeathered pterosaurs in the Mesozoic Daohugou biota in the Ningcheng region and its stratigraphic and biologic significances. *Geological Review* 48 (2): 221–224 (in Chinese).

Martill, D. M., and E. Frey. 1995. Color patterning preserved in Lower Cretaceous birds and insects: The Crato Formation of N.E. Brazil. *Neues Jahrbuch für Geologie und Paläontologie, Monatshefte* 2: 118–128.

Meyer, H. v. 1857. Beiträge zur näheren Kenntnis fossiler Reptilien. *Neues Jahrbuch für Mineralogie, Geologie und Paläontologie* 1857: 532.

———. 1861. *Archaeopteryx lithographica* (Vogel-Feder) und *Pterodactylus* von Solnhofen. *Neues Jahrbuch für Mineralogie, Geologie und Paläontologie* 1861: 678–679.

———. 1862. *Archaeopteryx lithographica* aus dem lithographischen Schiefer von Solnhofen. *Palaeontographica* A 10: 53–56.

Michard, J.-G. 1991. Description du *Compsognathus* (Saurischia, Theropoda) de Canjuers (Jurassique supérieur du Sud-Est de la France). Thèse de Doctorat de l'Université de Paris 7.

Norell, M. A., J. M. Clark, L. M. Chiappe, and D. Dazhzeved. 1995. A nesting dinosaur. *Nature* 378: 774–776.

Ostrom, J. H. 1969. Osteology of *Deinonychus antirrhopus,* an unusual theropod from the Lower Cretaceous of Montana. *Bulletin of the Peabody Museum of Natural History* 30: 1–165.

———. 1970. *Archaeopteryx:* Notice of a "new" specimen. *Science* 170: 537–538.

———. 1972. Description of the *Archaeopteryx* specimen in the Teyler Museum, Haarlem. *Proceedings of the Koninkljike Nederlandse Akademie van Wetenschappen* B 75(4): 289–305.

———. 1978. The osteology of *Compsognathus longipes* Wagner. *Zitteliana* 4: 73–118.

———. 1995. Wing biomechanics and the origin of bird flight. *Neues Jahrbuch für Geologie und Paläontologie,* Abhandlungen 195(1–3): 253–266.

Ostrom, J. H., S. O. Poore, and G. E. Goslow, Jr. 1999. Humeral rotation and wrist supination: important functional complex for the evolution of powered flight in birds? *Smithsonian Contributions to Paleobiology* 89: 301–309.

Padian, K., Ji Q., and Ji S.-A. 2001. Feathered dinosaurs and the origin of flight. In D. H. Tanke and K. Carpenter (eds.), *Mesozoic vertebrate life: new research inspired by the paleontology of Philip J. Currie,* pp. 117–135. Bloomington: Indiana University Press.

Rautian, A. S. 1978. A unique bird feather from the deposits of a Jurassic lake in the Karatau Range. *Paleontological Journal* 12(4): 520–528.

Reichel, M. 1941. L'*Archaeopteryx:* un ancêtre des oiseaux. *Nos Oiseaux* 16: 93–107.

Rietschel, S. 1976. *Archaeopteryx:* Tod und Einbettung. *Natur und Museum* 106: 280–286.
Rietschel, S. 1985. Feathers and wings of *Archaeopteryx* and the question of her flight ability. In M. K. Hecht, J. H. Ostrom, G. Viohl, and P. Wellnhofer (eds.), *The beginnings of birds,* pp. 251–260. Eichstätt: Freunde des Jura-Museums.
Sanz, J. L., L. M. Chiappe, B. P. Pérez-Moreno, A. D. Buscalioni, J. J. Moratalla, F. Ortega, and F. J. Poyato-Ariza. 1996. An Early Cretaceous bird from Spain and its implications for the evolution of avian flight. *Nature* 382: 442–445.
Sereno, P. C. 1999. The evolution of dinosaurs. *Science* 284: 2137–2147.
Sharov, A. G. 1971. New flying reptiles from the Mesozoic of Kazakhstan and Kirgisia. *Trudy of the Paleontological Institute, Akademia Nauk SSSR* 130: 104–113 (in Russian).
Sloan, C. P. 1999. Feathers for *T. rex*? *National Geographic Magazine* 196(5): 98–107.
Steiner, H. 1917. Das Problem der Diastataxie des Vogelflügels. *Jenaer Zeitschrift für Naturwissenschaften* 55: 221–596.
———. 1938. Über den "*Archaeopteryx*"-Schwanz der Vogelembryonen. *Vierteljahrschrift der Naturforschenden Gesellschaft in Zürich* 83: 279–300.
———. 1962. Befunde am dritten Exemplar des Urvogels Archaeopteryx. *Vierteljahrschrift der Naturforschenden Gesellschaft in Zürich* 107: 197–210.
Stephan, B. 1970. Eutaxie, Diastataxie und andere Probleme der Befiederung des Vogelflügels. *Mitteilungen des Zoologischen Museums Berlin* 46: 339–437.
———. 1985. Remarks on the reconstruction of *Archaeopteryx* wing. In M. K. Hecht, J. H. Ostrom, G. Viohl, and P. Wellnhofer (eds.), *The beginnings of birds,* pp. 261–265. Eichstätt: Freunde des Jura-Museums.
———. 1987. Urvögel. Archaeopterygiformes. 3. Auflage. *Neue Brehm-Bücherei* 465: 216 pp. Ziemsen, Wittenberg-Lutherstadt.
Stresemann, E. 1963. Variation in the number of primaries. *Condor* 65: 449–459.
Swisher III, C. C., Y.-Q. Wang, X.-L Wang, X. Xu, and Y. Wang. 1999. Cretaceous age for the feathered dinosaurs of Liaoning, China. *Nature* 400: 58–61.
Unwin, D. M., and N. N. Bakhurina. 1994. *Sordes pilosus* and the nature of the pterosaur flight apparatus. *Nature* 371: 62–64.
Vazquez, R. J. 1992. Functional osteology of the avian wrist and the evolution of flapping flight. *Journal of Morphology* 211: 259–268.
Wagner, A. 1861. Über ein neues, angeblich mit Vogelfedern versehenes Reptil aus dem Solnhofener lithographischen Schiefer. *Sitzungsberichte der Bayerischen Akademie der Wissenschaften, mathematisch-physikalische Classe:* 146–154.
Wang X.-L., Zhou Z.-H., Zhang F.-C., and Xu X. 2002. A nearly complete rhamphorhynchid pterosaur with exceptionally well preserved wings and "hairs" from Inner Mongolia, northeast China. *Chinese Science Bulletin* 47(3): 54–58.
Wellnhofer, P. 1974. Das fünfte Exemplar von *Archaeopteryx*. *Palaeontographica* A 147(4–6): 169–216.
———. 1975. Die Rhamphorhynchoidea (Pterosauria) der Oberjura-Plattenkalke Süddeutschlands. Teil III. Palökologie und Stammesgeschichte. *Palaeontographica* A 149: 1–30.

———. 1988a. A new specimen of *Archaeopteryx*. *Science* 240: 1790–1792.
———. 1988b. Ein neues Exemplar von *Archaeopteryx*. *Archaeopteryx* 6: 1–30.
———. 1991. *The illustrated encyclopedia of pterosaurs*. London: Salamander Books.
———. 1993. Das siebte Exemplar von *Archaeopteryx* aus den Solnhofener Schichten. *Archaeopteryx* 11: 1–48.
Xu X., Tang Z.-L., and Wang X.-L. 1999a. A therizinosaur dinosaur with integumentary structures from China. *Nature* 399: 350–354.
Xu X., Wang X.-L., and Wu X.-C. 1999b. A dromaeosaurid dinosaur with a filamentous integument from the Yixian Formation of China. *Nature* 401: 262–266.
Xu X., Zhou Z.-H., and R. O. Prum. 2001. Branched integumental structures in *Sinornithosaurus* and the origin of feathers. *Nature* 410: 200–204.
Yalden, D. W. 1985. Forelimb function in *Archaeopteryx*. In M. K. Hecht, J. H. Ostrom, G. Viohl, and P. Wellnhofer (eds.), *The beginnings of birds*, pp. 91–97. Eichstätt: Freunde des Jura-Museums.
Zhou Z.-H., and Wang X.-L. 2000. A new species of *Caudipteryx* from the Yixian Formation of Liaoning, Northeast China. *Vertebrata PalAsiatica* 38 (2): 113–127.

14. Dinosaur Crime Scene Investigations: Theropod Behavior at Como Bluff, Wyoming, and the Evolution of Birdness

Robert T. Bakker and Gary Bir

Abstract

Did dinosaurian predators feed their young? If you hatched out of the egg as a Late Jurassic allosaur, would your parents be there, ready to stuff meat into your maw? Or would you have to scrounge your first meal by yourself, the way hatchling lizards and turtles do today? If you lived within a theropod society, how would your pack choose hunting grounds and prey targets? How would your species deal with the other predators? How did differences among competing theropods drive the long-term evolutionary trends, as the top predator fauna shifted from ceratosaurids to megalosaurids to allosaurids and, finally, to tyrannosaurids?

A crime scene approach has been applied to Late Jurassic theropod behavior of Como Bluff, Wyoming. Shed teeth were used as "bullets" to map out the feeding habits of large theropods across a wide habitat mosaic of dry floodplains, wet floodplains, swamp and lake margins. Theropods with long, sinuous bodies—megalosaurids and ceratosaurids—specialized in large fish, turtles, and crocodiles, with a minor investment in large dinosaur carcasses. Allosaurids, the family with the stiffest, longest-legged body design, specialized in terrestrial dinosaur prey. Sauropods and stegosaurs were not year-round residents, and when the migratory herds disappeared, allosaurids shifted their feeding to aquatic sites. Hatchling allosaurids fed only on giant prey, surrounded by older juveniles and full-grown adults. The social behavior of allosaurs was much closer to that of large raptorial birds than to what is seen among crocodilians and lizards.

Introduction: Trying to Discover What Theropods Did Do, Instead of Theorizing about What They Should Do

To understand the behavior of carnivorous dinosaurs, we need to get inside the lives of theropods, to explore how the drama of competition and predation was played out across a shifting habitat mosaic of dry floodplains, broad swamps, and catastrophic floods. The Latest Jurassic theropods from the Morrison Formation at Como can be arranged in a sort of *Scala Naturae*, from least birdlike predators, close to a Late Triassic grade theropod, to the most birdlike genera that approach the conditions in the Cretaceous dromaeosaurids and tyrannosaurids (fig. 14.1). In general, long, sinuous bodies, short hindlimbs, and flexible ankles are the mark of a primitive theropod; stiff torso and tail, long limbs, and long, compressed ankles are the mark of a more avian construction. Here, then, is the theropod dramatis personae at Como:

Ceratosaurids

Ceratosaurus has many features in common with the dilophosaurs and *Coelophysis*-like theropods of the Late Triassic and Early Jurassic. Although I do not agree that these older theropods should be labeled "ceratosaurs" in a phyletic sense, Como ceratosaurids were living fossils in their day, retaining a very archaic body plan. Compared to the tyrannosaurids of the Late Cretaceous, the Como ceratosaurids possessed shorter shins and ankles, with less constriction of metatarsal III, a wider scapula blade, and longer and more flexible torsos with weaker development of interspinous ligaments—a flexibility continued through to the distal tail, where ceratosaurids lack the elongated prezygapophyses that could stiffen the distal half of the caudal appendage (fig. 14.1).

Megalosaurids

Como megalosaurids (including Jensen's torvosaurids) were also a venerable group that had been common in the Middle Jurassic. Megalosaurids were unique among Jurassic theropods in the acquisition of curiously short, thick forearm and metacarpal elements. Megalosaurids share with ceratosaurids a subdued neck curvature, wide scapular blades; short, wide metatarsals; long, low torsos; and sinuous tails. Megalosaurids are more advanced over ceratosaurids and approach the allosaur condition in the presence of huge hand claws.

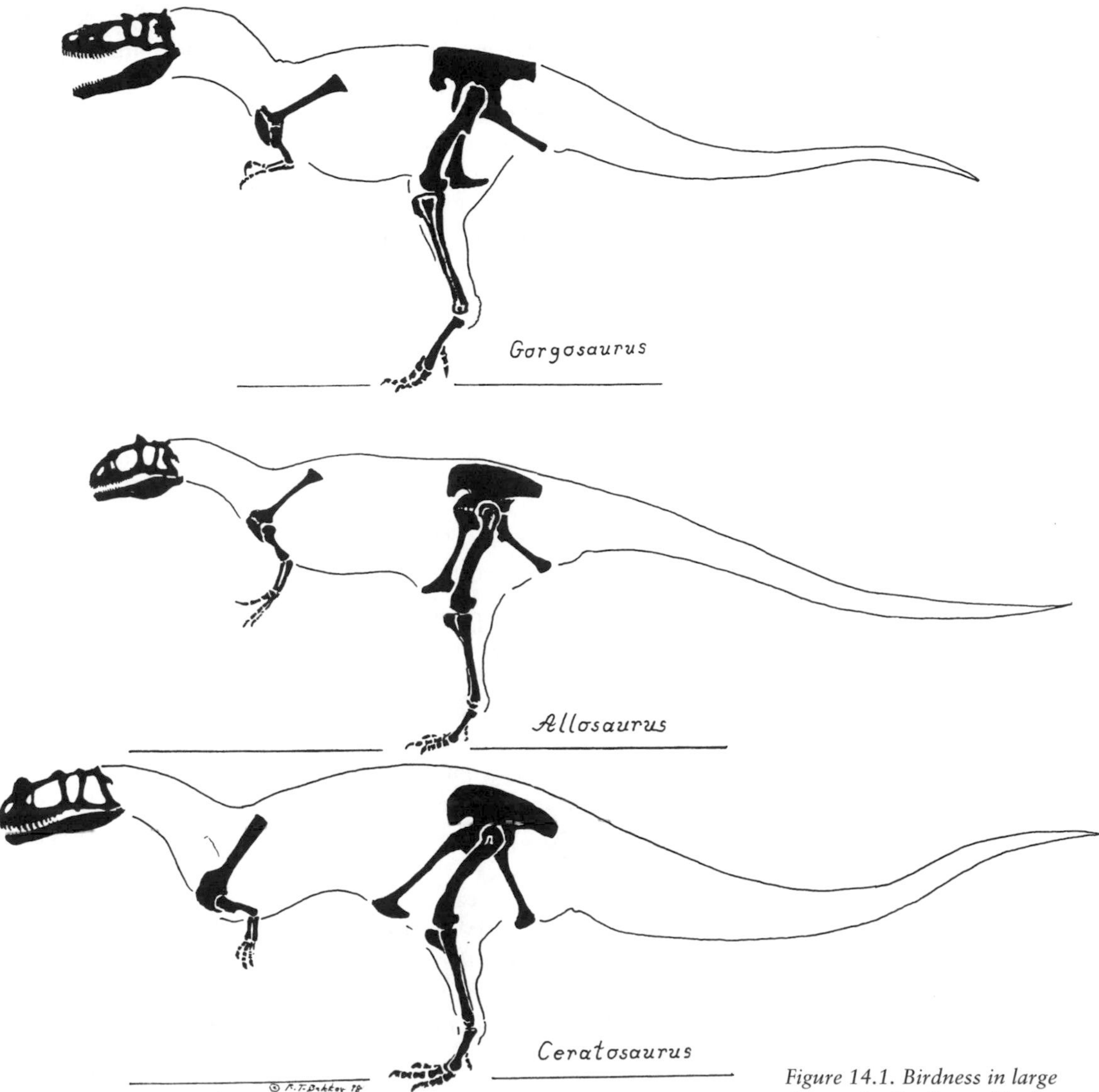

Figure 14.1. Birdness in large theropods. Outlines of body form, drawn to a constant femur length. Vertebral column arranged to give enough space between successive centra so the zygapophyseal joint surfaces are at the middle of their range of movement. Gorgosaurus n. sp. 1, *Indianapolis Children's Museum; femur length 823 mm.* Allosaurus fragilis, *topotype, USNM 4735; femur length 851 mm.* Ceratosaurus nasicornis *type, USNM 4734; femur length 624 mm. Note there is progressive shortening of the torso and increasingly sharp S-flexure in the neck from* Ceratosaurus *to* Gorgosaurus.

Allosaurids

Allosaurids have the most birdlike overall body shape of the large Como Late Jurassic theropods, with the shortest, stiffest torso, most sharply bent necks, distal tail with elongated prezygapophyses, thinnest scapula blade, longest hind legs relative to body length, and relatively longest metatarsals with some constriction of metatarsal III.

How did the spectrum of body form—from low and sinuous to compact, long-legged and birdlike—control the choice of habitat and hunting style? Allosaurids, with their long legs and compact torsos and stiff tails, seem best adapted for running in open terrain. Ceratosaurids and megalosaurids, with their lower, more sinuous bodies, seem far better shaped for snaking through dense forest and underbrush. A related question is: How did differences in behavior shape theropod fate? Megalosaurids and ceratosaurids, in the strict sense, make their last appearance at the top of the Morrison at Como. By the time of the earliest Cretaceous theropod faunas, these two archaic families are extinct. But allosaurids and a clan with similar features, the acrocanthosaurids, survive (Huti et al. 1996). Why? Was the allosaur design superior? The dominant top predators of the Late Cretaceous in Laurasia, the tyrannosaurids, have a body form that is allosauroid carried to extremes—the torso is very compact front-to-back and stiff, the distal tail is embraced by very long prezygapophyses, the neck is bent into a very tight S curve, the metatarsal bundle is very long and compressed, and the upper end of metatarsal II is severely squeezed between its neighbors. Whether or not tyrannosaurids are descendants of allosaurids or their kin, it appears that greater *birdness* was associated with top predator success. Why?

Jurassic Niches: Realized versus Potential

The ecological concept of "realized versus potential niche" is relevant here—it's all too often ignored in functional morphological studies. A potential niche is the complete array of resources a species could exploit if unhindered by the interference of other species. The realized niche is the full extent of resources actually exploited. As an excellent example, in Australia in the early 20th century the marsupial wolf (*Thylacinus cynocephalus*) was a fugitive species, driven out of most habitats by the very aggressive dingo (*Canis familiaris*) and restricted to the island of Tasmania. Before humans and their dingoes arrived, marsupial wolves had spread all over mainland Australia. Clearly the thylacines had the potential to be a top predator throughout the range of ground kangaroos. But the entry of a much faster, more social, more intelligent competitor constricted the realized niche.

We could ask: What happened to theropods with a shape like the ceratosaurids and megalosaurids, once dominant among top predators, when the allosaurids arrived in the latest Jurassic? Did allosaurs force the older theropod families into peripheral niches?

Limitations of Functional Morphology

The tradition of dinosaur functional morphology takes jaws, legs, and backbones and makes them understandable by analysis of muscle leverage and joint vectors. Usually we assumed that evolution made animal bodies "better" in some fundamental, biomechanical way, or permitted new species to exercise new behaviors so that the creatures could enter new niches. However, how could we be sure a dinosaur acted the way we anatomists said it should? Surely, a ceratosaur had a low, flexible body, but did living ceratosaurs really act like theropod equivalents of jaguars? There is a need for clues that come directly from the animal's behavior, such as footprints. Alas, footprints are rare at Como and are restricted to rock types that carry few bones (Bakker

1996, 1998a,b). Plus, the environment that preserves footprints—lime mudflats—probably was not the preferred hunting ground of most, or any common, Como predator species. Mapping the distribution of dinosaur bones through the diverse sedimentary environments offers tantalizing hints about availability of *prey*—haplocanthosaur sauropods, stegosaurs, and camptosaurs are restricted to flood sandstones sandwiched between well-drained floodplain sediments—but predator skeletons are so rare that patterns have not been obvious (Bakker 1996; 1998a,b; Connely 2002).

And then there's the specter of "Olson's Nightmare" (Olson 1962). Named for the late E. C. Olson of the Field Museum, a brilliant paleoecologist, the nightmare goes like this: Good skeletons of both predator and prey, of the sort anatomists yearn for, might represent rare and unusual events in the life of a species—carcasses floated into the area of deposition and buried without suffering damage from scavengers. And therefore, perfect skeletons could be hopelessly unreliable clues to the species's average preferred habits. Obvious examples are the mass deaths of zebra and wildebeest incurred when the herds cross swollen East African rivers. Hundreds of carcasses become embedded in sandbars, mixed with turtle and "croc" bones, providing a potential mass skeleton deposit that could persuade a future paleontologist that these herbivores were water lovers (Estes 1993). But, in fact, zebra and wildebeest are among the *least* aquatic of East African ungulates. Their behavior is designed for cropping grass on treeless plains and avoiding attacks by hyenas and lions, not feeding on riverine vegetation and coping with year-round assaults by crocodilians. It could be argued that a sample of many chewed-up carcasses, strewn over wide areas, is a better indicator of modal habits than an occasional pile of complete skeletons. But even such a dispersed sample of scavenged carcasses does not automatically provide conclusive evidence of which predator ate which herbivores.

Dinosaur "Bullets"

What we sought at Como was a set of clues to dinosaur feeding behavior that are independent of the dead carcasses. And we have found it in dinosaur "bullets." A bullet is a shed tooth crown left at a feeding site. Like crocodilians today, all Como dinosaurs were tooth-shedders throughout their lives, as long as they were healthy. Jawbones of both herbivorous and carnivorous dinosaurs and of Como crocodiles show that each socket was a dental magazine, filled with teeth stacked on top of each other like rounds inside the tubular magazine of a Henry rifle of the 1870s. Jaws in dinosaur embryos show that tooth replacement had begun before birth (personal observation of therizinosaur and Argentine sauropod embryos prepared by Mr. Terry Manning in 2000). Shed teeth are records of living Jurassic animals, not dead carcasses. An allosaur, ceratosaur, or megalosaur would start leaving bullets as soon as it hatched and ate its first meals at its nest, and the growing carnivore would continue to leave a trail of dental signatures for as long as it lived (fig. 14.2). If we could map the trail of shed crowns for a pack of allosaurs, we could—in theory—assemble a genuine predatory dossier showing how the theropod hatchlings, toddlers, adolescents, sub-adults, and adults used prey resources.

Unfortunately, shed teeth all too often fail to get the respect they deserve as crime scene clues. Notes in the American Museum tell of "*Coelurus*" crowns seen at Quarry 9 but not cataloged that were subsequently lost (Gilmore 1910). Shed teeth suffer the indignity of being treated as dino curios and petrified objets d'art—go to the Denver Gem and Mineral Show, and you will see hundreds of shed crowns on sale, alongside pretty quartz crystals. Dinosaur sites on public and private land are scoured by unscrupulous pothunters who scoop up the tooth crowns and conveniently lose the locality records.

Methods of Crime Scene Investigation

To explore the pattern of shed teeth, we sought a Morrison outcrop that had crime scene integrity—sites where the association between bones and shed teeth had not been disturbed by collectors picking up shed crowns from the surface. The western side of Como Bluff had been devastated by decades of amateurs and casual geology students. However, the eastern half of the anticline was remote, obscure, and hard to get to, and it had been protected by the landowners (Bakker 1996, 1998a,b; Connely 2002).

Sample Area

We have developed fifty sites here, representing a delightful range of paleoenvironments. Shed teeth are abundant on outcrops and in quarries; we gave special attention to recovering all the whole teeth and fragments.

To be useful for paleosociobiology, an ideal shed tooth sample would come from a single species lineage that did not evolve much during the sample time. Over 99 percent of our shed teeth at Como come from the upper half of the formation, the unit full of smectitic clays, roughly equivalent to what is labeled the Brushy Basin Member on the Colorado Plateau (Bakker 1996, 1998a,b; Connely 2002). How much diversity was there within big theropod genera in this section in our sample area? All the available evidence indicates that, at any one time, there was only one common species per genus and per family in our Como Morrison sites. There is no evidence of the supersized allosaur species, *Epanterias* (=? *Saurophagnax*). The size of allosaurids does not change in the sample interval; the length of proximal caudal centra in adults, where the centrum-arch suture is fused, is 110 mm to 120 mm from the earliest (Josef Sandstone) to the latest specimen (Claw Conglomerate). Quadrate shape documents two allosaurid taxa at Como, *Allosaurus sp. 1,* and another form with a thick, twisted quadrate (Bakker 1998a,b). However, in the sampling interval, true *Allosaurus* outnumbers the other form twenty to one, and so most shed teeth probably belong to *Allosaurus*. Several *Ceratosaurus* species have been named, including one from Como (*C. sulcatus*); the size and shape of our shed crowns fit the pattern in the skull with the best-preserved dentition, the tooth rows in the Fruita *Ceratosaurus* (Museum of Western Colorado 0001). Two distinct types of megalosaurid pelvis are present at Como, but all shed teeth agree with those in *Edmarka rex*.

It is intriguing that the large dinosaurs show little evolutionary change through the middle and upper parts of the Morrison at Como, but the mammals and turtles display an abrupt faunal event at the beginning of the upper swamp beds—the event signals the start of the Breakfast Bench Fauna. New families of mammals appear: the alticonodontid triconodontans, adapted for cutting vertebrate flesh, and the peculiar multituberculates of Family Zofiabaataridae, also highly carnivorous. (We separate alticonodontids as a family distinct from the European *Triconodon, Trioracodon* of England, and all the earlier Morrison triconodontans, which can be brigaded into *Priacodon;* see discussions in Fox 1969, 1976; Rasmussen and Callison 1981). Zofiabaatarids show some affinity to Cretaceous multituberculates in the reorientation of the jaw joint to allow greater fore-to-aft movement (Kielan-Jaworowska et al. 1987; Kielan-Jaworowska and Ensom 1992, 1994). Alticonodontids replace their more primitive triconodontid relatives in the Breakfast Bench and become the dominant mammalian carnivores all through the Cretaceous in North America but do not spread beyond the limits of the continent (Engelmann and Callison 1998; Simpson 1925, 1926, 1928, 1929; Fox 1969; Cifelli and Madsen 1998; Cifelli et al. 1998, 1999). Turtles also show an event at the beginning of the Breakfast Bench interval. What appear to be the first baenids appear in abundance (Bakker 1998a,b). Baenids become the dominant freshwater turtles in the Cretaceous and early Tertiary of North America but, as in the case of alticonodontid mammals, do not spread to other continents. In the Late Jurassic and earliest Cretaceous, North America appears to have been an island, sealed off from the rest of the world as far as turtles and triconodontan mammals were involved. Advanced, eucryptodire turtles were abundant and diverse in Asia and Europe but did not invade North America until the mid Cretaceous (Baur 1891; Brinkman and Peng 1993a,b; Peng and Brinkman 1993; Evans and Kemp 1975, 1976; Gaffney 1979). Most authors regard all the Como Morrison as Late Jurassic and earlier than the earliest Cretaceous of Purbeck, England (Kowalis et al. 1998; Ostrom and McIntosh 1966; Feist et al. 1995; Russell 1993). It is true that the Purbeck fauna has certain advanced elements missing from Como: large iguanodontids, very large ankylosaurians, and large, advanced pterodactyloids (Howse and Milner 1995; Wright 1998; Russell 1993). However, there are no reliable radiometric dates for the Breakfast Bench interval, and the advent of alticonodontid mammals plus baenid turtles gives a Cretaceous flavor to the fauna.

Identifying Shed Teeth

Gilmore (1920) claimed that Morrison theropods could not be distinguished on the basis of tooth crowns. He was wrong (Bakker 1996, and fig. 14.2 here). *Allosaurus* has rectangular crown bases with twisted cutting edges fore and aft; *Ceratosaurus* crowns have teardrop cross sections with cutting keels running up the midline fore and aft;

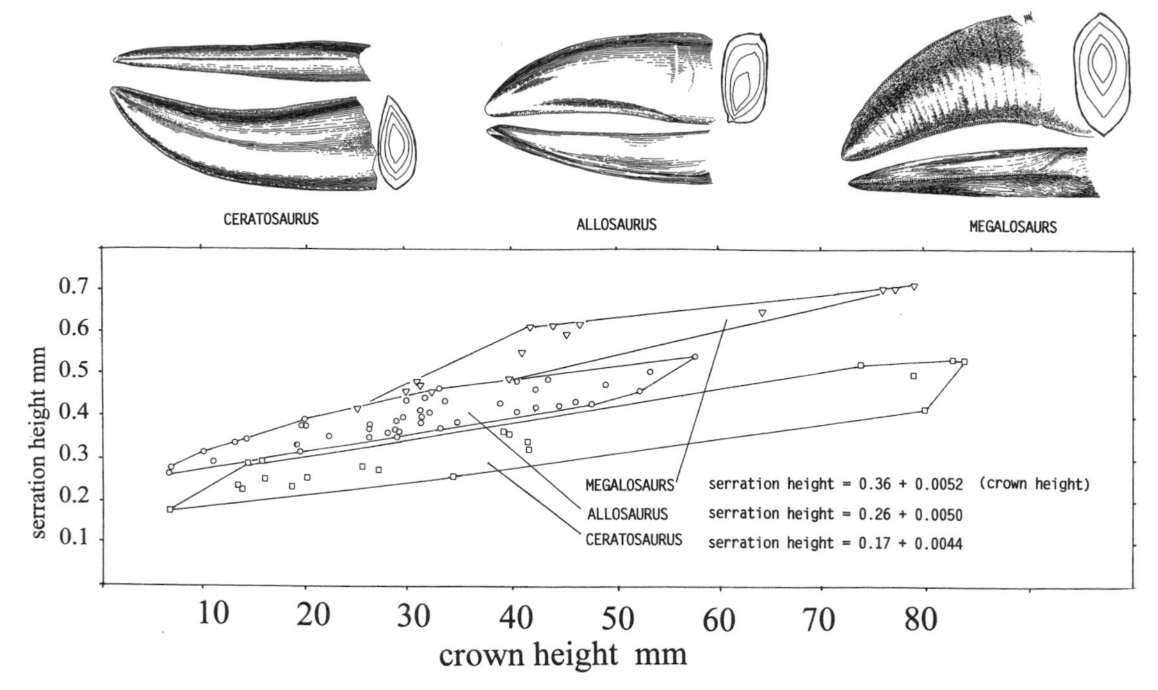

Figure 14.2. Identifying shed theropod teeth from Como. Y-axis: width of individual serration cusplet from the middle of the posterior row. Note: *in some theropods the serration size varies from crown base to tip. In megalosaurids the apical serrations are much larger than in the middle of the posterior row. X-axis: crown height, from base of enamel to apex. Large maxillary shed teeth shown at top.*

and the Como megalosaurid *Edmarka* has teardrop cross sections that are plumper than ceratosaurid's, wrinkled enamel, and very coarse serrations. The anterior incisiform teeth are also distinct: *Allosaurus* crowns have a thick elliptical cross section with two strong keels; *Ceratosaurus* crowns have a single keel and deep flutes on the inner (buccal) surface; the megalosaurid *Edmarka* has single-keeled crowns with smooth inner surfaces. Serrations in megalosaurids are blunt, swollen, and rounded. In ceratosaurids and allosaurids the serrations are laterally compressed, sharp, and have an acute apex directed toward the crown tip. Serration fineness has been used to identify theropods but this character requires a regression of serration size as a function of crown height (fig. 14.2). In all large Como theropods, as the crowns grow larger in ontogeny, the number of serrations per crown increases, and the size of the individual serration cusplet increases also (fig. 14.2). For all species there appears to be the same slope in this relationship. Ceratosaurids, allosaurids, and megalosaurids are distinguished by having different y-intercepts. And thus for any given crown height the ceratosaurids carry the finest serrations, the megalosaurids carry the coarsest, and allosaurids are intermediate.

Nearly all the shed crocodile teeth at Como (fig. 14.3) seem to belong to one species of goniopholid, judging from the size and shape of quadrates, vertebrae, and limb bones. It is close to the species represented by the complete skull recovered by commercial collectors, the Western Paleo Corporation, at Bone Cabin Quarry. This croc is very heterodont. Anterior tooth crowns are slender fangs, incurved and reinforced by strong anterior and posterior keels. Secondary keels and sulci run up the crown on the inner and outer surfaces, and between the minor keels is a fine, anastomosing pattern of very fine, sharp-edged ridges. The major keels carry fine knobs that give the appearance of protoserrations. Posterior crowns are acorn-shaped, with major anterior and posterior keels; anastomosing fine ridges are present but there are no secondary keels. Our Como *Goniopholis* seems to be heterodont at birth, unlike most crocodilians today. Even in the smallest size classes of shed crowns (1–2 mm) both tall fangs and acorn-shaped crowns are present. Three shed teeth indicate a second goniopholid species: The crowns are smooth except for sharply defined keels that have well developed, very fine serration-like nodulations.

Our detailed analysis of theropod habits is restricted to large top predators. Small theropods—skeletons or shed teeth—are usually very rare or absent in the Como Morrison, a striking contrast to the abundance and variety among Late Cretaceous carnivores in North America. In Como sites rich in aquatic prey, small theropod shed teeth do occur and represent *Ornitholestes*-like animals as well as dromaeosaurids and unidentified taxa (Bakker 1996).

Tooth Shedding Rates

To interpret fossil shed teeth in terms of number of individuals per unit time, we need a guide to the rate of shedding per individual animal per unit time. Fast-growing juveniles should shed teeth faster than old adults. Females growing eggs and embryos should shed teeth faster than indolent males of the same body mass. The common theropod predators and the Como crocodile *Goniopholis* sp. do not differ markedly in the number of tooth positions: *Ceratosaurus* has 3 premaxillary sockets + 13 maxillary sockets + 15 dentary sockets on each side = 31 total sockets on each side. Como-area megalosaurids probably have about 4 + 14 + 16 = 34; allosaurids have 5 + 15 + 15–17 = 35–37. *Goniopholis* sp. has 5 + 17 + 16–17 = 38–39. Extant crocodilians have slow-growth bone texture, and must shed teeth less rapidly than did theropods with fast-growth texture. However, Como *Goniopholis* has bone histology much like that of dinosaurs and birds, with a well-defined hollow core and open-weave histology through most of the cortical bone. Unlike the condition in Como turtles, the Como crocs do not display in their limb bones well-defined growth lines (LAGs—lines of arrested growth). Therefore we infer that growth rates of Como *Goniopholis* were faster than those of extant crocs and not greatly dissimilar to those of theropod dinosaurs. We use the anterior-posterior basal diameter of the crown to express size; the measurement is taken between the anterior and posterior keels. In all the theropod dinosaurs and crocodiles from Como, the basal diameter of the teeth changes along the row of sockets from anterior to posterior. The smallest premaxillary tooth is about half the size of the largest maxillary tooth, and the largest maxillary tooth is about 25 percent larger than the largest dentary tooth. Thus a single individual at any one time in its growth will be shedding teeth with a range of variation of about a factor of 2. Hence the distinction between age classes will be made less distinct in the frequency plots of size classes. Nevertheless, as described

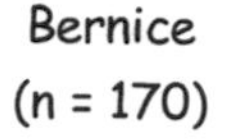

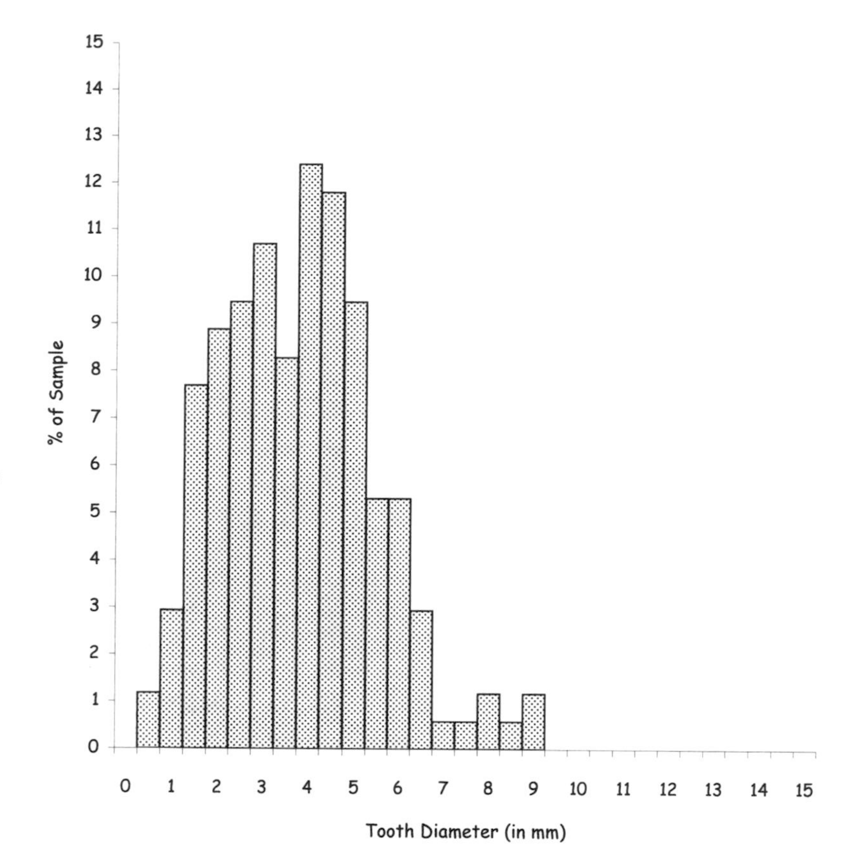

Figure 14.3. Diameters of Como crocodile shed teeth. Size-frequency distributions from a nursery site near shore and a more open water site. Bernice Quarry: no lungfish; abundant "amioid" fish, mammals, small dinosaurs. No large shed croc teeth from adult males. Many hatchling and yearling teeth.

below, the data set is tight enough to reveal strong differences among the species, and between members of the same species living in different environments.

The absolute growth rate of tooth size during ontogeny is not yet known, but it is knowable. Thin sections of crocodile and theropod shed crowns at Como show very distinct growth increments in the form of cone-shaped shells (Edmund 1962; Erickson 1996a,b; Erickson and Bochu 1999). Analysis of these growth lines has just begun. For example, two *Goniopholis* sp. blunt crowns, 7.8–7.9 mm wide, show four and five wide pale growth cones separated by thin dark LAGs. We interpret these lines as dry-season lines. We assume that the growth rate of crown width would slow down after an initially high hatchling-adolescent rate; therefore if the size-frequency plots of tooth crowns were converted to an age-frequency distribution, the right-hand part of the curves would be drawn out to the right.

Statistical Tests

Since the size-frequency distributions of shed teeth are obviously

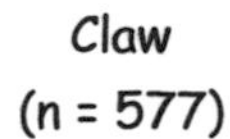

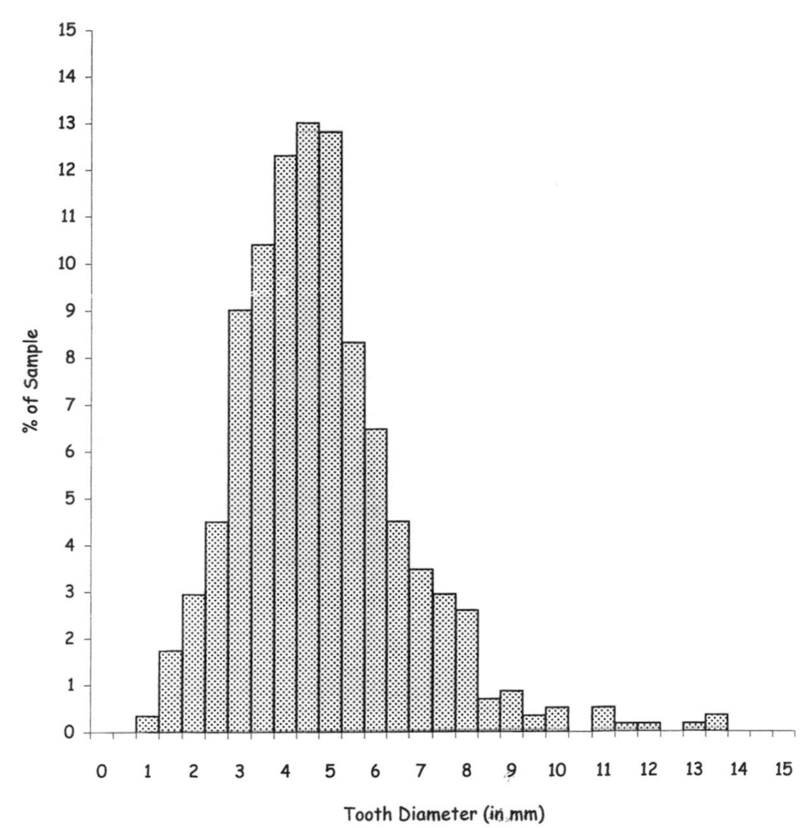

Figure 14.3. (cont.)
Claw Quarry: many lungfish; less common "amioid" fish, many turtle bones and crocodile bones. Few hatchling croc teeth. Very large shed teeth, from adult males.

not normal, to compare distributions we use a two-sample Komologorov-Smirnoff test (Siegel 1956); we accept as significant differences which are less probable than 0.05 of being ascribed to random variations within the same distribution. Identifying polymodalities is difficult, since random variations can produce gaps and modes in small samples; only one gap is found to be significant by the two-sample K-S test (fig 14.13—allosaur Turf; the test was against a theoretical distribution with the same sample size but a flat-topped shape across the size classes that record the "Lair Gap").

Excavation Protocols

Bulk washing of sediment over screens was not used, because at many locales teeth and small bones are too delicate to survive the treatment (Clemens 1963; Lofgren 1995; Estes and Berberian 1970). In the richest shed tooth sites, crowns break up into multiple fragments when exposed to water. Consequently, our field parties hand-quarry all sites, breaking down matrix into Ping-Pong ball–sized chunks. If any teeth or identifiable small bones are found, the rest of the matrix is

Figure 14.4. Scene from Bernice Quarry: a mother Goniopholis *and her hatchlings, playing with an alticonodont mammal.*

taken to the lab to be broken up under the microscope into clods 1 cm or smaller in diameter. Because small teeth and bones are stress concentrators within matrix clods, the small sediment lumps almost always break in planes that expose the fossils. Small clods that produce small bones under the 'scope are further disaggregated by being placed under a gentle water drip. The final water-drip treatment has exposed only a single identifiable bone—an ungual from a turtle. This process has revealed no complete teeth and only a few tooth fragments. Consequently we have confidence that few shed teeth are missed in our processing protocols. About 2,000 shed teeth and small bones have been recovered from all sites combined.

Usually, shed teeth are easy to recognize: The crown is abruptly truncated just below the enamel-dentine margin; there is a shallow depression on the underside of the truncation with a pebbly texture produced by the biochemical erosion of the root; and in crocodiles there is a raised rim around the depression. Crowns can and do break off their roots after the death of the animal. Teeth broken from jaws while the root was still in place have a jagged fracture surface without evidence of erosion. Isolated crowns also can come from very young developmental stages of a tooth that had not yet erupted before death. Tooth development in the living animal begins with the formation of an enamel shell that will become a crown. Replacement crowns that were deep inside the socket, underneath functional crowns, at the time of death do not have roots; the crown is a hollow cap with a conical space within the base, and there is no mark of chemical erosion.

Definition of the Mosaic of Habitats: Surf Sites, Wet Turf Sites, and Dry Turf Sites

We try to separate signals from four independent sources of information: 1) from the sediments—the physical crime scene; 2) from the ecological-indicator species—the biotic crime scene, turtles and crocs that were certainly linked to water; 3) from the condition of the "victims," the tooth-marked and dismembered carcasses of vertebrates; and 4) from the clues left by the "perps," the shed teeth of carnivores. The key sediment is the rock that forms the floor of the fossil accumulation, the surface where the victim's bones were dismembered and splintered and where the perpetrator's teeth were shed. The sediment that buried the crime scene does not shed light directly on the crime scene surface (fig. 14.4). Evidence at Como strongly suggests that many victims died, became desiccated, and sometimes were chewed up under the open air on wet floodplains and swamp margins, and then the vics' bones and the perps' teeth were moved a short distance when a layer of wet sediment moved across the crime scene.

Aquatic Vertebrate Indicators

The aquatic vertebrates—fish, turtles, and crocs—provide an independent indicator of dry versus wet crime scene conditions.

Fish

Como fish fall into two very different categories: (1) the "amioid" grade bony fish, similar to *Sinamia,* bowfin-like predators not exceeding 600 mm long as estimated from vertebrae and jaws (Hoyt 1962; Kirkland 1998); and (2) the ceratodontid lungfish, close to the living Australian lungfish *Neoceratodus fosteri* but with much blunter ridges on the tooth plates. The lungfish become varied and very large in the upper Morrison, probably approaching a half ton (fig. 14.5). The Australian lungfish does not make burrows and does not aestivate (Merrick and Schmida 1984; Kind et al. 1999). As we have found no evidence for burrows in the Como Morrison where lungfish tooth plates are common, these fish too probably did not construct refugia from desiccation. We infer that the large Como lungfish required larger bodies of water than the amiod grade fish did. Ceratodontid lungfish did not shed teeth; the crowns remained firmly attached to the supporting jawbones all through life, and the individual teeth grew by accretion. All our lungfish jawbones show predator damage; the anterior and posterior extensions of the jawbones are bitten off. At Quarry 13, on the west bluff at Como, lungfish jaws occur as complete bones without predator damage (Kirkland 1998). Clearly the mode of accumulation in our eastern bluff sites was different, with each and every fish jaw subjected to more tooth damage.

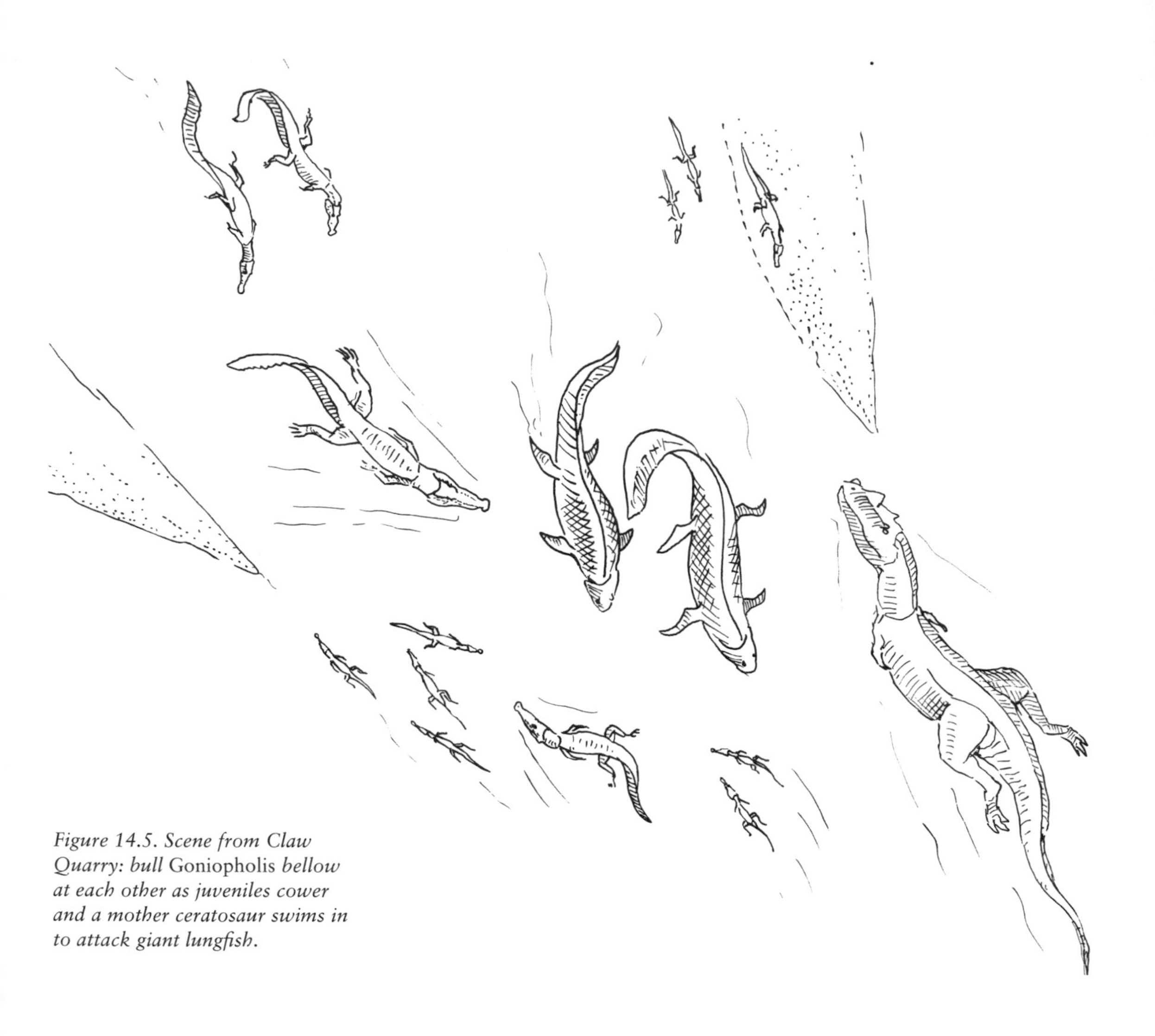

Figure 14.5. Scene from Claw Quarry: bull Goniopholis *bellow at each other as juveniles cower and a mother ceratosaur swims in to attack giant lungfish.*

Turtles

Morrison turtles fall into two family groups: the glyptopsids, *Glyptops* and the new genus "Peggops," and the more advanced baenids (Hay 1908; Bakker 1998a,b). Another, poorly defined category is "smooth turtle," *Dinochelys* and the new genus "Juliops," species with shells that lack the pebbly/sinuous ridge ornamentation of baenids and glyptopsids. All the Morrison turtles have the general shell and limb shape seen among extant pond and river species: The shells are depressed with the posterior margins pinched outwards and rolled up at the edges into a skirt. This shape is the rule among the large river turtles, such as the side-necked river turtles of the Amazon (*Podocnemis*), the highly aquatic Tobasco tortuga river turtle (*Dermatemys*) of Belize, and the chicken turtles and sliders of North America (*Deirochelys, Pseudemys,* and *Chrysemys*) (Earnst and Barbour 1989; Gibbons and Green 1978). The eye sockets in the glyptopsid turtles of the Morrison face strongly upward, as in the highly aquatic chicken turtle (*Deirochelys*). The baenid turtles of the Morrison (*Buzzops* and *Uluops*) have eye sockets that face laterally, but nonetheless the shell form agrees with that of river turtles. No Morrison turtles approach the body form of tortoises, box turtles, or any other terrestrial chelonian types of today. Therefore Como turtles should be linked to large bodies of water.

Crocodiles

The common Como crocodile *Goniopholis* sp. has an overall skull shape very like the river crocodiles of today, especially the American crocodile *Crocodilus acutus,* a species of large rivers and estuaries: The snout is sharply triangular, depressed, with moderate bulges near the front of the maxilla. Therefore the Morrison *Goniopholis* sp. can be taken as an aquatic feeder. It does differ from all modern crocodiles in having lighter bony armor and longer, more slender legs, indicating faster land-running habits.

Sediment Indicators

Two distinctly different types of sediment are involved in the formation of a crime scene (figs. 14.6, 14.7). First, there is the crime scene floor, the surface on which the victims' bodies were dismembered. Second, there's the crime scene sediment blanket, the layer that covered the floor. The covering blanket often is coarser sediment than the floor, and so carcasses and shed teeth of victims may have been moved by the current energy. The current movement is discussed below. Movement of carcasses and shed teeth does not destroy all useful data on predator behavior. In fact, as argued below, movement by turbidites provides a wide area survey of many feeding sites along a lake-swamp shore. The shape of the size-frequency graphs for turbidite-covered sites cannot be explained by hydraulic sorting: There are strong modes at several size classes, rather than the single mode that would be expected if hydraulics controlled tooth size. We have used the sediment indicators of habitat described in previous publications (Bakker 1996). Red and green mottled blocky mudstone with zones of caliche nodules records well-drained floodplains. Platy dark green to black mudstone with no caliche nodules records poorly drained floodplain and swamp. Green to gray blocky mudstone with zones of caliche nodules records an intermediate state of drained floodplains that retained reducing conditions between episodes of caliche formation. Feeding sites have been covered with a variety of coarse clastics. As described below, the most important are fining-upward sandstones that seem to be single-event floods, and graded rip-up-clast conglomerates that appear to be turbidites.

Categories of Feeding Sites

Using the aquatic species and the sediments, we define the following categories of predator-feeding sites:

Dry Surf

Crime Scene Floor: The Handle Mudstone: Blocky, green, non-smectitic mudstone with a few caliche nodules. Floor exceptionally wide—fossil bed extends for at least 500 m laterally, and perhaps for as much as 2 km.

Burial: By same sediment type.

Victims: Nearly all turtles, from complete skeletons to single shell plates, with dozens of partial shells disarticulated and piled up. Most individual turtle specimens occur in small zones, less than 1-m diameter, without mixing from other specimens. Most bones are horizontal but a few plates are vertical; evidently pushed into the surface by feet. A few tiny ornithopod dinosaurs and lungfish jaws; no crocodile parts. In one zone, 2-m diameter, hundreds of fragmented sauropod bones occur with no turtle pieces. Very well-preserved, thick-shelled freshwater clamshells are common. The clams are broken but the tissue is not recrystallized and shows internal fine structure.

Bullets: Rare. Three croc crowns. One small allosaur crown found within the fragmented sauropod bones. This fragmented sauropod may well be a regurgitated mass.

The Dry Turf–Handle surface is curious. It is the earliest bone-rich zone in the Como Morrison and lies below the earliest sandstone and above the lacustrine limestones of the Arrowtail Ranch Megafacies (Bakker 1996; 1998a,b). The sediments appear to be from a poorly drained floodplain, not a long-lived lake. There are no laminae. But the victims are nearly all aquatic turtles and large clams, as if some predator dragged or carried aquatic prey onto the floodplain and dismembered the carcasses there.

Surf

Crime Scene Floor: Most bones lie on wavy-bedded to platy dark mudstones without caliche nodules or roots or any other evidence of dry soil conditions.

Burial: A few large bones and many fish scales lie within the topmost 10 cm of the platy-to-wavy-bedded beds; most bones and teeth are buried within a parallel-bedded, graded conglomerate that appears to be a turbidity flow (fig. 14.6). The bed begins abruptly in a mass of mudclasts of many types—some are platy rip-up clasts derived from the underlying wavy-bedded bed, while some are rounded and angular clasts of pale mudstone containing sand and pebbles; the matrix between clasts is a mixture of clay and angular quartz sand with rounded lithic pebbles. Several zones of concentrated platy-mudclasts exist; one near the base of the conglomerate, and one or two more 10–60 cm higher. Nearly all the platy clasts lie horizontally in these zones. Between the zones, the platy clasts lie in every direction.

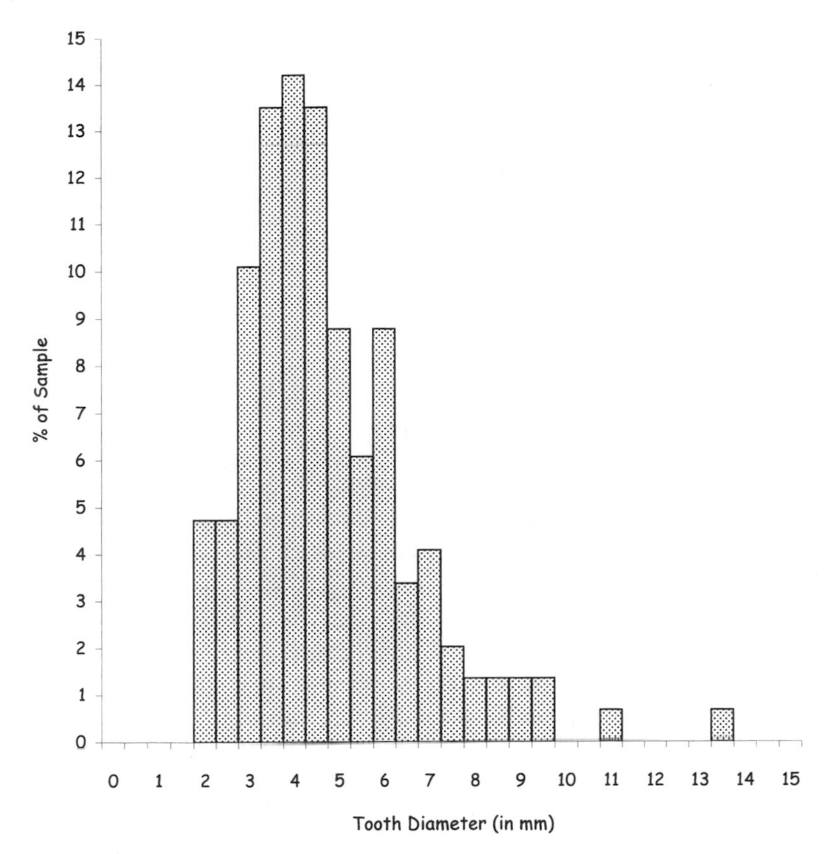

Figure 14.6. Como crocodile shed teeth. Size-frequency distribution from pooled lungfish quarries: Opkat Quarry, Arg Quarry, Truck Quarry, Ng Quarry, Godlungfish Quarry, BH Quarry, Pond Quarry. Very similar to Claw Quarry in lithology and fauna.

Long, thin fossils (dinosaur limb bones; unidentified long splinters) within the horizontal mudclast zones are also horizontal. Between the zones, thin fossils lie in every direction. Fossils are unsorted. A giant sauropod femur, 900 mm long, lies next to ten theropod teeth 6–10 mm tall and amioid fish spines 0.3 mm in diameter. Rounded balls of sauropod bone, 200-mm diameter, lie surrounded by fish fin bones 0.3 mm diameter and choristodire jaws 15 mm long. Disarticulated bones from single individuals lie next to one another. Examples: right and left *Goniopholis* femora, perfectly matched in size and shape; right and left first phalanges from the thumbs of a large crocodile; skull, jaws, vertebrae, limbs, and shell from a single baenid turtle; humerus and wing finger bones from a single pterodactyl. A few articulated specimens occur: a rib cage of a small dinosaur; the partial shell, skull, and limbs of a turtle. Fossils become rare as the bed fines upward into a wavy-platy sandy mudstone, and then into a silty, blocky mudstone with root traces.

The fossil-rich conglomerate must have been laid down suddenly in one hydraulic act, with little tumbling or mixing of prey skeletons and shed tooth bullets. The presence of many associated but disarticulated partial skeletons, with elements still close together in the rock, shows that individual carcasses were often carried and buried as units, without tumbling and mixing together of different carcasses. The extreme lack of hydraulic sorting indicates that lighter victims and shed teeth were not winnowed away to any large extent. The great variety of predator species represented by shed teeth, and the wide range of prey species, both terrestrial and aquatic, strongly suggest that remains from many feeding sites were swept onto the swamp bottom mud by a turbidite-like flow. A heuristic analogy is a shoreline in the Okavango swamp of southwest Africa. Large wading birds, hyenas, lions, crocodiles, and jackals feed on carcasses of fish, turtles, crocodiles, mammals, and birds (Bakker personal observation 1975). Mammals' fecal matter concentrates bones, as do regurgitates from birds. Dismembered and splintered carcasses from many predators lie on mudflats and on sandbars. A tectonic jolt or soft-sediment collapse of the delta could mobilize mud and sand and carry thousands of carcasses downslope onto the swamp floor.

Victims: Chewed vertebrate sample dominated by turtle shell parts, crocodile bones, and fish vertebrae, jaws, and fin spines.

Bullets: Very many crocs mixed with about 6 percent theropod, 1 percent choristodire.

Wet Turf

Crime Scene Floor: Same as above.

Burial: Same as above.

Victims: Articulated or associated parts of large adult sauropod dinosaurs, mixed with some turtle and crocodile parts.

Bullets: Large and small theropods mixed with some croc.

Wet Turf sites are local islands of terrestrial prey within a bed that has mostly aquatic prey. For example, at the Bern-Bronto Quarry, a partial adult apatosaur lay within a thin zone of dismembered vertebrate debris that extends for 400 m. In most places the debris is 99 percent crocodile and turtle, and the shed teeth are 95 percent crocodile. Within a 2-m radius of the apatosaur is a concentration of shed ceratosaur teeth.

Dry Turf

Crime Scene Floor: Blocky, smectite-rich, dark green claystone with many large and small caliche nodules.

Burial: By same sediment type. The burial layer is 2–2.5 m thick. Caliche nodules evidently have been reworked *upward* into the burial layer from the underlying floor (fig. 14.7). In the floor sediments the caliche nodules are arranged in orderly horizontal zones. Large nodules lie with their longest axis horizontal. In the burial layer the nodules lie in disordered arrays, with no horizontal zones, and the long axes of flattened nodules are vertical or oblique more often than horizontal. Many nodules in the burial layer were fractured and split apart before reaching their final resting place. At the top of the burial layer is a limestone, 10-20-cm thick, pinching out into limestone lumps here and there. Thin-shelled freshwater clams occur in the limestone. Fragments of the limestone have been worked *downward* into the burial layer. These angular fragments lie with their long axes at all angles to horizontal. Some clamshells also seemed to have been worked downward; the shells lie at various angles to the horizontal. Most articulated and associated skeletons lie with most bones horizontal on the floor. However, individual bones of some skeletons have been tilted vertically. Going upward through the burial layer, the orientation of bones becomes random, with many vertical elements. At the 1.2-m level, a partial sauropod skeleton lies with the long bones horizontal, as if to mark a second floor. Many bones on the floor are underlain by a halo of cushion-shaped limestone nodules; these nodules—"sapro-glaebules"—seem to have been induced by the presence of rotting flesh. The most likely explanation for the complex three-dimensional geometry is that the victims lay on a floor being dismembered. Then the burial layer arrived, in one or two installments. Carcasses continued to rot under the burial layer while the mud was still fluid. Bubbles of decomposition gases tilted bones up and lifted some bones and nodules. Because the limestone layer hardened before the burial mud, chunks of the limestone settled down through the burial layer.

Victims: Large sauropods, stegosaurs, allosaurs, megalosaurs. No small carcasses.

Bullets: All but one from allosaurids; hatchling bullets mixed with adults. This category contains the famous Nail Quarry.

Very Dry Turf

Crime Scene Floor: Sediments on quarry floor indicate some soil development under dry air: caliche nodular horizons, horizontal silici-

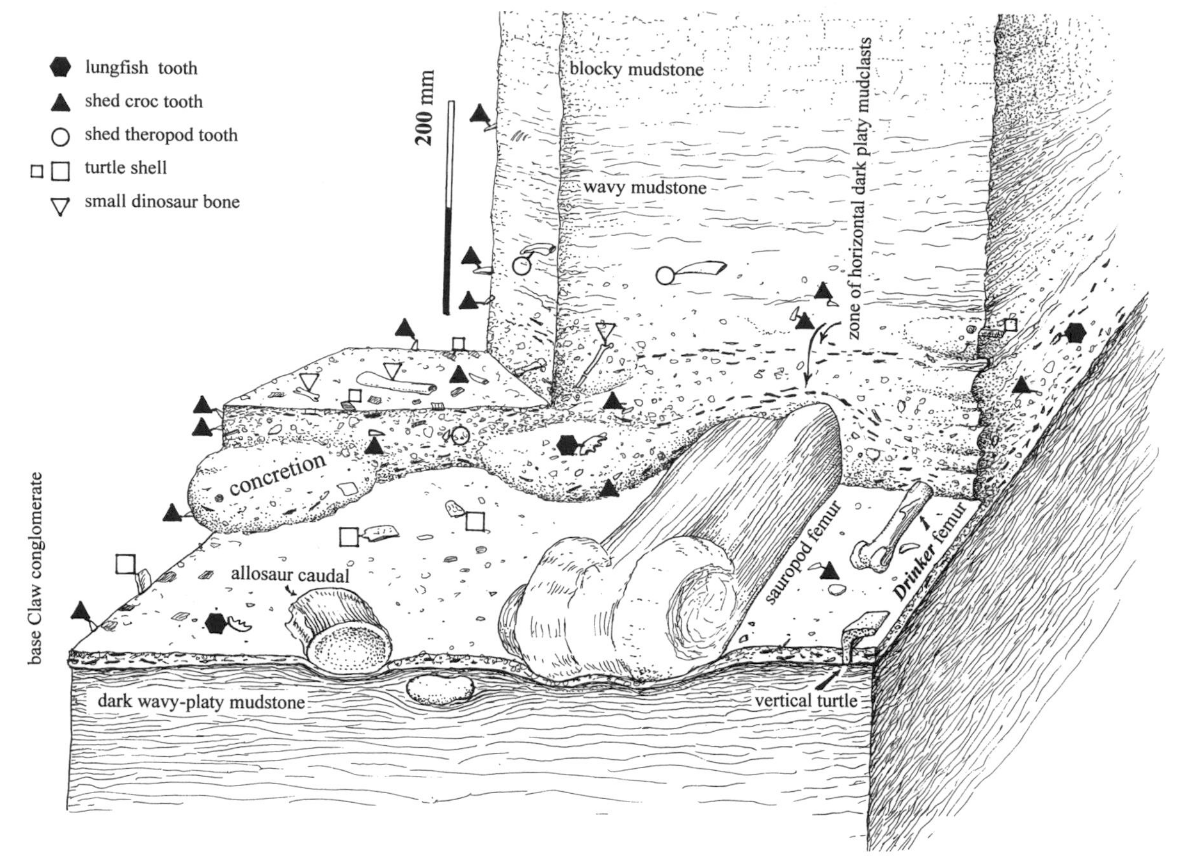

Figure 14.7. Quarry map of a small part of Claw Quarry, a Surf site. Note zones of platy rip-up clasts.

fied root systems, vertical root traces preserved in calcareous beads; red and green mottles developed in the mudstones; bones with deep desiccation cracks; calcified pellets from earthworms and soil arthropods.

Burial: By similar sediment.

Victims: Single sauropod skeletons; no mixture of aquatic vertebrates.

Bullets: No croc; a few theropod.

Very Dry Turf—Flood Burial

Crime Scene Floor: As in the Very Dry Turf.

Burial: By flood sandstone, beginning with basal conglomerate of reworked caliche nodules and mudclasts, fining upward through trough cross beds all trending in the same direction, and then climbing ripples. The sand bodies do not show lateral accretionary beds, and most of the sediment seems to be the result of one or two sudden depositional events. In any one sandstone body there usually is a single fining-upward sequence. These sandstones are incised into the underlying mudstone only in shallow, broad zones; these coarse clastic zones resemble a crevasse splay or, alternatively, an unconfined flood produced by a sudden flood. These burial sandstones are *not* normal river deposits left by a long-lived river system that carried water for many years.

Victims: Skeletons are sometimes concentrated at the boundary between the mudstone below and the conglomeratic base of the sand body; these specimens probably were lying on the floodplain just before the flood. Other concentrations occur in the middle and upper zones of the sand bodies. The mode of preservation is the same in all positions: many strings of vertebrae show the concave-up flexure of bodies that had dried out; many complete and partly articulated terrestrial dinosaur skeletons, plus many associated skeletons where bones from one individual are strewn downstream within a single trough. Bones are unsorted; small and large bones from a single skeleton lie disarticulated within 1–10 m, plus splinters. Faunal composition differs from that seen in other parts of the Morrison, and dinosaurs very rare elsewhere are very common—camptosaurs make up 10–50 percent of the sample and are virtually absent in other taphonomic modes; at one site, Quarry 13, stegosaurs make up 40 percent of the bones (Bakker 1996, 1998a,b). Aquatics very rare; a very few turtle plates and partial croc skeletons occur among hundreds of dinosaur bones and partial skeletons.

Bullets: Very few bullets of any kind. Theropod teeth and croc teeth much rarer, relative to victim bones, than in sites buried by mudstone or by parallel-bedded mudclast conglomerates.

These Flood Burial sites seem to represent "Olson's Nightmare," well-preserved carcasses that result from a most unusual event, repeated four times in 30 m of sediment. The lack of bullets and the high degree of articulation indicate that scavengers were prevented from thorough disarticulation of the corpses. The dried-death pose of vertebrae suggests that bodies were partly desiccated before burial. The lack of aquatics indicates that the floodwaters did not sample permanent bodies of water with large resident populations of turtles and crocodiles. One likely scenario is: a mass death of megafauna from drought, followed by a sudden flood that buried carcasses and sealed them off from scavengers.

Megafacies and Coexisting Habitats

There is an overall-change up-section through the Morrison at Como (Bakker 1996; 1998a,b). Four megafacies are recognized (fig. 14.8):

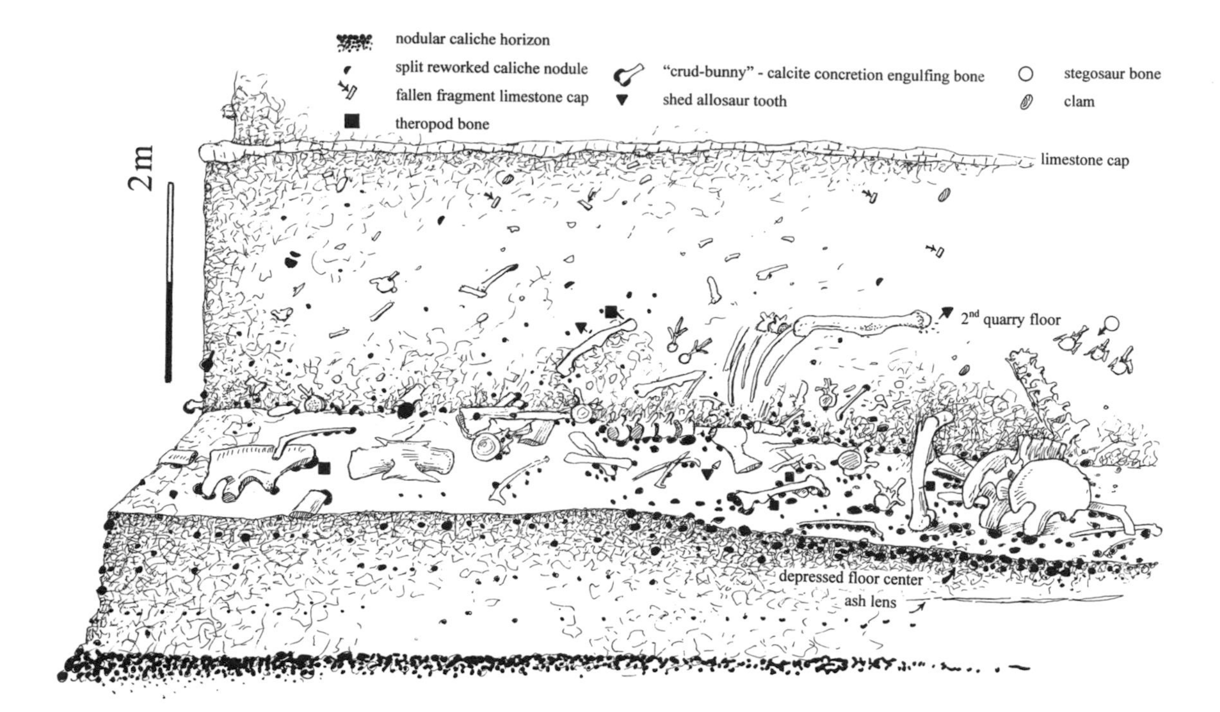

Figure 14.8. Quarry map of a small part of Nail Quarry, a Dry Turf site. Note basal concentration of horizontal bones, with more oblique orientations in higher levels.

1. The interval of the lowest 4 meters is the Arrowtail Ranch Megafacies: laterally continuous limestones 0.5 to 2 m thick, with abundant freshwater snails, charophytes, and plant parts, interbedded with blocky green mudstone, full of non-smectitic clays and few caliche nodules. Here is the Dry Surf feeding site.
2. The next 30 m of the Formation is the Nine Mile Megafacies: red and green blocky mottled mudstone, full of non-smectitic clays, with caliche zones, interrupted by flood sandstones. There are Dry Turf feeding sites.
3. The middle 30 m is the Carlin Ranch Megafacies: dark green/gray blocky mudstone, full of smectitic clays, with caliche zones, interbedded with some platy dark mudstone and thin, lenticular sandstones. There are Dry Turf feeding sites.
4. The upper 10–15 m is the Ward Ranch Megafacies: dark green/gray blocky mudstone with some caliche passing laterally and vertically into dark green/black/black platy mudstone, full of smectitic clay, interbedded with turbidite-like conglomeratic sheets and discontinuous limestones and cherts. There are Surf and Wet Turf feeding sites. As the Carlin Ranch Megafacies interfingers at its top with the Ward Ranch Megafacies, the Late Jurassic environments

represented habitats that coexisted. For example, a Dry Turf allosaur feeding site—Cam Bench Quarry—is at the top of the Carlin Ranch Megafacies at the same stratigraphic level as a Surf ceratosaurid-megalosaurid feeding site—Arg Quarry—at the bottom of the Carlin Ranch Megafacies. The coexistence of wet and dry feeding sites will be important in a later section, when we discuss seasonal shifts among allosaurids.

Results: Distribution of Shed Teeth

The keystone animal for understanding theropods at Como is the common riverine crocodile, *Goniopholis* sp. We assume that its sociobiology would be close to that of extant crocodiles, and hence we can use the shed teeth to test the faithfulness of the taphonomic process.

Como Crocodile Demographics

If fossil croc crowns show the pattern expected from the behavior of living crocodilian species, then we would have confidence that theropod shed teeth are reliable crime scene data as well. If the Como croc teeth show a pattern that makes no sense in the context of living crocodilians, then we would lose confidence in the theropod record. The Como crocodile data are the key challenge to the entire Jurassic Crime Scene Investigation project.

We make the following predictions about *Goniopholis* shed teeth at Como:

1. Croc bullets should be very rare in Dry Turf sites. Since *Goniopholis* is designed for aquatic feeding, its shed teeth should be rare or absent in dry floodplains with no aquatic prey. This prediction is fulfilled. Only one shed *Goniopholis* crown was recovered from the combined sample of all Dry Turf sites. These sites produced more than 80 shed theropod teeth. We conclude that the taphonomic process does not mix aquatic feeders with terrestrial feeders in any obfuscatory manner.

2. Croc bullets should be rare but present in Wet Turf sites. Modern day crocodilians do forage on land, near water, especially at night, and are aggressive competitors for land vertebrate carcasses that lie at or in the water. Therefore we predict that any large sauropod, stegosaur, or camptosaur should be a target for *Goniopholis* if the carcass is near the Jurassic waterways. This prediction is fulfilled. At all Wet Turf sites, where large dinosaur carcasses were dismembered in or near water, crocodile teeth are mixed with theropod teeth. In fact, although the largest bullets are from theropods at these Wet Turf sites, the total number of shed croc crowns exceeds the total theropod number. Clearly theropods and crocodiles were competitors for dinosaur carcasses in watery locales.

3. Shed croc teeth should be superabundant in all Surf sites. Crocodiles today dominate the predation and scavenging within permanent waterways, and so their shed teeth should outnumber those of theropods in fully aquatic sites. This prediction is fulfilled in an emphatic manner. In all Surf sites, crocodiles contribute 95 percent of the shed teeth.

4. Hatchling teeth should be most common at secluded, near-shore sites where a mother was guarding the nursery and where adult males were rare. In all modern crocodilian species the mother (and rarely, the father) guards the nest before the young hatch, and then guards the hatchlings for at least several months (Greer 1970; Singh and Bustard 1977; Rao and Singh 1994; Webb et al. 1987). The nests usually are located in protected locales, away from the population of aggressive adult males, often near shore in pools of quiet water. The mother does not usually feed the young; the hatchlings must capture small prey themselves (insect larvae, crayfish, small fish). Only after this period of maternal protection do the adolescents group together as a creche which tries to avoid big, aggressive adults. The adolescent creche eventually begins to filter into the general population in more open water. Therefore we would expect to find shed teeth of *Goniopholis* hatchlings only in sedimentary environments that were secluded and very near the shore in quiet water. For the purpose of exploring croc nurseries and adolescent creches, we can divide all Surf sites into two environmental categories:

   a) Lungfish—Poor, Secluded, Near-Shore (Bernice Quarry). Big lungfish are rare, most fish are small "amioid" grade gill-breathers, and small terrestrial species, the mammals and lizards, are common.
   b) Lungfish—Rich, Wide, Deep Water (Claw, OPK, Arg, Godlung Quarries, and other sites in pooled sample—see fig. 14.9). Big lungfish are common; mammals and lizards are rare. The samples from Claw Conglomerate contain at least three and perhaps as many as five species of large to very large lungfish; the largest specimens of *"Ceratodus" robustus* suggest a total fish length of 2 to 4 m, extrapolating from the size of the tooth plates. As described above, the big lungfish would require wide, deep, relatively permanent water.

If the Como pattern is concordant with predictions derived from modern crocodilians, hatchling shed teeth should be much more common and giant teeth from big males should be rarer at the Lungfish Poor Bernice locality than at the Lungfish Rich Claw locality. The prediction is confirmed (figs. 14.3, 14.9). The Bernice Quarry is the only locality that has a high percentage of shed teeth in the smallest five

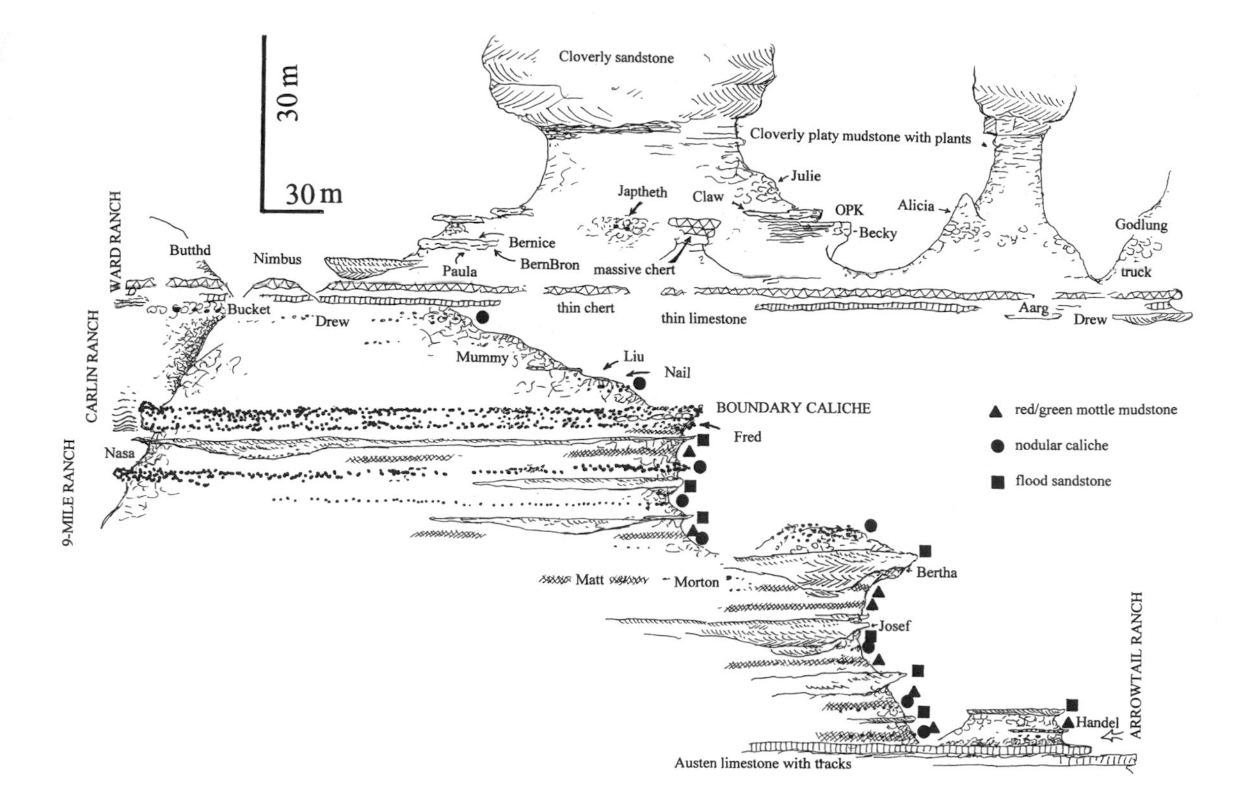

Figure 14.9. Stratigraphic distribution of shed tooth sites in the east half of Como. See map in Ostrom and McIntosh 1966 (outcrop east of Marshall Road); and measured sections in Bakker 1996, and Connely 2002. The localities have been compressed east-to-west; all sections are drawn to 70 m. The section depth varies from 70 m to 54 feet, but the sequence of facies is the same in all sections.

size categories (0.5–2 mm). The percentage in the smallest size classes is twice that of Claw locality and much greater than the combined total from the other sites with big lungfish (one-tailed K-S test, difference significant at 0.01 level). To put it in another context, there is a strong negative relation between large lungfish and baby croc teeth. Therefore Bernice locality must have been much closer to croc nurseries than were the big-lungfish/open water sites. Moreover, there is a highly significant lack of croc teeth bigger than 9.5 mm at Bernice Quarry, indicating that large males did not feed nearby.

5. Most shed teeth should be found where the main population foraged in wide, deep bodies of water. Lungfish-rich sites do provide the overwhelming majority of shed teeth, except for those from hatchlings (figs. 14.3, 14.9).

6. Shed teeth should show catastrophic mortality after the age of one to two years, and then a leveling off at the adult female and male sizes. When adolescent crocs move into the general population they suffer horrendous mortality, with only a small percent surviving to become breeding females, and fewer still becoming resident bulls. Since breeding females sacrifice their growth rates in favor of eggs and embryos, the average adult female is smaller than the domi-

nant bulls. Therefore the shed tooth distribution should level off at about two-thirds maximum size, the size that should represent sexual maturity of females. This prediction is met at Claw quarries (figs. 14.3, 14.9). The shed teeth do show a steep decline from a strong mode at 4–5 mm, probably 1.5 to 2.5 years old, followed by a leveling off at about 10 mm, and then a slow decline to the maximum of 15 mm.

7. Shed teeth should indicate residency all year long. Most if not all breeding populations of crocs today are resident; i.e., the adult females and big males stay in a small area and do not migrate long distances. Usually, a very few very large and aggressive bulls try to monopolize access to the breeding females, and young adult males may be driven out of their natal area. If breeding females and some bulls were resident in small sectors of the Como, then the distribution of shed teeth should be smooth, without gaps. Since shed teeth record the presence of live animals, the only way to generate gaps would be for large parts of the population to move out of the area and then return, or to have strong immigration. If crocs moved out seasonally and returned, there should be multiple gaps, e.g., if crocs hatched in the spring and the family left in the fall for half a year, the shed teeth would record 0-0.5-year-olds, 1-1.5-year-olds, 2-2.5-year-olds, and so forth. Our sampling interval of 0.5 mm should be fine enough to catch seasonal movement of young, fast-growing crocs. Tropical crocodilians today double their linear dimensions in 6–12 months; therefore, the interval from 0.5 mm to 1.5 mm should record the first year in *Goniopholis* sp. The interval from 0.5 to 4.5 mm should include several years of fast growth. In fact, the Claw Conglomerate croc shed tooth distribution is smooth, with no statistically significant gaps (fig. 14.3). When the class interval was reduced to 0.25 mm, no gaps appeared. Gaps do appear in the distribution at the righthand part of the curve, beyond the break in slope at about 8–9 mm. However, the sample sizes are so small here that the gaps probably are not significant. Therefore we conclude that the *Goniopholis* population was resident all year around.

8. Croc bones should be concentrated with croc teeth. Since extant crocodiles live and die in the same local habitats, shed teeth should be mingled with the crocodile bones. This prediction is fulfilled. All the sites with abundant crocodile shed teeth have abundant crocodile bones. No crocodile bones are known from Dry Turf sites. Two partially articulated skeletons are known from the Dry Turf sites covered with flood sandstones, but otherwise crocodilians are absent from this facies.

Figure 14.10. (opposite page) Size-frequency distributions of ceratosaurid (A), megalosaurid (B), and allosaurid (C). *(continued on p. 328)*

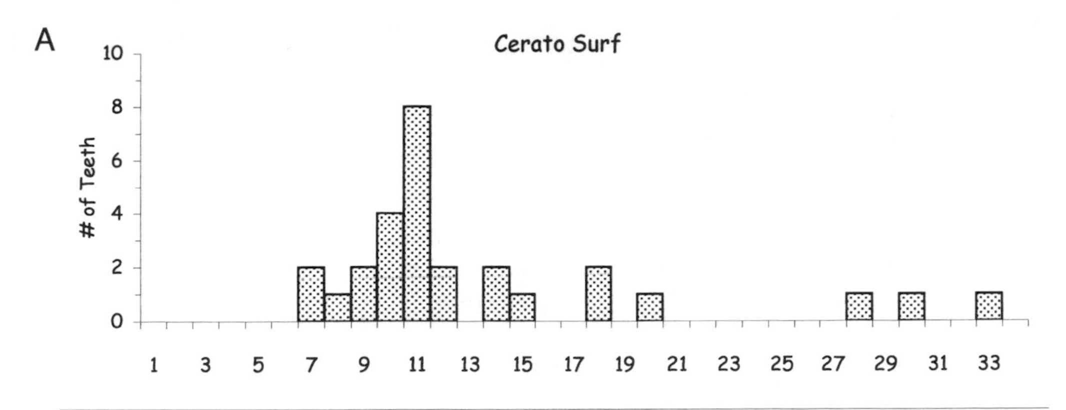
A
Cerato Surf
# of Teeth
0
2
4
6
8
10
1 3 5 7 9 11 13 15 17 19 21 23 25 27 29 31 33

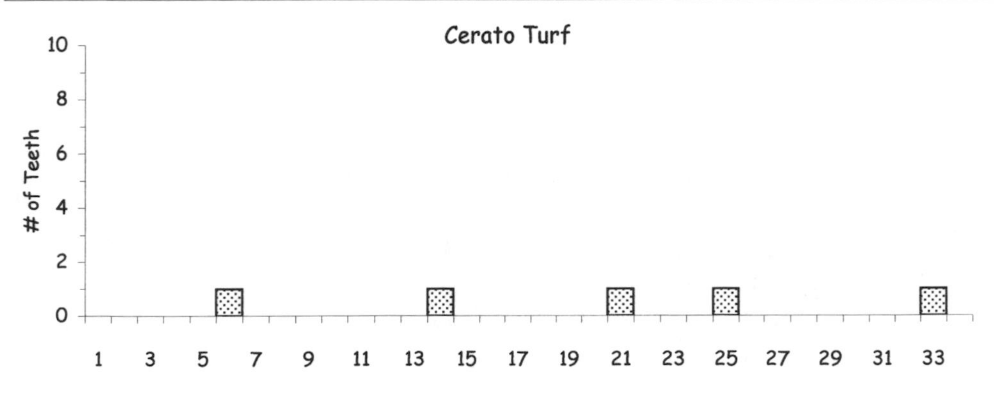
Cerato Turf
# of Teeth
0
2
4
6
8
10
1 3 5 7 9 11 13 15 17 19 21 23 25 27 29 31 33

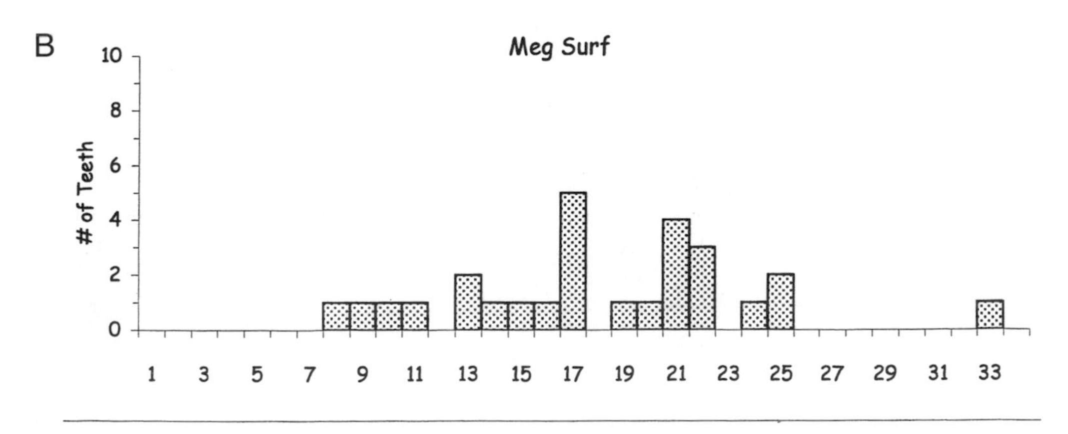
B
Meg Surf
# of Teeth
0
2
4
6
8
10
1 3 5 7 9 11 13 15 17 19 21 23 25 27 29 31 33

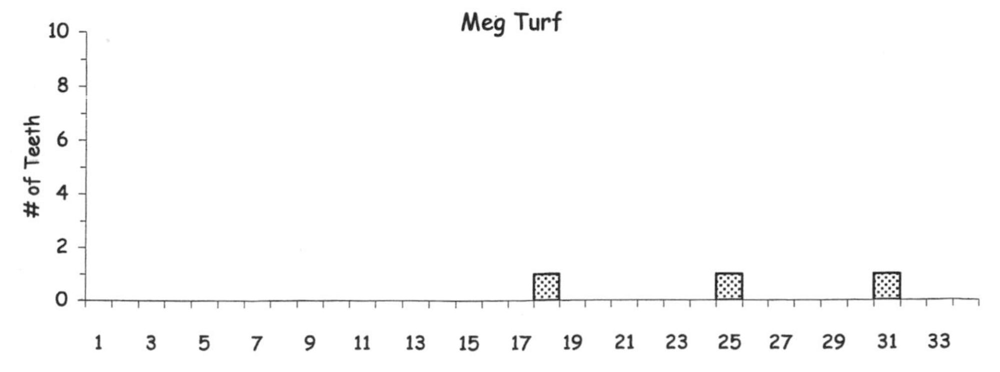
Meg Turf
# of Teeth
0
2
4
6
8
10
1 3 5 7 9 11 13 15 17 19 21 23 25 27 29 31 33

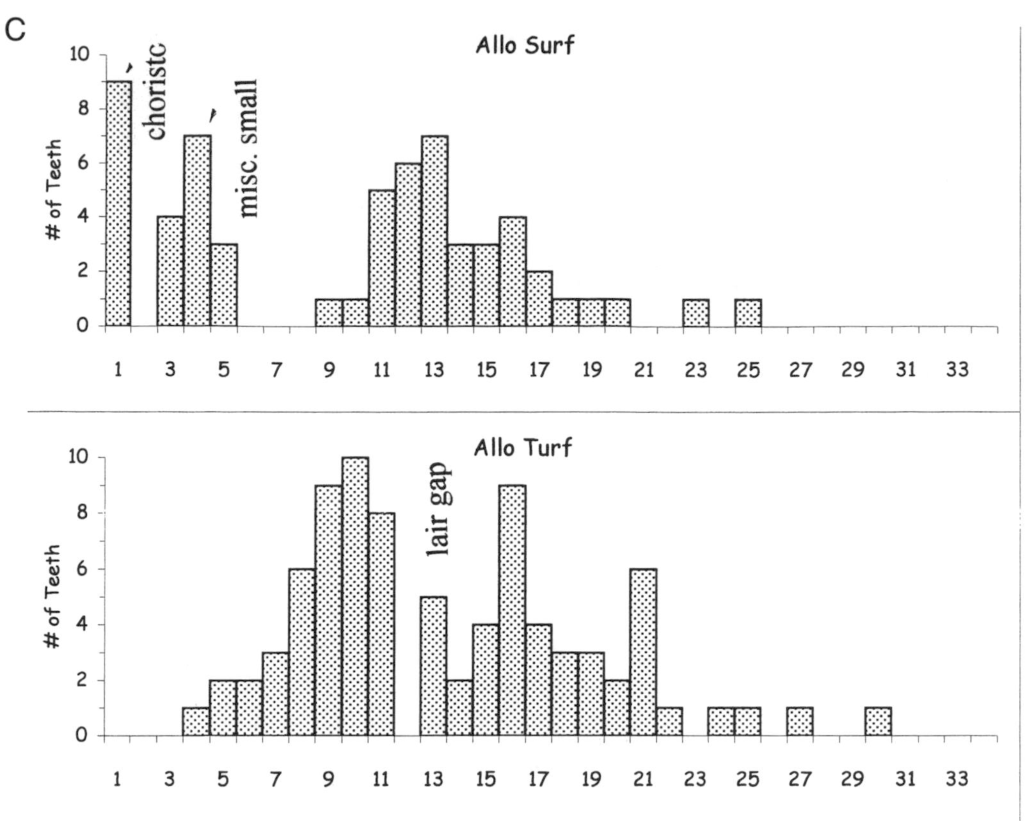

Figure 14.10. (continued)

9. Shed teeth and teeth lost from corpses should be preserved together. Since aquatic crocs should be year-round residents, teeth with roots, lost from corpses after death, should be found mingled with shed teeth. This prediction is fulfilled. All crocodile teeth with roots were found mixed with shed teeth. Teeth with roots equal 12 percent of the total shed sample at the Surf sites.

The croc sample is tremendously encouraging for paleoecology and, specifically, shed tooth crime scene investigations (fig. 14.10). Croc tooth patterns are largely concordant with the known behavior of extant freshwater crocodilians and their movement over micro habitats. Shed croc teeth occur in abundance where they should, and are rare or absent in sediment facies developed in habitats where crocs are excluded today. And, the ontogeny of Como crocs—from hatchling to adolescent to adult—is faithfully recorded in the shed tooth sample. We are encouraged to look at the distribution of shed theropod teeth.

Differences between Allosaur and Croc Behavior

If allosaur social behavior was more like a giant ground hawk than a crocodile, we can predict that the allosaur shed tooth sample would differ in several ways:

1. Hatchling allosaurs should be fed adult food at a lair.

CENSUS OF COMO CRIME SCENES, PERPS AND VICTIMS

type	quarry name	shed allo	shed cerato	shed meg	shed small therop	shed choris-todere	shed croc	lungfish jaw	turtle shell plates	sauropod skeleton	campto-saur skeleton	*Drinker* skeleton	shed sauropod
WARD RANCH FACIES													
turf	**Julie**											**7**	
wet turf	**Alicia**		**2**					**6**	**1**				
surf	**Claw**	**19**	**19**	**21**	**5**	**11**	**605**	**33**	**64**	**1**			
surf	**Opk**	**2**	**3**		**1**		**82**	**6**	**16**				
turf	**Becky**											**11**	
wet turf	**Japheth**		**2**						**5**	**1**			
surf	**Bernice**	**19**	**7**		**4**	**11**	**201**		**33**				**1**
wet turf	**BernBr**	**2**	**1**				**3**			**1**			
surf	**Truck**	**3**	**2**	**1**			**2**	**2**	**17**				
wet turf	**Paula**	**2**		**1**			**4**		**4**	**1**			
surf	**Godlun**	**1**		**1**			**11**	**2**	**9**				
surf	**Nimbus**			**1**			**6**	**3**	**5**				
surf	**Aarg**	**3**	**3**	**1**			**10**		**9**				
surf	**Drew**						**6**		**5**				
wet turf	**LastD**		**1**				**2**		**5**	**1**			
wet turf	**Bucket**	**2**					**2**		**3**	**1**			
surf	**Butthd**	**4**		**3**	**1**		**4**	**1**	**5**				
CARLIN RANCH FACIES													
dry turf	**CamB**	**10**	**1**	**1**	**1**		**1**		**2**	**2**			
dry turf	**Liu**	**2**								**1**			
dry turf	**Mummy**	**11**								**1**			**1**
dry turf	**Nail**	**47**			**1**					**10**			
9-MILE FACIES													
flood/turf	**Nasa**								**3**	**13**	**7**		
flood/turf	**Bertha**	**1**							**2**	**10**	**5**		
v dry turf	**Morton**	**2**								**1**			
flood/turf	**Josef**						**1**		**2**	**4**	**1**		
v dry turf	**Matt**									**1**			
ARROWTAIL FACIES													
dry turf	**Handel**	**1**					**2**	**2**	**43**			**?2**	

Figure 14.11. A census of the perps and victims at the Como crime scenes.

Carnivorous birds differ from crocs in providing prey to the hatchlings at the nest. A hatchling hawk or eagle is fed prey carcasses ripped into convenient fragments by the parent. So if allosaur parents fed their young à la eagle, then hatchling teeth should be mixed with adult teeth among the remains of large prey. This prediction is met (fig. 14.13). All allosaur shed teeth in the five smaller size classes were found in Dry Turf sites, where huge sauropod carcasses

Figure 14.12. An allosaur attacking a lungfish.

dominate the victim sample. At these sites the smallest prey carcass is a two-thirds juvenile allosaur, and 90 percent of the bones belong to adult sauropods. Tooth marks on the sauropods match the size of hatchling allosaur tooth size—the marks are like pin scratches drawn across the extremities of the prey bones. All shed hatchling teeth are found mingled with fully adult tooth crowns.

2. All growth stages of shed teeth should be together at allosaur lairs. Hawks, eagles, and birds in general have integrated extended families. Adults feed young until they are full grown; sometimes older siblings assist their parents in rearing the hatchlings. Consequently a hawk-style allosaur lair

Figure 14.13. Allosaur summer camp. A one-third grown allosaur attempts to eat a glyptopsid turtle (or vice versa).

should have shed teeth from every growth stage from smallest to full grown adult. This prediction is met (fig. 14.13). The Dry Turf sites have hatchling-size, adolescent, two-thirds grown, and full-grown crowns.

3. Intense parental care should reduce mortality from adolescent to full-grown adult. This prediction is met (fig. 14.13). It's useful to divide the shed teeth into 3 classes: (a) smallest to one-third grown; (b) one-third to two-thirds grown; and (c) two-thirds grown to maximum size. Among Como crocodiles, only 4 percent of the (b) class survive to grow into the final (c) size class. The distribution of allosaur crowns shows a much less catastrophic decrease; 33 percent of the (b) class survive to grow up to be the (c) class (difference significant at 0.05 level, two-sample K-S test). In other words, if you are an adolescent allosaur, your chances of surviving into adulthood are much greater than they would be if you were a Jurassic crocodile.

4. No resident sauropod/stegosaur prey herds. Predictions can be made about the seasonal availability of large dinosaur prey. Herbivorous dinosaurs differ from modern crocodiles

and lizards in having much faster growth rates, as indicated by the bone histology, and higher average foraging speed, as shown by the velocity calculated from footprints. Because the food requirements per year were higher than in modern reptiles, herbivorous dinosaurs would forage over a wider area and would be forced to migrate from one feeding area to another when severe dry seasons eliminated fodder in some locales. Migration of herds today is common among large herbivorous mammals in seasonally water-stressed environments. Such migration places great strain on the resident top carnivores. In retrospect, this result should not have been unexpected. Como Morrison sediments scream "seasonally dry" (Bakker, 1996, 1997, 1998a,b). The deep red/green soils, rarity of crocs and turtles compared with Cretaceous sections, and the many thick caliche layers speak of very well-drained, seasonally dry conditions in the Nine Mile Facies. Unlike the Late Cretaceous of Wyoming, the entire Como Morrison lacks any leaf beds, and fossil wood is very rare. The superabundance of turtles and croc shed teeth in the Ward Ranch Facies documents standing water, but that doesn't mean a rich, productive woodland was present. The total absence of frogs and salamanders indicates a high pH (amphibians are extirpated by alkalinity). The variety and thickness of local banded cherts also indicate times of high salinity when silica was mobilized and moved through the groundwater. The environment may well have been a series of alkaline lakes surrounded by wide mudflats that were inhospitable to most Jurassic trees and shrubs. The severity of the dry season leads to the prediction that the herds of large dinosaur herbivores were not year-round residents.

This prediction is fulfilled. Crime scene analysis provides strong evidence that the commonest Como herbivores, the sauropods, stegosaurs, and camptosaurids, were not residents in the Como area. Since all Como herbivores had tooth sockets stacked with replacement teeth, the rate of crown shedding must have been high where these dinosaurs fed for weeks or months at a time. Since mortality does occur where feeding is intense, herbivorous dinosaurs should leave their bones in places where their shed teeth are superabundant. Consequently the abundance of shed teeth/skeletons should distinguish between resident herbivores and those that were merely migrating through the area and not spending long periods feeding in the area of deposition. Studies in the Late Cretaceous of Wyoming and Montana show that the commonest herbivores were resident (Lofgren 1995; Estes and Berberian 1970). For example, at Glenrock, Wyoming, Paleon Museum workers have explored a thick section of Lance Formation that has yielded thirteen large disarticulated ceratopsian skulls and partial skeletons (*Triceratops* and *Torosaurus*). Three partial duckbill carcasses also have been recovered. Following the logic of Olson, such a sample of scattered, weathered, and scavenged carcasses should represent the resident fauna. Olson's logic works here. Micro-bone localities interbedded with the ceratopsian carcasses produce hundreds of herbivore crowns; commonest are ceratopsian, second commonest are duckbills. In this case, shed teeth confirm the bone sample. *Triceratops* and duckbills were resident.

The shed teeth/carcass situation at Como is quite astonishingly different. Sauropods of all species are the commonest herbivore skeletons in most facies. In the flood sandstones of the Nine Mile Facies stegosaurs and camptosaurs equal the sauropods and may exceed them (Quarry 13, NASA Quarries; Bakker 1996, 1997, 1998a,b). If these herbivores were resident, then their shed teeth should be abundant. They are not. Despite very careful excavation, and the large size of many herbivore teeth (camarasaur crowns, on average, are wider and thicker than those of large predators), we have recovered only a handful of shed herbivore crowns, and these few do not match the predictions from the bone sample. No stegosaur or camptosaur shed crowns are known from the flood sandstones where the bones are common. Two camptosaur shed crowns are from the swampy Ward Ranch Facies, where camptosaur bones are unknown. Sauropod shed teeth number only three from all locales. Stegosaur shed teeth number zero from all locales. Teeth with roots, dislodged after death, are much more common. We must emphasize that in the Cretaceous close kin of Como herbivores do leave abundant shed teeth. Worn camptosaur crowns resemble those of iguanodontids, whose shed teeth are common where iguanodontid bones are found (census of Wealden localities sampled by Oxford University Museum and British Museum, Natural History by senior author). Worn stegosaur crowns are close to those of ankylosaurians and thescelosaurs, whose shed crowns are common in Lance and Hell Creek microbeds (Estes and Berberian 1970). There is no biological or taphonomic reason why Como Morrison herbivores should not have left their shed teeth where the animals fed.

We are forced to conclude that in the Como Morrison the common large dinosaurian herbivores did not spend much time feeding in the area of deposition. Therefore resident large theropods that were dependent upon these migrating herds would be forced to leave or shift to another prey source.

5. Seasonal shifts in allosaur packs—summer at the lake. Allosaurs supply the overwhelming majority of bullets at sauropod carcass sites. Therefore, since they had food needs far higher than those of typical reptiles, the allosaurs should have been forced to migrate with the herds or shift to some sort of resident prey present during the dry season. This prediction is fulfilled (fig. 14.13). Shed teeth document a seasonal shifting of the allosaur feeding locations from Dry Turf locales to Wet Turf and Surf sites. If allosaurid packs stayed in one area all through the year, the shape of the size-frequency graph of shed teeth should be smooth, like the crocodile graph, without gaps, because all individuals shed teeth all through their ontogeny and all through the year. If the allosaurids shifted away from the dry floodplains during the dry season, then there should be gaps in the graph, representing the teeth that were shed in other locales. The allosaur Turf shed tooth distribution is not smooth; there is a highly significant gap at about one-third grown, the position at which the croc distributions show a very strong mode (two-tailed K-S test of the distribution against a hypothetical distribution with the same sample size but no gap, probability less than 0.05). This is the "Lair Gap" first noted seven years ago (Bakker 1996). There may be other gaps in the allosaur Turf distribution in size classes above one-third grown, but the sample size is far too small to be certain.

Where did the allosaurs go? Allosaurs traditionally have been portrayed as sauropod and stegosaur predators, and no previous study suggested that aquatic prey might be important. In fact allosaurs are the most common predator family in the two largest Surf samples, Claw and Bernice Localities, and are the most common in all the Surf localities combined. At these sites, shed allosaur crowns are found mingled with chewed and splintered lungfish jaws, "amiod" grade fish, and crocodile and turtle bones. If the Surf sites do represent the off-season activity of allosaurid packs, then the shape of the shed tooth distribution should match the gap in the Turf distribution. This prediction is fulfilled in remarkable fashion. The strong mode in the Surf distribution matches quite precisely the gap in the Turf distribution. (There may be other gaps in the Turf distribution and other modes in the Surf distribution, but small sample sizes make further analysis impossible at this time.) Therefore the Lair Gap seems to represent a time when the allosaur pack left the dry floodplains and moved to the lake and swamp margins, where the packs exploited aquatic prey alongside ceratosaurids and megalosaurids. Such seasonal shifts in a dominant top predator are common today, and many large cats do exploit aquatic reptiles at certain seasons. Jaguars ambush large Amazonian river turtles and caiman in Brazil, and tigers hunt mugger crocodiles in the Sundarban marshes and mangrove flats in India (Mountfort 1985).

Such a seasonal rhythm in prey switching means that investigation of functional morphology becomes more complex—and more interesting. It was not enough that allosaur jaws and teeth were adapted to deal with large, terrestrial prey; the form of the predator had to allow this family to compete successfully in a thoroughly different hunting milieu in the wetlands.

This picture of seasonally shifting allosaur packs is a revolutionary step away from the traditional portrait of allosaurs as dedicated, year-round sauropod killers. But the data from Como insists that the step be taken. Is the Como shed tooth record typical for all Morrison outcrops? We suspect so—though no comparable harvesting of shed teeth has been conducted. Shed sauropod and stegosaur teeth are uncommon in Utah and Colorado collections we have surveyed, compared to the abundance in Late Cretaceous sites.

Non-Allosaur Theropods

Ceratosaurids: Crocodile-Tailed Theropods

Ceratosaurids have been portrayed as hunters of sauropods, camptosaurs, and stegosaurs (Gilmore 1920). Como shed teeth indicate that these views are unbalanced. For two decades, the senior author has been puzzled by a sector of ceratosaur anatomy: the tail is shaped more like a crocodilian's than is any other well-known theropod. Ceratosaurids have very tall neural spines in the anterior caudal vertebrae, and the spines are nearly vertical, lacking the strong backward slant of allosaurs and megalosaurs. The chevrons too are very deep, so the tail profile agrees with that of freshwater crocodilians today (Gilmore 1920; fig. 14.1). All theropods could swim. Powerful kicks from the hind legs, plus sinuous swooshing from the tail could propel allosaurids through the water. But the shallow neural spines and chevrons, the stiffening effect of the posterior zygapophyses, and the short, stiff torso would make allosaurs poorer swimmers than ceratosaurids. Therefore we predict that ceratosaurid shed teeth would be most common mingled with aquatic carcasses.

This prediction is fulfilled (fig. 14.13). Most ceratosaurid shed teeth are from Surf sites. The abundance of small juvenile teeth, with a few full-grown teeth, suggests that ceratosaurid parents fed near the youngsters; perhaps the adults hunted the biggest lungfish and crocodiles and dragged the carcasses back to a shoreline lair. One adult shed tooth is from a Dry Turf sauropod site; two Wet Turf sites have juvenile and adult teeth mixed with sauropod carcasses. These data indicate that *Ceratosaurus* at Como had potential for feeding on large dinosaurs but was specializing in aquatic prey.

Megalosaurids as Mixed Terrestrial-Aquatic Predators

Since megalosaurids had longer, lower, more sinuous bodies than did allosaurids, we predict that megalosaur feeding sites would be more Surf and less Turf. This prediction is met (fig. 14.13). Two megalosaurid crowns were found with splintered sauropod bones at one site (Paula Quarry). Most megalosaurid crowns occur in lungfish-rich Surf sites. Megalosaurid shed teeth are absent from the lungfish-poor Bernice locality, the site we suspect was near crocodile nurseries. The percentage of the sample that is small juvenile (one-third grown and smaller) is smaller among megalosaurids than among ceratosaurids, suggesting a difference in social structure and survivorship—perhaps smaller clutch sizes. Certainly the portrayal of Morrison megalosaurids as specialized megasauropod killers is falsified by the shed teeth at Como.

Cannibalism and Intra-Guild Predation

Cannibalism is the rule, not the exception, among today's top predators. And, predators routinely kill other predator species. Lions kill hyenas and each other. Wolves kill and eat wolves and massacre coyotes. Therefore it would be expected that Como predators would feed on Como predators. This prediction is fulfilled. At Nail Quarry, Mummy Quarry, and Seamus Quarry, Dry Turf locales, five allosaurid skeletons were chewed by allosaurs, every tooth-marked bone surrounded by shed allosaur teeth ranging in size from hatchling (3.3 mm wide) to full-grown adult (31 mm wide). Megalosaurid bones are common at one floodplain site, Nail Quarry, where most of the megalosaurid bones are tooth-marked and surrounded by shed allosaurid teeth. But not a single megalosaurid shed tooth is from this or any Dry Turf locale, and so we infer that these locales were places where megalosaurids were dismembered by allosaurs but not where megalosaurids themselves fed.

Separation of Feeding from Death Sites

If Como theropods had wide foraging ranges and seasonal shifts in feeding sites, then the places where the predators left their shed teeth would not necessarily coincide with the places where they suffered mortality. In other words, unlike the situation with Como crocodiles, where shed teeth and bones are mixed together, there should be a disconnect between theropod carcasses (victims) and shed teeth (perp bullets). This prediction is fulfilled. Megalosaurid and ceratosaurid shed teeth are superabundant in the Claw Conglomerate but not a single recognizable bone from either family is found here, and so we infer that these sites were places where the predators fed but were not victims of predation themselves. As mentioned above, we have Dry Turf sites where megalosaurids were dismembered by allosaurids. Thus there is a suggestion that megalosaurids suffered mortality when allosaur foraging expeditions penetrated the outer limits of the megalosaurid range, or vice versa. Once again the shed tooth analysis reveals geographical-ecological complexity and order in the behavior of large Como theropods. The three common families of large theropods were not mixed indiscriminately across the habitat mosaic.

Family Extinction and Family Realized-Niche

As described in the Introduction, extinction and replacement at the Jurassic-Cretaceous boundary produced a net increase in the degree of birdness among large predators. The archaic, long-bodied, short-legged ceratosaurids and megalosaurids disappear; allosaurids persist, joined by acrocanthosaurids. In the later Cretaceous the very long-limbed, stiff-torsoed tyrannosaurids achieve a near monopoly in Laurasia within the top predator guilds. Why did the archaic theropods become extinct? A simple habitat shift does not explain the pattern. Tyrannosaurids seem less adapted to a dense, wet forest habitat than megalosaurids and ceratosaurids. The long-term trend toward stiffer bodies and longer limbs is independent of habitat. A shift in prey seems an inadequate explanation as well. Tyrannosaurids in the Western Interior Basin fed upon duckbills and horned dinosaurs, herbivores that were longer-limbed and certainly faster than sauropods and stegosaurs. However, in Mongolia, tyrannosaurids are found with sauropods as well as duckbills. In the entire interval from Early Jurassic to Late Cretaceous, there seems to have been a fundamental advantage to the long-limbed, short-torsoed body form, in all habitats and with all prey.

A very similar pattern occurs among top predators in the Cenozoic of North America and Europe (Osborn 1910; Bakker 1981). The Paleocene–Early Eocene predators are mostly sprawling, short-legged, flat-footed oxyaenids and primitive mesonychids. In the later Eocene and Oligocene, hyaenodontids with longer, stiffer limbs become common and diverse. In the Miocene and Pliocene, hyaenodontids are replaced by canids with even longer, even stiffer limbs. At the same time, modern-grade cats (neo-felids) with longer, stiffer limbs replace old-fashioned saber-toothed cats (nimravids). As a result of these replacements, in every New World habitat today the top predator has longer legs, more compressed metapodial bundles, and stiffer, more tightly locked ankle and wrist joints than did the top predators in the same environment in the Oligocene and Eocene. Why did longer, stiffer legs prove to be the winning geometry in mammalian top predators, even in habitats with dense underbrush?

Possibly the key connection is between stiff limbs/compact bodies and social behavior. Among extant mammals, the largest predators which are flat-footed and flexible predators today tend to be solitary—examples are Tasmanian devils (*Sarcophilus*), wolverines (*Gulo*), honey badgers (*Melivora),* and binturongs (*Arctitis*). The larger social predators tend to be long-limbed and digitigrade—examples are canids, hyaenids, and open-terrain cats such as lions and cheetahs (Estes 1993). Some long-limbed digitigrade predators are solitary—the tiger and jaguar being obvious examples. However, it can be argued that the modern-day canids, felids, and hyenids have been shaped by evolution in social settings, and that solitary habits are always secondary. If this is true, then long, digitigrade limbs are associated with a phase of natural selection that provides the potential for greater social complexity in any habitat.

Applying this tentative theory to the Late Jurassic, it could be that allosaurids were fundamentally more social than were the other large theropods, and this difference eventually proved to be an advantage even in wet, bushy terrain. Shed tooth analysis should, in time, provide tests for these ideas. If allosaurids did hunt and feed in larger social groups than the other large Morrison theropods did, then the allosaur bullet sample should occur in larger clumps. We look forward to expanding the crime scene analysis to other Morrison locales.

Summary

Shed teeth are magic bullets. The trail of dental signatures left by Jurassic crocodiles and dinosaurs throughout their lives in and around Como Bluff gives us an intimate look inside the society of these extinct creatures, their parental care, their search for hunting grounds, and their interactions with competitors. The following are the most salient conclusions:

1. Shed teeth record life activities—feeding sites—that do not necessarily correspond with burial sites of the same species. Megalosaur skeletons occur in habitats where megalosaurid shed teeth are absent. Ceratosaurid and megalosaurid shed teeth are superabundant in swampy facies where their bones do not occur.

2. Shed teeth of the aquatic Como crocodile *Goniopholis* sp. are distributed across environments as would be expected in a modern croc of comparable head/tooth design. Hatchling croc teeth are common only in secluded, shallow, nearshore habitats with small fish. Crowns from big adult males and females are most common in deep, wide, open waters with large lungfish. The 2-3-year-olds suffered precipitous mortality, and only a small percentage survived to become adults. Croc teeth are rare in dry floodplain habitats. Gap-free distribution of croc teeth indicates that the population was resident year-round. The faithfulness of croc shed teeth in mapping out croc behavior encourages the use of shed teeth to investigate dinosaurs.

3. Large herbivorous dinosaurs were not resident in the Como area in Morrison times. The abundance of skeletons and rarity of shed teeth indicate that sauropods, stegosaurs, and camptosaurs migrated through the area and suffered sporadic mass deaths, but did not linger to feed.

4. Predators with long, low, sinuous bodies—megalosaurids and ceratosaurids—fed mostly in and around permanent watercourses, preying upon large lungfish, crocodiles, and turtles. Ceratosaurids occasionally fed on sauropods in dry floodplains and wet floodplains. Megalosaurids very occasionally fed on sauropods on wet floodplains.

5. *Ceratosaurus* had the strongest swimming ability among Como Morrison theropods. Megalosaurids were also stronger swimmers than allosaurids.

6. Allosaurids were the fastest but least maneuverable among the three families of large theropods; allosaurids had overwhelming dominance in dry floodplain feeding sites with sauropod carcasses.

7. Nevertheless, allosaurids are the most common large preda-

tors in and around aquatic habitats. Allosaurids probably shifted their feeding sites from dry floodplains, where they fed mostly on sauropods, to lake margins and alkaline swamps, where they fed on aquatic vertebrates. Seasonal disappearance of migrating herbivore herds probably forced this dry-season shift among allosaurids.

8. Hatchling allosaurs fed on huge prey carcasses alongside their parents and older siblings. Allosaurids stayed in the family group until they were full-grown.

9. The long-bodied, primitive predators—the ceratosaurids and megalosaurids—were replaced by theropods with a more birdlike shape—the allosaurids and, in the Cretaceous, the tyrannosaurids. This long-term trend is repeated in the Cenozoic, as short-limbed creodonts with spreading feet are replaced by progressively longer-limbed, more digitigrade cats, dogs, and hyenas. The shift to longer, stiffer limbs may be associated with more complex social behavior.

Acknowledgments

Very special thanks go to the Harding Family and the Kirkbride Family, owners of the Carlin Ranch, Como Bluff, Wyoming. Their vigilance has preserved a superb outcrop of Morrison fossils from the depredations of sloppy university collectors and pothunters. By agreement with the owners of the Carlin Ranch, all fossils collected during this study must stay together and must stay at nonprofit museums within Wyoming. We thank Melissa Connely of Casper College for six years of intellectually vigorous assistance in analysis of the Como Morrison. Many volunteers from the Wyoming Dinosaur International Society contributed patient days excavating delicate fossils. Nancy Rufenacht, Ed Pulver, and Jordan Hand have run the WDIS field courses with extraordinary success. Sean Smith, Curator at the Glenrock Museum, has overseen the wonderfully complete sampling of shed teeth and skeletons from the Lance Formation. Two anonymous reviewers assisted in shaping the text.

This paper is dedicated to Dr. Martin Shugar. Marty is the indefatigable force behind this symposium. His selfless energy and vision brought together specimens and speakers into a unique combination. Without Marty, *Bambiraptor* never would have made its appearance in Florida, nor would this marvelous little dinosaur have stimulated the imagination of scientists and lay folks alike.

References

Bakker, R. T. 1981. The deer flees, the wolf pursues. In D. Futuyma and M. Slatkin (eds.), *Coevolution,* pp. 350–383. Sunderland, Mass.: Sinauer Associates.

———. 1996. The real Jurassic Park: dinosaurs and habitats at Como Bluff, Wyoming. *Museum of Northern Arizona Bulletin* 60: 35–49.

———. 1997. Raptor family values: Allosaur parents brought giant carcasses into their lair to feed their young. In *Dinofest International III*, pp. 51–63. Philadelphia: Academy of Natural Sciences.

———. 1998a. Brontosaur killers: Late Jurassic allosaurids as sabre-tooth cat analogues. *Gaia* 15: 145–158.

———. 1998b. Dinosaur mid-life crisis: the Jurassic/Cretaceous transition in Wyoming and Colorado. *New Mexico Museum of Natural History Bulletin* 14: 67–79.

Baur, G. 1891. On some little known American fossil tortoises. *Proceedings Academy of Natural Sciences of Philadelphia* 43: 411–430.

Brinkman, D. B., and Jiang-Hua Peng. 1993a. *Ordosemys leios*, n. gen., n. sp., a new turtle from the Early Cretaceous of the Ordos Basin, Inner Mongolia. *Canadian Journal of Earth Sciences* 30: 2128–2138.

———. 1993b. New Material of *Sinemys* (Testudines, Synemydidae) from the Early Cretaceous of China. *Canadian Journal of Earth Sciences* 30: 2139–2152.

———. 1996. A new species of *Zangrelia* from the Upper Cretaceous redbeds at Bayan Mandahu, Inner Mongolia. *Canadian Journal of Earth Sciences* 33: 526–540.

Cifelli, R. L., and S. K. Madsen. 1998. Triconodont mammals from the Medial Cretaceous of Utah. *Journal of Vertebrate Paleontology* 18: 403–411.

Cifelli, R. L., J. R. Wible, and F. A. Jenkins. 1998. Triconodont mammals from the Cloverly Formation (Lower Cretaceous), Montana and Wyoming. *Journal of Vertebrate Paleontology* 18(1): 237–241.

Cifelli, R. L., T. R. Lipka, C. R. Schaff, and T. B. Rowe. 1999. First Early Cretaceous mammal from the eastern seaboard of the United States. *Journal of Vertebrate Paleontology* 19: 199–203.

Clemens, W. A. 1963. Fossil mammals of the type Lance Formation. *University of California Publications in Geological Sciences* 48: 1–105.

Connely, M. V. 2002. Stratigraphy and paleontology of the Morrison Formation, Como Bluff, Wyoming. Master's thesis, Utah State University, Logan, Utah. 109 pp.

Earnst, C. H., and R. W. Barbour. 1989. *Turtles of the world*. Washington, D.C.: Smithsonian Press.

Edmund, A. G. 1962. Sequence and rate of tooth replacement in the Crocodilia. *Contributions, Life Science Division, Royal Ontario Museum* 56: 1–42.

Engelmann, G. F., and G. Callison. 1998. Mammalian faunas of the Morrison Formation. *Modern Geology* 23: 343–379.

Erickson, G. M. 1996a. Daily deposition of incremental lines in *Alligator* dentine and the assessment of tooth replacement rates using incremental line counts. *Journal of Morphology* 226: 189–194.

———. 1996b. Incremental lines of von Ebner in dinosaurs and the assessment of tooth replacement rates using growth line counts. *Proceedings National Academy of Sciences USA* 93: 144623–14627.

Erickson, G. M., and C. A. Bochu. 1999. How the terror crocodile grew so big. *Nature* 398: 205–206.

Estes, R. D. 1993. *The safari companion*. Port Mills, Vt.: Chelsea Green.

Estes, R., and P. Berberian. 1970. Paleoecology of a Late Cretaceous vertebrate community from Montana. *Breviora* 343: 11–35.

Evans, S. E., and T. S. Kemp. 1975. The cranial morphology of a new lower Cretaceous turtle from southern England. *Palaeontology* 18: 25–40.

———. 1976. A new turtle skull from the Purbeckian of England and a note on the early dichotomies of cryptodire turtles. *Palaeontology* 19: 317–324.

Feist, M., R. D. Lake, and C. J. Wood. 1995. Charophyte biostratigraphy of the Purbeck and Wealden of southern England. *Palaeontology* 36: 401–442.

Fox, R. C. 1969. Studies of Late Cretaceous vertebrates: III. A triconodont mammal from Alberta. *Canadian Journal of Zoology* 47: 1253–1256.

———. 1976. Additions to the mammalian local fauna from the Milk River Formation (Upper Cretaceous), Alberta. *Canadian Journal of Earth Sciences* 13: 1105–1118.

Gaffney, E. S. 1979. The Jurassic turtles of North America. *Bulletin American Museum of Natural History* 162: 95–135.

Gibbons, J. W., and J. L. Greene. 1978. Selected aspects of the ecology of the Chicken Turtle, *Deirocheyls reticularia. Journal of Herpetology* 12: 237–241.

Gilmore, C. W. 1910. A new rhynchocephalian reptile from Wyoming and notes on the fauna of "Quarry 9." *Proceedings U.S. National Museum* 37: 39–40.

———. 1920. Osteology of the carnivorous Dinosauria in the United States National Museum. *United States National Museum Bulletin* 110: 1–154.

Greer, A. 1970. Evolutionary and ecological significance of crocodilian nesting habits. *Nature* 5257: 523–524.

Hay, O. P. 1908. The fossil turtles of North America. *Carnegie Institution of Washington Publication* 75: 1–587.

Howse, S. C. B., and A. R. Milner. 1995. The pterodactyloids from the Purbeck Limestone Formation of Dorset. *Bulletin Natural History Museum London (Geology)* 51: 73–88.

Hoyt, E. P. 1983. *Bowfin.* New York: Van Nostrand.

Huti, S., D. M. Martill, and A. J. Barker. 1996. The first European allosaurid dinosaur. *Neues Jahrbuch Paläontologie Mineralogie* 1996: 635–644.

Kielan-Jaworowska, Z., D. Dasheveg, and B. A. Trofimov. 1987. Early Cretaceous multituberculates from Mongolia and a comparison with the Late Jurassic forms. *Palaeontologia Polonica* 32: 3–41.

Kielan-Jaworowska, Z., and P. C. Ensom. 1992. Multituberculate mammals from the Purbeck Limestone Formation of southern England. *Palaeontology* 35: 95–126.

———. 1994. Tiny plagiaulacoid multituberculate mammals from the Purbeck Limestone Formation of Dorset, England. *Palaeontology* 31: 17–31.

Kind, O., G. Grigg, K. Warburton, and C. Franklin. 1999. Habitat utilization and movement patterns of the Queensland lungfish (*Neoceratodus foster*). *Center for Conservation Biology* 22-33.14.12

Kirkland, J. A. 1998. Morrison fishes. *Modern Geology* 22: 503–533.

Kowalis, B. J., E. H. Christiansen, A. L. Deino, F. Peterson, C. E. Turner, M. J. Kunk, and J. D. Obradovich. 1998. The age of the Morrison Formation. *Moderen Geology* 22: 235–260.

Lofgren, D. L. 1995. The Bug Creek problem. *University of California Publications in Geological Sciences* 140: 1–185.

Merrick, J. R., and G. E. Schmida. 1984. *Australian freshwater fishes: biology and management.* North Ryde, N.S.W., Australia: Merrick.

Olson, E. C. 1962. Late Permian vertebrates, U.S. and U.S.S.R. *Transactions American Philosophical Society* New Series 52: 1–224.

Osborn, H. F. 1910. *The age of mammals*. New York: Macmillan.
Ostrom, J. H., and J. McIntosh. 1966. *Marsh's dinosaurs*. New Haven, Conn.: Yale University Press.
Peng, J.-H., and D. B. Brinkman. 1993. New material of *Xinjiangchelys* from the Late Jurassic Qigu Formation. *Canadian Journal of Earth Sciences* 30: 2013–2026.
Rao, R. J., and L. Singh. 1994. Status and conservation of the gharial in India. *Proceedings of the Working Meeting of the Crocodile Specialist Group* 12: 84–97.
Rasmussen, T., and G. Callison. 1981. A new species of triconodontid mammal from the Upper Jurassic of Colorado. *Journal of Paleontology* 55: 628–634.
Russell, D. A. 1993. The role of Central Asia in dinosaurian biogeography. *Canadian Journal of Earth Sciences* 30: 2002–2012.
Siegel, S. 1956. *Non-parametric statistics*. New York: McGraw Hill.
Simpson, G. G. 1925. Mesozoic mammalia: I. American triconodonts. Part 2. *American Journal of Science* 10: 334–358.
———. 1926. The fauna of Quarry Nine. *American Journal of Science* 67: 1–11.
———. 1928. *A catalogue of the Mesozoic Mammalia in the Geological Department of the British Museum*. London: British Museum.
———. 1929. American Mesozoic Mammalia. *Memoirs of the Peabody Museum of Natural History* 3.
Singh, L., and H. R. Bustard. 1977. Studies on the Indian Gharial, *Gavialis gangeticus* (Gmelin): V. Preliminary observations on maternal behavior. *Indian Forester* 103: 671–676.
Webb, G., S. Manolis, and P. Whitehead. 1987. *Wildlife management: crocodiles and alligators*. Sydney: Surrey Beatty.
Wright, J. 1998. Ichnological evidence for the use of the forelimb in iguanodontid locomotion. Special Papers in Palaeontology 60.
Wright, J. L., D. M. Unwin, M. G. Lockley, and E. C. Rainforth. 1997. Pterosaur tracks from the Purbeck Limestone Formation of Dorset, England. *Proceedings Geologists' Association* 108: 39–48.

Index